THE
MOBILE
COMMUNICATIONS
HANDBOOK

Editor-in-Chief

JERRY D. GIBSON
Texas A&M University
College Station, Texas

 CRC PRESS    IEEE PRESS

A CRC Handbook Published in Cooperation with IEEE Press

Library of Congress Cataloging-in-Publication Data

The mobile communications handbook / edited by Jerry D. Gibson.
    p.   cm. — (The electrical engineering handbook series)
    Includes bibliographical references and index.
    ISBN 0-8493-8573-3 (alk. paper)
    1. Mobile communication systems. I. Gibson, Jerry D.
II. Series.
TK6570.M6M5934 1996
621.3845—dc20

95-39771
CIP

© 1996 by CRC Press, Inc.

No claim to original U.S. Government works
International Standard Book Number 0-8493-8573-3
Library of Congress Card Number 95-39771
Printed in the United States of America   1 2 3 4 5 6 7 8 9 0
Printed on acid-free paper

Preface

The term *mobile communications* can include technologies ranging from cordless telephones, digital cellular mobile radio, and evolving personal communications systems to wireless data and networks, and this broad interpretation is the sense in which mobile communications is used here. To cover this range of topics, we present articles by 52 experts on various aspects of modulation, digital communications, and mobile communications applications. The articles are written to provide a succinct overview of each topic to quickly bring the reader up to date, but they also contain sufficient detail and references to encourage further study. The articles are more than a "just the facts" presentation, with many of the authors providing their personal insights and the evolutionary directions of the various fields.

The *Handbook* is separated into two parts. The first part is on the basic principles of analog and digital communications that pertain to mobile communications, and it consists of 14 tutorial/review articles that lay a solid groundwork for the wide-ranging aspects of mobile communications technologies. The second part of the *Handbook* contains 21 articles covering such topics as cellular mobile radio, personal communications systems, user location and addressing, wireless data, wireless local area networks (LANs), and prominent mobile communications standards in all parts of the world. Having the basic principles articles readily available gives the reader the ability to jump right to the mobile communications topic of interest and then efficiently fill in any gaps in his/her background by referring back to the basic principles section without searching through a textbook or shuffling off to the library.

Along these same lines, although there is an ordering to the articles, the sequence of articles does not have to be read from beginning to end. Each article was written to be an independent contribution but with intentional overlap between some articles to prevent the reader from having to flip back and forth between articles to extract the desired information. Interestingly, as the reader will discover, this overlap often admits alternative views of difficult issues.

The lead-off article for the wireless section is written by an acknowledged leader and innovator in wireless communications, Dr. Don Cox of Stanford University. Dr. Cox develops his unique perspective of the issues in wireless communications based on hard facts, concrete examples, and his interpretation of market and technology evolution. Besides being informative and easy to read, the article implicitly challenges those active in wireless communications to continually question the common knowledge in the field.

Personally, I must express how pleased I am to be associated with this project and these authors. The authors come from all parts of the world and constitute a who's who of workers in digital and mobile communications. Certainly, each article is an extraordinary contribution to the communications field, and the collection, I believe, is an important and unique reference for those interested in mobile communications research, development, and implementations. I must also acknowledge the exceptional efforts and talents of my wife, Elaine M. Gibson, who served as Managing Editor for the project, and was the principal contact point for the authors and the publisher. Her energy and organizational skills in coordinating each contribution and in organizing the *Handbook* were essential to its timely delivery and to its quality. Elaine and I both appreciate the patience, support, and guidance of Joel Claypool at CRC Press during all stages of developing and producing this *Handbook.*

Jerry D. Gibson
College Station, Texas

Editor-in-Chief

Jerry D. Gibson is a professor in the Department of Electrical Engineering at Texas A&M University, College Station. He received a B.S. degree in electrical engineering from the University of Texas at Austin in 1969 and M.S. and Ph.D. degrees from Southern Methodist University, Dallas, Texas in 1971 and 1973, respectively.

He has held positions at General Dynamics, Fort Worth (1969–72), the University of Notre Dame (1973–74), and the University of Nebraska, Lincoln (1974–76). In 1976 he joined Texas A&M University where he currently holds the J.W. Runyon, Jr. professorship in the Department of Electrical Engineering. During the fall of 1991, Dr. Gibson was on sabbatical with the Information Systems Laboratory and the Telecommunications Program in the Department of Electrical Engineering at Stanford University, California.

He is coauthor of the book *Introduction to Nonparametric Detection with Applications* (Academic Press, 1975 and IEEE Press, 1995) and author of the textbook, *Principles of Digital and Analog Communications* (Macmillan, 2nd edition, 1993). He was associate editor for Speech Processing of the *IEEE Transactions on Communications* from 1981 to 1985 and associate editor for Communications of the *IEEE Transactions on Information Theory* from 1988 to 1991. He is currently a member of the IEEE Information Theory Society Board of Governors (1990–1995), the Speech Technical Committee of the IEEE Signal Processing Society (1992–1995), and of the Editorial Board for the *Proceedings of the IEEE*.

Dr. Gibson is also editor-in-chief of *The Communications Handbook* (CRC Press, 1996), and the editor of the IEEE Press book series on signal processing.

In 1990, Dr. Gibson received The Fredrick Emmons Terman Award from The American Society for Engineering Education, and in 1992 was elected Fellow of the IEEE for contributions to the theory and practice of adaptive prediction and speech waveform coding. He was corecipient of the 1993 IEEE Signal Processing Society Senior Paper Award for the Speech Processing area.

His research interests include data, speech, image, and video compression, multimedia over networks, wireless communications, information theory, and digital signal processing.

Contributors

Saf Asghar
Advanced Micro Devices, Inc.
Austin, Texas

Melbourne Barton
Bellcore
Red Bank, New Jersey

V.K. Bhargava
University of Victoria
Victoria, British Columbia
Canada

Wai-Yip Chan
Illinois Institute of Technology
Chicago, Illinois

Giovanni Cherubini
IBM Zurich Research Laboratory
Ruschlikon, Switzerland

Stanley Chia
BT Laboratories
Ipswich, United Kingdom

Leon W. Couch II
University of Florida
Gainesville, Florida

Donald C. Cox
Stanford University
Stanford, California

Marc Delprat
Alcatel Mobile Communication
Columbes, France

Spiros Dimolitsas
COMSAT Laboratories
Clarksburg, Maryland

I.J. Fair
University of Victoria
Victoria, British Columbia
Canada

Ira Gerson
Motorola Corporate Systems
    Research Laboratories
Schaumburg, Illinois

Klein S. Gilhousen
QUALCOMM Incorporated
San Diego, California

Lajos Hanzo
University of Southampton
Southampton, United Kingdom

Tor Helleseth
University of Bergen
Bergen, Norway

Michael L. Honig
Northwestern University
Evanston, Illinois

Hwei P. Hsu
Fairleigh Dickinson University
Teaneck, New Jersey

Bijan Jabbari
George Mason University
Fairfax, Virginia

Ravi Jain
Bell Communications Research
Morristown, New Jersey

Varun Kapoor
Virginia Polytechnic Institute and
    State University
Blacksburg, Virginia

K. Kinoshita
NTT Do Co Mo

Andy D. Kucar
4U Communications Research Inc.
Ottawa, Ontario, Canada

P. Vijay Kumar
University of Southern California
Los Angeles, California

Vinod Kumar
Alcatel Mobile Communication
Columbes, France

Allen H. Levesque
GTE Laboratories, Inc.
Waltham, Massachusetts

Yi-Bing Lin
Bell Communications Research
Morristown, New Jersey

Joseph L. LoCicero
Illinois Institute of Technology
Chicago, Illinois

Paul Mermelstein
INRS-Télécommunications
Verdun, Quebec, Canada

Toshio Miki
NTT Mobile Communication
    Network, Inc.
Yokosuka-shi, Kanagawa, Japan

L.B. Milstein
University of California, San Diego
La Jolla, California

Seshadri Mohan
Bell Communications Research
Morristown, New Jersey

Rias Muhamed
Virginia Polytechnic Institute and
    State University
Blacksburg, Virginia

M. Nakagawa
Keio University
Yokohama, Japan

Geoffrey C. Orsak
George Mason University
Fairfax, Virginia

Kaveh Pahlavan
Worcester Polytechnic Institute
Worcester, Massachusetts

Bernd-Peter Paris
George Mason University
Fairfax, Virginia

Bhasker P. Patel
Illinois Institute of Technology
Chicago, Illinois

A. Paulraj
Stanford University
Stanford, California

Roman Pichna
University of Victoria
Victoria, British Columbia, Canada

John G. Proakis
Northeastern University
Boston, Massachusetts

Theodore S. Rappaport
Virginia Polytechnic Institute and
    State University
Blacksburg, Virginia

Arthur H.M. Ross
QUALCOMM Incorporated
San Diego, California

M.K. Simon
Jet Propulsion Laboratory
Pasadena, California

Suresh Singh
University of South Carolina
Columbia, South Carolina

Bernard Sklar
Communications Engineering
    Services
Tarzana, California

Raymond Steele
Southampton University and
    Multiple Access Communications
    Ltd.
Southampton, United Kingdom

Ferrel G. Stremler
University of Wisconsin
Madison, Wisconsin

Gordon L. Stüber
Georgia Institute of Technology
Atlanta, Georgia

Chong Kwan Un
Korea Advanced Institute of Science
    and Technology
Yusong-Ku, Taejon, Korea

Qiang Wang
University of Victoria
Victoria, British Columbia
Canada

Michel Daoud Yacoub
University of Campinas
Campinas, Sao Paolo, Brazil

Chong Ho Yoon
Hankook Aviation University
Koyang, Kyungki-Do, Korea

Contents

THE
MOBILE
COMMUNICATIONS
HANDBOOK

I

Basic Principles

1

Analog Modulation

Ferrel G. Stremler
University of Wisconsin

1.1 Introduction

Modulation is that process by which a property or a parameter of a given signal is varied in proportion to a second signal, which we term the input. Analog modulation generally refers to a modulation of the continuous complex exponential signal $a(t)\exp[j\theta(t)]$.[1] To make use of bandpass notation (and bandpass circuitry) we identify a carrier frequency ω_c about which we allow a bandwidth W (radians per second). Using the real-part notation, the modulated signal $\varphi(t)$ can be written as[2]

$$\varphi(t) = \text{Re}\{a(t)e^{j\theta(t)}\} = \text{Re}\{a(t)e^{j[\omega_c t + \gamma(t)]}\} \tag{1.1}$$

where $a(t)$ is the (time-varying) amplitude, ω_c the carrier frequency, and $\gamma(t)$ the (time-varying) angle with respect to the carrier phase $\omega_c t$. We assume that $a(t)$ and $\gamma(t)$ are slowly varying with respect to $\exp(j\omega_c t)$, i.e., we assume that $W \ll \omega_c$.

1.2 Amplitude Modulation

In **amplitude modulation** (**AM**), the angle term $\gamma(t)$ in Eq. (1.1) is a constant (often assumed to be zero) and $a(t)$ is made proportional to the input signal $f(t)$,

$$\varphi(t) = \text{Re}\{k_a f(t)e^{j\omega_c t}\} \tag{1.2}$$

where k_a is the proportionality constant of the modulator. The specific way in which this proportionality is achieved distinguishes several basic types of amplitude modulation.

---

[1] A relatively slow switching of a high-frequency sinusoidal signal on and off is sometimes referred to as continuous-wave (CW) modulation.

[2] The phasor notation is valid for the narrowband case $W \ll \omega_c$. Complex envelope notation and the concept of analytic signals can be used to extend this type of analysis to the wide bandwidth case; see, e.g., Proakis, J. G. 1989. *Digital Communications*, 2nd ed. McGraw-Hill, New York, Chap. 3.

Double-Sideband–Suppressed Carrier (DSB-SC)

If we assume that $f(t)$ is real valued, Eq. (1.2) simplifies to

$$\varphi(t) = k_a f(t) \cos \omega_c t \qquad (1.3)$$

An example of this type of modulation is shown in Fig. 1.1. Note that sign changes in the input $f(t)$ are conveyed as phase changes (of π radians) in carrier in the modulated waveform. Applying the frequency-shift property of the Fourier transform to Eq. (1.3), we find that the spectral density of $\varphi(t)$ is

$$\Phi(\omega) = \mathcal{F}\left\{\frac{1}{2}k_a f(t)e^{j\omega_c t} + \frac{1}{2}k_a f(t)e^{-j\omega_c t}\right\} = \frac{1}{2}F(\omega - \omega_c) + \frac{1}{2}F(\omega + \omega_c) \quad (1.4)$$

Thus, we see that this type of amplitude modulation translates the spectral density of $f(t)$ by $\pm\omega_c$ rad/s but leaves the spectral shape unaltered. This type of amplitude modulation is called **suppressed carrier** (SC) because the spectral density of $\varphi(t)$ has no (averaged) carrier in it, although the spectrum is centered at $\pm\omega_c$.

From the first term on the right-hand side of Eq. (1.4), we see that one-half of both the positive and the negative frequency content of $f(t)$ is shifted upward to positive frequencies in $\varphi(t)$, so that the

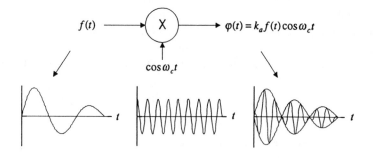

FIGURE 1.1   A suppressed-carrier amplitude modulation system.

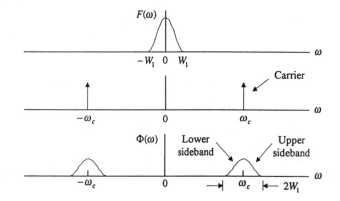

FIGURE 1.2   Spectral densities of signals shown in Fig. 1.1.

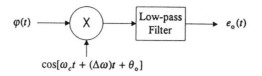

FIGURE 1.3 Demodulation of amplitude-modulated suppressed-carrier signals.

bandwidth is doubled.[3] This is illustrated in Fig. 1.2. The spectral content for positive frequencies above ω_c is called the *upper sideband* of $\varphi(t)$, and the spectral content for positive frequencies below ω_c is called the *lower sideband*. The fact that we have two sidebands and no identifiable averaged carrier present suggests the convenient designation: **double-sideband**, suppressed carrier (DSB-SC).

Demodulation of DSB-SC requires a multiplication operation to translate a portion of the spectral density down in frequency to baseband. A diagram of a basic receiver for DSB-SC is shown in Fig. 1.3. Introducing a frequency error $\Delta\omega$ and a phase error θ_0 in the locally generated reference in the receiver, the output signal $e_0(t)$ is

$$e_0(t) = \{f(t)\cos\omega_c t \cos[\omega_c t + (\Delta\omega)t + \theta_0]\}_{LP} = \frac{1}{2}f(t)\cos[(\Delta\omega)t + \theta_0] \quad (1.5)$$

The phase error θ_0 in the locally generated carrier causes a variable gain in the output signal; for small, fixed-phase errors this is tolerable. Any nonzero frequency error $\Delta\omega$, however, results in inacceptable multiplicative distortion. Thus, the use of synchronous detection of a DSB-SC modulated signal yields the original signal $f(t)$ with excellent fidelity.

Generation of a DSB-SC signal requires use of a nonlinear or a time-varying system. The former uses a nonlinear gain to generate the product-type term required; this nonlinearity may also introduce some carrier unless a signal balancing arrangement is provided to null out the carrier. The latter method generally uses a periodic switch-type multiplication that shifts a spectral replica of the input upward in frequency by the fundamental frequency of the switching waveform.

Double-Sideband–Large Carrier (DSB-LC)

Use of suppressed-carrier modulation requires provisions for acquiring and maintaining the necessary receiver synchronization. If we wish to use very inexpensive receivers at the expense of extra transmitter power, an alternative is to send a large enough carrier that no π jumps in carrier phase are necessary in the modulated waveform to convey sign changes of $f(t)$. To distinguish this case from the previous case, we designate this as double-sideband, **large carrier** (DSB-LC) modulation. Because commercial broadcast stations use this method of modulation, it is also commonly known as AM.

The DSB-LC modulated waveform can be described mathematically simply by adding a carrier term, $A\cos\omega_c t$, to the DSB-SC signal in Eq. (1.3),

$$\varphi_{AM}(t) = A\cos\omega_c t + k_a f(t)\cos\omega_c t = [A + k_a f(t)]\cos\omega_c t \quad (1.6)$$

If A is large enough, the envelope (magnitude) of the modulated waveform is proportional to $f(t)$. Demodulation in this case simply reduces to envelope detection (i.e., a diode and low-pass filter), plus deletion of an artificially generated average value caused by the added carrier term.

---

[3]By convention, the concepts of bandwidth, upper sideband, and lower sideband are defined for positive frequencies.

For the special case of a single-frequency sinusoid $f(t) = a \cos \omega_m t$, we define a dimensionless scale factor $m = ak_a/A$ to control the ratio of the peak sideband amplitude to the peak carrier amplitude. Using this sinusoidal input, Eq. (1.6) becomes

$$\varphi_{AM}(t) = A(1 + m \cos \omega_m t) \cos \omega_c t \tag{1.7}$$

The parameter m controls the relative proportions of peak sideband to peak carrier and is called the amplitude *modulation index*. A modulation index less than 100% on negative peaks of $f(t)$ does not result in a 180° phase change in the modulated waveform to convey a sign change in $f(t)$, and an inexpensive envelope detector can be used for demodulation. The economic advantages of use of an envelope detector for demodulation of DSB-LC instead of the synchronous detector necessary for DSB-SC come at the cost of a loss of low-frequency response (to provide for blocking the artificially generated nonzero average value) and a lowered power efficiency.

Generation of a DSB-LC signal can be accomplished by using gain nonlinearities to generate the product term required, or by a periodic switch-type time-varying system. The former is often used in low-power modulator applications, and the latter is used more in high-power modulator applications.

Commercial AM stations in the U.S. are assigned carrier frequencies at 10-kHz intervals from 520 to 1700 kHz. Required carrier stability is ±20 Hz, and the bandwidth of transmissions is nominally about 10 kHz. Stations in local proximity are usually assigned carrier frequencies that are separated by 30 kHz or more. Permissible transmitted (average) power levels range from 0.1 to 50 kW; licensed power output is for an unmodulated carrier. Transmission is primarily via ground-wave propagation during the day; sky-wave propagation may become more dominant at night in some circumstances. Interference between transmissions is controlled by a combination of frequency allocation, transmitter power, transmitting antenna pattern, and possible nighttime operating restrictions.

Quadrature Multiplexing (QM)

If we do not make the assumption that the input signal $f(t)$ is real valued in Eq. (1.2), we have

$$\varphi(t) = \text{Re}\{[x(t) - jy(t)]e^{j\omega_c t}\} \tag{1.8}$$

where $x(t)$ and $y(t)$ are the real and imaginary parts, respectively, of the input signal $f(t)$. Use of an identity for the real part of a product allows us to rewrite Eq. (1.8) as[4]

$$\varphi(t) = x(t) \cos \omega_c t + y(t) \sin \omega_c t = I(t) + Q(t) \tag{1.9}$$

where $I(t)$ is the in-phase component of $\varphi(t)$ and $Q(t)$ is the quadrature component. Because each is an example of DSB-SC modulation, the fact that the signals $\cos \omega_c t$ and $\sin \omega_c t$ are mutually orthogonal, and $x(t)$ and $y(t)$ are slowly varying with respect to $\exp(j\omega_c t)$, we can transmit (and recover) the two signals $x(t)$ and $y(t)$ simultaneously. This is referred to as **quadrature multiplexing** (QM). A quadrature multiplexed communication system is shown in Fig. 1.4.

Single Sideband (SSB)

The doubling of bandwidth in DSB modulation is a disadvantage if a given frequency band is crowded. Conversion to one sideband can be accomplished, in theory, by first generating DSB and then attenuating the undesired sideband. The **single-sideband** (SSB) filter requirements, however,

---

[4]Given two complex-valued quantities z_1, z_2, then $\text{Re}\{z_1 z_2\} = \text{Re}\{z_1\} \text{Re}\{z_2\} - \text{Im}\{z_1\} \text{Im}\{z_2\}$.

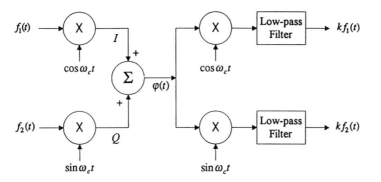

FIGURE 1.4 A quadrature-multiplexed communication system.

are stringent; ideally, the filter must pass all spectral components to one side of a carrier and attenuate all spectral components to the other side of the carrier. Using a complex-valued low-pass equivalent of the required bandpass filter, the frequency transfer function of the single-sideband filter is[5]

$$H(\omega) = \begin{cases} 2 & \omega > 0 \\ 0 & \omega < 0 \end{cases} \tag{1.10}$$

We designate the real and imaginary parts of this low-pass frequency transfer function in the following manner:

$$H(\omega) = H_r(\omega) + j H_i(\omega) \tag{1.11}$$

To maintain a phase characteristic that is an odd function of frequency and yet satisfies Eq. (1.10) requires[6]

$$H_r(\omega) = 1,$$
$$H_i(\omega) = \begin{cases} -j & \omega > 0 \\ j & \omega < 0 \end{cases} = -j\,\text{sgn}(\omega) \tag{1.12}$$

Using the known Fourier transform pair $\mathcal{F}\{\text{sgn}(\omega)\} = (-j\pi t)^{-1}$, we see that the required unit impulse response of the single-sideband filter is

$$h(t) = \delta(t) + \frac{1}{\pi t} \tag{1.13}$$

Thus, for an input $f(t)$, the corresponding filter output $g(t)$ is

$$g(t) = f(t) \otimes h(t) = f(t) + f(t) \otimes \frac{1}{\pi t} \tag{1.14}$$

where $\otimes$ is used to indicate the convolution operation.

The second term in the right-hand side of Eq. (1.14) is known as the *Hilbert transform* of $f(t)$, which we designate as $\hat{f}(t)$. Referring to Eq. (1.12) we see that each spectral component in $\hat{f}(t)$ is

---

[5]A gain factor of 2 is used here for convenience in the notation.
[6]Inequalities shown are for an upper-sideband filter; if the inequalities are interchanged, then we have a lower sideband filter.

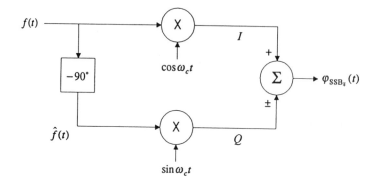

FIGURE 1.5   The phase-shift method of generating SSB modulation.

in phase quadrature ($-90°$) with that of $f(t)$. The Hilbert transform operation in time is not well defined at $t = 0$, and this causes peaks of undefined amplitude at points of finite discontinuity in $f(t)$. As a result, the use of SSB modulation is restricted to smooth waveforms to retain reasonable control of the overall signal envelope.

An interesting application of the preceding results and quadrature multiplexing leads to the phase shift method of generating SSB shown in Fig. 1.5. A major problem in the design of a phase-shift SSB system is the practical realization of the 90° phase-shift filter for $f(t)$, because all spectral components in the band must be shifted by exactly 90°. On the other hand, the SSB filter method must allow some finite attenuation rate at the band edge. Mathematically, the SSB filter method and the SSB phase-shift method are equivalent. In either case, the resultant SSB-SC signal can be written as

$$\varphi_{SSB_\mp}(t) = \mathrm{Re}\{[f(t) \pm j\hat{f}(t)]e^{j\omega_c t}\} \tag{1.15}$$

where SSB_+ designates upper sideband and SSB_- designates lower sideband. If the SSB-SC signal described by Eq. (1.15) is applied to the input of the synchronous detector shown in Fig. 1.3, the output can be written in the form

$$e_0(t) = \mathrm{Re}\{[f(t) \pm j\hat{f}(t)]e^{j\omega_c t}e^{-j(\omega_c t+\Delta\omega t+\theta_0)}\} \tag{1.16}$$

where $\Delta\omega$ and θ_0 are the frequency and phase errors, respectively, between transmitter and receiver. Phase distortion, i.e., a mix of $f(t)$ and $\hat{f}(t)$, results if $\theta_0 \neq 0$ and, in addition, a spectral shift results if $\Delta\omega \neq 0$. Although the fidelity is not very good, the SSB-SC receiver does not have the serious multiplicative distortion effect that occurs in DSB-SC reception using synchronous detection, and intelligible speech can be received using only half of the bandwidth of DSB systems, but at a cost of reduced fidelity. Single-sideband modulation is used in many of the amateur radio bands as a primary method for transmission of voice signals for communication.

Vestigial-Sideband (VSB) Modulation

To conserve spectral occupancy, a compromise can be made between the bandwidth requirements of SSB and DSB modulation by transmission of most of one sideband and only a portion or vestige of the second sideband. This **vestigial-sideband** (VSB) modulation is transmitted in such a way that synchronous detection of the modulated signal reproduces the original input signal. The VSB modulator can be represented by a DSB modulator followed by a VSB filter with frequency transfer function $H_V(\omega)$; the spectral density of the filter output is

$$\Phi_{VSB}(\omega) = \Phi_{DSB}(\omega)H_V(\omega) = \left[\frac{1}{2}F(\omega - \omega_c) + \frac{1}{2}F(\omega + \omega_c)\right]H_V(\omega) \qquad (1.17)$$

The output of a synchronous detector with the input signal $\varphi_{VSB}(t)$ is

$$e_0(t) = [\varphi_{VSB}(t)\cos\omega_c t]_{LP}$$
$$E_0(\omega) = \frac{1}{4}F(\omega)H_V(\omega + \omega_c) + \frac{1}{4}F(\omega)H_V(\omega - \omega_c) \qquad (1.18)$$

Thus, for a faithful reproduction of $f(t)$ we require that

$$[H_V(\omega - \omega_c) + H_V(\omega + \omega_c)]_{LP} = \text{const} \qquad (|\omega| - \omega_c) < \omega_m \qquad (1.19)$$

Choosing the constant in Eq. (1.19) to be $2H_V(\omega_c)$, we find we require that $|H_V(\omega)|$ be anti-symmetric about the carrier frequency ω_c. Although we have used synchronous detection in the preceding results, these principles also hold if a large carrier is present and if envelope detection is used.

Commercial television stations in the U.S. use VSB-LC modulation for transmission of the luminance (black and white) video signal, and quadrature-multiplexed DSB-SC and VSB-SC modulation for the color video signals. The color signal subcarrier is 3.579545 MHz above the video carrier, and is sent during a brief burst in the blanked portion of each horizontal trace line. The audio is sent using **frequency modulation** (FM) (cf. next section) with a carrier frequency 4.5 MHz above the frequency of the video carrier and with a peak frequency deviation of 25 kHz. The (audio) stereo is sent using DSB-SC/FM with a subcarrier frequency equal to twice the horizontal trace frequency of the video signal. Stations are assigned video carrier frequencies at 6-MHz intervals from 55.25 to 83.25 MHz and 175.25 to 211.25 MHz in the VHF band, and from 471.25 to 801.25 MHz in the UHF band. Maximum effective radiated (average) power is 100–316 kW in the VHF band and 5 MW in the UHF band. The audio carrier is transmitted 7–10 dB lower average power level than the video carrier.[7]

1.3 Angle Modulation

A general relation between the instantaneous angular rate $\omega_i(t)$ and the angle $\theta(t)$ in circular motion is

$$\theta(t) = \int_0^t \omega_i(\tau)\, d\tau + \theta_0 \qquad (1.20)$$

Taking the derivative of both sides of Eq. (1.20), we have

$$\omega_i(t) = \frac{d\theta}{dt} \qquad (1.21)$$

When applied to the phasor signal representation in Eq. (1.1), $\omega_i(t)$ is called the instantaneous frequency of the signal $\varphi(t)$. This definition of frequency differs, in general, from that associated with spectral content in Fourier analysis. The latter is time averaged and, thus, cannot be time dependent; in contrast, the instantaneous frequency can be time dependent.

---

[7]For more information on television systems, see, e.g., Benson, K. B. ed. 1986. *Television Engineering Handbook*. McGraw-Hill, New York.

In **angle modulation** we assume a constant amplitude in Eq. (1.1) and modulate the phase angle with respect to the carrier in proportion to the input signal $f(t)$. For a direct proportionality, we have

$$\theta(t) = \omega_c t + k_p f(t) + \theta_0 \qquad \text{(PM)} \tag{1.22}$$

and this type of angle modulation is called **phase modulation** (PM). Another possibility to make the instantaneous frequency $\omega_i(t)$ proportional to $f(t)$ is

$$\omega_i(t) = \omega_c + k_f f(t), \qquad \text{(FM)} \tag{1.23}$$

and this type is called **frequency modulation** (FM).

FM

Because the angle vs amplitude transformation is not linear—and thus superposition cannot be used—we choose a specific signal for the input,

$$f(t) = a \cos \omega_m t \tag{1.24}$$

The resulting instantaneous frequency [cf. Eqs. (1.23) and (1.24)] is

$$\omega_i(t) = \omega_c + a k_f \cos \omega_m t \tag{1.25}$$

where k_f is the frequency modulator constant; typical units are radians per second per volt.

Defining a new constant $\Delta\omega = a k_f$, called the *peak frequency deviation* (from carrier), we rewrite Eq. (1.25) as

$$\omega_i(t) = \omega_c + \Delta\omega \cos \omega_m t \tag{1.26}$$

Using Eq. (1.26) in Eq. (1.20), the phase angle of the FM signal with sinusoidal input is

$$\theta(t) = \omega_c t + \beta \sin \omega_m t + \theta_0 \tag{1.27}$$

where

$$\beta = \frac{\Delta\omega}{\omega_m} \tag{1.28}$$

is called the frequency *modulation index.* Combining Eqs. (1.1) and (1.27), we have[8]

$$\varphi_{\text{FM}}(t) = \text{Re}\{A e^{j\omega_c t} e^{j\beta \sin \omega_m t}\} \tag{1.29}$$

To obtain a result in terms of averaged spectral content, we note that the second exponential term in Eq. (1.29) is a periodic function (with a fundamental frequency of ω_m rad/s). Thus, it can be expanded in a Fourier series; doing this gives the result[9]

$$\varphi_{\text{FM}}(t) = \text{Re}\left\{ A e^{j\omega_c t} \sum_{n=-\infty}^{\infty} J_n(\beta) e^{jn\omega_m t} \right\} \tag{1.30}$$

---

[8]The constant phase θ_0 has been absorbed into A for convenience in this equation.

[9]Notation differs in describing Eq. (1.30). Some authors refer to the $J_n(\beta)$ as sidebands, and then we can say that $\varphi_{\text{FM}}(t)$ has an infinite number of sidebands. Others refer to only two sidebands, one above the carrier and one below, and then each sideband contains multiple side frequencies.

where $J_n(\beta)$ is the Bessel function of the first kind, of order n and argument β. Based on properties of these Bessel functions, an approximate relation for the bandwidth of $\varphi_{FM}(t)$, known as Carson's rule, is

$$W \approx 2(\Delta\omega + \omega_m) = 2\omega_m(1 + \beta) \qquad (1.31)$$

Tabulated numerical values of Bessel functions or a direct numerical computation can be used for more accurate measures of bandwidth.

The analytical methods used here in applying the Fourier series for a spectral representation of angle-modulated signals is a very powerful one and can be used for other choices of periodic signals. Another approach is to use the fast Fourier transform (FFT) algorithm for a direct computation of the spectral components. Both approaches measure time-averaged spectral components, in contrast to the concept of instantaneous frequency.

Demodulation of FM requires circuitry that produces an output amplitude, which is proportional to the instantaneous frequency from carrier. Many modern receivers use a feedback method, called the phase-locked loop (PLL), to provide good demodulation without the use of tuned circuits. Because instantaneous frequency is the derivative of phase [cf. Eq. (1.21)], demodulation of FM tends to emphasize high-frequency noise, a disadvantage overcome by use of PM. An effect of the nonlinear modulation in FM and PM is to suppress weak signals (or noise); this helps to define service areas for commercial broadcasting, and provides a noise quieting effect in suppressing additive noise in the output if the signal-to-noise ratio is high.

Commercial FM stations in the U.S. are assigned carrier frequencies at 200-kHz intervals from 88.1 to 107.9 MHz. The peak frequency deviation is 75 kHz. Minimum required carrier frequency stability is $\pm 0.002\%$ (i.e., ± 2 kHz at 100 MHz). Stereo audio signals are sent using DSB-SC modulation at a subcarrier frequency of 38 kHz; a pilot tone not exceeding 10% of the peak frequency deviation is sent at 19 kHz. Permissible transmitted (average) power levels range from 0.25 to 100 kW. Transmissions received beyond the line-of-sight distance are weak and subject to fading.

PM

The basic difference between PM and FM is that the phase (with respect to the carrier) in the modulated waveform is proportional to the input signal $f(t)$ in PM and to the integral of the input $f(t)$ in FM. For sinusoidal analyses this introduces only a slight modification. For the sinusoidal input described in Eq. (1.24), the resulting PM signal has a phase

$$\theta(t) = \omega_c t + \Delta\theta \cos\omega_m t + \theta_0 \qquad (1.32)$$

where $\Delta\theta = ak_p$ is the *peak phase deviation* from the carrier and k_p is the phase modulator constant (in radians per volt). Taking a derivative of Eq. (1.32) and comparing the result with Eq. (1.26), we see that the peak frequency deviation in PM is proportional not only to the amplitude of the modulating signal but also to its frequency; that is,

$$\Delta\omega = \begin{cases} ak_f & \text{for FM} \\ ak_p\omega_m = (\Delta\theta)\omega_m & \text{for PM} \end{cases} \qquad (1.33)$$

This makes PM less desirable to transmit when $\Delta\omega$ is fixed (as in commercial FM). There are some advantages in the demodulation of PM, however, which make its use desirable, as noted in the preceding section. The role of the modulation index β and the analysis in terms of Bessel functions for PM are essentially the same as in the analysis of FM [cf. Eq. (1.28)]. Note that the numerical value of β is equal to the peak phase deviation $\Delta\theta$ in the PM case [cf. Eqs. (1.27) and (1.32)].

Defining Terms

Amplitude modulation: A continuous-wave modulation using amplitude variations in proportion to the amplitude of the input signal.

Angle modulation: A continuous-wave modulation using angle variations in proportion to the amplitude of the input signal.

Double Sideband (DSB): An amplitude-modulated signal (LC or SC) having two spectral sidebands symmetrically about a carrier frequency.

Frequency modulation (FM): A continuous-wave modulation using instantaneous frequency variations from carrier in proportion to the amplitude of the input signal.

Large Carrier (LC): A modulated signal in which a relatively large proportion of the spectrum is concentrated at the carrier frequency (used with DSB, SSB, VSB).

Phase modulation (PM): A continuous-wave modulation using phase variations from carrier in proportion to the amplitude of the input signal.

Quadrature Multiplexing: Use of both in-phase (cosine reference) and quadrature (sine reference) modulation.

Suppressed carrier (SC): A modulated signal in which a relatively small proportion (ideally, zero) of the spectrum is concentrated at the carrier frequency (used with DSB, SSB, VSB).

Single sideband (SSB): An amplitude-modulated signal (LC or SC) having one spectral sideband either above (upper) or below (lower) the carrier frequency.

Vestigial sideband (VSB): An amplitude-modulated signal for which the spectral density has the major portion of one sideband and a vestige of the second sideband.

References

Gregg, W.D. 1977. *Analog and Digital Communication*, John Wiley & Sons, New York.

Gibson, J.D. 1993. *Principles of Digital and Analog Communications*, 2nd ed., Macmillan, New York, NY.

Haykin, S. 1994. *Communication Systems*, 3rd ed., John Wiley & Sons, New York, NY.

Panter, P.F. 1965. *Modulation, Noise, and Spectral Analysis*, McGraw-Hill, New York.

Stremler, F.G. 1990. *Introduction to Communication Systems*, 3rd ed., Addison-Wesley, Reading, MA.

Taub, H. and Schilling, D.L. 1986. *Principles of Communication Systems*, 2nd ed., McGraw-Hill, New York.

Further Information

Some practical aspects of amplitude modulation and demodulation are described in R.S. Carson's *Radio Communications Concepts: Analog*, published in 1990 by John Wiley & Sons, New York, NY.

For explanations of some of the popular circuits used for angle demodulation, see, e.g., H. Stark, F.B. Tuteur, and J.B. Anderson's *Modern Electrical Communications*, 1988, 2nd ed., published by Prentice Hall, Englewood Cliffs, NJ.

Circuits design considerations for analog modulation systems are covered by K.K. Clarke and D.T. Hess, in *Communication Circuits: Analysis and Design*, 1971, Addison-Wesley, Reading, MA. (Reprinted in 1994 by Krieger Publishing Co., Melbourne, FL.)

2

Sampling

Hwei P. Hsu
Fairleigh Dickinson University

2.1 Introduction

To transmit analog message signals, such as speech signals or video signals, by digital means, the signal has to be converted into digital form. This process is known as analog-to-digital conversion. The sampling process is the first process performed in this conversion, and it converts a continuous-time signal into a discrete-time signal or a sequence of numbers. Digital transmission of analog signals is possible by virtue of the sampling theorem, and the sampling operation is performed in accordance with the sampling theorem.

In this chapter, using the Fourier transform technique, we present this remarkable sampling theorem and discuss the operation of sampling and practical aspects of sampling.

2.2 Instantaneous Sampling

Suppose we sample an arbitrary analog signal $m(t)$ shown in Fig. 2.1(a) instantaneously at a uniform rate, once every T_s seconds. As a result of this sampling process, we obtain an infinite sequence of samples $\{m(nT_s)\}$, where n takes on all possible integers. This form of sampling is called *instantaneous sampling*. We refer to T_s as the **sampling interval**, and its reciprocal $1/T_s = f_s$ as the **sampling rate**. Sampling rate (samples per second) is often cited in terms of sampling frequency expressed in hertz.

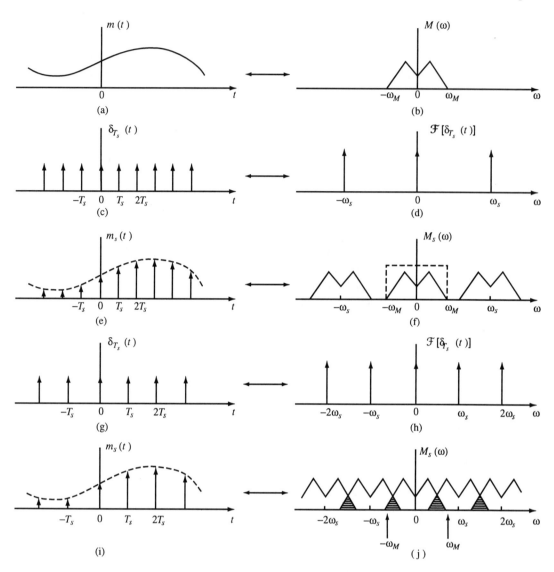

FIGURE 2.1   Illustration of instantaneous sampling and sampling theorem.

Ideal Sampled Signal

Let $m_s(t)$ be obtained by multiplication of $m(t)$ by the unit impulse train $\delta_T(t)$ with period T_s [Fig. 2.1(c)], that is,

$$m_s(t) = m(t)\delta_{T_s}(t) = m(t) \sum_{n=-\infty}^{\infty} \delta(t - nT_s)$$

$$= \sum_{n=-\infty}^{\infty} m(t)\delta(t - nT_s) = \sum_{n=-\infty}^{\infty} m(nT_s)\delta(t - nT_s) \qquad (2.1)$$

where we used the property of the δ function, $m(t)\delta(t - t_0) = m(t_0)\delta(t - t_0)$. The signal $m_s(t)$ [Fig. 2.1(e)] is referred to as the **ideal sampled signal**.

Band-Limited Signals

A real-valued signal $m(t)$ is called a **band-limited signal** if its Fourier transform $M(\omega)$ satisfies the condition

$$M(\omega) = 0 \qquad \text{for } |\omega| > \omega_M \tag{2.2}$$

where $\omega_M = 2\pi f_M$ [Fig. 2.1(b)]. A band-limited signal specified by Eq. (2.2) is often referred to as a *low-pass signal*.

2.3 Sampling Theorem

The sampling theorem states that a band-limited signal $m(t)$ specified by Eq. (2.2) can be uniquely determined from its values $m(nT_s)$ sampled at uniform interval T_s if $T_s \leq \pi/\omega_M = 1/(2f_M)$. In fact, when $T_s = \pi/\omega_M$, $m(t)$ is given by

$$m(t) = \sum_{n=-\infty}^{\infty} m(nT_s) \frac{\sin \omega_M(t - nT_s)}{\omega_M(t - nT_s)} \tag{2.3}$$

which is known as the **Nyquist–Shannon interpolation formula** and it is also sometimes called the *cardinal series*. The sampling interval $T_s = 1/(2f_M)$ is called the *Nyquist interval* and the minimum rate $f_s = 1/T_s = 2f_M$ is known as the **Nyquist rate**.

Illustration of the instantaneous sampling process and the sampling theorem is shown in Fig. 2.1. The Fourier transform of the unit impulse train is given by [Fig. 2.1(d)]

$$\mathcal{F}\{\delta_{T_s}(t)\} = \omega_s \sum_{n=-\infty}^{\infty} \delta(\omega - n\omega_s) \qquad \omega_s = 2\pi/T_s \tag{2.4}$$

Then, by the convolution property of the Fourier transform, the Fourier transform $M_s(\omega)$ of the ideal sampled signal $m_s(t)$ is given by

$$M_s(\omega) = \frac{1}{2\pi} \left[M(\omega) * \omega_s \sum_{n=-\infty}^{\infty} \delta(\omega - n\omega_s) \right]$$

$$= \frac{1}{T_s} \sum_{n=-\infty}^{\infty} M(\omega - n\omega_s) \tag{2.5}$$

where $*$ denotes convolution and we used the convolution property of the δ-function $M(\omega) * \delta(\omega - \omega_0) = M(\omega - \omega_0)$. Thus, the sampling has produced images of $M(\omega)$ along the frequency axis. Note that $M_s(\omega)$ will repeat periodically without overlap as long as $\omega_s \geq 2\omega_M$ or $f_s \geq 2f_M$ [Fig. 2.1(f)]. It is clear from Fig. 2.1(f) that we can recover $M(\omega)$ and, hence, $m(t)$ by passing the sampled signal $m_s(t)$ through an ideal low-pass filter having frequency response

$$H(\omega) = \begin{cases} T_s, & |\omega| \leq \omega_M \\ 0, & \text{otherwise} \end{cases} \tag{2.6}$$

where $\omega_M = \pi/T_s$. Then

$$M(\omega) = M_s(\omega)H(\omega) \tag{2.7}$$

Taking the inverse Fourier transform of Eq. (2.6), we obtain the impulse response $h(t)$ of the ideal low-pass filter as

$$h(t) = \frac{\sin \omega_M t}{\omega_M t} \tag{2.8}$$

Taking the inverse Fourier transform of Eq. (2.7), we obtain

$$m(t) = m_s(t) * h(t)$$

$$= \sum_{n=-\infty}^{\infty} m(nT_s)\delta(t - nT_s) * \frac{\sin \omega_M t}{\omega_M t}$$

$$= \sum_{n=-\infty}^{\infty} m(nT_s)\frac{\sin \omega_M (t - nT_s)}{\omega_M (t - nT_s)} \tag{2.9}$$

which is Eq. (2.3).

The situation shown in Fig. 2.1(j) corresponds to the case where $f_s < 2f_M$. In this case there is an overlap between $M(\omega)$ and $M(\omega - \omega_M)$. This overlap of the spectra is known as *aliasing* or *foldover*. When this aliasing occurs, the signal is distorted and it is impossible to recover the original signal $m(t)$ from the sampled signal. To avoid aliasing, in practice, the signal is sampled at a rate slightly higher than the Nyquist rate. If $f_s > 2f_M$, then as shown in Fig. 2.1(f), there is a gap between the upper limit ω_M of $M(\omega)$ and the lower limit $\omega_s - \omega_M$ of $M(\omega - \omega_s)$. This range from ω_M to $\omega_s - \omega_M$ is called a *guard band*. As an example, speech transmitted via telephone is generally limited to $f_M = 3.3$ kHz (by passing the sampled signal through a low-pass filter). The Nyquist rate is, thus, 6.6 kHz. For digital transmission, the speech is normally sampled at the rate $f_s = 8$ kHz. The guard band is then $f_s - 2f_M = 1.4$ kHz. The use of a sampling rate higher than the Nyquist rate also has the desirable effect of making it somewhat easier to design the low-pass reconstruction filter so as to recover the original signal from the sampled signal.

2.4 Sampling of Sinusoidal Signals

A special case is the sampling of a sinusoidal signal having the frequency f_M. In this case we require that $f_s > 2f_M$ rather that $f_s \geq 2f_M$. To see that this condition is necessary, let $f_s = 2f_M$. Now, if an initial sample is taken at the instant the sinusoidal signal is zero, then all successive samples will also be zero. This situation is avoided by requiring $f_s > 2f_M$.

2.5 Sampling of Bandpass Signals

A real-valued signal $m(t)$ is called a **bandpass signal** if its Fourier transform $M(\omega)$ satisfies the condition

$$M(\omega) = 0 \quad \text{except for} \quad \begin{cases} \omega_1 < \omega < \omega_2 \\ -\omega_2 < \omega < -\omega_1 \end{cases} \tag{2.10}$$

where $\omega_1 = 2\pi f_1$ and $\omega_2 = 2\pi f_2$ [Fig. 2.2(a)].

The sampling theorem for a band-limited signal has shown that a sampling rate of $2f_2$ or greater is adequate for a low-pass signal having the highest frequency f_2. Therefore, treating $m(t)$ specified by Eq. (2.10) as a special case of such a low-pass signal, we conclude that a sampling rate of $2f_2$ is adequate for the sampling of the bandpass signal $m(t)$. But it is not necessary to sample this fast. The minimum allowable sampling rate depends on f_1, f_2, and the bandwidth $f_B = f_2 - f_1$.

Let us consider the direct sampling of the bandpass signal specified by Eq. (2.10). The spectrum of the sampled signal is periodic with the period $\omega_s = 2\pi f_s$, where f_s is the sampling frequency,

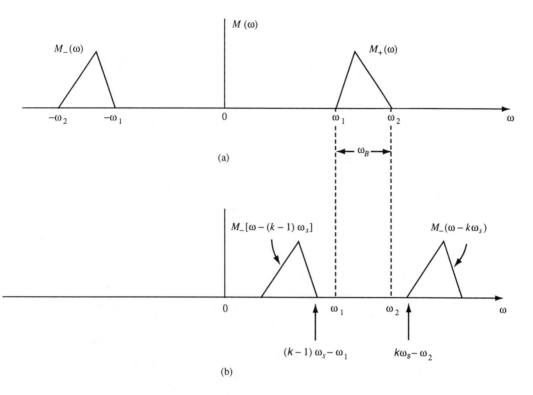

FIGURE 2.2 (a) Spectrum of a bandpass signal; (b) Shifted spectra of $M_-(\omega)$.

as in Eq. (2.4). Shown in Fig. 2.2(b) are the two right shifted spectra of the negative side spectrum $M_-(\omega)$. If the recovering of the bandpass signal is achieved by passing the sampled signal through an ideal bandpass filter covering the frequency bands $(-\omega_2, -\omega_1)$ and (ω_1, ω_2), it is necessary that there be no aliasing problem. From Fig. 2.2(b), it is clear that to avoid overlap it is necessary that

$$\omega_s \geq 2(\omega_2 - \omega_1) \tag{2.11}$$

$$(k-1)\omega_s - \omega_1 \leq \omega_1 \tag{2.12}$$

and

$$k\omega_s - \omega_2 \geq \omega_2 \tag{2.13}$$

where $\omega_1 = 2\pi f_1$, $\omega_2 = 2\pi f_2$, and k is an integer ($k = 1, 2, \ldots$). Since $f_1 = f_2 - f_B$, these constraints can be expressed as

$$1 \leq k \leq \frac{f_2}{f_B} \leq \frac{k}{2}\frac{f_s}{f_B} \tag{2.14}$$

and

$$\frac{k-1}{2}\frac{f_s}{f_B} \leq \frac{f_2}{f_B} - 1 \tag{2.15}$$

A graphical description of Eqs. (2.14) and (2.15) is illustrated in Fig. 2.3. The unshaded regions represent where the constraints are satisfied, whereas the shaded regions represent the regions where the constraints are not satisfied and overlap will occur. The solid line in Fig. 2.3 shows the locus of

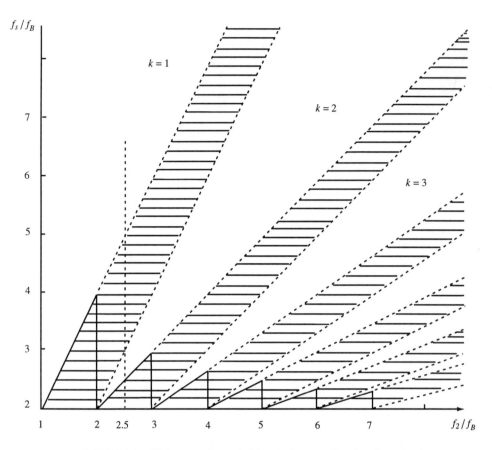

FIGURE 2.3   Minimum and permissible sampling rates for a bandpass signal.

the minimum sampling rate. The minimum sampling rate is given by

$$\min\{f_s\} = \frac{2f_2}{m} \tag{2.16}$$

where m is the largest integer not exceeding f_2/f_B. Note that if the ratio f_2/f_B is an integer, then the minimum sampling rate is $2f_B$. As an example, consider a bandpass signal with $f_1 = 1.5$ kHz and $f_2 = 2.5$ kHz. Here $f_B = f_2 - f_1 = 1$ kHz, and $f_2/f_B = 2.5$. Then from Eq. (2.16) and Fig. 2.3 we see that the minimum sampling rate is $2f_2/2 = f_2 = 2.5$ kHz, and allowable ranges of sampling rate are 2.5 kHz $\leq f_s \leq 3$ kHz and $f_s \geq 5$ kHz $(= 2f_2)$.

2.6 Practical Sampling

In practice, the sampling of an analog signal is performed by means of high-speed switching circuits, and the sampling process takes the form of *natural sampling* or **flat-top sampling**.

Natural Sampling

Natural sampling of a band-limited signal $m(t)$ is shown in Fig. 2.4. The sampled signal $m_{ns}(t)$ can be expressed as

$$m_{ns}(t) = m(t)x_p(t) \tag{2.17}$$

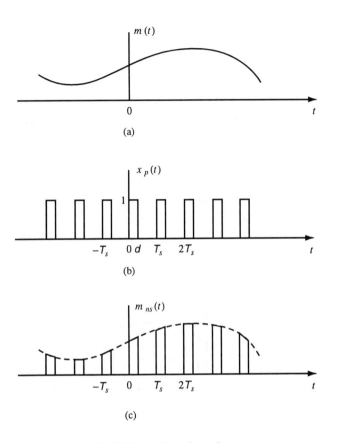

FIGURE 2.4 Natural sampling.

where $x_p(t)$ is the periodic train of rectangular pulses with fundamental period T_s, and each rectangular pulse in $x_p(t)$ has duration d and unit amplitude [Fig. 2.4(b)]. Observe that the sampled signal $m_{ns}(t)$ consists of a sequence of pulses of varying amplitude whose tops follow the waveform of the signal $m(t)$ [Fig. 2.4(c)].

The Fourier transform of $x_p(t)$ is

$$X_p(\omega) = \sum_{n=-\infty}^{\infty} c_n \delta(\omega - n\omega_s) \qquad \omega_s = 2\pi/T_s \tag{2.18}$$

where

$$c_n = \frac{d}{T_s}\frac{\sin(n\omega_s d/2)}{n\omega_s d/2}e^{-jn\omega_s d/2} \tag{2.19}$$

Then the Fourier transform of $m_{ns}(t)$ is given by

$$M_{ns}(\omega) = M(\omega) * X_p(\omega) = \sum_{n=-\infty}^{\infty} c_n M(\omega - n\omega_s) \tag{2.20}$$

from which we see that the effect of the natural sampling is to multiply the nth shifted spectrum $M(\omega - n\omega_s)$ by a constant c_n. Thus, the original signal $m(t)$ can be reconstructed from $m_{ns}(t)$ with no distortion by passing $m_{ns}(t)$ through an ideal low-pass filter if the sampling rate f_s is equal to or greater than the Nyquist rate $2f_M$.

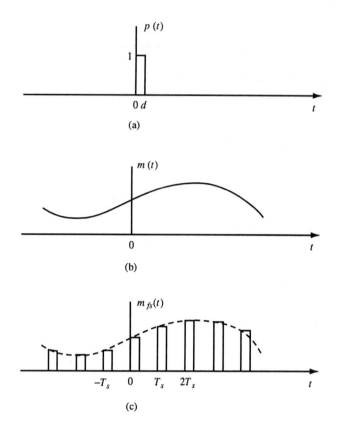

FIGURE 2.5 Flat-top sampling.

Flat-Top Sampling

The sampled waveform, produced by practical sampling devices that are the sample and hold types, has the form [Fig. 2.5(c)]

$$m_{fs}(t) = \sum_{n=-\infty}^{\infty} m(nT_s)p(t - nT_s) \tag{2.21}$$

where $p(t)$ is a rectangular pulse of duration d with unit amplitude [Fig. 2.5(a)]. This type of sampling is known as **flat-top sampling**. Using the ideal sampled signal $m_s(t)$ of Eq. (2.1), $m_{fs}(t)$ can be expressed as

$$m_{fs}(t) = p(t) * \left[\sum_{n=-\infty}^{\infty} m(nT_s)\delta(t - nT_s) \right] = p(t) * m_s(t) \tag{2.22}$$

Using the convolution property of the Fourier transform and Eq. (2.4), the Fourier transform of $m_{fs}(t)$ is given by

$$M_{fs}(\omega) = P(\omega)M_s(\omega) = \frac{1}{T_s} \sum_{n=-\infty}^{\infty} P(\omega)M(\omega - n\omega_s) \tag{2.23}$$

where

$$P(\omega) = d\frac{\sin(\omega d/2)}{\omega d/2}e^{-j\omega d/2} \qquad (2.24)$$

From Eq. (2.23) we see that by using flat-top sampling we have introduced amplitude distortion and time delay, and the primary effect is an attenuation of high-frequency components. This effect is known as the *aperture effect*. The aperture effect can be compensated by an equalizing filter with a frequency response $H_{eq}(\omega) = 1/P(\omega)$. If the pulse duration d is chosen such that $d \ll T_s$, however, then $P(\omega)$ is essentially constant over the baseband and no equalization may be needed.

2.7 Sampling Theorem in the Frequency Domain

The sampling theorem expressed in Eq. (2.4) is the time-domain sampling theorem. There is a dual to this time-domain sampling theorem, i.e., the sampling theorem in the frequency domain.

Time-limited signals: A continuous-time signal $m(t)$ is called **time limited** if

$$m(t) = 0 \qquad \text{for} \quad |t| > |T_0| \qquad (2.25)$$

Frequency-domain sampling theorem: The frequency-domain sampling theorem states that the Fourier transform $M(\omega)$ of a time-limited signal $m(t)$ specified by Eq. (2.25) can be uniquely determined from its values $M(n\omega_s)$ sampled at a uniform rate ω_s if $\omega_s \leq \pi/T_0$. In fact, when $\omega_s = \pi/T_0$, then $M(\omega)$ is given by

$$M(\omega) = \sum_{n=-\infty}^{\infty} M(n\omega_s)\frac{\sin T_0(\omega - n\omega_s)}{T_0(\omega - n\omega_s)} \qquad (2.26)$$

2.8 Summary and Discussion

The sampling theorem is the fundamental principle of digital communications. We state the sampling theorem in two parts.

Theorem 2.1. If the signal contains no frequency higher than f_M Hz, it is completely described by specifying its samples taken at instants of time spaced $1/2f_M$ s.

Theorem 2.2. The signal can be completely recovered from its samples taken at the rate of $2f_M$ samples per second or higher.

The preceding sampling theorem assumes that the signal is strictly band limited. It is known that if a signal is band limited it cannot be time limited and vice versa. In many practical applications, the signal to be sampled is time limited and, consequently, it cannot be strictly band limited. Nevertheless, we know that the frequency components of physically occurring signals attenuate rapidly beyond some defined bandwidth, and for practical purposes we consider these signals are band limited. This approximation of real signals by band limited ones introduces no significant error in the application of the sampling theorem. When such a signal is sampled, we band limit the signal by filtering before sampling and sample at a rate slightly higher than the nominal Nyquist rate.

Defining Terms

Band-Limited Signal: A signal whose frequency content (Fourier transform) is equal to zero above some specified frequency.

Bandpass Signal: A signal whose frequency content (Fourier transform) is nonzero only in a band of frequencies not including the origin.

Flat-top Sampling: Sampling with finite width pulses that maintain a constant value for a time period less than or equal to the sampling interval. The constant value is the amplitude of the signal at the desired sampling instant.

Ideal Sampled Signal: A signal sampled using an ideal impulse train.

Nyquist Rate: The minimum allowable sampling rate of $2 f_M$ samples per second, to reconstruct a signal band limited to f_M hertz.

Nyquist-Shannon Interpolation Formula: The infinite series representing a time domain waveform in terms of its ideal samples taken at uniform intervals.

Sampling Interval: The time between samples in uniform sampling.

Sampling Rate: The number of samples taken per second (expressed in Hertz and equal to the reciprocal of the sampling interval).

Time-limited: A signal that is zero outside of some specified time interval.

References

Brown, J.L. Jr. 1980. First order sampling of bandpass signals—A new approach. *IEEE Trans. Information Theory.* IT-26(5):613–615.

Byrne, C.L. and Fitzgerald, R. M. 1982. Time-limited sampling theorem for band-limited signals, *IEEE Trans. Information Theory.* IT-28(5):807–809.

Hsu, H.P. 1984. *Applied Fourier Analysis,* Harcourt Brace Jovanovich, San Diego, CA.

Hsu, H.P. 1993. *Analog and Digital Communications,* McGraw-Hill, New York.

Hulthén, R. 1983. Restoring causal signals by analytical continuation: A generalized sampling theorem for causal signals. *IEEE Trans. Acoustics, Speech, and Signal Processing.* ASSP-31(5):1294–1298.

Jerri, A.J. 1977. The Shannon sampling theorem—Its various extensions and applications: A tutorial review, *Proc. IEEE.* 65(11):1565–1596.

Further Information

For a tutorial review of the sampling theorem, historical notes, and earlier references see Jerri (1977).

3

Pulse Code Modulation[1]

Leon W. Couch II
University of Florida

3.1 Introduction

Pulse code modulation (PCM) is analog-to-digital conversion of a special type where the information contained in the instantaneous samples of an analog signal is represented by digital words in a serial bit stream.

If we assume that each of the digital words has n binary digits, there are $M = 2^n$ unique code words that are possible, each code word corresponding to a certain amplitude level. Each sample value from the analog signal, however, can be any one of an infinite number of levels, so that the digital word that represents the amplitude closest to the actual sampled value is used. This is called **quantizing**. That is, instead of using the exact sample value of the analog waveform, the sample is replaced by the closest allowed value, where there are M allowed values, and each allowed value corresponds to one of the code words.

PCM is very popular because of the many advantages it offers. Some of these advantages are as follows.

- Relatively inexpensive digital circuitry may be used extensively in the system.
- PCM signals derived from all types of analog sources (audio, video, etc.) may be time-division multiplexed with data signals (e.g., from digital computers) and transmitted over a common high-speed digital communication system.
- In long-distance digital telephone systems requiring repeaters, a *clean* PCM waveform can be regenerated at the output of each repeater, where the input consists of a noisy PCM waveform. The noise at the input, however, may cause bit errors in the regenerated PCM output signal.

[1] *Source:* Leon W. Couch, II 1993. *Digital and Analog Communication Systems,* 4th ed., Macmillan Publishing Co., New York. With permission.

0-8493-8573-3/96/$0.00+$.50

• The noise performance of a digital system can be superior to that of an analog system. In addition, the probability of error for the system output can be reduced even further by the use of appropriate coding techniques.

These advantages usually outweigh the main disadvantage of PCM: a much wider bandwidth than that of the corresponding analog signal.

3.2 Generation of PCM

The PCM signal is generated by carrying out three basic operations: sampling, quantizing, and encoding (see Fig. 3.1). The sampling operation generates an instantaneously-sampled flat-top **pulse-amplitude modulated** (PAM) signal.

The quantizing operation is illustrated in Fig. 3.2 for the $M = 8$ level case. This quantizer is said to be *uniform* since all of the steps are of equal size. Since we are approximating the analog sample values by using a finite number of levels ($M = 8$ in this illustration), *error* is introduced into the recovered output analog signal because of the quantizing effect. The error waveform is illustrated in Fig. 3.2c. The quantizing error consists of the difference between the analog signal at the sampler input and the output of the quantizer. Note that the peak value of the error (± 1) is one-half of the quantizer step size (2). If we sample at the Nyquist rate ($2B$, where B is the absolute bandwidth, in hertz, of the input analog signal) or faster and there is negligible channel noise, there will still be noise, called *quantizing noise*, on the recovered analog waveform due to this error. The quantizing noise can also be thought of as a round-off error. The quantizer output is a *quantized* (i.e., only M possible amplitude values) PAM signal.

TABLE 3.1   3-b Gray Code for $M = 8$ levels

Quantized Sample Voltage	Gray Code Word (PCM Output)	
+7	110	
+5	111	
+3	101	
+1	100	
		Mirror image except for sign bit
−1	000	
−3	001	
−5	011	
−7	010	

Source: Couch, L.W. II 1993. *Digital and Analog Communication Systems*, 4th ed. Macmillan Publishing Co., New York, p. 145. With permission.

The PCM signal is obtained from the quantized PAM signal by encoding each quantized sample value into a digital word. It is up to the system designer to specify the exact code word that will represent a particular quantized level. If a Gray code of Table 3.1 is used, the resulting PCM signal is shown in Fig. 3.2d where the PCM word for each quantized sample is strobed out of the encoder by the next clock pulse. The Gray code was chosen because it has only 1-b change for each step change in the quantized level. Consequently, single errors in the received PCM code word will cause minimum errors in the recovered analog level, provided that the sign bit is not in error.

Here we have described PCM systems that represent the quantized analog sample values by *binary* code words. Of course, it is possible to represent the quantized analog samples by digital words using other than

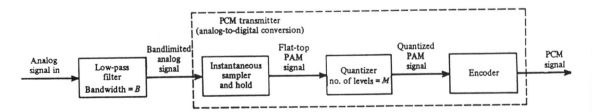

FIGURE 3.1   A PCM transmitter. *Source:* Couch, L.W. II 1993. *Digital and Analog Communication Systems,* 4th ed., Macmillan Publishing Co., New York, p. 143. With permission.

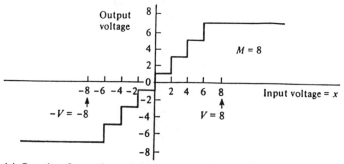

(a) Quantizer Output-Input Characteristics

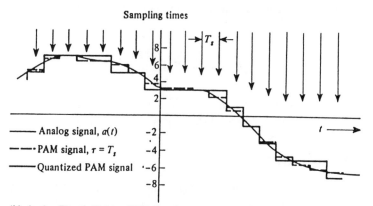

(b) Analog Signal, Flat-top PAM Signal, and Quantized PAM Signal

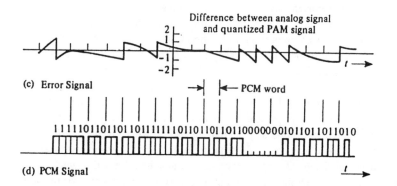

FIGURE 3.2   Illustration of waveforms in a PCM system. *Source:* Couch, L.W. II 1993. *Digital and Analog Communication Systems*, 4th ed., Macmillan Publishing Co., New York, p. 144. With permission.

base 2. That is, for base q, the number of quantized levels allowed is $M = q^n$, where n is the number of q base digits in the code word. We will not pursue this topic since binary ($q = 2$) digital circuits are most commonly used.

3.3 Percent Quantizing Noise

The quantizer at the PCM encoder produces an error signal at the PCM decoder output as illustrated in Fig. 3.2c. The peak value of this error signal may be expressed as a percentage of the maximum possible analog signal amplitude. Referring to Fig. 3.2c, a peak error of 1 V occurs for a maximum analog signal amplitude of $M = 8$ V as shown in Fig. 3.1c. Thus, in general,

$$\frac{2P}{100} = \frac{1}{M} = \frac{1}{2^n}$$

or

$$2^n = \frac{50}{P} \tag{3.1}$$

where P is the peak percentage error for a PCM system that uses n bit code words. The design value of n needed in order to have less than P percent error is obtained by taking the base 2 logarithm of both sides of Eq. (3.1), where it is realized that $\log_2(x) = [\log_{10}(x)]/\log_{10}(2) = 3.32 \log_{10}(x)$. That is,

$$n \geq 3.32 \log_{10}\left(\frac{50}{P}\right) \tag{3.2}$$

where n is the number of bits needed in the PCM word in order to obtain less than P percent error in the recovered analog signal (i.e., decoded PCM signal).

3.4 Practical PCM Circuits

Three techniques are used to implement the analog-to-digital converter (ADC) encoding operation. These are the *counting* or *ramp, serial* or *successive approximation,* and *parallel* or *flash* encoders.

In the counting encoder, at the same time that the sample is taken, a ramp generator is energized and a binary counter is started. The output of the ramp generator is continuously compared to the sample value; when the value of the ramp becomes equal to the sample value, the binary value of the counter is read. This count is taken to be the PCM word. The binary counter and the ramp generator are then reset to zero and are ready to be reenergized at the next sampling time. This technique requires only a few components, but the speed of this type of ADC is usually limited by the speed of the counter. The Intersil ICL7126 CMOS ADC integrated circuit uses this technique.

The serial encoder compares the value of the sample with trial quantized values. Successive trials depend on whether the past comparator outputs are positive or negative. The trial values are chosen first in large steps and then in small steps so that the process will converge rapidly. The trial voltages are generated by a series of voltage dividers that are configured by (on-off) switches. These switches are controlled by digital logic. After the process converges, the value of the switch settings is read out as the PCM word. This technique requires more precision components (for the voltage dividers) than the ramp technique. The speed of the feedback ADC technique is determined by the speed of the switches. The National Semiconductor ADC0804 8-b ADC uses this technique.

The parallel encoder uses a set of parallel comparators with reference levels that are the permitted quantized values. The sample value is fed into all of the parallel comparators simultaneously. The high or low level of the comparator outputs determines the binary PCM word with the aid of some digital logic. This is a fast ADC technique but requires more hardware than the other two methods. The RCA CA3318 8-b ADC integrated circuit is an example of the technique.

All of the integrated circuits listed as examples have parallel digital outputs that correspond to the digital word that represents the analog sample value. For generation of PCM, the parallel output (digital word) needs to be converted to serial form for transmission over a two-wire channel. This is accomplished by using a parallel-to-serial converter integrated circuit, which is also known as a **serial-input-output** (SIO) chip. The SIO chip includes a shift register that is set to contain the parallel data (usually, from 8 or 16 input lines). Then the data are shifted out of the last stage of the shift register bit by bit onto a single output line to produce the serial format. Furthermore, the SIO chips are usually full duplex; that is, they have two sets of shift registers, one that functions for data flowing in each direction. One shift register converts parallel input data to serial output data for transmission over the channel, and, simultaneously, the other shift register converts received serial data from another input to parallel data that are available at another output. Three types of SIO chips are available: the *universal asynchronous receiver/transmitter* (UART), the *universal synchronous receiver/transmitter* (USRT), and the *universal synchronous/asynchronous receiver transmitter* (USART). The UART transmits and receives asynchronous serial data, the USRT transmits and receives synchronous serial data, and the USART combines both a UART and a USRT on one chip.

At the receiving end the PCM signal is decoded back into an analog signal by using a digital-to-analog converter (DAC) chip. If the DAC chip has a parallel data input, the received serial PCM data are first converted to a parallel form using a SIO chip as described in the preceding paragraph. The parallel data are then converted to an approximation of the analog sample value by the DAC chip. This conversion is usually accomplished by using the parallel digital word to set the configuration of electronic switches on a resistive current (or voltage) divider network so that the analog output is produced. This is called a *multiplying* DAC since the analog output voltage is directly proportional to the divider reference voltage multiplied by the value of the digital word. The Motorola MC1408 and the National Semiconductor DAC0808 8-b DAC chips are examples of this technique. The DAC chip outputs samples of the quantized analog signal that approximates the analog sample values. This may be smoothed by a low-pass reconstruction filter to produce the analog output.

The Electrical Engineering Handbook [Dorf, 1993, pp. 771–782] gives more details on ADC, DAC, and PCM circuits.

3.5 Bandwidth of PCM

A good question to ask is: What is the spectrum of a PCM signal? For the case of PAM signalling, the spectrum of the PAM signal could be obtained as a function of the spectrum of the input analog signal because the PAM signal is a linear function of the analog signal. This is not the case for PCM. As shown in Figs. 3.1 and 3.2, the PCM signal is a nonlinear function of the input signal. Consequently, the spectrum of the PCM signal is not directly related to the spectrum of the input analog signal. It can be shown that the spectrum of the PCM signal depends on the bit rate, the correlation of the PCM data, and on the PCM waveform pulse shape (usually rectangular) used to describe the bits [Couch, 1993; Couch, 1995]. From Fig. 3.2, the bit rate is

$$R = nf_s \tag{3.3}$$

where n is the number of bits in the PCM word ($M = 2^n$) and f_s is the sampling rate. For no aliasing we require $f_s \geq 2B$ where B is the bandwidth of the analog signal (that is to be converted to the PCM signal). The dimensionality theorem [Couch, 1993; Couch, 1995] shows that the bandwidth of the PCM waveform is bounded by

$$B_{\text{PCM}} \geq \frac{1}{2}R = \frac{1}{2}nf_s \tag{3.4}$$

where equality is obtained if a $(\sin x)/x$ type of pulse shape is used to generate the PCM waveform. The exact spectrum for the PCM waveform will depend on the pulse shape that is used as well as on the type of line encoding. For example, if one uses a rectangular pulse shape with polar nonreturn to zero (NRZ) line coding, the first null bandwidth is simply

$$B_{\text{PCM}} = R = nf_s \text{ Hz} \tag{3.5}$$

TABLE 3.2   Performance of a PCM System with Uniform
Quantizing and No Channel Noise

Number of Quantizer Levels Used, M	Length of the PCM Word, n (bits)	Bandwidth of PCM Signal (First Null Bandwidth)[a]	Recovered Analog Signal Power-to-Quantizing Noise Power Ratios (dB) $(S/N)_{out}$
2	1	2B	6.0
4	2	4B	12.0
8	3	6B	18.1
16	4	8B	24.1
32	5	10B	30.1
64	6	12B	36.1
128	7	14B	42.1
256	8	16B	48.2
512	9	18B	54.2
1,024	10	20B	60.2
2,048	11	22B	66.2
4,096	12	24B	72.2
8,192	13	26B	78.3
16,384	14	28B	84.3
32,768	15	30B	90.3
65,536	16	32B	96.3

[a]B is the absolute bandwidth of the input analog signal. *Source:* Couch, L.W. II 1993. *Digital and Analog Communication Systems,* 4th ed. Macmillan Publishing Co., New York, p. 148. With permission.

Table 3.2 presents a tabulation of this result for the case of the minimum sampling rate, $f_s = 2B$. Note that Eq. (3.4) demonstrates that the bandwidth of the PCM signal has a lower bound given by

$$B_{PCM} \geq nB \qquad\qquad (3.6)$$

where $f_s > 2B$ and B is the bandwidth of the corresponding analog signal. Thus, for reasonable values of n, the bandwidth of the PCM signal will be significantly larger than the bandwidth of the corresponding analog signal that it represents. For the example shown in Fig. 3.2 where $n = 3$, the PCM signal bandwidth will be at least three times wider than that of the corresponding analog signal. Furthermore, if the bandwidth of the PCM signal is reduced by improper filtering or by passing the PCM signal through a system that has a poor frequency response, the filtered pulses will be elongated (stretched in width) so that pulses corresponding to any one bit will smear into adjacent bit slots. If this condition becomes too serious, it will cause errors in the detected bits. This pulse smearing effect is called **intersymbol interference** (ISI).

3.6 Effects of Noise

The analog signal that is recovered at the PCM system output is corrupted by noise. Two main effects produce this noise or distortion: 1) quantizing noise that is caused by the M-step quantizer at the PCM transmitter and 2) bit errors in the recovered PCM signal. The bit errors are caused by *channel noise* as well as improper channel filtering, which causes ISI. In addition, if the input analog signal is not strictly band limited, there will be some aliasing noise on the recovered analog signal [Spilker, 1977]. Under certain assumptions, it can be shown that the recovered analog *average* signal power to the average noise power [Couch, 1993] is

$$\left(\frac{S}{N}\right)_{\text{out}} = \frac{M^2}{1 + 4(M^2 - 1)P_e} \tag{3.7}$$

where M is the number of uniformly spaced quantizer levels used in the PCM transmitter and P_e is the probability of bit error in the recovered binary PCM signal at the receiver DAC before it is converted back into an analog signal. Most practical systems are designed so that P_e is negligible. Consequently, if we assume that there are no bit errors due to channel noise (i.e., $P_e = 0$), the S/N due only to quantizing errors is

$$\left(\frac{S}{N}\right)_{\text{out}} = M^2 \tag{3.8}$$

Numerical values for these S/N ratios are given in Table 3.2.

To realize these S/N ratios, one critical assumption is that the peak-to-peak level of the analog waveform at the input to the PCM encoder is set to the design level of the quantizer. For example, referring to Fig. 3.2, this corresponds to the input traversing the range $-V$ to $+V$ volts where $V = 8$ V is the design level of the quantizer. Equation (3.7) was derived for waveforms with equally likely values, such as a triangle waveshape, that have a peak-to-peak value of $2V$ and an rms value of $V/\sqrt{3}$, where V is the design peak level of the quantizer.

From a practical viewpoint, the quantizing noise at the output of the PCM decoder can be categorized into four types depending on the operating conditions. The four types are overload noise, random noise, granular noise, and hunting noise. As discussed earlier, the level of the analog waveform at the input of the PCM encoder needs to be set so that its peak level does not exceed the design peak of V volts. If the peak input does exceed V, the recovered analog waveform at the output of the PCM system will have flat tops near the peak values. This produces *overload noise*. The flat tops are easily seen on an oscilloscope, and the recovered analog waveform sounds distorted since the flat topping produces unwanted harmonic components. For example, this type of distortion can be heard on PCM telephone systems when there are high levels such as dial tones, busy signals, or off-hook warning signals.

The second type of noise, *random noise*, is produced by the random quantization errors in the PCM system under normal operating conditions when the input level is properly set. This type of condition is assumed in Eq. (3.8). Random noise has a white hissing sound. If the input level is not sufficiently large, the S/N will deteriorate from that given by Eq. (3.8); the quantizing noise will still remain more or less random.

If the input level is reduced further to a relatively small value with respect to the design level, the error values are not equally likely from sample to sample, and the noise has a harsh sound resembling gravel being poured into a barrel. This is called *granular noise*. This type of noise can be randomized (noise power decreased) by increasing the number of quantization levels and, consequently, increasing the PCM bit rate. Alternatively, granular can be reduced by using a nonuniform quantizer, such as the μ-law or A-law quantizers that are described in the Sec. 3.7.

The fourth type of quantizing noise that may occur at the output of a PCM system is *hunting noise*. It can occur when the input analog waveform is nearly constant, including when there is no signal (i.e., zero level). For these conditions the sample values at the quantizer output (see Fig. 3.2) can oscillate between two adjacent quantization levels, causing an undesired sinusoidal type tone of frequency $1/2 f_s$ at the output of the PCM system. Hunting noise can be reduced by filtering out the tone or by designing the quantizer so that there is no vertical step at the constant value of the inputs, such as at 0-V input for the no signal case. For the no signal case, the hunting noise is also called *idle channel noise*. Idle channel noise can be reduced by using a horizontal step at the origin of the quantizer output–input characteristic instead of a vertical step as shown in Fig. 3.2.

Recalling that $M = 2^n$, we may express Eq. (3.8) in decibels by taking $10 \log_{10}(\cdot)$ of both sides of the equation,

$$\left(\frac{S}{N}\right)_{\text{dB}} = 6.02n + \alpha \tag{3.9}$$

where n is the number of bits in the PCM word and $\alpha = 0$. This equation—called the 6-dB rule—points out the significant performance characteristic for PCM: an additional 6-dB improvement in S/N is obtained for each bit added to the PCM word. This is illustrated in Table 3.2. Equation (3.9) is valid for a wide variety of assumptions (such as various types of input waveshapes and quantification characteristics), although the value of α will depend on these assumptions [Jayant and Noll, 1984]. Of course, it is assumed that there are no bit errors and that the input signal level is large enough to range over a significant number of quantizing levels.

One may use Table 3.2 to examine the design requirements in a proposed PCM system. For example, high fidelity enthusiasts are turning to digital audio recording techniques. Here PCM signals are recorded instead of the analog audio signal to produce superb sound reproduction. For a dynamic range of 90 dB, it is seen that at least 15-b PCM words would be required. Furthermore, if the analog signal had a bandwidth of 20 kHz, the first null bandwidth for rectangular bit-shape PCM would be $2 \times 20\,\text{kHz} \times 15 = 600\,\text{kHz}$. Consequently, video-type tape recorders are needed to record and reproduce high-quality digital audio signals. Although this type of recording technique might seem ridiculous at first, it is realized that expensive high-quality analog recording devices are hard pressed to reproduce a dynamic range of 70 dB. Thus, digital audio is one way to achieve improved performance. This is being proven in the marketplace with the popularity of the digital compact disk (CD). The CD uses a 16-b PCM word and a sampling rate of 44.1 kHz on each stereo channel [Miyaoka, 1984; Peek, 1985]. Reed–Solomon coding with interleaving is used to correct burst errors that occur as a result of scratches and fingerprints on the compact disk.

3.7 Nonuniform Quantizing: μ-Law and A-Law Companding

Voice analog signals are more likely to have amplitude values near zero than at the extreme peak values allowed. For example, when digitizing voice signals, if the peak value allowed is 1 V, weak passages may have voltage levels on the order of 0.1 V (20 dB down). For signals such as these with nonuniform amplitude distribution, the granular quantizing noise will be a serious problem if the step size is not reduced for amplitude values near zero and increased for extremely large values. This is called nonuniform quantizing since a variable step size is used. An example of a nonuniform quantizing characteristic is shown in Fig. 3.3.

The effect of nonuniform quantizing can be obtained by first passing the analog signal through a compression (nonlinear) amplifier and then into the PCM circuit that uses a uniform quantizer. In the U.S., a μ-law type of compression characteristic is used. It is defined [Smith, 1957] by

$$|w_2(t)| = \frac{\ln(1 + \mu|w_1(t)|)}{\ln(1 + \mu)} \tag{3.10}$$

where the allowed peak values of $w_1(t)$ are ± 1 (i.e., $|w_1(t)| \le 1$), μ is a positive constant that is a parameter. This compression characteristic is shown in Fig. 3.3(b) for several values of μ, and it is noted that $\mu \to 0$ corresponds to linear amplification (uniform quantization overall). In the United States, Canada, and Japan, the telephone companies use a $\mu = 255$ compression characteristic in their PCM systems [Dammann, McDaniel, and Maddox, 1972].

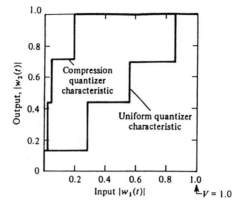

(a) $M = 8$ Quantizer Characteristic

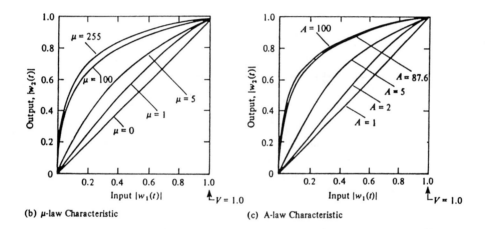

(b) μ-law Characteristic

(c) A-law Characteristic

FIGURE 3.3 Compression characteristics (first quadrant shown). *Source:* Couch, L.W. II 1993. *Digital and Analog Communication Systems,* 4th ed., Macmillan Publishing Co., New York, p. 153. With permission.

Another compression law, used mainly in Europe, is the A-law characteristic. It is defined [Cattermole, 1969] by

$$|w_2(t)| = \begin{cases} \dfrac{A|w_1(t)|}{1 + \ln A}, & 0 \leq |w_1(t)| \leq \dfrac{1}{A} \\[2ex] \dfrac{1 + \ln(A|w_1(t)|)}{1 + \ln A}, & \dfrac{1}{A} \leq |w_1(t)| \leq 1 \end{cases} \tag{3.11}$$

where $|w_1(t)| < 1$ and A is a positive constant. The A-law compression characteristic is shown in Fig. 3.3(c). The typical value for A is 87.6.

When compression is used at the transmitter, *expansion* (i.e., decompression) must be used at the receiver output to restore signal levels to their correct relative values. The *expandor* characteristic is the inverse of the compression characteristic, and the combination of a compressor and an expandor is called a *compandor*.

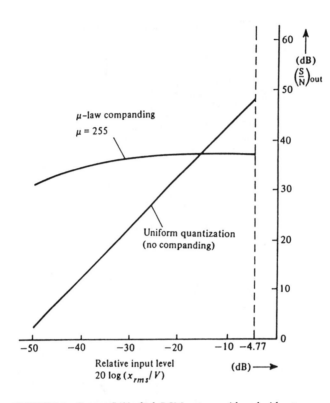

FIGURE 3.4   Output S/N of 8-b PCM systems with and without companding. *Source:* Couch, L.W. II 1993. *Digital and Analog Communication Systems,* 4th ed., Macmillan Publishing Co., New York, p. 155. With permission.

Once again, it can be shown that the output S/N follows the 6-dB law [Couch, 1993]

$$\left(\frac{S}{N}\right)_{dB} = 6.02 + \alpha \tag{3.12}$$

where for uniform quantizing

$$\alpha = 4.77 - 20\log(V/x_{\text{rms}}) \tag{3.13}$$

and for sufficiently large input levels[2] for μ-law companding

$$\alpha \approx 4.77 - 20\log[\ln(1+\mu)] \tag{3.14}$$

and for A-law companding [Jayant and Noll, 1984]

$$\alpha \approx 4.77 - 20\log[1 + \ln A] \tag{3.15}$$

n is the number of bits used in the PCM word, V is the peak design level of the quantizer, and x_{rms} is the rms value of the input analog signal. Notice that the output S/N is a function of the

---

[2]See Couch, 1993 or Lathi, 1989 for a more complicated expression that is valid for any input level.

input level for the uniform quantizing (no companding) case but is relatively insensitive to input level for μ-law and A-law companding, as shown in Fig. 3.4. The ratio V/x_{rms} is called the *loading factor*. The input level is often set for a loading factor of 4 (12 dB) to ensure that the overload quantizing noise will be negligible. In practice this gives $\alpha = -7.3$ for the case of uniform encoding as compared to $\alpha = 0$, which was obtained for the ideal conditions associated with Eq. (3.8).

3.8 Example: Design of a PCM System

Assume that an analog voice-frequency signal, which occupies a band from 300 to 3400 Hz, is to be transmitted over a binary PCM system. The minimum sampling frequency would be $2 \times 3.4 = 6.8$ kHz. In practice the signal is oversampled, and in the U.S. a sampling frequency of 8 kHz is the standard used for voice-frequency signals in telephone communication systems. Assume that each sample value is represented by 8 b; then the bit rate of the PCM signal is

$$R = (f_s \text{ samples/s})(n \text{ b/s})$$
$$= (8\,k \text{ samples/s})(8 \text{ b/s}) = 64\,\text{kb/s} \tag{3.16}$$

Referring to the dimensionality theorem [Eq. (3.4)], we realize that the theoretically minimum absolute bandwidth of the PCM signal is

$$B_{min} = \frac{1}{2}D = 32\,\text{kHz} \tag{3.17}$$

and this is realized if the PCM waveform consists of $(\sin x)/x$ pulse shapes. If rectangular pulse shaping is used, the absolute bandwidth is infinity, and the first null bandwidth [Eq. (3.5)] is

$$B_{null} = R = \frac{1}{T_b} = 64\,\text{kHz} \tag{3.18}$$

That is, we require a bandwidth of 64 kHz to transmit this digital voice PCM signal where the bandwidth of the original analog voice signal was, at most, 4 kHz. Using $n = 8$ in Eq. (3.1), the error on the recovered analog signal is $\pm0.2\%$. Using Eqs. (3.12) and (3.13) for the case of uniform quantizing with a loading factor, V/x_{rms}, of 10 (20 dB), we get for uniform quantizing

$$\left(\frac{S}{N}\right)_{dB} = 32.9\,\text{dB} \tag{3.19}$$

Using Eqs. (3.12) and (3.14) for the case of $\mu = 255$ companding, we get

$$\left(\frac{S}{N}\right) = 38.05\,\text{dB} \tag{3.20}$$

These results are illustrated in Fig. 3.4.

Defining Terms

Intersymbol interference: Filtering of a digital waveform so that a pulse corresponding to 1 b will smear (stretch in width) into adjacent bit slots.

Pulse amplitude modulation: An analog signal is represented by a train of pulses where the pulse amplitudes are proportional to the analog signal amplitude.

Pulse code modulation: A serial bit stream that consists of binary words which represent quantized sample values of an analog signal.

Quantizing: Replacing a sample value with the closest allowed value.

References

Cattermole, K.W. 1969. *Principles of Pulse-code Modulation*, American Elsevier, New York, NY.

Couch, L.W. 1993. *Digital and Analog Communication Systems*, 4th ed., Macmillan Publishing Co., New York, NY.

Couch, L.W. 1995. *Modern Communication Systems: Principles and Applications*, Macmillan Publishing Co., New York, NY.

Dammann, C.L., McDaniel, L.D., and Maddox, C.L. 1972. D2 Channel Bank—Multiplexing and Coding. *B. S. T. J.* 12(10):1675–1700.

Dorf, R.C. 1993. *The Electrical Engineering Handbook*, CRC Press, Inc., Boca Raton, FL.

Jayant, N.S. and Noll, P. 1984. *Digital Coding of Waveforms*, Prentice Hall, Englewood Cliffs, NJ.

Lathi, B.P. 1989. *Modern Digital and Analog Communication Systems*, 2nd ed, Holt, Rinehart and Winston, New York, NY.

Miyaoka, S. 1984. Digital Audio Is Compact and Rugged. *IEEE Spectrum.* 21(3):35–39.

Peek, J.B.H. 1985. Communication Aspects of the Compact Disk Digital Audio System. *IEEE Comm. Mag.* 23(2):7–15.

Smith, B. 1957. Instantaneous Companding of Quantized Signals. *B. S. T. J.* 36(5):653–709.

Spilker, J.J. 1977. *Digital Communications by Satellite*, Prentice Hall, Englewood Cliffs, NJ.

Further Information

Further Information Many practical design situations and applications of PCM transmission via twisted-pair T-1 telephone lines, fiber optic cable, microwave relay, and satellite systems are given in Couch, 1993 and Couch, 1995.

4

Baseband Signalling and Pulse Shaping

Michael L. Honig
Northwestern University

Melbourne Barton
Bellcore

Many physical communications channels, such as radio channels, accept a continuous-time wave-form as input. Consequently, a sequence of source bits, representing data or a digitized analog signal, must be converted to a continuous-time waveform at the transmitter. In general, each successive group of bits taken from this sequence is mapped to a particular continuous-time pulse. In this chapter we discuss the basic principles involved in selecting such a pulse for channels that can be characterized as linear and time invariant with finite bandwidth.

4.1 Communications System Model

Figure 4.1(a) shows a simple block diagram of a communications system. The sequence of source bits $\{b_i\}$ are grouped into sequential blocks (vectors) of m bits $\{\boldsymbol{b}_i\}$, and each binary vector $\boldsymbol{b}_i$ is mapped to one of 2^m pulses, $p(\boldsymbol{b}_i; t)$, which is transmitted over the channel. The transmitted signal as a function of time can be written as

$$s(t) = \sum_i p(\boldsymbol{b}_i; t - iT) \tag{4.1}$$

FIGURE 4.1(a)   Communication system model.   The source bits are grouped into binary vectors, which are mapped to a sequence of pulse shapes.

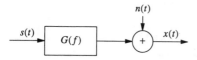

FIGURE 4.1(b)   Channel model consisting of a linear, time-invariant system (transfer function) followed by additive noise.

where $1/T$ is the rate at which each group of m bits, or pulses, is introduced to the channel. The information (bit) rate is therefore m/T.

The channel in Fig. 4.1(a) can be a radio link, which may distort the input signal $s(t)$ in a variety of ways. For example, it may introduce pulse dispersion (due to finite bandwidth) and multipath, as well as additive background noise. The output of the channel is denoted as $x(t)$, which is processed by the receiver to determine estimates of the source bits. The receiver can be quite complicated; however, for the purpose of this discussion, it is sufficient to assume only that it contains a front-end filter and a sampler, as shown in Fig. 4.1(a). This assumption is valid for a wide variety of detection strategies. The purpose of the receiver filter is to remove noise outside of the transmitted frequency band and to compensate for the channel frequency response.

A commonly used channel model is shown in Fig. 4.1(b) and consists of a linear, time-invariant filter, denoted as $G(f)$, followed by additive noise $n(t)$. The channel output is, therefore,

$$x(t) = [g(t) * s(t)] + n(t) \qquad (4.2)$$

where $g(t)$ is the channel impulse response associated with $G(f)$, and the asterisk denotes convolution,

$$g(t) * s(t) = \int_{-\infty}^{\infty} g(t - \tau) s(\tau) \, d\tau$$

This channel model accounts for all linear, time-invariant channel impairments, such as finite bandwidth and time-invariant multipath. It does not account for time-varying impairments, such as rapid fading due to time-varying multipath. Nevertheless, this model can be considered valid over short time periods during which the multipath parameters remain constant.

In Fig. 4.1 it is assumed that all signals are **baseband signals**, which means that the frequency content is centered around $f = 0$ (DC). The channel passband, therefore, partially coincides with the transmitted spectrum. In general, this condition requires that the transmitted signal be modulated by an appropriate carrier frequency and demodulated at the receiver. In that case, the model in Fig. 4.1 still applies; however, *baseband-equivalent* signals must be derived from their modulated (passband) counterparts. *Baseband signalling* and *pulse shaping* refers to the way in which a group of source bits is mapped to a baseband transmitted pulse.

As a simple example of baseband signalling, we can take $m = 1$ (map each source bit to a pulse), assign a 0 bit to a pulse $p(t)$, and a 1 bit to the pulse $-p(t)$. Perhaps the simplest example of a baseband pulse is the *rectangular* pulse given by $p(t) = 1$, $0 < t \leq T$, and $p(t) = 0$ elsewhere. In this case, we can write the transmitted signal as

$$s(t) = \sum_i A_i p(t - iT) \qquad (4.3)$$

where each symbol A_i takes on a value of $+1$ or -1, depending on the value of the ith bit, and $1/T$ is the *symbol rate*, namely, the rate at which the symbols A_i are introduced to the channel.

The preceding example is called *binary* **pulse amplitude modulation (PAM)**, since the data symbols A_i are binary valued, and they amplitude modulate the transmitted pulse $p(t)$. The information rate (bits per second) in this case is the same as the symbol rate $1/T$. As a simple extension of this signalling technique, we can increase m and choose A_i from one of $M = 2^m$ values to transmit at bit rate m/T. This is known as M-ary PAM. For example, letting $m = 2$, each pair of bits can be mapped to a pulse in the set $\{p(t), -p(t), 3p(t), -3p(t)\}$.

In general, the transmitted symbols $\{A_i\}$, the baseband pulse $p(t)$, and channel impulse response $g(t)$ can be *complex valued*. For example, each successive pair of bits might select a symbol from the set $\{1, -1, j, -j\}$, where $j = \sqrt{-1}$. This is a consequence of considering the baseband equivalent of passband modulation. (That is, generating a transmitted spectrum which is centered around a carrier frequency f_c.) Here we are not concerned with the relation between the passband and baseband equivalent models and simply point out that the discussion and results in this chapter apply to complex-valued symbols and pulse shapes.

As an example of a signalling technique which is not PAM, let $m = 1$ and

$$
\begin{aligned}
p(0; t) &= \begin{cases} \sqrt{2}\sin(2\pi f_1 t) & 0 < t < T \\ 0 & \text{elsewhere} \end{cases} \\[2mm]
p(1; t) &= \begin{cases} \sqrt{2}\sin(2\pi f_2 t) & 0 < t < T \\ 0 & \text{elsewhere} \end{cases}
\end{aligned}
\tag{4.4}
$$

where f_1 and $f_2 \neq f_1$ are fixed frequencies selected so that $f_1 T$ and $f_2 T$ (number of cycles for each bit) are multiples of $1/2$. These pulses are *orthogonal*, namely,

$$
\int_0^T p(1; t)p(0; t)\, \mathrm{d}t = 0
$$

This choice of pulse shapes is called binary **frequency-shift keying (FSK)**.

Another example of a set of orthogonal pulse shapes for $m = 2$ bits$/T$ is shown in Fig. 4.2. Because these pulses may have as many as three transitions within a symbol period, the transmitted spectrum occupies roughly four times the transmitted spectrum of binary PAM with a rectangular pulse shape. The spectrum is, therefore, spread across a much larger band than the smallest required for reliable transmission, assuming a data rate of $2/T$. This type of signalling is referred

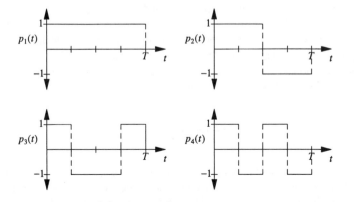

FIGURE 4.2   Four orthogonal spread-spectrum pulse shapes.

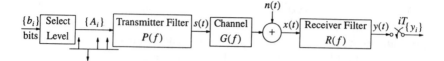

FIGURE 4.3   Baseband model of a pulse amplitude modulation system.

to as **spread-spectrum**. Spread-spectrum signals are more robust with respect to interference from other transmitted signals than are narrowband signals.[1]

4.2 Intersymbol Interference and the Nyquist Criterion

Consider the transmission of a PAM signal illustrated in Fig. 4.3. The source bits $\{b_i\}$ are mapped to a sequence of levels $\{A_i\}$, which modulate the transmitter pulse $p(t)$. The channel input is, therefore, given by Eq. (4.3) where $p(t)$ is the impulse response of the transmitter *pulse-shaping filter* $P(f)$ shown in Fig. 4.3. The input to the transmitter filter $P(f)$ is the modulated sequence of delta functions $\sum_i A_i \delta(t - iT)$. The channel is represented by the transfer function $G(f)$ (plus noise), which has impulse response $g(t)$, and the receiver filter has transfer function $R(f)$ with associated impulse response $r(t)$.

Let $h(t)$ be the overall impulse response of the combined transmitter, channel, and receiver, which has transfer function $H(f) = P(f)G(f)R(f)$. We can write $h(t) = p(t) * g(t) * r(t)$. The output of the receiver filter is then

$$y(t) = \sum_i A_i h(t - iT) + \tilde{n}(t) \tag{4.5}$$

where $\tilde{n}(t) = r(t) * n(t)$ is the output of the filter $R(f)$ with input $n(t)$. Assuming that samples are collected at the output of the filter $R(f)$ at the symbol rate $1/T$, we can write the kth sample of $y(t)$ as

$$y(kT) = \sum_i A_i h(kT - iT) + \tilde{n}(kT)$$

$$= A_k h(0) + \sum_{i \neq k} A_i h(kT - iT) + \tilde{n}(kT) \tag{4.6}$$

The first term on the right-hand side of Eq. (4.6) is the kth transmitted symbol scaled by the system impulse response at $t = 0$. If this were the only term on the right side of Eq. (4.6), we could obtain the source bits without error by scaling the received samples by $1/h(0)$. The second term on the right-hand side of Eq. (4.6) is called **intersymbol interference**, which reflects the view that neighboring symbols interfere with the detection of each desired symbol.

One possible criterion for choosing the transmitter and receiver filters is to minimize intersymbol interference. Specifically, if we choose $p(t)$ and $r(t)$ so that

$$h(kT) = \begin{cases} 1 & k = 0 \\ 0 & k \neq 0 \end{cases} \tag{4.7}$$

---

[1]This example can also be viewed as coded binary PAM. Namely, each pair of two source bits are mapped to 4 coded bits, which are transmitted via binary PAM with a rectangular pulse. The current IS-95 air interface uses an extension of this signalling method in which groups of 6 b are mapped to 64 orthogonal pulse shapes with as many as 63 transitions during a symbol.

then the kth received sample is

$$y(kT) = A_k + \tilde{n}(kT) \tag{4.8}$$

In this case, the intersymbol interference has been eliminated. This choice of $p(t)$ and $r(t)$ is called a **zero-forcing** solution, since it forces the intersymbol interference to zero. Depending on the type of detection scheme used, a zero-forcing solution may not be desirable. This is because the probability of error also depends on the noise intensity, which generally increases when intersymbol interference is suppressed. It is instructive, however, to examine the properties of the zero-forcing solution.

We now view Eq. (4.7) in the frequency domain. Since $h(t)$ has Fourier transform

$$H(f) = P(f)G(f)R(f) \tag{4.9}$$

where $P(f)$ is the Fourier transform of $p(t)$, the bandwidth of $H(f)$ is limited by the bandwidth of the channel $G(f)$. We will assume that $G(f) = 0$, $|f| > W$. The sampled impulse response $h(kT)$ can, therefore, be written as the inverse Fourier transform

$$h(kT) = \int_{-W}^{W} H(f)e^{j2\pi fkT} \, df$$

Through a series of manipulations, this integral can be rewritten as an inverse discrete Fourier transform,

$$h(kT) = T \int_{-1/(2T)}^{1/(2T)} H_{eq}(e^{j2\pi fT})e^{j2\pi fkT} \, df \tag{4.10a}$$

where

$$H_{eq}(e^{j2\pi fT}) = \frac{1}{T} \sum_k H\left(f + \frac{k}{T}\right)$$

$$= \frac{1}{T} \sum_k P\left(f + \frac{k}{T}\right)G\left(f + \frac{k}{T}\right)R\left(f + \frac{k}{T}\right) \tag{4.10b}$$

This relation states that $H_{eq}(z)$, $z = e^{j2\pi fT}$, is the discrete Fourier transform of the sequence $\{h_k\}$, where $h_k = h(kT)$. Sampling the impulse response $h(t)$ therefore changes the transfer function $H(f)$ to the *aliased* frequency response $H_{eq}(e^{j2\pi fT})$. From Eqs. (4.10) and (4.6) we conclude that $H_{eq}(z)$ is the transfer function that relates the sequence of input data symbols $\{A_i\}$ to the sequence of received samples $\{y_i\}$, where $y_i = y(iT)$, in the absence of noise. This is illustrated in Fig. 4.4. For this reason, $H_{eq}(z)$ is called the **equivalent discrete-time transfer function** for the overall system transfer function $H(f)$.

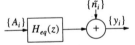

FIGURE 4.4 Equivalent discrete-time channel for the PAM system shown in Fig. 4.3 $[y_i = y(iT), \tilde{n}_i = \tilde{n}(iT)]$

Since $H_{eq}(e^{j2\pi fT})$ is the discrete Fourier transform of the sequence $\{h_k\}$, the time-domain, or sequence condition (4.7) is equivalent to the frequency-domain condition

$$H_{eq}(e^{j2\pi fT}) = 1 \tag{4.11}$$

This relation is called the **Nyquist criterion**. From Eqs. (4.10b) and (4.11) we make the following observations.

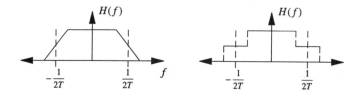

FIGURE 4.5 Two examples of frequency responses that satisfy the Nyquist criterion.

1. To satisfy the Nyquist criterion, the channel bandwidth W must be at least $1/(2T)$. Otherwise, $G(f + n/T) = 0$ for f in some interval of positive length for all n, which implies that $H_{\text{eq}}(e^{j2\pi fT}) = 0$ for f in the same interval.

2. For the minimum bandwidth $W = 1/(2T)$, Eqs. (4.10b) and (4.11) imply that $H(f) = T$ for $|f| < 1/(2T)$ and $H(f) = 0$ elsewhere. This implies that the system impulse response is given by

$$h(t) = \frac{\sin(\pi t/T)}{\pi t/T} \tag{4.12}$$

(Since $\int_{-\infty}^{\infty} h^2(t)\,dt = T$, the transmitted signal $s(t) = \sum_i A_i h(t - iT)$ has power equal to the symbol variance $E[|A_i|^2]$.) The impulse response in Eq. (4.12) is called a *minimum bandwidth* or Nyquist pulse. The frequency band $[-1/(2T), 1/(2T)]$ [i.e., the passband of $H(f)$] is called the **Nyquist band**.

3. Suppose that the channel is bandlimited to twice the Nyquist bandwidth. That is, $G(f) = 0$ for $|f| > 1/T$. The condition (4.11) then becomes

$$H(f) + H\left(f - \frac{1}{T}\right) + H\left(f + \frac{1}{T}\right) = T \tag{4.13}$$

Assume for the moment that $H(f)$ and $h(t)$ are both real valued, so that $H(f)$ is an even function of $f [H(f) = H(-f)]$. This is the case when the receiver filter is the matched filter (see Sec. 4.3). We can then rewrite Eq. (4.13) as

$$H(f) + H\left(\frac{1}{T} - f\right) = T, \qquad 0 < f < \frac{1}{2T} \tag{4.14}$$

which states that $H(f)$ must have odd symmetry about $f = 1/(2T)$. This is illustrated in Fig. 4.5, which shows two different transfer functions $H(f)$ that satisfy the Nyquist criterion.

4. The pulse shape $p(t)$ enters into Eq. (4.11) only through the product $P(f)R(f)$. Consequently, either $P(f)$ or $R(f)$ can be fixed, and the other filter can be adjusted or adapted to the particular channel. Typically, the pulse shape $p(t)$ is fixed, and the receiver filter is adapted to the (possibly time-varying) channel.

Raised Cosine Pulse

Suppose that the channel is ideal with transfer function

$$G(f) = \begin{cases} 1, & |f| < W \\ 0, & |f| > W \end{cases} \tag{4.15}$$

To maximize bandwidth efficiency, Nyquist pulses given by Eq. (4.12) should be used where $W = 1/(2T)$. This type of signalling, however, has two major drawbacks. First, Nyquist pulses are noncausal and of infinite duration. They can be approximated in practice by introducing an appropriate delay, and truncating the pulse. The pulse, however, decays very slowly, namely, as $1/t$, so that the truncation window must be wide. This is equivalent to observing that the ideal bandlimited frequency response given by Eq. (4.15) is difficult to approximate closely. The second drawback, which is more important, is the fact that this type of signalling is not robust with respect to sampling jitter. Namely, a small sampling offset ε produces the output sample

$$y(kT + \varepsilon) = \sum_i A_i \frac{\sin[\pi(k - i + \varepsilon/T)]}{\pi(k - i + \varepsilon/T)} \tag{4.16}$$

Since the Nyquist pulse decays as $1/t$, this sum is not guaranteed to converge. A particular choice of symbols $\{A_i\}$ can, therefore, lead to very large intersymbol interference, no matter how small the offset. Minimum bandwidth signalling is therefore impractical.

The preceding problem is generally solved in one of two ways in practice:

1. The pulse bandwidth is increased to provide a faster pulse decay than $1/t$.
2. A *controlled* amount of intersymbol interference is introduced at the transmitter, which can be subtracted out at the receiver.

The former approach sacrifices bandwidth efficiency, whereas the latter approach sacrifices power efficiency. We will examine the latter approach in Sec. 4.5. The most common example of a pulse, which illustrates the first technique, is the **raised cosine pulse**, given by

$$h(t) = \left[\frac{\sin(\pi t/T)}{\pi t/T} \right] \left[\frac{\cos(\alpha \pi t/T)}{1 - (2\alpha t/T)^2} \right] \tag{4.17}$$

which has Fourier transform

$$H(f) = \begin{cases} T & 0 \le |f| \le \dfrac{1 - \alpha}{2T} \\[2ex] \dfrac{T}{2}\left\{ 1 + \cos\left[\dfrac{\pi T}{\alpha}\left(|f| - \dfrac{1 - \alpha}{2T} \right) \right] \right\} & \dfrac{1 - \alpha}{2T} \le |f| \le \dfrac{1 + \alpha}{2T} \\[2ex] 0 & |f| > \dfrac{1 + \alpha}{2T} \end{cases} \tag{4.18}$$

where $0 \le \alpha \le 1$.

Plots of $p(t)$ and $P(f)$ are shown in Fig. 4.6 for different values of α. It is easily verified that $h(t)$ satisfies the Nyquist criterion (4.7) and, consequently, $H(f)$ satisfies Eq. (4.11). When $\alpha = 0$, $H(f)$ is the Nyquist pulse with minimum bandwidth $1/(2T)$, and when $\alpha > 0$, $H(f)$ has bandwidth $(1 + \alpha)/(2T)$ with a raised cosine rolloff. The parameter α, therefore, represents the additional, or **excess bandwidth** as a fraction of the minimum bandwidth $1/(2T)$. For example, when $\alpha = 1$, we say that that the pulse is a raised cosine pulse with 100% excess bandwidth. This is because the pulse bandwidth $1/T$ is twice the minimum bandwidth. Because the raised cosine pulse decays as $1/t^3$, performance is robust with respect to sampling offsets.

The raised cosine frequency response (4.18) applies to the combination of transmitter, channel, and receiver. If the transmitted pulse shape $p(t)$ is a raised cosine pulse, then $h(t)$ is a raised cosine pulse only if the combined receiver and channel frequency response is constant. Even with an ideal (transparent) channel, however, the optimum (matched) receiver filter response is generally not constant in the presence of additive Gaussian noise. An alternative is to transmit the *square-root raised cosine* pulse shape, which has frequency response $P(f)$ given by the square-root of the raised cosine frequency response in Eq. (4.18). Assuming an ideal channel, setting the receiver frequency response $R(f) = P(f)$ then results in an overall raised cosine system response $H(f)$.

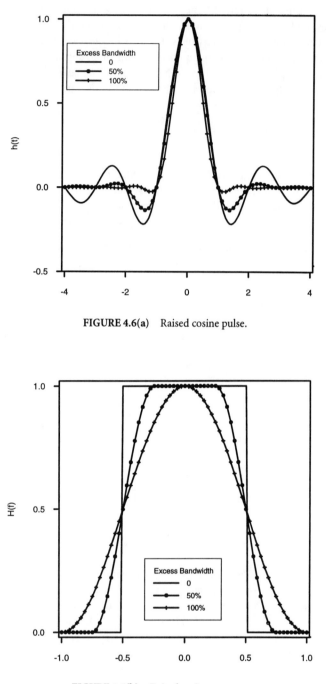

FIGURE 4.6(a)   Raised cosine pulse.

FIGURE 4.6(b)   Raised cosine spectrum.

4.3 Nyquist Criterion with Matched Filtering

Consider the transmission of an isolated pulse $A_0\delta(t)$. In this case the input to the receiver in Fig. 4.3 is

$$x(t) = A_0\tilde{g}(t) + n(t) \tag{4.19}$$

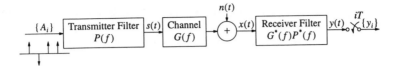

FIGURE 4.7 Baseband PAM model with a matched filter at the receiver.

where $\tilde{g}(t)$ is the inverse Fourier transform of the combined transmitter-channel transfer function $\tilde{G}(f) = P(f)G(f)$. We will assume that the noise $n(t)$ is white with spectrum $N_0/2$. The output of the receiver filter is then

$$y(t) = r(t) * x(t) = A_0[r(t) * \tilde{g}(t)] + [r(t) * n(t)] \qquad (4.20)$$

The first term on the right-hand side is the desired signal, and the second term is noise. Assuming that $y(t)$ is sampled at $t = 0$, the ratio of signal energy to noise energy, or signal-to-noise ratio (*SNR*) at the sampling instant, is

$$SNR = \frac{E[|A_0|^2] \left| \int_{-\infty}^{\infty} r(-t)\tilde{g}(t)\,dt \right|^2}{\frac{N_0}{2} \int_{-\infty}^{\infty} |r(t)|^2\,dt} \qquad (4.21)$$

The receiver impulse response that maximizes this expression is $r(t) = \tilde{g}^*(-t)$ [complex conjugate of $\tilde{g}(-t)$], which is known as the **matched filter** impulse response. The associated transfer function is $R(f) = \tilde{G}^*(f)$.

Choosing the receiver filter to be the matched filter is optimal in more general situations, such as when detecting a sequence of channel symbols with intersymbol interference (assuming the additive noise is Gaussian). We, therefore, reconsider the Nyquist criterion when the receiver filter is the matched filter. In this case, the baseband model is shown in Fig. 4.7, and the output of the receiver filter is given by

$$y(t) = \sum_i A_i h(t - iT) + \tilde{n}(t) \qquad (4.22)$$

where the baseband pulse $h(t)$ is now the impulse response of the filter with transfer function $|\tilde{G}(f)|^2 = |P(f)G(f)|^2$. This impulse response is the *autocorrelation* of the impulse response of the combined transmitter-channel filter $\tilde{G}(f)$,

$$h(t) = \int_{-\infty}^{\infty} \tilde{g}^*(s)\tilde{g}(s+t)\,ds \qquad (4.23)$$

With a matched filter at the receiver, the equivalent discrete-time transfer function is

$$H_{eq}(e^{j2\pi fT}) = \frac{1}{T}\sum_k \left| \tilde{G}\left(f - \frac{k}{T}\right) \right|^2$$

$$= \frac{1}{T}\sum_k \left| P\left(f - \frac{k}{T}\right)G\left(f - \frac{k}{T}\right) \right|^2 \qquad (4.24)$$

which relates the sequence of transmitted symbols $\{A_k\}$ to the sequence of received samples $\{y_k\}$ in the absence of noise. Note that $H_{eq}(e^{j2\pi fT})$ is positive, real valued, and an even function of f. If the channel is bandlimited to twice the Nyquist bandwidth, then $H(f) = 0$ for $|f| > 1/T$, and the Nyquist condition is given by Eq. (4.14) where $H(f) = |G(f)P(f)|^2$. The aliasing sum in Eq. (4.10b) can therefore be described as a folding operation in which the channel response $|H(f)|^2$ is folded around the Nyquist frequency $1/(2T)$. For this reason, $H_{eq}(e^{j2\pi fT})$ with a matched receiver filter is often referred to as the folded channel spectrum.

4.4 Eye Diagrams

One way to assess the severity of distortion due to intersymbol interference in a digital communications system is to examine the **eye diagram**. The eye diagram is illustrated in Fig. 4.8 for a raised cosine pulse shape with 25% excess bandwidth and an ideal bandlimited channel. Figure 4.8(a) shows the data signal at the receiver

$$y(t) = \sum_i A_i h(t - iT) + n(t) \tag{4.25}$$

where $h(t)$ is given by Eq. (4.17), $\alpha = 1/4$, each symbol A_i is independently chosen from the set $\{\pm 1, \pm 3\}$, where each symbol is equally likely, and $n(t)$ is bandlimited white Gaussian noise. (The received *SNR* is 30 dB.) The eye diagram is constructed from the time-domain data signal $y(t)$ as follows (assuming nominal sampling times at $kT, k = 0, 1, 2, \ldots$):

1. Partition the waveform $y(t)$ into successive segments of length T starting from $t = T/2$.
2. Translate each of these waveform segments $[y(t), (k + 1/2)T \leq t \leq (k + 3/2)T, k = 0, 1, 2, \ldots]$ to the interval $[-T/2, T/2]$, and superimpose.

The resulting picture is shown in Fig. 4.8(b) for the $y(t)$ shown in Fig. 4.8(a). (Partitioning $y(t)$ into successive segments of length $iT, i > 1$, is also possible. This would result in i successive eye diagrams.) The number of eye openings is one less than the number of transmitted signal levels. In practice, the eye diagram is easily viewed on an oscilloscope by applying the received waveform $y(t)$ to the vertical deflection plates of the oscilloscope and applying a sawtooth waveform at the

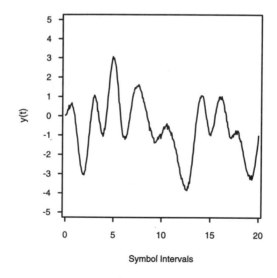

FIGURE 4.8(a)   Received signal $y(t)$.

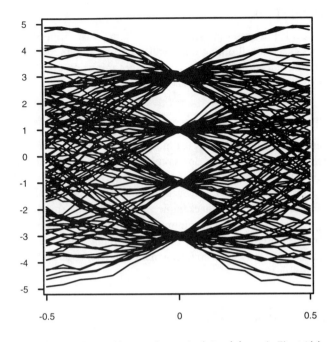

FIGURE 4.8(b)   Eye diagram for received signal shown in Fig. 4.8(a).

symbol rate $1/T$ to the horizontal deflection plates. This causes successive symbol intervals to be translated into one interval on the oscilloscope display.

Each waveform segment $y(t)$, $(k+1/2)T \leq t \leq (k+3/2)T$, depends on the particular sequence of channel symbols surrounding A_k. The number of channel symbols that affects a particular waveform segment depends on the extent of the intersymbol interference, shown in Eq. (4.6). This, in turn, depends on the duration of the impulse response $h(t)$. For example, if $h(t)$ has most of its energy in the interval $0 < t < mT$, then each waveform segment depends on approximately m symbols. Assuming binary transmission, this implies that there are a total of 2^m waveform segments that can be superimposed in the eye diagram. (It is possible that only one sequence of channel symbols causes significant intersymbol interference, and this sequence occurs with very low probability.) In current digital wireless applications the impulse response typically spans only a few symbols.

The eye diagram has the following important features which measure the performance of a digital communications system.

Vertical Eye Opening

The vertical openings at any time t_0, $-T/2 \leq t_0 \leq T/2$, represent the separation between signal levels with worst-case intersymbol interference, assuming that $y(t)$ is sampled at times $t = kT + t_0$, $k = 0, 1, 2, \ldots$. It is possible for the intersymbol interference to be large enough so that this vertical opening between some, or all, signal levels disappears altogether. In that case, the eye is said to be closed. Otherwise, the eye is said to be open. A closed eye implies that if the estimated bits are obtained by thresholding the samples $y(kT)$, then the decisions will depend primarily on the intersymbol interference rather than on the desired symbol. The probability of error will, therefore, be close to $1/2$. Conversely, wide vertical spacings between signal levels imply a large degree of immunity to additive noise. In general, $y(t)$ should be sampled at the times $kT + t_0$, $k = 0, 1, 2, \ldots$, where t_0 is chosen to maximize the vertical eye opening.

Horizontal Eye Opening

The width of each opening indicates the sensitivity to timing offset. Specifically, a very narrow eye opening indicates that a small timing offset will result in sampling where the eye is closed. Conversely, a wide horizontal opening indicates that a large timing offset can be tolerated, although the error probability will depend on the vertical opening.

Slope of the Inner Eye

The slope of the inner eye indicates sensitivity to timing jitter or variance in the timing offset. Specifically, a very steep slope means that the eye closes rapidly as the timing offset increases. In this case, a significant amount of jitter in the sampling times significantly increases the probability of error.

The shape of the eye diagram is determined by the pulse shape. In general, the faster the baseband pulse decays, the wider the eye opening. For example, a rectangular pulse produces a box-shaped eye diagram (assuming binary signalling). The minimum bandwidth pulse shape Eq. (4.12) produces an eye diagram which is closed for all t except for $t = 0$. This is because, as shown earlier, an arbitrarily small timing offset can lead to an intersymbol interference term that is arbitrarily large, depending on the data sequence.

4.5 Partial-Response Signalling

To avoid the problems associated with Nyquist signalling over an ideal bandlimited channel, bandwidth and/or power efficiency must be compromised. Raised cosine pulses compromise bandwidth efficiency to gain robustness with respect to timing errors. Another possibility is to introduce a controlled amount of intersymbol at the transmitter, which can be removed at the receiver. This approach is called **partial-response (PR) signalling**. The terminology reflects the fact that the sampled system impulse response does not have the full response given by the Nyquist condition Eq. (4.7).

To illustrate PR signalling, suppose that the Nyquist condition Eq. (4.7) is replaced by the condition

$$h_k = \begin{cases} 1 & k = 0, 1 \\ 0 & \text{all other } k \end{cases} \tag{4.26}$$

The kth received sample is then

$$y_k = A_k + A_{k-1} + \tilde{n}_k \tag{4.27}$$

so that there is intersymbol interference from one neighboring transmitted symbol. For now we focus on the spectral characteristics of PR signalling and defer discussion of how to detect the transmitted sequence $\{A_k\}$ in the presence of intersymbol interference. The equivalent discrete-time transfer function in this case is the discrete Fourier transform of the sequence in Eq. (4.26),

$$H_{\text{eq}}(e^{j2\pi fT}) = \frac{1}{T} \sum_k H\left(f + \frac{k}{T}\right)$$

$$= 1 + e^{-j2\pi fT} = 2e^{-j\pi fT} \cos(\pi fT) \tag{4.28}$$

As in the full-response case, for Eq. (4.28) to be satisfied, the *minimum* bandwidth of the channel $G(f)$ and transmitter filter $P(f)$ is $W = 1/(2T)$. Assuming $P(f)$ has this minimum bandwidth implies

$$H(f) = \begin{cases} 2T e^{-j\pi fT} \cos(\pi fT) & |f| < 1/(2T) \\ 0 & |f| > 1/(2T) \end{cases} \tag{4.29a}$$

and

$$h(t) = T\{\text{sinc}(t/T) + \text{sinc}[(t - T)/T]\} \tag{4.29b}$$

where $\text{sinc}\, x = (\sin \pi x)/(\pi x)$. This pulse is called a *duobinary* pulse and is shown along with the associated $H(f)$ in Fig. 4.9. [Notice that $h(t)$ satisfies Eq. (4.26).] Unlike the ideal bandlimited frequency response, the transfer function $H(f)$ in Eq. (4.29a) is continuous and is, therefore, easily approximated by a physically realizable filter. Duobinary PR was first proposed by Lender, 1963, and later generalized by Kretzmer, 1966.

The main advantage of the duobinary pulse Eq. (4.29b), relative to the minimum bandwidth pulse Eq. (4.12), is that signalling at the Nyquist symbol rate is feasible with zero excess bandwidth. Because the pulse decays much more rapidly than a Nyquist pulse, it is robust with respect to timing errors. Selecting the transmitter and receiver filters so that the overall system response is duobinary is appropriate in situations where the channel frequency response $G(f)$ is near zero or has a rapid rolloff at the Nyquist band edge $f = 1/(2T)$.

As another example of PR signaling, consider the *modified* duobinary partial response

$$h_k = \begin{cases} 1 & k = -1 \\ -1 & k = 1 \\ 0 & \text{all other } k \end{cases} \tag{4.30}$$

which has equivalent discrete-time transfer function

$$\begin{aligned} H_{\text{eq}}(e^{j2\pi fT}) &= e^{j2\pi fT} - e^{-j2\pi fT} \\ &= j2 \sin(2\pi fT) \end{aligned} \tag{4.31}$$

With zero excess bandwidth, the overall system response is

$$H(f) = \begin{cases} j2T \sin(2\pi fT) & |f| < 1/(2T) \\ 0 & |f| > 1/(2T) \end{cases} \tag{4.32a}$$

and

$$h(t) = T\{\text{sinc}[(t + T)/T] - \text{sinc}[(t - T)/T]\} \tag{4.32b}$$

These functions are plotted in Fig. 4.10. This pulse shape is appropriate when the channel response $G(f)$ is near zero at both DC ($f = 0$) and at the Nyquist band edge. This is often the case for wire (twisted-pair) channels where the transmitted signal is coupled to the channel through a transformer. Like duobinary PR, modified duobinary allows minimum bandwidth signalling at the Nyquist rate.

A particular partial response is often identified by the polynomial

$$\sum_{k=0}^{K} h_k D^k$$

where D (for delay) takes the place of the usual z^{-1} in the z transform of the sequence $\{h_k\}$. For example, duobinary is also referred to as $1 + D$ partial response.

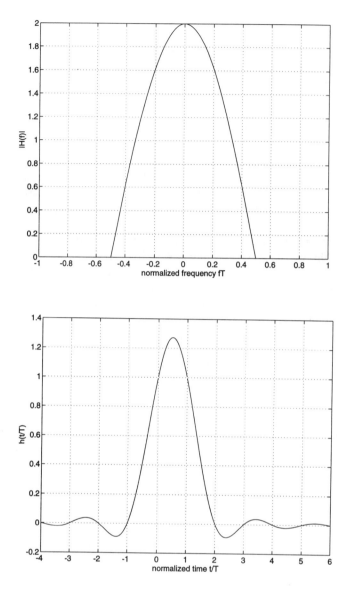

FIGURE 4.9   Duobinary frequency response and minimum bandwidth pulse.

In general, more complicated system responses than those shown in Figs. 4.9 and 4.10 can be generated by choosing more nonzero coefficients in the sequence $\{h_k\}$. This complicates detection, however, because of the additional intersymbol interference that is generated.

Rather than modulating a PR pulse $h(t)$, a PR signal can also be generated by filtering the sequence of transmitted levels $\{A_i\}$. This is shown in Fig. 4.11. Namely, the transmitted levels are first passed through a discrete-time (digital) filter with transfer function $P_d(e^{j2\pi fT})$ (where the subscript d indicates discrete). [Note that $P_d(e^{j2\pi fT})$ can be selected to be $H_{eq}(e^{j2\pi fT})$.] The outputs of this filter form the PAM signal, where the pulse shaping filter $P(f) = 1$, $|f| < 1/(2T)$ and is zero elsewhere. If the transmitted levels $\{A_k\}$ are selected independently and are identically distributed, then the transmitted spectrum is $\sigma_A^2 |P_d(e^{j2\pi fT})|^2$ for $|f| < 1/(2T)$ and is zero for $|f| > 1/(2T)$, where $\sigma_A^2 = E[|A_k|^2]$.

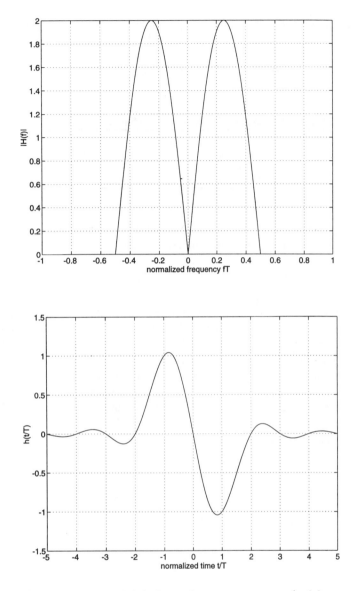

FIGURE 4.10 Modified duobinary frequency response and minimum bandwidth pulse.

FIGURE 4.11 Generation of PR signal.

Shaping the transmitted spectrum to have nulls coincident with nulls in the channel response potentially offers significant performance advantages. By introducing intersymbol interference, however, PR signalling increases the number of received signal levels, which increases the complexity of the detector and may reduce immunity to noise. For example, the set of received signal levels for duobinary signalling is $\{0, \pm 2\}$ from which the transmitted levels $\{\pm 1\}$ must be estimated. The

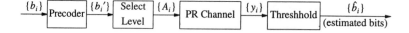

FIGURE 4.12   Precoding for a PR channel.

TABLE 4.1   Example of Precoding for Duobinary PR.

$\{b_i\}$:		1	0	0	1	1	1	0	0	1	0
$\{b_i'\}$:	0	1	1	1	0	1	0	0	0	1	1
$\{A_i\}$:	−1	1	1	1	−1	1	−1	−1	−1	1	1
$\{y_i\}$:		0	2	2	0	0	0	−2	−2	0	2

performance of a particular PR scheme depends on the channel characteristics, as well as the type of detector used at the receiver. We now describe a simple suboptimal detection strategy.

Precoding

Consider the received signal sample Eq. (4.27) with duobinary signalling. If the receiver has correctly decoded the symbol A_{k-1}, then in the absence of noise A_k can be decoded by subtracting A_{k-1} from the received sample y_k. If an error occurs, however, then subtracting the preceding symbol estimate from the received sample will cause the error to propogate to successive detected symbols. To avoid this problem, the transmitted levels can be **precoded** in such a way as to compensate for the intersymbol interference introduced by the overall partial response.

We first illustrate precoding for duobinary PR. The sequence of operations is illustrated in Fig. 4.12. Let $\{b_k\}$ denote the sequence of source bits where $b_k \in \{0, 1\}$. This sequence is transformed to the sequence $\{b_k'\}$ by the operation

$$b_k' = b_k \oplus b_{k-1}' \tag{4.33}$$

where $\oplus$ denotes modulo 2 addition (exclusive OR). The sequence $\{b_k'\}$ is mapped to the sequence of binary transmitted signal levels $\{A_k\}$ according to

$$A_k = 2b_k' - 1 \tag{4.34}$$

That is, $b_k' = 0$ $(b_k = 1)$ is mapped to the transmitted level $A_k = -1$ $(A_k = 1)$. In the absence of noise, the received symbol is then

$$y_k = A_k + A_{k-1} = 2(b_k' + b_{k-1}' - 1) \tag{4.35}$$

and combining Eqs. (4.33) and (4.35) gives

$$b_k = \left(\frac{1}{2}y_k + 1\right) \bmod 2 \tag{4.36}$$

That is, if $y_k = \pm 2$, then $b_k = 0$, and if $y_k = 0$, then $b_k = 1$. Precoding, therefore, enables the detector to make *symbol-by-symbol* decisions that do not depend on previous decisions. Table 4.1 shows a sequence of transmitted bits $\{b_i\}$, precoded bits $\{b_i'\}$, transmitted signal levels $\{A_i\}$, and received samples $\{y_i\}$.

The preceding precoding technique can be extended to multilevel PAM and to other PR channels. Suppose that the PR is specified by

$$H_{eq}(D) = \sum_{k=0}^{K} h_k D^k$$

where the coefficients are integers and that the source symbols $\{b_k\}$ are selected from the set $\{0, 1, \ldots, M-1\}$. These symbols are transformed to the sequence $\{b'_k\}$ via the precoding operation

$$b'_k = \left(b_k - \sum_{i=1}^{K} h_i b'_{k-i} \right) \bmod M \tag{4.37}$$

Because of the modulo operation, each symbol b'_k is also in the set $\{0, 1, \ldots, M-1\}$. The kth transmitted signal level is given by

$$A_k = 2b'_k - (M-1) \tag{4.38}$$

so that the set of transmitted levels is $\{-(M-1), \ldots, (M-1)\}$ (i.e., a shifted version of the set of values assumed by b_k). In the absence of noise the received sample is

$$y_k = \sum_{i=0}^{K} h_i A_{k-i} \tag{4.39}$$

and it can be shown that the kth source symbol is given by

$$b_k = \frac{1}{2}(y_k + (M-1) \cdot H_{eq}(1)) \bmod M \tag{4.40}$$

Precoding the symbols $\{b_k\}$ in this manner, therefore, enables symbol-by-symbol decisions at the receiver. In the presence of noise, more sophisticated detection schemes (e.g., maximum likelihood) can be used with PR signalling to obtain improvements in performance.

4.6 Additional Considerations

In many applications, bandwidth and intersymbol interference are not the only important considerations for selecting baseband pulses. Here we give a brief discussion of additional practical constraints that may influence this selection.

Average Transmitted Power and Spectral Constraints

The constraint on average transmitted power varies according to the application. For example, low-average power is highly desirable for mobile wireless applications that use battery-powered transmitters. In many applications (e.g., digital subscriber loops, as well as digital radio), constraints are imposed to limit the amount of interference, or crosstalk, radiated into neighboring receivers and communications systems. Because this type of interference is frequency dependent, the constraint may take the form of a spectral mask that specifies the maximum allowable transmitted power as a function of frequency. For example, crosstalk in wireline channels is generally caused by capacitive coupling and increases as a function of frequency. Consequently, to reduce the amount of crosstalk generated at a particular transmitter, the pulse shaping filter generally attenuates high frequencies more than low frequencies.

In radio applications where signals are assigned different frequency bands, constraints on the transmitted spectrum are imposed to limit *adjacent-channel interference*. This interference is generated by transmitters assigned to adjacent frequency bands. Therefore, a constraint is needed to limit the amount of *out-of-band power* generated by each transmitter, in addition to an overall average power constraint. To meet this constraint, the transmitter filter in Fig. 4.3 must have a sufficiently steep rolloff at the edges of the assigned frequency band. (Conversely, if the transmitted signals are time multiplexed, then the duration of the system impulse response must be contained within the assigned time slot.)

Peak-to-Average Power

In addition to a constraint on average transmitted power, a *peak-power* constraint is often imposed as well. This constraint is important in practice for the following reasons:

1. The dynamic range of the transmitter is limited. In particular, saturation of the output amplifier will "clip" the transmitted waveform.
2. Rapid fades can severely distort signals with high peak-to-average power.
3. The transmitted signal may be subjected to nonlinearities. Saturation of the output amplifier is one example. Another example that pertains to wireline applications is the companding process in the voice telephone network [Kalet and Saltzberg, 1994]. Namely, the compander used to reduce quantization noise for pulse-code modulated voice signals introduces amplitude-dependent distortion in data signals.

The preceding impairments or constraints indicate that the transmitted waveform should have a low peak-to-average power ratio. The peak-to-average power ratio is minimized by using binary signalling with rectangular pulse shapes. However, this compromises bandwidth efficiency. In applications where peak-to-average ratio should be low, binary signalling with rounded pulses are often used.

Channel and Receiver Characteristics

The type of channel impairments encountered and the type of detection scheme used at the receiver can also influence the choice of a transmitted pulse shape. For example, a constant amplitude pulse is appropriate for a fast fading environment with noncoherent detection. The ability to track channel characteristics, such as phase, may allow more bandwidth efficient pulse shapes in addition to multilevel signalling.

High-speed data communications over time-varying channels requires that the transmitter and/or receiver adapt to the changing channel characteristics. Adapting the transmitter to compensate for a time-varying channel requires a feedback channel through which the receiver can notify the transmitter of changes in channel characteristics. Because of this extra complication, adapting the receiver is often preferred to adapting the transmitter pulse shape. However, the following examples are notable exceptions.

1. The current IS-95 air interface for direct-sequence code-division multiple access adapts the transmitter power to control the amount of interference generated and to compensate for channel fades. This can be viewed as a simple form of adaptive transmitter pulse shaping in which a single parameter associated with the pulse shape is varied.
2. Multitone modulation divides the channel bandwidth into small subbands, and the transmitted power and source bits are distributed among these subbands to maximize the information rate. The received signal-to-noise ratio for each subband must be transmitted back to the transmitter to guide the allocation of transmitted bits and power [Bingham, 1990].

In addition to multitone modulation, *adaptive precoding* (also known as Tomlinson–Harashima precoding [Tomlinson, 1971; Harashima and Miyakawa, 1972]) is another way in which the transmitter can adapt to the channel frequency response. Adaptive precoding is an extension of the technique described earlier for partial-response channels. Namely, the equivalent discrete-time channel impulse response is measured at the receiver and sent back to the transmitter, where it is used in a precoder. The precoder compensates for the intersymbol interference introduced by the channel, allowing the receiver to detect the data by a simple threshhold operation. Both multitone modulation and precoding have been used with wireline channels (voiceband modems and digital subscriber loops).

Complexity

Generation of a bandwidth-efficient signal requires a filter with a sharp cutoff. In addition, bandwidth-efficient pulse shapes can complicate other system functions, such as timing and carrier recovery. If sufficient bandwidth is available, the cost can be reduced by using a rectangular pulse shape with a simple detection strategy (low-pass filter and threshold).

Tolerance to Interference

Interference is one of the primary channel impairments associated with digital radio. In addition to adjacent-channel interference described earlier, *cochannel interference* may be generated by other transmitters assigned to the same frequency band as the desired signal. Co-channel interference can be controlled through frequency (and perhaps time slot) assignments and by pulse shaping. For example, assuming fixed average power, increasing the bandwidth occupied by the signal lowers the power spectral density and decreases the amount of interference into a narrowband system that occupies part of the available bandwidth. Sufficient bandwidth spreading, therefore, enables wideband signals to be overlaid on top of narrowband signals without disrupting either service.

Probability of Intercept and Detection

The broadcast nature of wireless channels generally makes eavesdropping easier than for wired channels. A requirement for most commercial, as well as military applications, is to guarantee the privacy of user conversations (low probability of intercept). An additional requirement, in some applications, is that determining whether or not communications is taking place must be difficult (low probability of detection). Spread-spectrum waveforms are attractive in these applications since spreading the pulse energy over a wide frequency band decreases the power spectral density and, hence, makes the signal less visible. Power-efficient modulation combined with coding enables a further reduction in transmitted power for a target error rate.

4.7 Examples

We conclude this chapter with a brief description of baseband pulse shapes used in existing and emerging standards for digital mobile cellular and Personal Communications Services (PCS).

Global System for Mobile Communications (GSM)

The European GSM standard for digital mobile cellular communications operates in the 900-MHz frequency band, and is based on time-division multiple access (TDMA) [Rahnema, 1993]. A special variant of binary FSK is used called *Gaussian minimum-shift keying (GMSK)*. The GMSK modulator is illustrated in Fig. 4.13. The input to the modulator is a binary PAM signal $s(t)$, given by Eq. (4.3), where the pulse $p(t)$ is a Gaussian function and $|s(t)| < 1$. This waveform frequency modulates the carrier f_c, so that the (passband) transmitted signal is

$$w(t) = K \cos\left[2\pi f_c t + 2\pi f_d \int_{-\infty}^{t} s(\tau)\, d\tau \right]$$

FIGURE 4.13   Generation of GMSK signal; LPF is low-pass filter.

The maximum frequency deviation from the carrier is $f_d = 1/(2T)$, which characterizes minimum-shift keying. This technique can be used with a noncoherent receiver that is easy to implement. Because the transmitted signal has a constant envelope, the data can be reliably detected in the presence of rapid fades that are characteristic of mobile radio channels.

U.S. Digital Cellular (IS-54)

The IS-54 air interface operates in the 800-MHz band and is based on TDMA [EIA/TIA, 1991]. The baseband signal is given by Eq. (4.3) where the symbols are complex-valued, corresponding to quadrature phase modulation. The pulse has a square-root raised cosine spectrum with 35% excess bandwidth.

Interim Standard-95

The IS-95 air interface for digital mobile cellular uses spread-spectrum signalling (CDMA) in the 800-MHz band [TIA, 1993]. The baseband transmitted pulse shapes are analogous to those shown in Fig. 4.2, where the number of square pulses (chips) per bit is 128. To improve spectral efficiency the (wideband) transmitted signal is filtered by an approximation to an ideal low-pass response with a small amount of excess bandwidth. This shapes the chips so that they resemble minimum bandwidth pulses.

Personal Access Communications System (PACS)

Both PACS and the Japanese personal handy phone (PHP) system are TDMA systems which have been proposed for personal communications systems (PCS), and operate near 2 Ghz [Cox, 1995]. The baseband signal is given by Eq. (4.3) with four complex symbols representing four-phase quadrature modulation. The baseband pulse has a square-root raised cosine spectrum with 50% excess bandwidth.

Defining Terms

Baseband signal: A signal with frequency content centered around DC.

Equivalent discrete-time transfer function: A discrete-time transfer function (z transform) that relates the transmitted amplitudes to received samples in the absence of noise.

Excess bandwidth: That part of the baseband transmitted spectrum which is not contained within the Nyquist band.

Eye diagram: Superposition of segments of a received PAM signal that indicates the amount of intersymbol interference present.

Frequency-shift keying: A digital modulation technique in which the transmitted pulse is sinusoidal, where the frequency is determined by the source bits.

Intersymbol interference: The additive contribution (interference) to a received sample from transmitted symbols other than the symbol to be detected.

Matched filter: The receiver filter with impulse response equal to the time-reversed, complex conjugate impulse response of the combined transmitter filter-channel impulse response.

Nyquist band: The narrowest frequency band that can support a PAM signal without intersymbol interference (the interval $[-1/(2T), 1/(2T)]$ where $1/T$ is the symbol rate).

Nyquist criterion: A condition on the overall frequency response of a PAM system that ensures the absence of intersymbol interference.

Partial-response signalling: A signalling technique in which a controlled amount of intersymbol interference is introduced at the transmitter in order to shape the transmitted spectrum.

Precoding: A transformation of source symbols at the transmitter that compensates for intersymbol interference introduced by the channel.

Pulse amplitude modulation (PAM): A digital modulation technique in which the source bits are mapped to a sequence of amplitudes that modulate a transmitted pulse.

Raised cosine pulse: A pulse shape with Fourier transform that decays to zero according to a raised cosine; see Eq. (4.18). The amount of excess bandwidth is conveniently determined by a single parameter (α).

Spread spectrum: A signalling technique in which the pulse bandwidth is many times wider than the Nyquist bandwidth.

Zero-forcing criterion: A design constraint which specifies that intersymbol interference be eliminated.

References

Bingham, J.A.C. 1990. Multicarrier modulation for data transmission: an idea whose time has come. *IEEE Commun. Mag.* 28(May):5–14.

Cox, D.C. 1995. Wireless personal communications: what is it? *IEEE Personal Comm.* 2(2):20–35.

Electronic Industries Association/Telecommunications Industry Association. 1991. Recommended minimum performance standards for 800 MHz dual-mode mobile stations. Incorp. EIA/TIA 19B, EIA/TIA Project No. 2216, March.

Harashima, H. and Miyakawa, H. 1972. Matched-transmission technique for channels with intersymbol interference. *IEEE Trans. on Commun.* COM-20(Aug.):774–780.

Kalet, I. and Saltzberg, B.R. 1994. QAM transmission through a companding channel—signal constellations and detection. *IEEE Trans. on Comm.* 42(2–4):417–429.

Kretzmer, E.R. 1966. Generalization of a technique for binary data communication. *IEEE Trans. Comm. Tech.* COM-14 (Feb.):67, 68.

Lender, A. 1963. The duobinary technique for high-speed data Transmission. *AIEE Trans. on Comm. Electronics*, 82 (March):214–218.

Rahnema, M. 1993. Overview of the GSM system and protocol architecture. *IEEE Commun. Mag.* (April):92–100.

Telecommunication Industry Association. 1993. Mobile station-base station compatibility standard for dual-mode wideband spread spectrum cellular system. TIA/EIA/IS-95. July.

Tomlinson, M. 1971. New automatic equalizer employing modulo arithmetic. *Electron. Lett.* 7(March):138, 139.

Further Information

Baseband signalling and pulse shaping is fundamental to the design of any digital communications system and is, therefore, covered in numerous texts on digital communications. For more advanced treatments see E.A. Lee and D.G. Messerschmitt, *Digital Communication*, Kluwer 1994, and J.G. Proakis, *Digital Communications*, McGraw-Hill 1995.

5

Channel Equalization

John G. Proakis
Northeastern University

5.1 Characterization of Channel Distortion

Many communication channels, including telephone channels, and some radio channels, may be generally characterized as band-limited linear filters. Consequently, such channels are described by their frequency response $C(f)$, which may be expressed as

$$C(f) = A(f)e^{j\theta(f)} \qquad (5.1)$$

where $A(f)$ is called the *amplitude response* and $\theta(f)$ is called the *phase response*. Another characteristic that is sometimes used in place of the phase response is the *envelope delay* or *group delay*, which is defined as

$$\tau(f) = -\frac{1}{2\pi}\frac{d\theta(f)}{df} \qquad (5.2)$$

A channel is said to be nondistorting or ideal if, within the bandwidth W occupied by the transmitted signal, $A(f) = \text{const}$ and $\theta(f)$ is a linear function of frequency [or the envelope delay $\tau(f) = \text{const}$]. On the other hand, if $A(f)$ and $\tau(f)$ are not constant within the bandwidth occupied by the transmitted signal, the channel distorts the signal. If $A(f)$ is not constant, the distortion is called *amplitude distortion* and if $\tau(f)$ is not constant, the distortion on the transmitted signal is called *delay distortion*.

As a result of the amplitude and delay distortion caused by the nonideal channel frequency response characteristic $C(f)$, a succession of pulses transmitted through the channel at rates comparable to the bandwidth W are smeared to the point that they are no longer distinguishable as well-defined pulses at the receiving terminal. Instead, they overlap and, thus, we have **intersymbol interference (ISI)**. As an example of the effect of delay distortion on a transmitted pulse, Fig. 5.1(a) illustrates a band-limited pulse having zeros periodically spaced in time at points labeled $\pm T$, $\pm 2T$,

0-8493-8573-3/96/$0.00+$.50

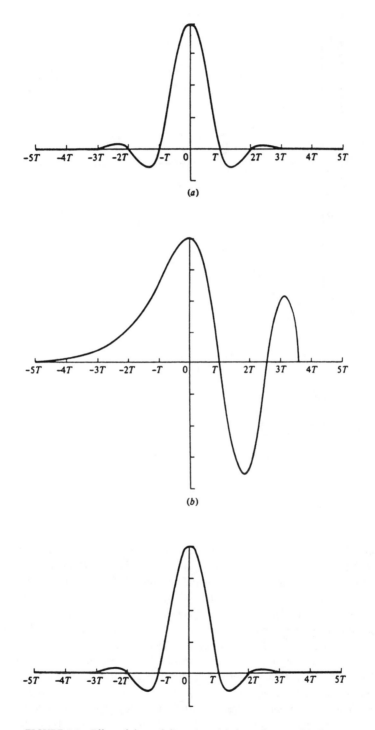

FIGURE 5.1   Effect of channel distortion: (a) channel input, (b) channel
output, (c) equalizer output.

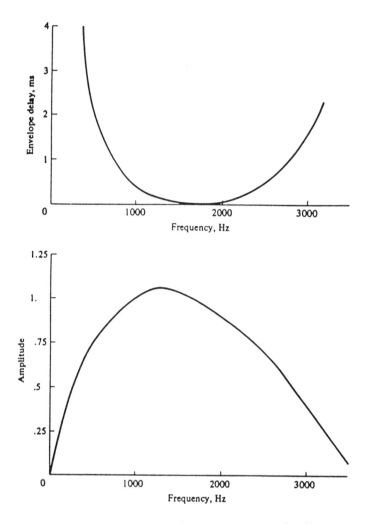

FIGURE 5.2 Average amplitude and delay characteristics of medium-range telephone channel.

etc. If information is conveyed by the pulse amplitude, as in pulse amplitude modulation (PAM), for example, then one can transmit a sequence of pulses, each of which has a peak at the periodic zeros of the other pulses. Transmission of the pulse through a channel modeled as having a linear envelope delay characteristic $\tau(f)$ [quadratic phase $\theta(f)$], however, results in the received pulse shown in Fig. 5.1(b) having zero crossings that are no longer periodically spaced. Consequently a sequence of successive pulses would be smeared into one another, and the peaks of the pulses would no longer be distinguishable. Thus, the channel delay distortion results in intersymbol interference. As will be discussed in this chapter, it is possible to compensate for the nonideal frequency response characteristic of the channel by use of a filter or equalizer at the demodulator. Figure 5.1(c) illustrates the output of a linear equalizer that compensates for the linear distortion in the channel.

The extent of the intersymbol interference on a telephone channel can be appreciated by observing a frequency response characteristic of the channel. Figure 5.2 illustrates the measured average amplitude and delay as a function of frequency for a medium-range (180–725 mi) telephone channel of the switched telecommunications network as given by Duffy and Tratcher, 1971. We observe that the usable band of the channel extends from about 300 Hz to about 3000 Hz. The corresponding impulse response of the average channel is shown in Fig. 5.3. Its duration is about 10 ms. In

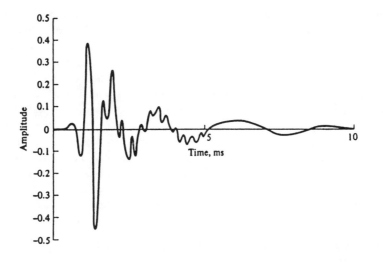

FIGURE 5.3 Impulse response of average channel with amplitude and delay shown in Fig. 5.2.

comparison, the transmitted symbol rates on such a channel may be of the order of 2500 pulses or symbols per second. Hence, intersymbol interference might extend over 20–30 symbols.

Besides telephone channels, there are other physical channels that exhibit some form of time dispersion and, thus, introduce intersymbol interference. Radio channels, such as short-wave ionospheric propagation (HF), tropospheric scatter, and mobile cellular radio are three examples of time-dispersive wireless channels. In these channels, time dispersion and, hence, intersymbol interference is the result of multiple propagation paths with different path delays. The number

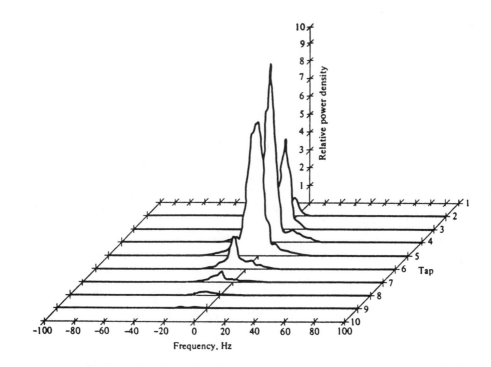

FIGURE 5.4 Scattering function of a medium-range tropospheric scatter channel.

of paths and the relative time delays among the paths vary with time and, for this reason, these radio channels are usually called time-variant multipath channels. The time-variant multipath conditions give rise to a wide variety of frequency response characteristics. Consequently, the frequency response characterization that is used for telephone channels is inappropriate for time-variant multipath channels. Instead, these radio channels are characterized statistically in terms of the scattering function, which, in brief, is a two-dimensional representation of the average received signal power as a function of relative time delay and Doppler frequency (see Proakis, 1995).

For illustrative purposes, a scattering function measured on a medium-range (150 mi) tropospheric scatter channel is shown in Fig. 5.4. The total time duration (multipath spread) of the channel response is approximately 0.7 μs on the average, and the spread between half-power points in Doppler frequency is a little less than 1 Hz on the strongest path and somewhat larger on the other paths. Typically, if one is transmitting at a rate of 10^7 symbols/s over such a channel, the multipath spread of 0.7 μs will result in intersymbol interference that spans about seven symbols.

5.2 Characterization of Intersymbol Interference

In a digital communication system, channel distortion causes intersymbol interference, as illustrated in the preceding section. In this section, we shall present a model that characterizes the ISI. The digital modulation methods to which this treatment applies are PAM, phase-shift keying (PSK) and quadrature amplitude modulation (QAM). The transmitted signal for these three types of modulation may be expressed as

$$\begin{aligned} s(t) &= v_c(t) \cos 2\pi f_c t - v_s(t) \sin 2\pi f_c t \\ &= \text{Re}[v(t) e^{j2\pi f_c t}] \end{aligned} \tag{5.3}$$

where $v(t) = v_c(t) + j v_s(t)$ is called the *equivalent low-pass signal*, f_c is the carrier frequency, and Re[ ] denotes the real part of the quantity in brackets.

In general, the equivalent low-pass signal is expressed as

$$v(t) = \sum_{n=0}^{\infty} I_n g_T(t - nT) \tag{5.4}$$

where $g_T(t)$ is the basic pulse shape that is selected to control the spectral characteristics of the transmitted signal, $\{I_n\}$ the sequence of transmitted information symbols selected from a signal constellation consisting of M points, and T the signal interval ($1/T$ is the symbol rate). For PAM, PSK, and QAM, the values of I_n are points from M-ary signal constellations. Figure 5.5 illustrates the signal constellations for the case of $M = 8$ signal points. Note that for PAM, the signal constellation is one dimensional. Hence, the equivalent low-pass signal $v(t)$ is real valued, i.e., $v_s(t) = 0$ and $v_c(t) = v(t)$. For M-ary ($M > 2$) PSK and QAM, the signal constellations are two dimensional and, hence, $v(t)$ is complex valued.

The signal $s(t)$ is transmitted over a bandpass channel that may be characterized by an equivalent low-pass frequency response $C(f)$. Consequently, the equivalent low-pass received signal can be represented as

$$r(t) = \sum_{n=0}^{\infty} I_n h(t - nT) + w(t) \tag{5.5}$$

where $h(t) = g_T(t) * c(t)$, and $c(t)$ is the impulse response of the equivalent low-pass channel, the asterisk denotes convolution, and $w(t)$ represents the additive noise in the channel.

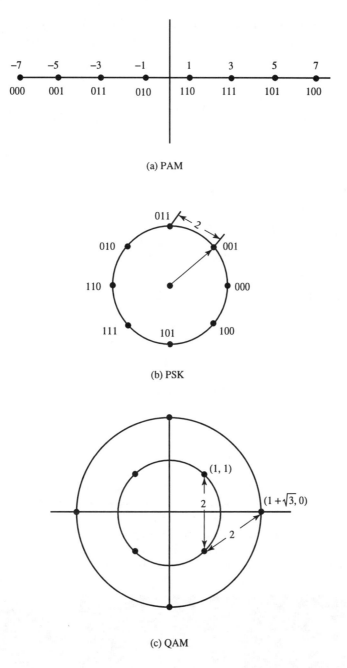

(a) PAM

(b) PSK

(c) QAM

FIGURE 5.5 $M = 8$ signal constellations for PAM, PSK, and QAM.

To characterize the ISI, suppose that the received signal is passed through a receiving filter and then sampled at the rate $1/T$ samples/s. In general, the optimum filter at the receiver is matched to the received signal pulse $h(t)$. Hence, the frequency response of this filter is $H^*(f)$. We denote its output as

$$y(t) = \sum_{n=0}^{\infty} I_n \, x(t - nT) + v(t) \tag{5.6}$$

where $x(t)$ is the signal pulse response of the receiving filter, i.e., $X(f) = H(f)H^*(f) = |H(f)|^2$, and $v(t)$ is the response of the receiving filter to the noise $w(t)$. Now, if $y(t)$ is sampled at times $t = kT$, $k = 0, 1, 2, \ldots$, we have

$$y(kT) \equiv y_k = \sum_{n=0}^{\infty} I_n x(kT - nT) + v(kT)$$

$$= \sum_{n=0}^{\infty} I_n x_{k-n} + v_k, \qquad k = 0, 1, \ldots \tag{5.7}$$

The sample values $\{y_k\}$ can be expressed as

$$y_k = x_0 \left(I_k + \frac{1}{x_0} \sum_{\substack{n=0 \\ n \neq k}}^{\infty} I_n x_{k-n} \right) + v_k, \qquad k = 0, 1, \ldots \tag{5.8}$$

The term x_0 is an arbitrary scale factor, which we arbitrarily set equal to unity for convenience. Then

$$y_k = I_k + \sum_{\substack{n=0 \\ n \neq k}}^{\infty} I_n x_{k-n} + v_k \tag{5.9}$$

The term I_k represents the desired information symbol at the kth sampling instant, the term

$$\sum_{\substack{n=0 \\ n \neq k}}^{\infty} I_n x_{k-n} \tag{5.10}$$

represents the ISI, and v_k is the additive noise variable at the kth sampling instant.

The amount of ISI, and noise in a digital communications system can be viewed on an oscilloscope. For PAM signals, we can display the received signal $y(t)$ on the vertical input with the horizontal sweep rate set at $1/T$. The resulting oscilloscope display is called an *eye pattern* because of its resemblance to the human eye. For example, Fig. 5.6 illustrates the eye patterns for binary and four-level PAM modulation. The effect of ISI is to cause the eye to close, thereby reducing the margin

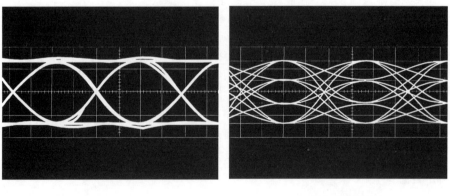

BINARY   QUATERNARY

FIGURE 5.6 Examples of eye patterns for binary and quaternary amplitude shift keying (or PAM).

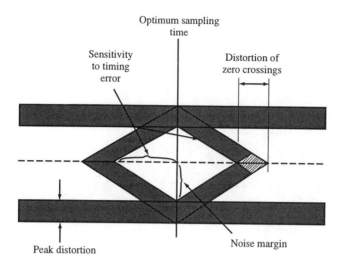

FIGURE 5.7 Effect of intersymbol interference on eye opening.

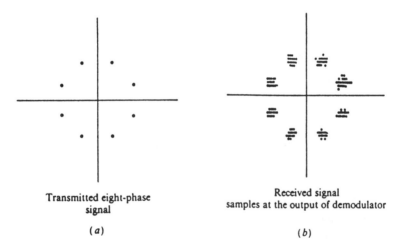

Transmitted eight-phase signal

(*a*)

Received signal samples at the output of demodulator

(*b*)

FIGURE 5.8 Two-dimensional digital eye patterns.

for additive noise to cause errors. Figure 5.7 graphically illustrates the effect of ISI in reducing the opening of a binary eye. Note that intersymbol interference distorts the position of the zero crossings and causes a reduction in the eye opening. Thus, it causes the system to be more sensitive to a synchronization error.

For PSK and QAM it is customary to display the eye pattern as a two-dimensional scatter diagram illustrating the sampled values $\{y_k\}$ that represent the decision variables at the sampling instants. Figure 5.8 illustrates such an eye pattern for an 8-PSK signal. In the absence of intersymbol interference and noise, the superimposed signals at the sampling instants would result in eight distinct points corresponding to the eight transmitted signal phases. Intersymbol interference and noise result in a deviation of the received samples $\{y_k\}$ from the desired 8-PSK signal. The larger the intersymbol interference and noise, the larger the scattering of the received signal samples relative to the transmitted signal points.

In practice, the transmitter and receiver filters are designed for zero ISI at the desired sampling times $t = kT$. Thus, if $G_T(f)$ is the frequency response of the transmitter filter and $G_R(f)$ is the

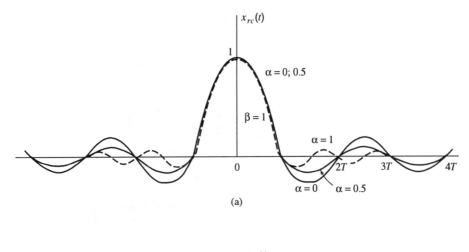

(a)

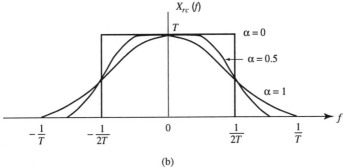

(b)

FIGURE 5.9   Pulses having a raised cosine spectrum.

frequency response of the receiver filter, then the product $G_T(f) G_R(f)$ is designed to yield zero ISI. For example, the product $G_T(f) G_R(f)$ may be selected as

$$G_T(f)G_R(f) = X_{rc}(f) \tag{5.11}$$

where $X_{rc}(f)$ is the raised-cosine frequency response characteristic, defined as

$$X_{rc}(f) = \begin{cases} T, & 0 \le |f| \le (1-\alpha)/2T \\ \dfrac{T}{2}\lfloor 1 + \cos\dfrac{\pi T}{\alpha}(|f| - \dfrac{1-\alpha}{2T}), & \dfrac{1-\alpha}{2T} \le |f| \le \dfrac{1+\alpha}{2T} \\ 0, & |f| > \dfrac{1+\alpha}{2T} \end{cases} \tag{5.12}$$

where α is called the *rolloff* factor, which takes values in the range $0 \le \alpha \le 1$, and $1/T$ is the symbol rate. The frequency response $X_{rc}(f)$ is illustrated in Fig. 5.9(a) for $\alpha = 0$, $1/2$, and 1. Note that when $\alpha = 0$, $X_{rc}(f)$ reduces to an ideal brick wall physically nonrealizable frequency response with bandwidth occupancy $1/2T$. The frequency $1/2T$ is called the *Nyquist frequency*. For $\alpha > 0$, the bandwidth occupied by the desired signal $X_{rc}(f)$ beyond the Nyquist frequency $1/2T$ is called the *excess bandwidth*, and is usually expressed as a percentage of the Nyquist frequency. For example, when $\alpha = 1/2$, the excess bandwidth is 50% and when $\alpha = 1$, the excess bandwidth is 100%. The signal pulse $x_{rc}(t)$ having the raised-cosine spectrum is

$$x_{rc}(t) = \frac{\sin \pi t/T}{\pi t/T} \frac{\cos(\pi \alpha t/T)}{1 - 4\alpha^2 t^2/T^2} \tag{5.13}$$

Figure 5.9(b) illustrates $x_{rc}(t)$ for $\alpha = 0$, $1/2$, and 1. Note that $x_{rc}(t) = 1$ at $t = 0$ and $x_{rc}(t) = 0$ at $t = kT$, $k = \pm 1, \pm 2, \ldots$. Consequently, at the sampling instants $t = kT$, $k \neq 0$, there is no ISI from adjacent symbols when there is no channel distortion. In the presence of channel distortion, however, the ISI given by Eq. (5.10) is no longer zero, and a channel equalizer is needed to minimize its effect on system performance.

5.3 Linear Equalizers

The most common type of channel equalizer used in practice to reduce ISI is a linear transversal filter with adjustable coefficients $\{c_i\}$, as shown in Fig. 5.10.

On channels whose frequency response characteristics are unknown, but time invariant, we may measure the channel characteristics and adjust the parameters of the equalizer; once adjusted, the parameters remain fixed during the transmission of data. Such equalizers are called **preset equalizers**. On the other hand, **adaptive equalizers** update their parameters on a periodic basis during the transmission of data and, thus, they are capable of tracking a slowly time-varying channel response.

First, let us consider the design characteristics for a linear equalizer from a frequency domain viewpoint. Figure 5.11 shows a block diagram of a system that employs a linear filter as a **channel equalizer**.

The demodulator consists of a receiver filter with frequency response $G_R(f)$ in cascade with a channel equalizing filter that has a frequency response $G_E(f)$. As indicated in the preceding section, the receiver filter response $G_R(f)$ is matched to the transmitter response, i.e., $G_R(f) = G_T^*(f)$, and the product $G_R(f)G_T(f)$ is usually designed so that there is zero ISI at the sampling instants as, for example, when $G_R(t)G_T(f) = X_{rc}(f)$.

For the system shown in Fig. 5.11, in which the channel frequency response is not ideal, the desired condition for zero ISI is

$$G_T(f)C(f)G_R(f)G_E(f) = X_{rc}(f) \tag{5.14}$$

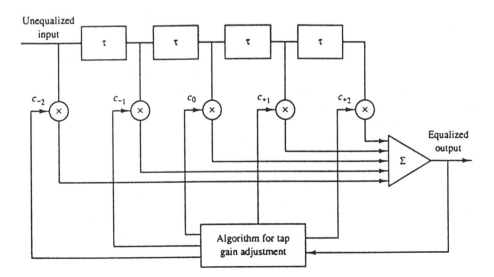

FIGURE 5.10 Linear transversal filter.

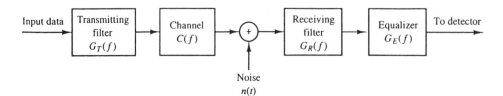

FIGURE 5.11   Block diagram of a system with an equalizer.

where $X_{rc}(f)$ is the desired raised-cosine spectral characteristic. Since $G_T(f)G_R(f) = X_{rc}(f)$ by design, the frequency response of the equalizer that compensates for the channel distortion is

$$G_E(f) = \frac{1}{C(f)} = \frac{1}{|C(f)|} e^{-j\theta_c(f)} \tag{5.15}$$

Thus, the amplitude response of the equalizer is $|G_E(f)| = 1/|C(f)|$ and its phase response is $\theta_E(f) = -\theta_c(f)$. In this case, the equalizer is said to be the *inverse channel filter* to the channel response.

We note that the inverse channel filter completely eliminates ISI caused by the channel. Since it forces the ISI to be zero at the sampling instants $t = kT$, $k = 0, 1, \ldots$, the equalizer is called a **zero-forcing equalizer.** Hence, the input to the detector is simply

$$z_k = I_k + \eta_k, \qquad k = 0, 1, \ldots \tag{5.16}$$

where η_k represents the additive noise and I_k is the desired symbol.

In practice, the ISI caused by channel distortion is usually limited to a finite number of symbols on either side of the desired symbol. Hence, the number of terms that constitute the ISI in the summation given by Eq. (5.10) is finite. As a consequence, in practice the channel equalizer is implemented as a finite duration impulse response (FIR) filter, or transversal filter, with adjustable tap coefficients $\{c_n\}$, as illustrated in Fig. 5.10. The time delay τ between adjacent taps may be selected as large as T, the symbol interval, in which case the FIR equalizer is called a **symbol-spaced equalizer.** In this case, the input to the equalizer is the sampled sequence given by Eq. (5.7). We note that when the symbol rate $1/T < 2W$, however, frequencies in the received signal above the folding frequency $1/T$ are aliased into frequencies below $1/T$. In this case, the equalizer compensates for the aliased channel-distorted signal.

On the other hand, when the time delay τ between adjacent taps is selected such that $1/\tau \geq 2W > 1/T$, no aliasing occurs and, hence, the inverse channel equalizer compensates for the true channel distortion. Since $\tau < T$, the channel equalizer is said to have *fractionally spaced taps* and it is called a **fractionally spaced equalizer.** In practice, τ is often selected as $\tau = T/2$. Notice that, in this case, the sampling rate at the output of the filter $G_R(f)$ is $2/T$.

The impulse response of the FIR equalizer is

$$g_E(t) = \sum_{n=-N}^{N} c_n \delta(t - n\tau) \tag{5.17}$$

and the corresponding frequency response is

$$G_E(f) = \sum_{n=-N}^{N} c_n e^{-j2\pi f n\tau} \tag{5.18}$$

where $\{c_n\}$ are the $(2N + 1)$ equalizer coefficients and N is chosen sufficiently large so that the equalizer spans the length of the ISI, i.e., $2N + 1 \geq L$, where L is the number of signal samples spanned by the ISI. Since $X(f) = G_T(f)C(f)G_R(f)$ and $x(t)$ is the signal pulse corresponding to $X(f)$, then the equalized output signal pulse is

$$q(t) = \sum_{n=-N}^{N} c_n x(t - n\tau) \tag{5.19}$$

The zero-forcing condition can now be applied to the samples of $q(t)$ taken at times $t = mT$. These samples are

$$q(mT) = \sum_{n=-N}^{N} c_n x(mT - n\tau), \qquad m = 0, \pm 1, \ldots, \pm N \tag{5.20}$$

Since there are $2N + 1$ equalizer coefficients, we can control only $2N + 1$ sampled values of $q(t)$. Specifically, we may force the conditions

$$q(mT) = \sum_{n=-N}^{N} c_n x(mT - n\tau) = \begin{cases} 1, & m = 0 \\ 0, & m = \pm 1, \pm 2, \ldots, \pm N \end{cases} \tag{5.21}$$

which may be expressed in matrix form as $Xc = q$, where X is a $(2N + 1) \times (2N + 1)$ matrix with elements $\{x(mT - n\tau)\}$, c is the $(2N + 1)$ coefficient vector and q is the $(2N + 1)$ column vector with one nonzero element. Thus, we obtain a set of $2N + 1$ linear equations for the coefficients of the zero-forcing equalizer.

We should emphasize that the FIR zero-forcing equalizer does not completely eliminate ISI because it has a finite length. As N is increased, however, the residual ISI can be reduced, and in the limit as $N \to \infty$, the ISI is completely eliminated.

Example 5.1 Consider a channel distorted pulse $x(t)$, at the input to the equalizer, given by the expression

$$x(t) = \frac{1}{1 + \left(\dfrac{2t}{T}\right)^2}$$

where $1/T$ is the symbol rate. The pulse is sampled at the rate $2/T$ and equalized by a zero-forcing equalizer. Determine the coefficients of a five-tap zero-forcing equalizer.

Solution 5.1 According to Eq. (5.21), the zero-forcing equalizer must satisfy the equations

$$q(mT) = \sum_{n=-2}^{2} c_n x(mT - nT/2) = \begin{cases} 1, & m = 0 \\ 0, & m = \pm 1, \pm 2 \end{cases}$$

The matrix X with elements $x(mT - nT/2)$ is given as

$$X = \begin{bmatrix} \dfrac{1}{5} & \dfrac{1}{10} & \dfrac{1}{17} & \dfrac{1}{26} & \dfrac{1}{37} \\[2mm] 1 & \dfrac{1}{2} & \dfrac{1}{5} & \dfrac{1}{10} & \dfrac{1}{17} \\[2mm] \dfrac{1}{5} & \dfrac{1}{2} & 1 & \dfrac{1}{2} & \dfrac{1}{5} \\[2mm] \dfrac{1}{17} & \dfrac{1}{10} & \dfrac{1}{5} & \dfrac{1}{2} & 1 \\[2mm] \dfrac{1}{37} & \dfrac{1}{26} & \dfrac{1}{17} & \dfrac{1}{10} & \dfrac{1}{5} \end{bmatrix} \tag{5.22}$$

The coefficient vector c and the vector q are given as

$$c = \begin{bmatrix} c_{-2} \\ c_{-1} \\ c_0 \\ c_1 \\ c_2 \end{bmatrix} \qquad q = \begin{bmatrix} 0 \\ 0 \\ 1 \\ 0 \\ 0 \end{bmatrix} \tag{5.23}$$

Then, the linear equations $Xc = q$ can be solved by inverting the matrix X. Thus, we obtain

$$c_{\text{opt}} = X^{-1}q = \begin{bmatrix} -2.2 \\ 4.9 \\ -3 \\ 4.9 \\ -2.2 \end{bmatrix} \tag{5.24}$$

One drawback to the zero-forcing equalizer is that it ignores the presence of additive noise. As a consequence, its use may result in significant noise enhancement. This is easily seen by noting that in a frequency range where $C(f)$ is small, the channel equalizer $G_E(f) = 1/C(f)$ compensates by placing a large gain in that frequency range. Consequently, the noise in that frequency range is greatly enhanced. An alternative is to relax the zero ISI condition and select the channel equalizer characteristic such that the combined power in the residual ISI and the additive noise at the output of the equalizer is minimized. A channel equalizer that is optimized based on the minimum mean square error (MMSE) criterion accomplishes the desired goal.

To elaborate, let us consider the noise corrupted output of the FIR equalizer, which is

$$z(t) = \sum_{n=-N}^{N} c_n y(t - n\tau) \tag{5.25}$$

where $y(t)$ is the input to the equalizer, given by Eq. (5.6). The equalizer output is sampled at times $t = mT$. Thus, we obtain

$$z(mT) = \sum_{n=-N}^{N} c_n y(mT - n\tau) \tag{5.26}$$

The desired response at the output of the equalizer at $t = mT$ is the transmitted symbol I_m. The error is defined as the difference between I_m and $z(mT)$. Then, the mean square error (*MSE*) between the actual output sample $z(mT)$ and the desired values I_m is

$$MSE = E|z(mT) - I_m|^2$$

$$= E\left[\left|\sum_{n=-N}^{N} c_n y(mT - n\tau) - I_m\right|^2\right]$$

$$= \sum_{n=-N}^{N} \sum_{k=-N}^{N} c_n c_k R_Y(n - k)$$

$$- 2\sum_{k=-N}^{N} c_k R_{IY}(k) + E(|I_m|^2) \tag{5.27}$$

where the correlations are defined as

$$R_Y(n - k) = E[y^*(mT - n\tau)y(mT - k\tau)]$$

$$R_{IY}(k) = E[y(mT - k\tau)I_m^*] \tag{5.28}$$

and the expectation is taken with respect to the random information sequence $\{I_m\}$ and the additive noise.

The minimum *MSE* solution is obtained by differentiating Eq. (5.27) with respect to the equalizer coefficients $\{c_n\}$. Thus, we obtain the necessary conditions for the minimum *MSE* as

$$\sum_{n=-N}^{N} c_n R_Y(n - k) = R_{IY}(k), \qquad k = 0, \pm1, 2, \ldots, \pm N \tag{5.29}$$

These are the $(2N + 1)$ linear equations for the equalizer coefficients. In contrast to the zero-forcing solution already described, these equations depend on the statistical properties (the autocorrelation) of the noise as well as the ISI through the autocorrelation $R_Y(n)$.

In practice, the autocorrelation matrix $R_Y(n)$ and the crosscorrelation vector $R_{IY}(n)$ are unknown a priori. These correlation sequences can be estimated, however, by transmitting a test signal over the channel and using the time-average estimates

$$\hat{R}_Y(n) = \frac{1}{K}\sum_{k=1}^{K} y^*(kT - n\tau)y(kT)$$

$$\hat{R}_{IY}(n) = \frac{1}{K}\sum_{k=1}^{K} y(kT - n\tau)I_k^* \tag{5.30}$$

in place of the ensemble averages to solve for the equalizer coefficients given by Eq. (5.29).

Adaptive Linear Equalizers

We have shown that the tap coefficients of a linear equalizer can be determined by solving a set of linear equations. In the zero-forcing optimization criterion, the linear equations are given by Eq. (5.21). On the other hand, if the optimization criterion is based on minimizing the *MSE*, the optimum equalizer coefficients are determined by solving the set of linear equations given by Eq. (5.29).

In both cases, we may express the set of linear equations in the general matrix form

$$\boldsymbol{Bc} = \boldsymbol{d} \qquad\qquad (5.31)$$

where $\boldsymbol{B}$ is a $(2N + 1) \times (2N + 1)$ matrix, $\boldsymbol{c}$ is a column vector representing the $2N + 1$ equalizer coefficients, and $\boldsymbol{d}$ a $(2N + 1)$-dimensional column vector. The solution of Eq. (5.31) yields

$$\boldsymbol{c}_{\text{opt}} = \boldsymbol{B}^{-1}\boldsymbol{d} \qquad\qquad (5.32)$$

In practical implementations of equalizers, the solution of Eq. (5.31) for the optimum coefficient vector is usually obtained by an iterative procedure that avoids the explicit computation of the inverse of the matrix $\boldsymbol{B}$. The simplest iterative procedure is the method of steepest descent, in which one begins by choosing arbitrarily the coefficient vector $\boldsymbol{c}$, say $\boldsymbol{c}_0$. This initial choice of coefficients corresponds to a point on the criterion function that is being optimized. For example, in the case of the *MSE* criterion, the initial guess $\boldsymbol{c}_0$ corresponds to a point on the quadratic *MSE* surface in the $(2N + 1)$-dimensional space of coefficients. The gradient vector, defined as $\boldsymbol{g}_0$, which is the derivative of the *MSE* with respect to the $2N + 1$ filter coefficients, is then computed at this point on the criterion surface, and each tap coefficient is changed in the direction opposite to its corresponding gradient component. The change in the jth tap coefficient is proportional to the size of the jth gradient component.

For example, the gradient vector denoted as $\boldsymbol{g}_k$, for the MSE criterion, found by taking the derivatives of the MSE with respect to each of the $2N + 1$ coefficients, is

$$\boldsymbol{g}_k = \boldsymbol{Bc}_k - \boldsymbol{d}, \qquad k = 0, 1, 2, \ldots \qquad\qquad (5.33)$$

Then the coefficient vector $\boldsymbol{c}_k$ is updated according to the relation

$$\boldsymbol{c}_{k+1} = \boldsymbol{c}_k - \Delta \boldsymbol{g}_k \qquad\qquad (5.34)$$

where Δ is the *step-size parameter* for the iterative procedure. To ensure convergence of the iterative procedure, Δ is chosen to be a small positive number. In such a case, the gradient vector $\boldsymbol{g}_k$ converges toward zero, i.e., $\boldsymbol{g}_k \to 0$ as $k \to \infty$, and the coefficient vector $\boldsymbol{c}_k \to \boldsymbol{c}_{\text{opt}}$ as illustrated in Fig. 5.12 based on two-dimensional optimization. In general, convergence of the equalizer tap coefficients to $\boldsymbol{c}_{\text{opt}}$ cannot be attained in a finite number of iterations with the steepest-descent method. The optimum solution $\boldsymbol{c}_{\text{opt}}$, however, can be approached as closely as desired in a few hundred iterations. In digital communication systems that employ channel equalizers, each iteration corresponds to a time interval for sending one symbol and, hence, a few hundred iterations to achieve convergence to $\boldsymbol{c}_{\text{opt}}$ corresponds to a fraction of a second.

Adaptive channel equalization is required for channels whose characteristics change with time. In such a case, the ISI varies with time. The channel equalizer must track such time variations in the channel response and adapt its coefficients to reduce the ISI. In the context of the preceding discussion, the optimum coefficient vector $\boldsymbol{c}_{\text{opt}}$ varies with time due to time variations in the matrix $\boldsymbol{B}$ and, for the case of the *MSE* criterion, time variations in the vector $\boldsymbol{d}$. Under these conditions, the iterative method described can be modified to use estimates of the gradient components. Thus, the algorithm for adjusting the equalizer tap coefficients may be expressed as

$$\hat{\boldsymbol{c}}_{k+1} = \hat{\boldsymbol{c}}_k - \Delta \hat{\boldsymbol{g}}_k \qquad\qquad (5.35)$$

where $\hat{\boldsymbol{g}}_k$ denotes an estimate of the gradient vector $\boldsymbol{g}_k$ and $\hat{\boldsymbol{c}}_k$ denotes the estimate of the tap coefficient vector.

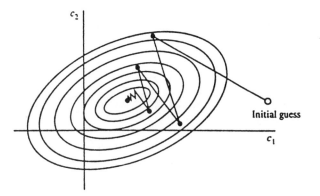

FIGURE 5.12 Examples of convergence characteristics of a gradient algorithm.

In the case of the *MSE* criterion, the gradient vector g_k given by Eq. (5.33) may also be expressed as

$$g_k = -E\left(e_k y_k^*\right)$$

An estimate $\hat{g}_k$ of the gradient vector at the kth iteration is computed as

$$\hat{g}_k = -e_k y_k^* \tag{5.36}$$

where e_k denotes the difference between the desired output from the equalizer at the kth time instant and the actual output $z(kT)$, and y_k denotes the column vector of $2N + 1$ received signal values contained in the equalizer at time instant k. The *error signal* e_k is expressed as

$$e_k = I_k - z_k \tag{5.37}$$

where $z_k = z(kT)$ is the equalizer output given by Eq. (5.26) and I_k is the desired symbol. Hence, by substituting Eq. (5.36) into Eq. (5.35), we obtain the adaptive algorithm for optimizing the tap coefficients (based on the *MSE* criterion) as

$$\hat{c}_{k+1} = \hat{c}_k + \Delta e_k y_k^* \tag{5.38}$$

Since an estimate of the gradient vector is used in Eq. (5.38) the algorithm is called a **stochastic gradient algorithm**; it is also known as the **LMS algorithm**.

A block diagram of an adaptive equalizer that adapts its tap coefficients according to Eq. (5.38) is illustrated in Fig. 5.13. Note that the difference between the desired output I_k and the actual output z_k from the equalizer is used to form the error signal e_k. This error is scaled by the step-size parameter Δ, and the scaled error signal Δe_k multiplies the received signal values $\{y(kT - n\tau)\}$ at the $2N + 1$ taps. The products $\Delta e_k y^*(kT - n\tau)$ at the $(2N + 1)$ taps are then added to the previous values of the tap coefficients to obtain the updated tap coefficients, according to Eq. (5.38). This computation is repeated as each new symbol is received. Thus, the equalizer coefficients are updated at the symbol rate.

Initially, the adaptive equalizer is trained by the transmission of a known pseudo-random sequence $\{I_m\}$ over the channel. At the demodulator, the equalizer employs the known sequence to adjust its coefficients. Upon initial adjustment, the adaptive equalizer switches from a **training mode** to a **decision-directed mode**, in which case the decisions at the output of the detector are sufficiently reliable so that the error signal is formed by computing the difference between the

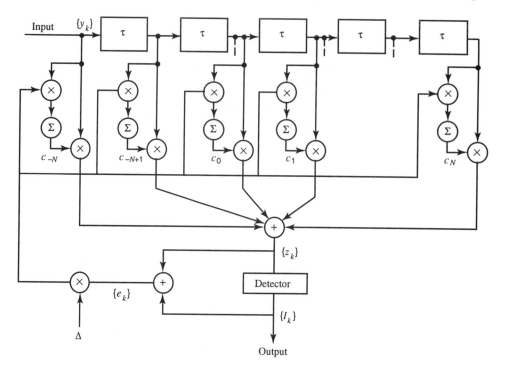

FIGURE 5.13   Linear adaptive equalizer based on the *MSE* criterion.

detector output and the equalizer output, i.e.,

$$e_k = \tilde{I}_k - z_k \tag{5.39}$$

where $\tilde{I}_k$ is the output of the detector. In general, decision errors at the output of the detector occur infrequently and, consequently, such errors have little effect on the performance of the tracking algorithm given by Eq. (5.38).

A rule of thumb for selecting the step-size parameter so as to ensure convergence and good tracking capabilities in slowly varying channels is

$$\Delta = \frac{1}{5(2N+1)P_R} \tag{5.40}$$

where P_R denotes the received signal-plus-noise power, which can be estimated from the received signal (see Proakis, 1995).

The convergence characteristic of the stochastic gradient algorithm in Eq. (5.38) is illustrated in Fig. 5.14. These graphs were obtained from a computer simulation of an 11-tap adaptive equalizer operating on a channel with a rather modest amount of ISI. The input signal-plus-noise power P_R was normalized to unity. The rule of thumb given in Eq. (5.40) for selecting the step size gives $\Delta = 0.018$. The effect of making Δ too large is illustrated by the large jumps in *MSE* as shown for $\Delta = 0.115$. As Δ is decreased, the convergence is slowed somewhat, but a lower MSE is achieved, indicating that the estimated coefficients are closer to c_{opt}.

Although we have described in some detail the operation of an adaptive equalizer that is optimized on the basis of the MSE criterion, the operation of an adaptive equalizer based on the zero-forcing method is very similar. The major difference lies in the method for estimating the gradient vectors g_k at each iteration. A block diagram of an adaptive zero-forcing equalizer is shown in Fig. 5.15.

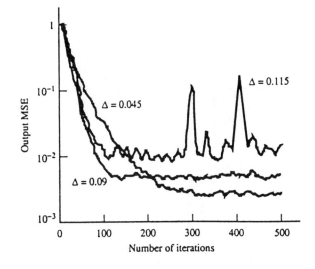

FIGURE 5.14 Initial convergence characteristics of the *LMS* algorithm with different step sizes.

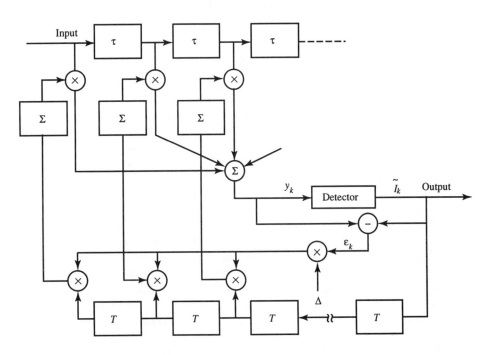

FIGURE 5.15 An adaptive zero-forcing equalizer.

For more details on the tap coefficient update method for a zero-forcing equalizer, the reader is referred to the papers by Lucky, 1965 and 1966, and the text by Proakis, 1995.

5.4 Decision-Feedback Equalizer

The linear filter equalizers described in the preceding section are very effective on channels, such as wire line telephone channels, where the ISI is not severe. The severity of the ISI is directly related

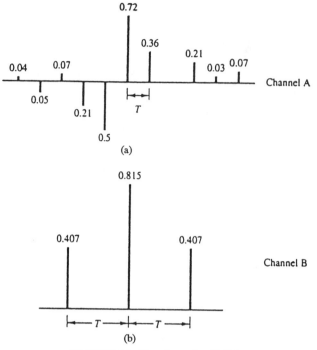

FIGURE 5.16   Two channels with ISI.

to the spectral characteristics and not necessarily to the time span of the ISI. For example, consider the ISI resulting from the two channels that are illustrated in Fig. 5.16. The time span for the ISI in channel A is 5 symbol intervals on each side of the desired signal component, which has a value of 0.72. On the other hand, the time span for the ISI in channel B is one symbol interval on each side of the desired signal component, which has a value of 0.815. The energy of the total response is normalized to unity for both channels.

In spite of the shorter ISI span, channel B results in more severe ISI. This is evidenced in the frequency response characteristics of these channels, which are shown in Fig. 5.17. We observe that channel B has a spectral null [the frequency response $C(f) = 0$ for some frequencies in the band $|f| \leq W$] at $f = 1/2T$, whereas this does not occur in the case of channel A. Consequently, a linear equalizer will introduce a large gain in its frequency response to compensate for the channel null. Thus, the noise in channel B will be enhanced much more than in channel A. This implies that the performance of the linear equalizer for channel B will be sufficiently poorer than that for channel A. This fact is borne out by the computer simulation results for the performance of the two linear equalizers shown in Fig. 5.18. Hence, the basic limitation of a linear equalizer is that it performs poorly on channels having spectral nulls. Such channels are often encountered in radio communications, such as ionospheric transmission at frequencies below 30 MHz and mobile radio channels, such as those used for cellular radio communications.

A **decision-feedback equalizer** (DFE) is a nonlinear equalizer that employs previous decisions to eliminate the ISI caused by previously detected symbols on the current symbol to be detected. A simple block diagram for a DFE is shown in Fig. 5.19. The DFE consists of two filters. The first filter is called a *feedforward filter* and it is generally a fractionally spaced FIR filter with adjustable tap coefficients. This filter is identical in form to the linear equalizer already described. Its input is the received filtered signal $y(t)$ sampled at some rate that is a multiple of the symbol rate, e.g., at rate $2/T$. The second filter is a *feedback filter*. It is implemented as an FIR filter with symbol-spaced taps having adjustable coefficients. Its input is the set of previously detected symbols. The output of the feedback filter is subtracted from the output of the feedforward filter to form the input to the

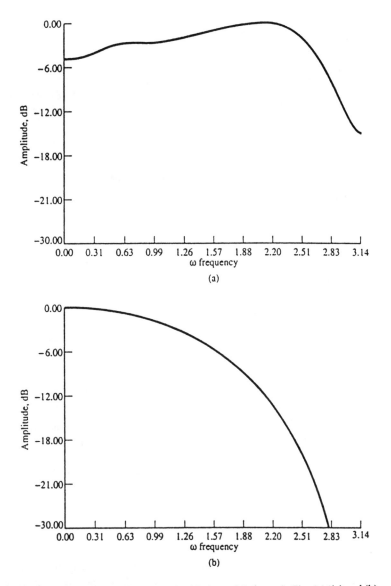

FIGURE 5.17 Amplitude spectra for (a) channel A shown in Fig. 5.16(a) and (b) channel B shown in Fig. 5.16(b).

detector. Thus, we have

$$z_m = \sum_{n=-N_1}^{0} c_n y(mT - n\tau) - \sum_{n=1}^{N_2} b_n \tilde{I}_{m-n} \qquad (5.41)$$

where $\{c_n\}$ and $\{b_n\}$ are the adjustable coefficients of the feedforward and feedback filters, respectively, $\tilde{I}_{m-n}, n = 1, 2, \ldots, N_2$ are the previously detected symbols, $N_1 + 1$ is the length of the feedforward filter, and N_2 is the length of the feedback filter. Based on the input z_m, the detector determines which of the possible transmitted symbols is closest in distance to the input signal I_m. Thus, it makes its decision and outputs $\tilde{I}_m$. What makes the DFE nonlinear is the nonlinear characteristic of the detector that provides the input to the feedback filter.

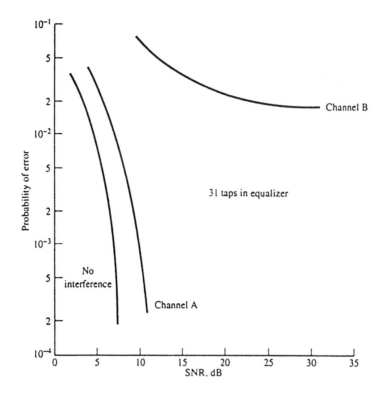

FIGURE 5.18   Error-rate performance of linear *MSE* equalizer.

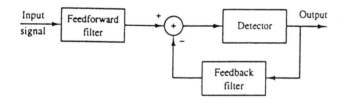

FIGURE 5.19   Block diagram of DFE.

The tap coefficients of the feedforward and feedback filters are selected to optimize some desired performance measure. For mathematical simplicity, the *MSE* criterion is usually applied, and a stochastic gradient algorithm is commonly used to implement an adaptive DFE. Figure 5.20 illustrates the block diagram of an adaptive DFE whose tap coefficients are adjusted by means of the LMS stochastic gradient algorithm. Figure 5.21 illustrates the probability of error performance of the DFE, obtained by computer simulation, for binary PAM transmission over channel B. The gain in performance relative to that of a linear equalizer is clearly evident.

We should mention that decision errors from the detector that are fed to the feedback filter have a small effect on the performance of the DFE. In general, a small loss in performance of one to two decibels is possible at error rates below 10^{-2}, as illustrated in Fig. 5.21, but the decision errors in the feedback filters are not catastrophic.

5.5 Maximum-Likelihood Sequence Detection

Although the DFE outperforms a linear equalizer, it is not the optimum equalizer from the viewpoint of minimizing the probability of error in the detection of the information sequence $\{I_k\}$ from the

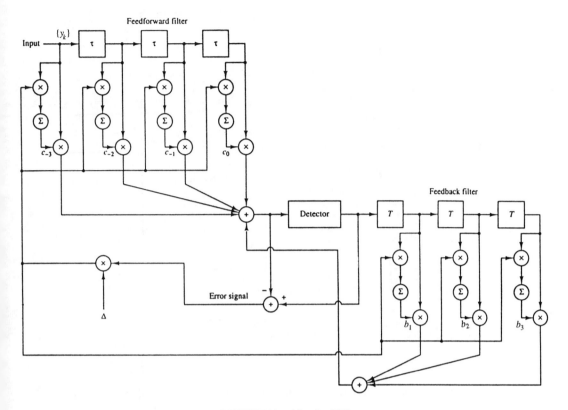

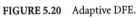

FIGURE 5.20 Adaptive DFE.

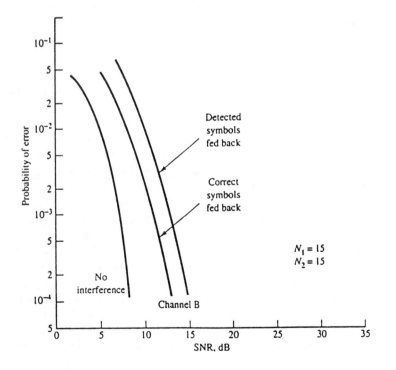

FIGURE 5.21 Performance of DFE with and without error propagation.

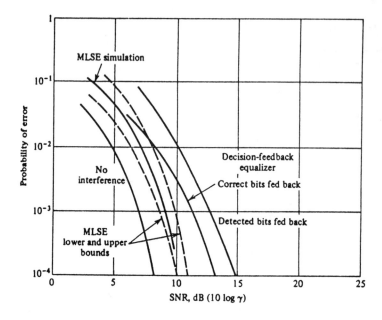

FIGURE 5.22 Comparison of performance between MLSE and decision-feedback equalization for channel B of Fig. 5.16.

received signal samples $\{y_k\}$ given in Eq. (5). In a digital communication system that transmits information over a channel that causes ISI, the optimum detector is a maximum-likelihood symbol sequence detector which produces at its output the most probable symbol sequence $\{\tilde{I}_k\}$ for the given received sampled sequence $\{y_k\}$. That is, the detector finds the sequence $\{\tilde{I}_k\}$ that maximizes the *likelihood function*

$$\Lambda(\{I_k\}) = \ln p(\{y_k\} \mid \{I_k\}) \tag{5.42}$$

where $p(\{y_k\} \mid \{I_k\})$ is the joint probability of the received sequence $\{y_k\}$ conditioned on $\{I_k\}$. The sequence of symbols $\{\tilde{I}_k\}$ that maximizes this joint conditional probability is called the **maximum-likelihood sequence detector**.

An algorithm that implements maximum-likelihood sequence detection (MLSD) is the Viterbi algorithm, which was originally devised for decoding convolutional codes. For a description of this algorithm in the context of sequence detection in the presence of ISI, the reader is referred to the paper by Forney, 1972 and the text by Proakis, 1995.

The major drawback of MLSD for channels with ISI is the exponential behavior in computational complexity as a function of the span of the ISI. Consequently, MLSD is practical only for channels where the ISI spans only a few symbols and the ISI is severe, in the sense that it causes a severe degradation in the performance of a linear equalizer or a decision-feedback equalizer. For example, Fig. 5.22 illustrates the error probability performance of the Viterbi algorithm for a binary PAM signal transmitted through channel B (see Fig. 5.16). For purposes of comparison, we also illustrate the probability of error for a DFE. Both results were obtained by computer simulation. We observe that the performance of the maximum likelihood sequence detector is about 4.5 dB better than that of the DFE at an error probability of 10^{-4}. Hence, this is one example where the ML sequence detector provides a significant performance gain on a channel with a relatively short ISI span.

5.6 Conclusions

Channel equalizers are widely used in digital communication systems to mitigate the effects of ISI cause by channel distortion. Linear equalizers are widely used for high-speed modems that transmit data over telephone channels. For wireless (radio) transmission, such as in mobile cellular communications and interoffice communications, the multipath propagation of the transmitted signal results in severe ISI. Such channels require more powerful equalizers to combat the severe ISI. The decision-feedback equalizer and the MLSD are two nonlinear channel equalizers that are suitable for radio channels with severe ISI.

Defining Terms

Adaptive equalizer: A channel equalizer whose parameters are updated automatically and adaptively during transmission of data.

Channel equalizer: A device that is used to reduce the effects of channel distortion in a received signal.

Decision-directed mode: Mode for adjustment of the equalizer coefficient adaptively based on the use of the detected symbols at the output of the detector.

Decision-feedback equalizer (DFE): An adaptive equalizer that consists of a feedforward filter and a feedback filter, where the latter is fed with previously detected symbols that are used to eliminate the intersymbol interference due to the tail in the channel impulse response.

Fractionally spaced equalizer: A tapped-delay line channel equalizer in which the delay between adjacent taps is less than the duration of a transmitted symbol.

Intersymbol interference: Interference in a received symbol from adjacent (nearby) transmitted symbols caused by channel distortion in data transmission.

LMS algorithm: See stochastic gradient algorithm.

Maximum-likelihood sequence detector: A detector for estimating the most probable sequence of data symbols by maximizing the likelihood function of the received signal.

Preset equalizer: A channel equalizer whose parameters are fixed (time-invariant) during transmission of data.

Stochastic gradient algorithm: An algorithm for adaptively adjusting the coefficients of an equalizer based on the use of (noise-corrupted) estimates of the gradients.

Symbol-spaced equalizer: A tapped-delay line channel equalizer in which the delay between adjacent taps is equal to the duration of a transmitted symbol.

Training mode: Mode for adjustment of the equalizer coefficients based on the transmission of a known sequence of transmitted symbols.

Zero-forcing equalizer: A channel equalizer whose parameters are adjusted to completely eliminate intersymbol interference in a sequence of transmitted data symbols.

References

Lucky, R.W. 1965. Automatic equalization for digital communications. *Bell Syst. Tech. J.*, 44(April): 547–588.

Lucky, R.W. 1966. Techniques for adaptive equalization of digital communication. *Bell Syst. Tech. J.* 45(Feb.):255–286.

Forney, G.D., Jr. 1972. Maximum-likelihood sequence estimation of digital sequences in the presence of intersymbol interference. *IEEE Trans. Inform. Theory*, IT-18(May):363–378.

Proakis, J.G. 1995. *Digital Communications*, 3rd ed. McGraw-Hill, New York.

Further Information

For a comprehensive treatment of adaptive equalization techniques and their performance characteristics, the reader may refer to the book by Proakis, 1995. The two papers by Lucky, 1965 and 1966, provide a treatment on linear equalizers based on the zero-forcing criterion. Additional information on decision-feedback equalizers may be found in the journal papers "An Adaptive Decision-Feedback Equalizer" by D.A. George, R.R. Bowen, and J.R. Storey, *IEEE Transactions on Communications Technology*, Vol. COM-19, pp. 281–293, June 1971, and "Feedback Equalization for Fading Dispersive Channels" by P. Monsen, *IEEE Transactions on Information Theory*, Vol. IT-17, pp. 56–64, January 1971. A through treatment of channel equalization based on maximum-likelihood sequence detection is given in the paper by Forney, 1972.

6

Line Coding

Joseph L. LoCicero
Illinois Institute of Technology

Bhasker P. Patel
Illinois Institute of Technology

6.1 Introduction

The terminology **line coding** originated in telephony with the need to transmit digital information across a copper telephone *line;* more specifically, binary data over a digital repeatered line. The concept of line coding, however, readily applies to any transmission line or channel. In a digital communication system, there exists a known set of symbols to be transmitted. These can be designated as $\{m_i\}$, $i = 1, 2, \ldots, N$, with a probability of occurrence $\{p_i\}$, $i = 1, 2, \ldots, N$, where the sequentially transmitted symbols are generally assumed to be statistically independent. The conversion or *coding* of these abstract symbols into real, temporal waveforms to be transmitted in baseband is the process of line coding. Since the most common type of line coding is for binary data, such a waveform can be succinctly termed a direct format for serial bits. The concentration in this section will be line coding for binary data.

Different channel characteristics, as well as different applications and performance requirements, have provided the impetus for the development and study of various types of line coding [Bellamy, 1991; Bell Telephone Laboratries (BTL), 1970]. For example, the channel might be ac coupled and, thus, could not support a line code with a dc component or large dc content. Synchronization or timing recovery requirements might necessitate a discrete component at the data rate. The channel bandwidth and **crosstalk** limitations might dictate the type of line coding employed. Even such factors as the complexity of the encoder and the economy of the decoder could determine the line code chosen. Each line code has its own distinct properties. Depending on the application, one property may be more important than the other. In what follows, we describe, in general, the most desirable features that are considered when choosing a line code.

It is commonly accepted [Bellamy, 1991; BTL, 1970; Couch, 1994; Lathi, 1989] that the dominant considerations effecting the choice of a line code are: 1) timing, 2) dc content, 3) power spectrum, 4) performance monitoring, 5) probability of error, and 6) transparency. Each of these are detailed in the following paragraphs.

1) *Timing:* The waveform produced by a line code should contain enough timing information such that the receiver can synchronize with the transmitter and decode the received signal properly. The timing content should be relatively independent of source statistics, i.e., a long string of 1s or 0s should not result in loss of timing or jitter at the receiver.

2) *DC content:* Since the repeaters used in telephony are ac coupled, it is desirable to have zero dc in the waveform produced by a given line code. If a signal with significant dc content is used in ac coupled lines, it will cause **dc wander** in the received waveform. That is, the received signal baseline will vary with time. Telephone lines do not pass dc due to ac coupling with transformers and capacitors to eliminate dc ground loops. Because of this, the telephone channel causes a droop in constant signals. This causes dc wander. It can be eliminated by dc restoration circuits, feedback systems, or with specially designed line codes.

3) *Power spectrum:* The power spectrum and bandwidth of the transmitted signal should be matched to the frequency response of the channel to avoid significant distortion. Also, the power spectrum should be such that most of the energy is contained in as small bandwidth as possible. The smaller is the bandwidth, the higher is the transmission efficiency.

4) *Performance monitoring:* It is very desirable to detect errors caused by a noisy transmission channel. The error detection capability in turn allows performance monitoring while the channel is in use (i.e., without elaborate testing procedures that require suspending use of the channel).

5) *Probability of error:* The average error probability should be as small as possible for a given transmitter power. This reflects the reliability of the line code.

6) *Transparency:* A line code should allow all the possible patterns of 1s and 0s. If a certain pattern is undesirable due to other considerations, it should be mapped to a unique alternative pattern.

6.2 Common Line Coding Formats

A line coding format consists of a formal definition of the line code that specifies how a string of binary digits are converted to a line code waveform. There are two major classes of binary line codes: **level codes** and **transition codes**. Level codes carry information in their voltage level, which may be high or low for a full bit period or part of the bit period. Level codes are usually instantaneous since they typically encode a binary digit into a distinct waveform, independent of any past binary data. However, some level codes do exhibit memory. Transition codes carry information in the change in level appearing in the line code waveform. Transition codes may be instantaneous, but they generally have memory, using past binary data to dictate the present waveform. There are two common forms of level line codes: one is called **return to zero (RZ)** and the other is called **nonreturn to zero (NRZ)**. In RZ coding, the level of the pulse returns to zero for a portion of the bit interval. In NRZ coding, the level of the pulse is maintained during the entire bit interval.

Line coding formats are further classified according to the polarity of the voltage levels used to represent the data. If only one polarity of voltage level is used, i.e., positive or negative (in addition to the zero level) then it is called **unipolar** signalling. If both positive and negative voltage levels are being used, with or without a zero voltage level, then it is called **polar** signalling. The term **bipolar** signalling is used by some authors to designate a specific line coding scheme with positive, negative, and zero voltage levels. This will be described in detail later in this section. The formal definition of five common line codes is given in the following along with a representative waveform, the *power spectral density* (PSD), the probability of error, and a discussion of advantages and disadvantages. In some cases specific applications are noted.

Unipolar NRZ (Binary On-Off Keying)

In this line code, a binary **1** is represented by a non- zero voltage level and a binary **0** is represented by a zero voltage level as shown in Fig. 6.1(a). This is an instantaneous level code. The PSD of this code with equally likely **1**s and **0**s is given by [Couch, 1994; Lathi, 1989]

$$S_1(f) = \frac{V^2 T}{4} \left(\frac{\sin \pi f T}{\pi f T} \right)^2 + \frac{V^2}{4} \delta(f) \tag{6.1}$$

where V is the binary **1** voltage level, $T = 1/R$ is the bit duration, and R is the bit rate in bits per second. The spectrum of unipolar NRZ is plotted in Fig. 6.2(a). This PSD is a two-sided even spectrum, although only half of the plot is shown for efficiency of presentation. If the probability of a binary **1** is p, and the probability of a binary **0** is $(1 - p)$, then the PSD of this code, in the most general case, is $4p(1 - p) S_1(f)$. Considering the frequency of the first spectral null as the bandwidth of the waveform, the bandwidth of unipolar NRZ is R in hertz. The error rate performance of this code, for equally likely data, with additive white Gaussian noise (AWGN) and optimum, i.e., matched filter, detection is given by [Bellamy, 1991; Couch, 1994]

$$P_e = \frac{1}{2} \text{erfc} \left(\sqrt{\frac{E_b}{2N_0}} \right) \tag{6.2}$$

where E_b/N_0 is a measure of the signal-to-noise ratio (SNR) of the received signal. In general, E_b is the energy per bit and $N_0/2$ is the two-sided PSD of the AWGN. More specifically, for unipolar NRZ, E_b is the energy in a binary **1**, which is $V^2 T$. The performance of the unipolar NRZ code is plotted in Fig. 6.3.

The principle advantages of unipolar NRZ are ease of generation, since it requires only a single power supply, and a relatively low bandwidth of R Hz. There are quite a few disadvantages of this line code. A loss of synchronization and timing jitter can result with a long sequence of **1**s or **0**s because no pulse transition is present. The code has no error detection capability and, hence, performance cannot be monitored. There is a significant dc component as well as a dc content. The error rate performance is not as good as that of polar line codes.

Unipolar RZ

In this line code, a binary **1** is represented by a nonzero voltage level during a portion of the bit duration, usually for half of the bit period, and a zero voltage level for rest of the bit duration. A binary **0** is represented by a zero voltage level during the entire bit duration. Thus, this is an instantaneous level code. Figure 6.1(b) illustrates a unipolar RZ waveform in which the **1** is represented by a nonzero voltage level for half the bit period. The PSD of this line code, with equally likely binary digits, is given by [Couch, 1994; Feher, 1977; Lathi, 1989]

$$S_2(f) = \frac{V^2 T}{16} \left(\frac{\sin \pi f T/2}{\pi f T/2} \right)^2$$
$$+ \frac{V^2}{4\pi^2} \left[\frac{\pi^2}{4} \delta(f) + \sum_{n=-\infty}^{\infty} \frac{1}{(2n+1)^2} \delta(f - (2n+1)R) \right] \tag{6.3}$$

where again V is the binary **1** voltage level, and $T = 1/R$ is the bit period. The spectrum of this code is drawn in Fig. 6.2(a). In the most general case, when the probability of a **1** is p, the continuous portion of the PSD in Eq. (6.3) is scaled by the factor $4p(1 - p)$ and the discrete portion is scaled by the factor $4p^2$. The first null bandwidth of unipolar RZ is $2R$ Hz. The error rate performance

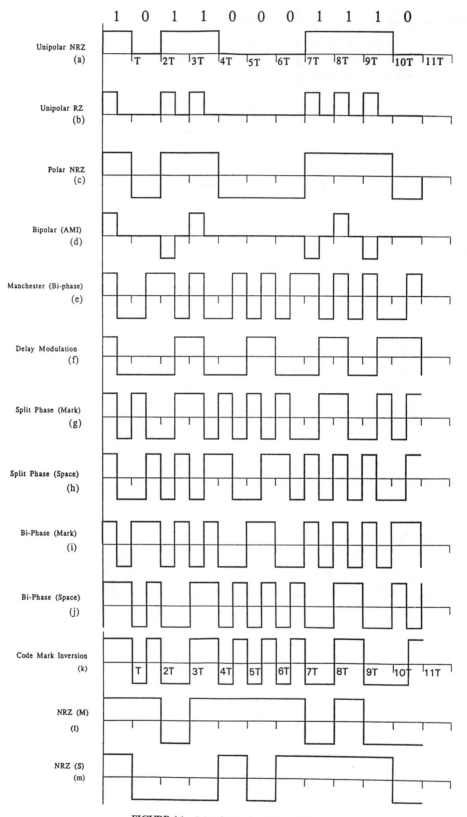

FIGURE 6.1 Waveforms for different line codes.

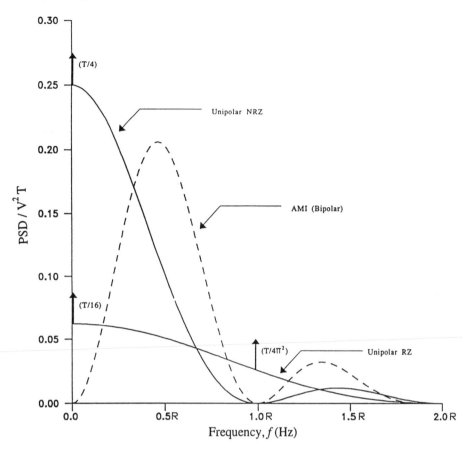

FIGURE 6.2(a) Power spectral density of different line codes, where $R = 1/T$ is the bit rate.

of this line code is the same as that of the unipolar NRZ provided we increase the voltage level of this code such that the energy in binary **1**, E_b, is the same for both codes. The probability of error is given by Eq. (6.2) and identified in Fig. 6.3. If the voltage level and bit period are the same for unipolar NRZ and unipolar RZ, then the energy in a binary **1** for unipolar RZ will be $V^2T/2$ and the probability of error is worse by 3 dB.

The main advantages of unipolar RZ are, again, ease of generation since it requires a single power supply and the presence of a discrete spectral component at the symbol rate, which allows simple timing recovery. A number of disadvantages exist for this line code. It has a nonzero dc component and nonzero dc content, which can lead to dc wander. A long string of **0**s will lack pulse transitions and could lead to loss of synchronization. There is no error detection capability and, hence, performance monitoring is not possible. The bandwidth requirement ($2R$ Hz) is higher than that of NRZ signals. The error rate performance is worse than that of polar line codes.

Unipolar NRZ as well as unipolar RZ are examples of pulse/no-pulse type of signalling. In this type of signalling, the pulse for a binary **0**, $g_2(t)$, is zero and the pulse for a binary **1** is specified generically as $g_1(t) = g(t)$. Using $G(f)$ as the Fourier transform of $g(t)$, the PSD of pulse/no-pulse signalling is given as [Feher, 1977; Gibson, 1993; Lindsey, 1973]

$$S_{\text{PNP}}(f) = p(1-p)R|G(f)|^2 + p^2R^2 \sum_{n=-\infty}^{\infty} |G(nR)|^2 \delta(f - nR) \qquad (6.4)$$

where p is the probability of a binary **1**, and R is the bit rate.

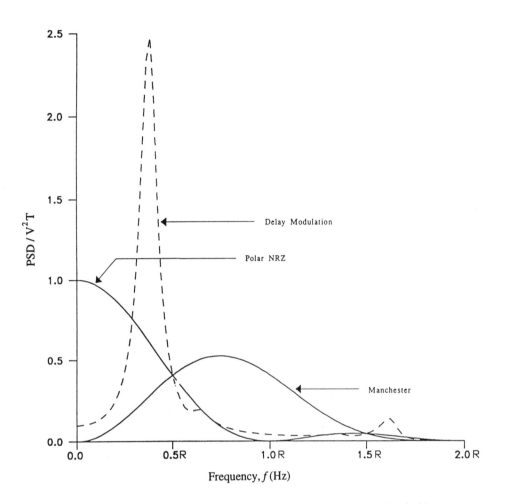

FIGURE 6.2(b)   Power spectral density of different line codes, where $R = 1/T$ is the bit rate.

Polar NRZ

In this line code, a binary **1** is represented by a positive voltage $+V$ and a binary **0** is represented by a negative voltage $-V$ over the full bit period. This code is also referred to as NRZ (L), since a bit is represented by maintaining a level (L) during its entire period. A polar NRZ waveform is shown in Fig. 6.1(c). This is again an instantaneous level code. Alternatively, a **1** may be represented by a $-V$ voltage level and a **0** by a $+V$ voltage level, without changing the spectral characteristics and performance of the line code. The PSD of this line code with equally likely bits is given by [Couch, 1994; Lathi, 1989]

$$S_3(f) = V^2T \left(\frac{\sin \pi fT}{\pi fT} \right)^2 \tag{6.5}$$

This is plotted in Fig. 6.2(b). When the probability of a **1** is p, and p is not 0.5, a dc component exists, and the PSD becomes [Lindsey, 1973]

$$S_{3p}(f) = 4V^2Tp(1-p) \left(\frac{\sin \pi fT}{\pi fT} \right)^2 + V^2(1-2p)^2\delta(f) \tag{6.6}$$

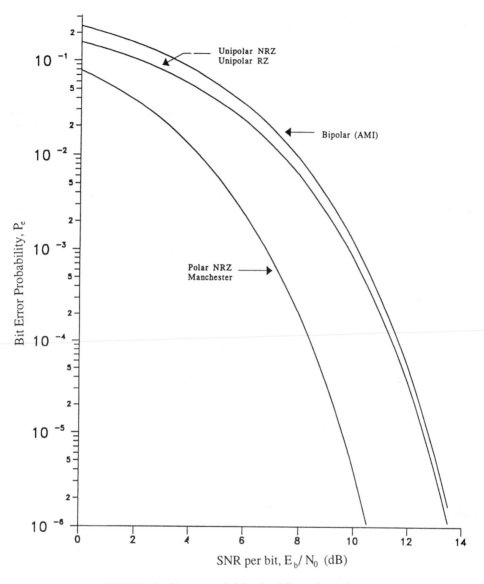

FIGURE 6.3   Bit error probability for different line codes.

The first null bandwidth for this line code is again R Hz, independent of p. The probability of error of this line code when $p = 0.5$ is given by [Bellamy, 1991; Couch, 1994]

$$P_e = \frac{1}{2}\mathrm{erfc}\left(\sqrt{\frac{E_b}{N_0}}\right) \tag{6.7}$$

The performance of polar NRZ is plotted in Fig. 6.3. This is better than the error performance of the unipolar codes by 3 dB.

The advantages of polar NRZ include a low-bandwidth requirement, R Hz, comparable to unipolar NRZ, very good error probability, and greatly reduced dc because the waveform has a zero dc component when $p = 0.5$ even though the dc content is never zero. A few notable disadvantages are that there is no error detection capability, and that a long string of **1**s or **0**s could result in loss of synchronization, since there are no transitions during the string duration. Two power supplies are required to generate this code.

Polar RZ [Bipolar, Alternate Mark Inversion (AMI), or Pseudoternary]

In this scheme, a binary **1** is represented by alternating the positive and negative voltage levels, which return to zero for a portion of the bit duration, generally half the bit period. A binary **0** is represented by a zero voltage level during the entire bit duration. This line coding scheme is often called **alternate mark inversion** (AMI) since 1s (marks) are represented by alternating positive and negative pulses. It is also called *pseudoternary* since three different voltage levels are used to represent binary data. Some authors designate this line code as bipolar RZ (BRZ). An AMI waveform is shown in Fig. 6.1(d). Note that this is a level code with memory. The AMI code is well known for its use in telephony. The PSD of this line code with memory is given by [Bellamy, 1991; BTL, 1970; Gibson, 1993]

$$S_{4p}(f) = 2p(1-p)R|G(f)|^2 \left(\frac{1 - \cos 2\pi f T}{1 + (2p-1)^2 + 2(2p-1)\cos 2\pi f T} \right) \quad (6.8)$$

where $G(f)$ is the Fourier transform of the pulse used to represent a binary **1**, and p is the probability of a binary **1**. When $p = 0.5$ and square pulses with amplitude $\pm V$ and duration $T/2$ are used to represent binary 1s, the PSD becomes

$$S_4(f) = \frac{V^2 T}{4} \left(\frac{\sin \pi f T/2}{\pi f T/2} \right)^2 \sin^2(\pi f T) \quad (6.9)$$

This PSD is plotted in Fig. 6.2(a). The first null bandwidth of this waveform is R Hz. This is true for RZ rectangular pulses, independent of the value of p in Eq. (6.8). The error rate performance of this line code for equally likely binary data is given by [Couch, 1994]

$$P_e \approx \frac{3}{4} \text{erfc}\left(\sqrt{\frac{E_b}{2N_0}} \right), \qquad E_b/N_0 > 2 \quad (6.10)$$

This curve is plotted in Fig. 6.3 and is seen to be no more than 0.5 dB worse than the unipolar codes.

The advantages of polar RZ (or AMI, as it is most commonly called) outweigh the disadvantages. This code has no dc component and zero dc content, completely avoiding the dc wander problem. Timing recovery is rather easy since squaring, or full-wave rectifying, this type of signal yields a unipolar RZ waveform with a discrete component at the bit rate, R Hz. Because of the alternating polarity pulses for binary 1s, this code has error detection and, hence, performance monitoring capability. It has a low-bandwidth requirement, R Hz, comparable to unipolar NRZ. The obvious disadvantage is that the error rate performance is worse than that of the unipolar and polar waveforms. A long string of 0s could result in loss of synchronization, and two power supplies are required for this code.

Manchester Coding (Split Phase or Digital Biphase)

In this coding, a binary **1** is represented by a pulse that has positive voltage during the first-half of the bit duration and negative voltage during second-half of the bit duration. A binary **0** is represented by a pulse that is negative during the first-half of the bit duration and positive during the second-half of the bit duration. The negative or positive midbit transition indicates a binary **1** or binary **0**, respectively. Thus, a Manchester code is classified as an instantaneous transition code; it has no memory. The code is also called diphase because a square wave with a 0° phase is used to represent a binary **1** and a square wave with a phase of 180° used to represent a binary **0**; or vice versa. This line code is used in Ethernet local area networks (LANs). The waveform for Manchester coding is shown in Fig. 6.1(e). The PSD of a Manchester waveform with equally likely bits is given by [Couch, 1994; Lathi, 1989]

$$S_5(f) = V^2 T \left(\frac{\sin \pi f T/2}{\pi f T/2} \right)^2 \sin^2(\pi f T/2) \tag{6.11}$$

where $\pm V$ are used as the positive/negative voltage levels for this code. Its spectrum is plotted in Fig. 6.2(b). When the probability p of a binary **1**, is not equal to one-half, the continuous portion of the PSD is reduced in amplitude and discrete components appear at integer multiples of the bit rate, $R = 1/T$. The resulting PSD is [Feher, 1977; Lindsey, 1973]

$$S_{5p}(f) = V^2 T 4p(1-p) \left(\frac{\sin \pi f T/2}{\pi f T/2} \right)^2 \sin^2 \frac{\pi f T}{2}$$

$$+ V^2 (1-2p)^2 \sum_{n=-\infty, n \neq 0}^{\infty} \left(\frac{2}{n\pi} \right)^2 \delta(f - nR) \tag{6.12}$$

The first null bandwidth of the waveform generated by a Manchester code is $2R$ Hz. The error rate performance of this waveform when $p = 0.5$ is the same as that of polar NRZ, given by Eq. (6.9), and plotted in Fig. 6.3.

The advantages of this code include a zero dc content on an individual pulse basis, so no pattern of bits can cause dc buildup; midbit transitions are always present making it is easy to extract timing information; and it has good error rate performance, identical to polar NRZ. The main disadvantage of this code is a larger bandwidth than any of the other common codes. Also, it has no error detection capability and, hence, performance monitoring is not possible.

Polar NRZ and Manchester coding are examples of the use of pure polar signalling where the pulse for a binary **0**, $g_2(t)$ is the negative of the pulse for a binary **1**, i.e., $g_2(t) = -g_1(t)$. This is also referred to as an antipodal signal set. For this broad type of polar binary line code, the PSD is given by [Lindsey, 1973]

$$S_{\text{BP}}(f) = 4p(1-p)R|G(f)|^2 + (2p-1)^2 R^2 \sum_{n=-\infty}^{\infty} |G(nR)|^2 \delta(f - nR) \tag{6.13}$$

where $|G(f)|$ is the magnitude of the Fourier transform of either $g_1(t)$ or $g_2(t)$.

A further generalization of the PSD of binary line codes can be given, wherein a continuous spectrum and a discrete spectrum is evident. Let a binary **1**, with probability p, be represented by $g_1(t)$ over the $T = 1/R$ second bit interval; and let a binary **0**, with probability $1 - p$, be represented by $g_2(t)$ over the same T second bit interval. The two-sided PSD for this general binary line code is [Lindsey, 1973]

$$S_{\text{GB}}(f) = p(1-p)R|G_1(f) - G_2(f)|^2$$

$$+ R^2 \sum_{n=-\infty}^{\infty} |pG_1(nR) + (1-p)G_2(nR)|^2 \delta(f - nR) \tag{6.14}$$

where the Fourier transform of $g_1(t)$ and $g_2(t)$ are given by $G_1(f)$ and $G_2(f)$, respectively.

6.3 Alternate Line Codes

Most of the line codes discussed thus far were instantaneous level codes. Only AMI had memory, and Manchester was an instantaneous transition code. The alternate line codes presented in this section all have memory. The first four are transition codes, where binary data is represented as the presence or absence of a transition, or by the direction of transition, i.e., positive to negative or vice versa. The last four codes described in this section are level line codes with memory.

Delay Modulation (Miller Code)

In this line code, a binary **1** is represented by a transition at the midbit position, and a binary **0** is represented by no transition at the midbit position. If a **0** is followed by another **0**, however, the signal transition also occurs at the end of the bit interval, that is, between the two **0**s. An example of delay modulation is shown in Fig. 6.1(f). It is clear that delay modulation is a transition code with memory. This code achieves the goal of providing good timing content without sacrificing bandwidth. The PSD of the Miller code for equally likely data is given by [Lindsey, 1973]

$$
\begin{aligned}
S_6(f) = \ & \frac{V^2 T}{2(\pi f T)^2 (17 + 8\cos 2\pi f T)} \\
& \times (23 - 2\cos \pi f T - 22\cos 2\pi f T \\
& - 12\cos 3\pi f T + 5\cos 4\pi f T + 12\cos 5\pi f T \\
& + 2\cos 6\pi f T - 8\cos 7\pi f T + 2\cos 8\pi f T)
\end{aligned} \tag{6.15}
$$

This spectrum is plotted in Fig. 6.2(b). The advantages of this code are that it requires relatively low bandwidth, most of the energy is contained in less than $0.5R$. However, there is no distinct spectral null within the $2R$-Hz band. It has low dc content and no dc component. It has very good timing content, and carrier tracking is easier than Manchester coding. Error rate performance is comparable to that of the common line codes. One important disadvantage is that it has no error detection capability and, hence, performance cannot be monitored.

Split Phase (Mark)

This code is similar to Manchester in the sense that there are always midbit transitions. Hence, this code is relatively easy to synchronize and has no dc. Unlike Manchester, however, split phase (mark) encodes a binary digit into a midbit transition dependent on the midbit transition in the previous bit period [Stremler, 1980]. Specifically, a binary **1** produces a reversal of midbit transition relative to the previous midbit transition. A binary **0** produces no reversal of the midbit transition. Certainly this is a transition code with memory. An example of a split phase (mark) coded waveform is shown in Fig. 6.1(g), where the waveform in the first bit period is chosen arbitrarily. Since this method encodes bits differentially, there is no 180°-phase ambiguity associated with some line codes. This phase ambiguity may not be an issue in most baseband links but is important if the line code is modulated. Split phase (space) is very similar to split phase (mark), where the role of the binary **1** and binary **0** are interchanged. An example of a split phase (space) coded waveform is given in Fig. 6.1(h); again, the first bit waveform is arbitrary.

Biphase (Mark)

This code, designated as Bi ϕ-M, is similar to a Miller code in that a binary **1** is represented by a midbit transition, and a binary **0** has no midbit transition. However, this code always has a transition at the beginning of a bit period [Lindsey, 1973]. Thus, the code is easy to synchronize and has no dc. An example of Bi ϕ-M is given in Fig. 6.1(i), where the direction of the transition at $t = 0$ is arbitrarily chosen. Biphase (space) or Bi ϕ-S is similar to Bi ϕ-M, except the role of the binary data is reversed. Here a binary **0** (space) produces a midbit transition, and a binary **1** does not have a midbit transition. A waveform example of Bi ϕ-S is shown in Fig. 6.1(j). Both Bi ϕ-S and Bi ϕ-M are transition codes with memory.

Code Mark Inversion (CMI)

This line code is used as the interface to a Consultative Committee on International Telegraphy and Telephony (CCITT) multiplexer and is very similar to Bi ϕ-S. A binary **1** is encoded as an NRZ pulse with alternate polarity, $+V$ or $-V$. A binary **0** is encoded with a definitive midbit transition (or square wave phase) [Bellamy, 1991]. An example of this waveform is shown in Fig. 6.1(k) where a negative to positive transition (or 180° phase) is used for a binary **0**. The voltage level of the first binary **1** in this example is chosen arbitrarily. This example waveform is identical to Bi ϕ-S shown in Fig. 6.1(j), except for the last bit. CMI has good synchronization properties and has no dc.

NRZ (I)

This type of line code uses an inversion (I) to designate binary digits, specifically, a change in level or no change in level. There are two variants of this code, NRZ mark (M) and NRZ space (S) [Couch, 1994, Stremler, 1980]. In NRZ (M), a change of level is used to indicate a binary **1**, and no change of level is used to indicate a binary **0**. In NRZ (S) a change of level is used to indicate a binary **0**, and no change of level is used to indicate a binary **1**. Waveforms for NRZ (M) and NRZ (S) are depicted in Fig. 6.1(l) and Fig. 6.1(m), respectively, where the voltage level of the first binary **1** in the example is chosen arbitrarily. These codes are level codes with memory. In general, line codes that use differential encoding, like NRZ (I), are insensitive to 180° phase ambiguity. Clock recovery with NRZ (I) is not particularly good, and dc wander is a problem as well. Its bandwidth is comparable to polar NRZ.

Binary *N* Zero Substitution (BNZS)

The common bipolar code AMI has many desirable properties of a line code. Its major limitation, however, is that a long string of zeros can lead to loss of synchronization and timing jitter because there are no pulses in the waveform for relatively long periods of time. **Binary *N* zero substitution (BNZS)** attempts to improve AMI by substituting a special code of length *N* for all strings of *N* zeros. This special code contains pulses that look like binary 1s but purposely produce violations of the AMI pulse convention. Two consecutive pulses of the same polarity violate the AMI pulse convention, independent of the number of zeros between the two consecutive pulses. These violations can be detected at the receiver, and the special code replaced by *N* zeros. The special code contains pulses facilitating synchronization even when the original data has long string of zeros. The special code is chosen such that the desirable properties of AMI coding are retained despite the AMI pulse convention violations, i.e., dc balance and error detection capability. The only disadvantage of BNZS compared to AMI is a slight increase in crosstalk due to the increased number of pulses and, hence, an increase in the average energy in the code.

Choosing different values of *N* yields different BNZS codes. The value of *N* is chosen to meet the timing requirements of the application. In telephony, there are three commonly used BNZS codes: B6ZS, B3ZS, and B8ZS. All BNZS codes are level codes with memory.

In a B6ZS code, a string of six consecutive zeros is replaced by one of two the special codes according to the rule:

If the last pulse was positive $(+)$, the special code is: $0 + - 0 - +$.
If the last pulse was negative $(-)$, the special code is: $0 - + 0 + -$.

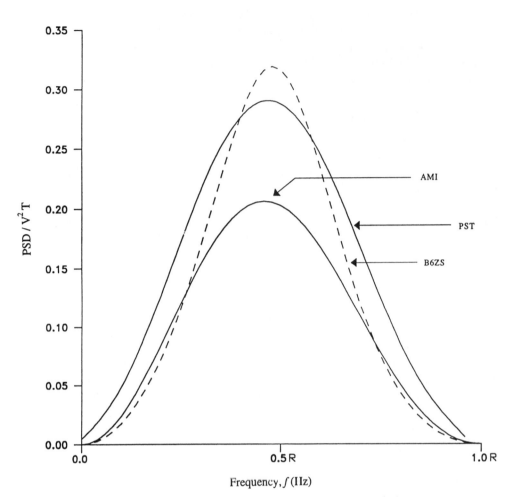

FIGURE 6.4   Power spectral density of different line codes, where $R = 1/T$ is the bit rate.

Here a zero indicates a zero voltage level for the bit period; a plus designates a positive pulse; and a minus indicates a negative pulse.

This special code causes two AMI pulse violations: in its second bit position and in its fifth bit position. These violations are easily detected at the receiver and zeros resubstituted. If the number of consecutive zeros is 12, 18, 24, ..., the substitution is repeated 2, 3, 4, ... times. Since the number of violations is even, the B6ZS waveform is the same as the AMI waveform outside the special code, i.e., between special code sequences.

There are four pulses introduced by the special code that facilitates timing recovery. Also, note that the special code is dc balanced. An example of the B6ZS code is given as follows, where the special code is indicated by the bold characters.

Original data: 0 1 0 0 0 0 0 0 1 1 0 1 0 0 0 0 0 0 1 1
B6ZS format: 0 + **0** **+** **−** **0** **−** **+** − + 0 − **0** **−** **+** **0** **+** **−** + −

The computation of the PSD of a B6ZS code is tedious. Its shape is given in Fig. 6.4, for comparison purposes with AMI, for the case of equally likely data.

In a B3ZS code, a string of three consecutive zeros is replaced by either $B0V$ or **00V**, where B denotes a pulse obeying the AMI (bipolar) convention and V denotes a pulse violating the

TABLE 6.1   B3ZS Substitution Rules

Number of B Pulses Since Last Violation	Polarity of Last B Pulse	Substitution Code	Substitution Code Form
Odd	Negative $(-)$	$0\,0\,-$	$00V$
Odd	Positive $(+)$	$0\,0\,+$	$00V$
Even	Negative $(-)$	$+\,0\,+$	$B0V$
Even	Positive $(+)$	$-\,0\,-$	$B0V$

AMI convention. $B0V$ or $00V$ is chosen such that the number of bipolar (B) pulses between the violations is odd. The B3ZS rules are summarized in Table 6.1.

Observe that the violation always occurs in the third bit position of the substitution code, and so it can be easily detected and zero replacement made at the receiver. Also, the substitution code selection maintains dc balance. There is either one or two pulses in the substitution code, facilitating synchronization. The error detection capability of AMI is retained in B3ZS because a single channel error would make the number of bipolar pulses between violations even instead of being odd. Unlike B6ZS, the B3ZS waveform between violations may not be the same as the AMI waveform. B3ZS is used in the digital signal-3 (DS-3) signal interface in North America and also in the long distance-4 (LD-4) coaxial transmission system in Canada. Next is an example of a B3ZS code, using the same symbol meaning as in the B6ZS code.

Original data:	1 0 0 1 0 0 0 1 1 0 0 0 0 1 0 0 0 1
B3ZS format:	
Even No. of B pulses:	$+$ 0 0 $-$ $+$ 0 $+$ $-$ $+$ $-$ **0** $-$ 0 $+$ **0 0** $+$ $-$
Odd No. of B pulses:	$+$ 0 0 $-$ **0 0** $-$ $+$ $-$ $+$ **0** $+$ 0 $-$ **0 0** $-$ $+$

The last BNZS code considered here uses $N = 8$. A B8ZS code is used to provide transparent channels for the Integrated Services Digital Network (ISDN) on T1 lines and is similar to the B6ZS code. Here a string of eight consecutive zeros is replaced by one of two special codes according to the following rule:

If the last pulse was positive $(+)$, the special code is: $0\ \ 0\ \ 0\ \ +\ \ -\ \ 0\ \ -\ \ +$.

If the last pulse was negative $(-)$, the special code is: $0\ \ 0\ \ 0\ \ -\ \ +\ \ 0\ \ +\ \ -$.

There are two bipolar violations in the special codes, at the fourth and seventh bit positions. The code is dc balanced, and the error detection capability of AMI is retained. The waveform between substitutions is the same as that of AMI. If the number of consecutive zeros is 16, 24, ..., then the substitution is repeated 2, 3, ..., times.

High-Density Bipolar N (HDBN)

This coding algorithm is a CCITT standard recommended by the Conference of European Posts and Telecommunications Administrations (CEPT), a European standards body. It is quite similar to BNZS coding. It is thus a level code with memory. Whenever there is a string of $N + 1$ consecutive zeros, they are replaced by a special code of length $N + 1$ containing AMI violations. Specific codes can be constructed for different values of N. A specific **high-density bipolar N (HDBN)** code, HDB3, is implemented as a CEPT primary digital signal. It is very similar to the B3ZS code. In this code, a string of four consecutive zeros is replaced by either $B00V$ or $000V$. $B00V$ or $000V$ is chosen such that the number of bipolar (B) pulses between violations is odd. The HDB3 rules are summarized in Table 6.2.

TABLE 6.2   HDB3 Substitution Rules

Number of *B* Pulses Since Last Violation	Polarity of Last *B* Pulse	Substitution Code	Substitution Code Form
Odd	Negative (−)	**0 0 0 −**	**000***V*
Odd	Positive (+)	**0 0 0 +**	**000***V*
Even	Negative (−)	**+ 0 0 +**	***B*00***V*
Even	Positive (+)	**− 0 0 −**	***B*00***V*

Here the violation always occurs in the fourth bit position of the substitution code, so that it can be easily detected and zero replacement made at the receiver. Also, the substitution code selection maintains dc balance. There is either one or two pulses in the substitution code facilitating synchronization. The error detection capability of AMI is retained in HDB3 because a single channel error would make the number of bipolar pulses between violations even instead of being odd.

Ternary Coding

Many line coding schemes employ three symbols or levels to represent only one bit of information, like AMI. Theoretically, it should be possible to transmit information more efficiently with three symbols, specifically the maximum efficiency is $\log_2 3 = 1.58$ bits per symbol. Alternatively, the redundancy in the code signal space can be used to provide better error control. Two examples of ternary coding are described next [Bellamy, 1991; BTL, 1970]: **pair selected ternary** (**PST**) and **4 binary 3 ternary** (**4B3T**). The PST code has many of the desirable properties of line codes, but its transmission efficiency is still 1 bit per symbol. The 4B3T code also has many of the desirable properties of line codes, and it has increased transmission efficiency.

In the PST code, two consecutive bits, termed a binary pair, are grouped together to form a word. These binary pairs are assigned codewords consisting of two ternary symbols, where each ternary symbol can be +, −, or 0, just as in AMI. There are nine possible ternary codewords. Ternary codewords with identical elements, however, are avoided, i.e., ++, −−, and 00. The remaining six codewords are transmitted using two modes called + mode and − mode. The modes are switched whenever a codeword with a single pulse is transmitted. The PST code and mode switching rules are summarized in Table 6.3.

PST is designed to maintain dc balance and include a strong timing component. One drawback of this code is that the bits must be framed into pairs. At the receiver, an *out-of-frame* condition is signalled when unused ternary codewords (++, −−, and 00) are detected. The mode switching property of PST provides error detection capability. PST can be classified as a level code with memory.

If the original data for PST coding contains only 1s or 0s, an alternating sequence of +− +− ⋯ is transmitted. As a result, an out-of-frame condition can not be detected. This problem can be minimized by using the modified PST code as shown in Table 6.4.

It is tedious to derive the PSD of a PST coded waveform. Again, Fig. 6.4 shows the PSD of the PST code along with the PSD of AMI and B6ZS for comparison purposes, all for equally likely binary data. Observe that PST has more power than AMI and, thus, a larger amount of energy per bit, which translates into slightly increased crosstalk.

In 4B3T coding, words consisting of four binary digits are mapped into three ternary symbols. Four bits imply $2^4 = 16$ possible binary words, whereas three ternary symbols allow $3^3 = 27$ possible ternary codewords. The binary-to-ternary conversion in 4B3T in-

TABLE 6.3   PST Codeword Assignment and Mode Switching Rules

Binary Pair	Ternary Codewords + Mode	− Mode	Mode Switching
11	+ −	+ −	No
10	+ 0	− 0	Yes
01	0 +	0 −	Yes
00	− +	− +	No

TABLE 6.4   Modified PST Codeword Assignment and Mode Switching Rules

Binary Pair	Ternary Codewords + Mode	− Mode	Mode Switching
11	+ 0	0 −	Yes
10	+ −	+ −	No
01	− +	− +	No
00	0 +	− 0	Yes

TABLE 6.5 4B3T Codeword Assignment

Binary Words	Column 1	Column 2	Column 3
		Ternary Codewords	
0000	$-\,-\,-$		$+\,+\,+$
0001	$-\,-\,0$		$+\,+\,0$
0010	$-\,0\,-$		$+\,0\,+$
0011	$0\,-\,-$		$0\,+\,+$
0100	$-\,-\,+$		$+\,+\,-$
0101	$-\,+\,-$		$+\,-\,+$
0110	$+\,-\,-$		$-\,+\,+$
0111	$-\,0\,0$		$+\,0\,0$
1000	$0\,-\,0$		$0\,+\,0$
1001	$0\,0\,-$		$0\,0\,+$
1010		$0\,+\,-$	
1011		$0\,-\,+$	
1100		$+\,0\,-$	
1101		$-\,0\,+$	
1110		$+\,-\,0$	
1111		$-\,+\,0$	

sures dc balance and a strong timing component. The specific codeword assignment is as shown in Table 6.5.

There are three types of codewords in Table 6.5, organized into three columns. The codewords in the first column have negative dc, codewords in the second column have zero dc, and those in the third column have positive dc. The encoder monitors the integer variable

$$I = N_p - N_n, \qquad (6.16)$$

where N_p is the number of positive pulses transmitted and N_n are the number of negative pulses transmitted. Codewords are chosen according to following rule:

If $I < 0$, choose the ternary codeword from columns 1 and 2.

If $I > 0$, choose the ternary codeword from columns 2 and 3.

If $I = 0$, choose the ternary word from column 2, and from column 1
if the previous $I > 0$ or from column 3 if the previous $I < 0$.

Note that the ternary codeword 000 is not used, but the remaining 26 codewords are used in a complementary manner. For example, the column 1 codeword for 0001 is $-\,-\,0$, whereas the column 3 codeword is $+\,+\,0$. The maximum transmission efficiency for the 4B3T code is 1.33 bits per symbol compared to 1 bit per symbol for the other line codes. The disadvantages of 4B3T are that framing is required and that performance monitoring is complicated. The 4B3T code is used in the T148 span line developed by ITT Telecommunications. This code allows transmission of 48 channels using only 50% more bandwidth than required by T1 lines, instead of 100% more bandwidth.

6.4 Multilevel Signalling, Partial Response Signalling, and Duobinary Coding

Ternary coding, such as 4B3T, is an example of the use of more than two levels to improve the transmission efficiency. To increase the transmission efficiency further, more levels and/or more signal processing is needed. Multilevel signalling allows an improvement in the transmission efficiency at the expense of an increase in the error rate, i.e., more transmitter power will be required to maintain a given probability of error. In partial response signalling, intersymbol interference is deliberately introduced by using pulses that are wider and, hence, require less bandwidth. The controlled amount of interference from each pulse can be removed at the receiver. This improves the transmission efficiency, at the expense of increased complexity. **Duobinary coding**, a special case of partial response signalling, requires only the minimum theoretical bandwidth of $0.5R$ Hz. In what follows these techniques are discussed in slightly more detail.

Multilevel Signalling

The number of levels that can be used for a line code is not restricted to two or three. Since more levels or symbols allow higher transmission efficiency, multilevel signalling can be considered in bandwidth-limited applications. Specifically, if the signalling rate or baud rate is R_s and the number of levels used is L, the equivalent transmission bit rate R_b is given by

$$R_b = R_s \log_2[L]. \qquad (6.17)$$

Alternatively, multilevel signalling can be used to reduce the baud rate, which in turn can reduce crosstalk for the same equivalent bit rate. The penalty, however, is that the SNR must increase to achieve the same error rate. The T1G carrier system of AT&T uses multilevel signalling with $L = 4$ and a baud rate of 3.152 mega-symbols/s to double the capacity of the T1C system from 48 channels to 96 channels. Also, a four level signalling scheme at 80-kB is used to achieve 160 kb/s as a basic rate in a digital subscriber loop (DSL) for ISDN.

Partial Response Signalling and Duobinary Coding

This class of signalling is also called *correlative* coding because it purposely introduces a controlled or correlated amount of intersymbol interference in each symbol. At the receiver, the known amount of interference is effectively removed from each symbol. The advantage of this signalling is that wider pulses can be used requiring less bandwidth, but the SNR must be increased to realize a given error rate. Also, errors can propagate unless *precoding* is used.

There are many commonly used partial response signalling schemes, often described in terms of the delay operator D, which represents one signalling interval delay. For example, in $(1 + D)$ signalling the current pulse and the previous pulse are added. The T1D system of AT&T uses $(1+D)$ signalling with precoding, referred to as duobinary signalling, to convert binary (two level) data into ternary (three level) data at the same rate. This requires the minimum theoretical channel bandwidth without the deleterious effects of intersymbol interference and avoids error propagation. Complete details regarding duobinary coding are found in Lender, 1963 and Schwartz, 1980. Some partial response signalling schemes, such as $(1 - D)$, are used to shape the bandwidth rather than control it. Another interesting example of duobinary coding is a $(1 - D^2)$, which can be analyzed as the product $(1-D)(1+D)$. It is used by GTE in its modified T carrier system. AT&T also uses $(1-D^2)$ with four input levels to achieve an equivalent data rate of 1.544 Mb/s in only a 0.5-MHz bandwidth.

6.5 Bandwidth Comparison

We have provided the PSD expressions for most of the commonly used line codes. The actual bandwidth requirement, however, depends on the pulse shape used and the definition of bandwidth itself. There are many ways to define bandwidth, for example, as a percentage of the total power or the sidelobe suppression relative to the main lobe. Using the first null of the PSD of the code as the definition of bandwidth, Table 6.6 provides a useful bandwidth comparison.

TABLE 6.6 First Null Bandwidth Comparison

Bandwidth	Codes	
R	Unipolar NRZ	BNZS
	Polar NRZ	HDBN
	Polar RZ (AMI)	PST
$2R$	Unipolar RZ	Split Phase
	Manchester	CMI

The notable omission in Table 6.6 is delay modulation (Miller code). It does not have a first null in the $2R$-Hz band, but most of its power is contained in less than $0.5R$ Hz.

6.6 Concluding Remarks

An in-depth presentation of line coding, particularly applicable to telephony, has been included in this chapter. The most desirable characteristics of line codes were discussed. We introduced five common line codes and eight alternate line codes. Each line code was illustrated by an example waveform. In most cases expressions for the PSD and the probability of error were given and plotted. Advantages and disadvantages of all codes were included in the discussion, and some specific applications were noted. Line codes for optical fiber channels and networks built around them, such as fiber distributed data interface (FDDI) were not included in this section. A discussion of line codes for optical fiber channels, and other new developments in this topic area can be found in Bellamy, 1991, Bic, Duponteil, and Imbeaux, 1991, and Bylanski, 1976.

Defining Terms

Alternate mark inversion (AMI): A popular name for bipolar line coding using three levels: zero, positive, and negative.

Bipolar: A particular line coding scheme using three levels: zero, positive, and negative.

Binary N zero substitution (BNZS): A class of coding schemes that attempts to improve AMI line coding.

Crosstalk: An unwanted signal from an adjacent channel.

DC Wander: The dc level variation in the received signal due to a channel that cannot support dc.

Duobinary coding: A coding scheme with binary input and ternary output requiring the minimum theoretical channel bandwidth.

4 Binary 3 Ternary (4B3T): A line coding scheme that maps four binary digits into three ternary symbols.

High-density bipolar N (HDBN): A class of coding schemes that attempts to improve AMI.

Level codes: Line codes carrying information in their voltage levels.

Line coding: The process of converting abstract symbols into real, temporal waveforms to be transmitted through a baseband channel.

Nonreturn to zero (NRZ): A signal that stays at a nonzero level for the entire bit duration.

Polar: A line coding scheme using both polarity of voltages, with or without a zero level.

Pair selected ternary (PST): A coding scheme based on selecting a pair of three level symbols.

Return to zero (RZ): A signal that returns to zero for a portion of the bit duration.

Transition codes: Line codes carrying information in voltage level transitions.

Unipolar: A line coding scheme using only one polarity of voltage, in addition to a zero level.

References

Bellamy, J. 1991. *Digital Telephony*, John Wiley & Sons, Inc., New York, NY.

Bell Telephone Laboratories Technical Staff Members. 1970. *Transmission Systems for Communications*, 4th ed. Western Electric Company, Inc., Technical Publications, Winston-Salem, NC.

Bic, J.C., Duponteil, D., and Imbeaux, J.C. 1991. *Elements of Digital Communication*, John Wiley & Sons, Inc., New York, NY.

Bylanski, P. 1976. *Digital Transmission Systems*, Peter Peregrinus Ltd., Herts, England.

Couch, L.W. 1994. *Modern Communication Systems: Principles and Applications*, Prentice-Hall, Inc., Englewood Cliffs, NJ.

Feher, K. 1977. *Digital Modulation Techniques in an Interference Environment*, EMC Encyclopedia Series, Vol. IX. Don White Consultants, Inc., Germantown, MD.

Gibson, J.D. 1993. *Principles of Analog and Digital Communications*, MacMillan Publishing, Inc., New York, NY.

Lathi, B.P. 1989. *Modern Digital and Analog Communication Systems*, Holt, Rinehart and Winston, Inc., Philadelphia, PA.

Lender, A. 1963. Duobinary Techniques for High Speed Data Transmission, *IEEE Trans. Commun. Electron.*, CE-82(May):214–218.

Lindsey, W.C. and Simon, M.K. 1973. *Telecommunication Systems Engineering*, Prentice-Hall, Inc., Englewood Cliffs, NJ.

Schwartz, M. 1980. *Information Transmission, Modulation, and Noise*, McGraw-Hill Book Co., Inc., New York, NY.

Stremler, F.G. 1990. *Introduction to Communication Systems*, Addison-Wesley Publishing, Co., Reading, MA.

7

Echo Cancellation

Giovanni Cherubini
IBM Zurich Research Laboratory

7.1 Introduction

Full-duplex data transmission over a single twisted-pair cable permits the simultaneous flow of information in two directions when the same frequency band is used. Examples of this technique are digital communication systems that operate over the telephone network. In a digital subscriber loop, at each end of the full-duplex link, a circuit known as a hybrid separates the two directions of transmission. To avoid signal reflections at the near- and far-end hybrid, a precise knowledge of the line impedance would be required. Since the line impedance depends on line parameters that, in general, are not exactly known, however, an attenuated and distorted replica of the transmit signal leaks to the receiver input as an echo signal. Data-driven adaptive echo cancellation mitigates the effects of impedance mismatch.

A similar problem is caused by crosstalk in transmission systems over voice-grade unshielded twisted-pair cables for local-area network applications, where multipair cables are used to physically separate the two directions of transmission. Crosstalk is a statistical phenomenon due to randomly varying differential capacitive and inductive coupling between adjacent two-wire transmission lines. At the rates of several megabits per second that are usually considered for local-area network applications, near-end crosstalk represents the dominant disturbance; hence near-end crosstalk cancellation must be performed to ensure reliable communication.

In voiceband data modems, the model for the echo channel is considerably different from the echo model adopted in baseband transmission. In fact, since the transmitted passband signal is obtained by modulating a complex-valued baseband signal, the far-end echo signal may experience significant jitter and frequency shift, which are caused by signal processing at intermediate points in the telephone network. Therefore, a digital adaptive echo canceller for passband transmission needs to embody algorithms that account for the presence of such additional impairments.

In this chapter, we describe the echo channel models and the structure of digital echo cancellers for baseband and passband transmission and address the tradeoffs between complexity, speed of adaptation, and accuracy of cancellation in adaptive echo cancellers.

7.2 Baseband Transmission

The model of a full-duplex data transmission system with adaptive echo cancellation is shown in Fig. 7.1. To describe system operations, we consider one end of the full-duplex link. The configuration for a baseband channel digital echo canceller is shown in Fig. 7.2. The transmitted data consist of a sequence $\{a_n\}$ of independent and identically distributed (i.i.d.) real-valued symbols from the M-ary alphabet $\mathcal{A} = \{\pm 1, \ldots, \pm(M-1)\}$. The sequence $\{a_n\}$ is converted into an analog signal by a digital-to-analog (D/A) converter. The conversion to a staircase signal by a zero-order hold D/A converter is described by the frequency response $H_{D/A}(f) = T \sin(\pi f T)/(\pi f T)$, where T is the modulation interval. The D/A converter output is filtered by the analog transmit filter and is input to the channel through the hybrid.

The signal $x(t)$ at the output of the low-pass analog receive filter has three components, namely, the signal from the far-end transmitter $r(t)$, the echo $u(t)$, and additive Gaussian noise $w(t)$. The signal $x(t)$ is given by

$$x(t) = r(t) + u(t) + w(t)$$

$$= \sum_{n=-\infty}^{\infty} a_n^{\text{FE}} h(t - nT) + \sum_{n=-\infty}^{\infty} a_n h_E(t - nT) + w(t) \qquad (7.1)$$

where $\{a_n^{\text{FE}}\}$ is the sequence of symbols from the far-end transmitter, and $h(t)$ and $h_E(t) = \{h_{D/A} \otimes g_E\}(t)$ are the impulse responses of the overall channel and the echo channel, respectively. In the expression of $h_E(t)$, the function $h_{D/A}(t)$ is the inverse Fourier transform of $H_{D/A}(f)$, and the operator $\otimes$ denotes convolution. The signal obtained after echo cancellation is processed by a detector that outputs the sequence of estimated symbols $\{\hat{a}_n^{\text{FE}}\}$. In the case of data transmission for local-area network applications, where near-end crosstalk represents the main disturbance, the configuration of a digital near-end crosstalk canceller is obtained from Fig. 7.2, with the echo channel replaced by the crosstalk channel.

In general, we consider baseband signalling techniques such that the signal at the output of the overall channel has nonnegligible excess bandwidth, i.e., nonnegligible spectral components at frequencies larger than half of the modulation rate, $|f| \geq 1/2T$. Therefore, to avoid aliasing, the signal $x(t)$ is sampled at twice the modulation rate or at a higher sampling rate. Assuming a

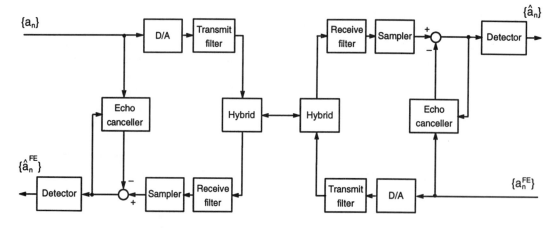

FIGURE 7.1 Model of a full-duplex transmission system.

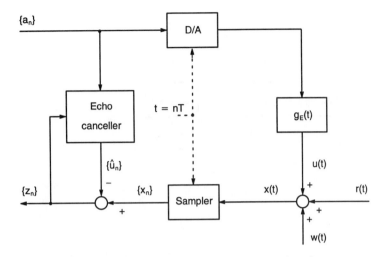

FIGURE 7.2 Configuration for a baseband channel echo canceller.

sampling rate equal to m/T, $m > 1$, the ith sample during the nth modulation interval is given by

$$x\left[(nm+i)\frac{T}{m}\right] = x_{nm+i} = r_{nm+i} + u_{nm+i} + w_{nm+i}, \qquad i = 0, \ldots, m-1$$

$$= \sum_{k=-\infty}^{\infty} h_{km+i}^{\text{FE}} a_{n-k} + \sum_{k=-\infty}^{\infty} h_{km+i}^{E} a_{n-k} + w_{nm+i} \qquad (7.2)$$

where $\{h_{nm+i}, \ i = 0, \ldots, m-1\}$ and $\{h_{nm+i}^{E}, \ i = 0, \ldots, m-1\}$ are the discrete-time impulse responses of the overall channel and the echo channel, respectively, and $\{w_{nm+i}, i = 0, \ldots, m-1\}$ is a sequence of Gaussian noise samples with zero mean and variance σ_w^2. Equation (7.2) suggests that the sequence of samples $\{x_{nm+i}, \ i = 0, \ldots, m-1\}$ be regarded as a set of m interleaved sequences, each with a sampling rate equal to the modulation rate. Similarly, the sequence of echo samples $\{u_{nm+i}, \ i = 0, \ldots, m-1\}$ can be regarded as a set of m interleaved sequences that are output by m independent echo channels with discrete-time impulse responses $\{h_{nm+i}^{E}\}$, $i = 0, \ldots, m-1$, and an identical sequence $\{a_n\}$ of input symbols [Lee and Messerschmitt, 1994]. Hence, echo cancellation can be performed by m interleaved echo cancellers, as shown in Fig. 7.3. Since the performance of each canceller is independent of the other $m-1$ units, in the remaining part of this section we will consider the operations of a single echo canceller.

The echo canceller generates an estimate $\hat{u}_n$ of the echo signal. If we consider a transversal filter realization, $\hat{u}_n$ is obtained as the inner product of the vector of filter coefficients at time $t = nT$, $c_n' = (c_{n,0}, \ldots, c_{n,N-1})$ and the vector of signals stored in the echo canceller delay line at the same instant, $a_n' = (a_n, \ldots, a_{n-N+1})$

$$\hat{u}_n = c_n' a_n = \sum_{k=0}^{N-1} c_{n,k} a_{n-k} \qquad (7.3)$$

where c_n' denotes the transpose of the vector c_n. The estimate of the echo is subtracted from the received signal. The result is defined as the cancellation error signal

$$z_n = x_n - \hat{u}_n = x_n - c_n' a_n \qquad (7.4)$$

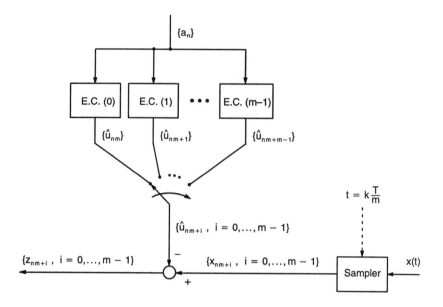

FIGURE 7.3   A set of m interleaved echo cancellers.

The echo attenuation that must be provided by the echo canceller to achieve proper system operation depends on the application. For example, for the Integrated Services Digital Network (ISDN) U-Interface transceiver, the echo attenuation must be larger than 55 dB [Messerschmitt, 1986]. It is then required that the echo signals outside of the time span of the echo canceller delay line be negligible, i.e., $h_n^E \approx 0$ for $n < 0$ and $n > N - 1$. As a measure of system performance, we consider the mean square error ε_n^2 at the output of the echo canceller at time $t = nT$, defined by

$$\varepsilon_n^2 = E\left\{z_n^2\right\} \tag{7.5}$$

where $\{z_n\}$ is the error sequence and $E\{\cdot\}$ denotes the expectation operator. For a particular coefficient vector c_n, substitution of Eq. (7.4) into Eq. (7.5) yields

$$\varepsilon_n^2 = E\left\{x_n^2\right\} - 2c_n'q + c_n'R\,c_n \tag{7.6}$$

where $q = E\{x_n a_n\}$ and $R = E\{a_n a_n'\}$. With the assumption of i.i.d. transmitted symbols, the correlation matrix R is diagonal. The elements on the diagonal are equal to the variance of the transmitted symbols, $\sigma_a^2 = (M^2 - 1)/3$. The minimum mean square error is given by

$$\varepsilon_{\min}^2 = E\left\{x_n^2\right\} - c_{\text{opt}}'R\,c_{\text{opt}} \tag{7.7}$$

where the optimum coefficient vector is $c_{\text{opt}} = R^{-1}q$. We note that proper system operation is achieved only if the transmitted symbols are uncorrelated with the symbols from the far-end transmitter. If this condition is satisfied, the optimum filter coefficients are given by the values of the discrete-time echo channel impulse response, i.e., $c_{n,k} = h_k^E$, $k = 0, \ldots, N - 1$.

By the decision-directed stochastic gradient algorithm, also known as the least mean square algorithm, the coefficients of the echo canceller converge in the mean to c_{opt}. The stochastic gradient algorithm for an N-tap adaptive linear transversal filter is formulated as follows:

$$c_{n+1} = c_n - \frac{1}{2}\alpha\nabla_c\left\{z_n^2\right\} = c_n + \alpha z_n a_n \tag{7.8}$$

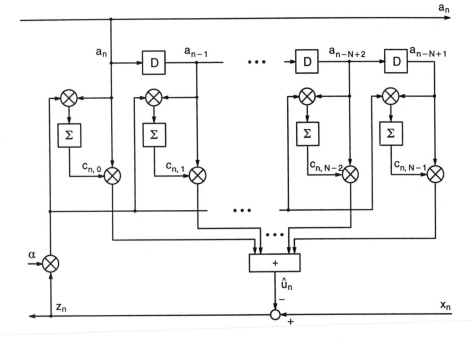

FIGURE 7.4   Block diagram of an adaptive transversal filter echo canceller.

where α is the adaptation gain and

$$\nabla_c\{z_n^2\} = \left(\frac{\partial z_n^2}{\partial c_{n,0}}, \ldots, \frac{\partial z_n^2}{\partial c_{n,N-1}}\right) = -2z_n a_n$$

is the gradient of the squared error with respect to the vector of coefficients. The block diagram of an adaptive transversal filter echo canceller is shown in Fig. 7.4.

If we define the vector $p_n = c_{opt} - c_n$, the mean square error can be expressed as

$$\varepsilon_n^2 = \varepsilon_{min}^2 + p_n' R p_n \tag{7.9}$$

where the term $p_n' R p_n$ represents an excess mean square distortion due to the misadjustment of the filter settings. The analysis of the convergence behavior of the excess mean square distortion was first proposed for adaptive equalizers [Ungerboeck, 1972] and later extended to adaptive echo cancellers [Messerschmitt, 1984]. Under the assumption that the vectors p_n and a_n are statistically independent, the dynamics of the mean square error are given by

$$E\{\varepsilon_n^2\} = \varepsilon_0^2\left[1 - \alpha\sigma_a^2\left(2 - \alpha N\sigma_a^2\right)\right]^n + \frac{2\varepsilon_{min}^2}{2 - \alpha N\sigma_a^2} \tag{7.10}$$

where ε_0^2 is determined by the initial conditions. The mean square error converges to a finite steady-state value ε_∞^2 if the stability condition $0 < \alpha < 2/(N\sigma_a^2)$ is satisfied. The optimum adaptation gain that yields fastest convergence at the beginning of the adaptation process is $\alpha_{opt} = 1/(N\sigma_a^2)$. The corresponding time constant and asymptotic mean square error are $\tau_{opt} = N$ and $\varepsilon_\infty^2 = 2\varepsilon_{min}^2$, respectively.

We note that a fixed adaptation gain equal to α_{opt} could not be adopted in practice, since after echo cancellation the signal from the far-end transmitter would be embedded in a residual echo having approximately the same power. If the time constant of the convergence mode is not a critical system parameter, an adaptation gain smaller than α_{opt} will be adopted to achieve an asymptotic mean square error close to ε_{min}^2. On the other hand, if fast convergence is required, a variable gain will be chosen.

Several techniques have been proposed to increase the speed of convergence of the stochastic gradient algorithm. In particular, for echo cancellation in data transmission, the speed of adaptation is reduced by the presence of the signal from the far-end transmitter in the cancellation error. To mitigate this problem, the data signal can be adaptively removed from the cancellation error by a decision-directed algorithm [Falconer, 1982].

Modified versions of the stochastic gradient algorithm have been also proposed to reduce system complexity. For example, the sign algorithm suggests that only the sign of the error signal be used to compute an approximation of the stochastic gradient [Duttweiler, 1982]. An alternative means to reduce the implementation complexity of an adaptive echo canceller consists in the choice of a filter structure with a lower computational complexity than the transversal filter.

At high data rates, very large scale integration (VLSI) technology is needed for the implementation of transceivers for full-duplex data transmission. High-speed echo cancellers and near-end crosstalk cancellers that do not require multiplications represent an attractive solution because of their low complexity. As an example of an architecture suitable for VLSI implementation, we consider echo cancellation by a distributed-arithmetic filter, where multiplications are replaced by table lookup and shift-and-add operations [Smith, Cowan, and Adams, 1988]. By segmenting the echo canceller into filter sections of shorter lengths, various tradeoffs concerning the number of operations per modulation interval and the number of memory locations needed to store the lookup tables are possible. Adaptivity is achieved by updating the values stored in the lookup tables by the stochastic gradient algorithm.

To describe the principles of operations of a distributed-arithmetic echo canceller, we assume that the number of elements in the alphabet of input symbols is a power of two, $M = 2^W$. Therefore, each symbol is represented by the vector $(a_n^{(0)}, \ldots, a_n^{(W-1)})$, where $a_n^{(i)}, i = 0, \ldots, W - 1$, are independent binary random variables, i.e.,

$$a_n = \sum_{w=0}^{W-1} \left(2a_n^{(w)} - 1\right)2^w = \sum_{w=0}^{W-1} b_n^{(w)}2^w \tag{7.11}$$

where $b_n^{(w)} = (2a_n^{(w)} - 1) \in \{-1, +1\}$. If we recall the expression (7.1) of the output of a transversal filter, by substituting Eq. (7.11) into Eq. (7.1) and segmenting the delay line of the echo canceller into L sections with $K = N/L$ delay elements each, we obtain

$$\hat{u}_n = \sum_{\ell=0}^{L-1} \sum_{w=0}^{W-1} 2^w \left[\sum_{k=0}^{K-1} b_{n-\ell K-k}^{(w)} c_{n,\ell K+k} \right] \tag{7.12}$$

Equation (7.12) suggests that the filter output can be computed using a set of $L2^K$ lookup values that are stored in L lookup tables with 2^K memory locations each. The binary vectors $a_{n,\ell}^{(w)} = (a_{n-(\ell+1)K-1}^{(w)}, \ldots, a_{n-\ell K}^{(w)})$, $w = 0, \ldots, W - 1$, $\ell = 0, \ldots, L - 1$, determine the addresses of the memory locations where the lookup values that are needed to compute the filter output are stored. The filter output is obtained by WL table lookup and shift-and-add operations.

We observe that $a_{n,\ell}^{(w)}$ and its binary complement $\bar{a}_{n,\ell}^{(w)}$ select two values that differ only in their sign. This symmetry is exploited to halve the number of lookup values to be stored. To determine the output of a distributed-arithmetic filter with reduced memory size, we use the identity

$$\hat{u}_n = \sum_{\ell=0}^{L-1} \sum_{w=0}^{W-1} 2^w b_{n-\ell K-k_0}^{(w)} \left[c_{\ell K+k_0} + b_{n-\ell K-k_0}^{(w)} \sum_{\substack{k=0 \\ k \neq k_0}}^{K-1} b_{n-\ell K-k}^{(w)} c_{n,\ell K+k} \right] \tag{7.13}$$

where k_0 can be any element of the set $\{0, \ldots, K - 1\}$. In the following, we take $k_0 = 0$. Then the binary symbols $b_{n-\ell K}^{(w)}$ determine whether the selected lookup values are to be added or subtracted. Each lookup table has now 2^{K-1} memory locations, and the filter output is given by

$$\hat{u}_n = \sum_{\ell=0}^{L-1} \sum_{w=0}^{W-1} 2^w b_{n-\ell K}^{(w)} d_n\left(i_{n,\ell}^{(w)}, \ell\right) \tag{7.14}$$

where $d_n(k, \ell)$, $k = 0, \ldots, 2^{K-1} - 1$, $\ell = 0, \ldots, L - 1$, are the lookup values, and $i_{n,\ell}^{(w)}$, $w = 0, \ldots, W - 1, \ell = 0, \ldots, L - 1$, are the lookup indices computed as follows:

$$i_{n,\ell}^{(w)} = \begin{cases} \sum_{k=1}^{K-1} a_{n-\ell K-k}^{(w)} 2^{k-1} & \text{if } a_{n-\ell K}^{(w)} = 1 \\ \sum_{k=1}^{K-1} \bar{a}_{n-\ell K-k}^{(w)} 2^{k-1} & \text{if } a_{n-\ell K}^{(w)} = 0 \end{cases} \tag{7.15}$$

We note that, as long as Eqs. (7.12) and (7.13) hold for some coefficient vector $(c_{n,0}, \ldots, c_{n,N-1})$, the distributed-arithmetic filter emulates the operation of a linear transversal filter. For arbitrary values $d_n(k, \ell)$, however, a nonlinear filtering operation results.

The expression of the stochastic gradient algorithm to update the lookup values of a distributed-arithmetic echo canceller takes the form

$$d_{n+1} = d_n - \frac{1}{2}\alpha \nabla_d\{z_n^2\} = d_n + \alpha z_n y_n \tag{7.16}$$

where $d_n' = [d_n'(0), \ldots, d_n'(L-1)]$, with $d_n'(\ell) = [d_n(0, \ell), \ldots, d_n(2^{K-1}-1, \ell)]$, and $y_n' = [y_n'(0), \ldots, y_n'(L-1)]$, with

$$y_n'(\ell) = \sum_{w=0}^{W-1} 2^w b_{n-\ell K}^{(w)} \left(\delta_{0, i_{n,\ell}^{(w)}}, \ldots, \delta_{2^{K-1}-1, i_{n,\ell}^{(w)}}\right)$$

are $L2^{K-1} \times 1$ vectors and where $\delta_{i,j}$ is the Kronecker delta. We note that at each iteration only those lookup values that are selected to generate the filter output are updated. The block diagram of an adaptive distributed-arithmetic echo canceller is shown in Fig. 7.5.

The analysis of the mean square error convergence behavior and steady-state performance has been extended to adaptive distributed-arithmetic echo cancellers [Cherubini, 1993]. The dynamics of the mean square error are given by

$$E\{\varepsilon_n^2\} = \varepsilon_0^2 \left[\frac{\alpha\sigma_a^2}{1-2^{K-1}}(2 - \alpha L\sigma_a^2)\right]^n + \frac{2\varepsilon_{\min}^2}{2 - \alpha L\sigma_a^2} \tag{7.17}$$

The stability condition for the echo canceller is $0 < \alpha < 2/(L\sigma_a^2)$. For a given adaptation gain, echo canceller stability depends on the number of lookup tables and on the variance of the transmitted symbols. Therefore, the time span of the echo canceller can be increased without affecting system stability, provided that the number L of lookup tables is kept constant. In that case, however, mean square error convergence will be slower. From Eq. (7.17), one finds that the optimum adaptation gain that permits the fastest mean square error convergence at the beginning of the adaptation process is $\alpha_{opt} = 1/(L\sigma_a^2)$. The time constant of the convergence mode is $\tau_{opt} = L2^{K-1}$. The smallest achievable time constant is thus proportional to the total number of lookup values. The realization of a distributed-arithmetic echo canceller can be further simplified by updating at each iteration only the lookup values that are addressed by the most significant bits of the symbols stored in the delay line. If the number of signal levels is large, the complexity required

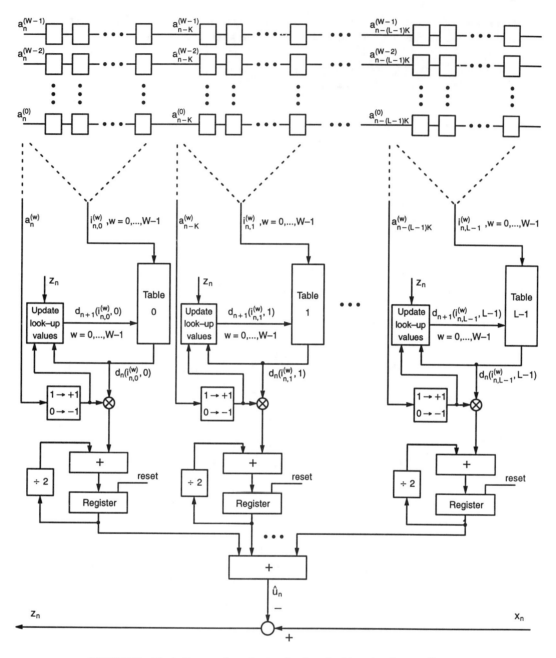

FIGURE 7.5 Block diagram of an adaptive distributed-arithmetic echo canceller.

for adaptation can be considerably reduced at the price of a small increase of the time constant of the convergence mode.

7.3 Passband Transmission

Although most of the concepts presented in the preceding sections can be readily extended to echo cancellation for passband transmission, the case of full-duplex transmission over a voiceband data channel requires a specific discussion. We consider the passband channel model shown in

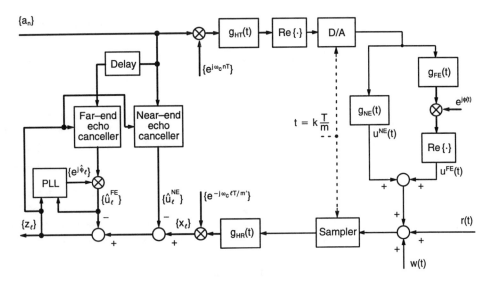

FIGURE 7.6 Configuration for a passband channel echo canceller.

Fig. 7.6. The transmitter generates a sequence $\{a_n\}$ of i.i.d. complex-valued symbols, which are modulated by the carrier $e^{j\omega_c nT}$, where T and ω_c denote the modulation interval and the carrier frequency, respectively. The discrete-time signal at the output of the transmit Hilbert filter may be regarded as an analytic signal, which is generated at the rate of m/T samples/s, $m > 1$. The real part of the analytic signal is converted into an analog signal by a D/A converter and input to the channel. We note that by transmitting the real part of a complex-valued signal positive- and negative-frequency components become folded. The image band attenuation of the transmit Hilbert filter thus determines the achievable echo suppression. In fact, the receiver cannot extract aliasing image-band components from desired passband frequency components, and the echo canceller is able to suppress only echo arising from transmitted passband components.

The output of the echo channel is represented as the sum of two contributions. The near-end echo $u^{NE}(t)$ arises from the impedance mismatch between the hybrid and the transmission line, as in the case of baseband transmission. The far-end echo $u^{FE}(t)$ represents the contribution due to echos that are generated at intermediate points in the telephone network. These echos are characterized by additional impairments, such as jitter and frequency shift, which are accounted for by introducing a carrier-phase rotation of an angle $\phi(t)$ in the model of the far-end echo.

At the receiver, samples of the signal at the channel output are obtained synchronously with the transmitter timing, at the sampling rate of m/T samples/s. The discrete-time received signal is converted to a complex-valued baseband signal $\{x_{nm'+i}, i = 0, \ldots, m' - 1\}$, at the rate of m'/T samples/s, $1 < m' < m$, through filtering by the receive Hilbert filter, decimation, and demodulation. From delayed transmit symbols, estimates of the near- and far-end echo signals after demodulation, $\{\hat{u}^{NE}_{nm'+i}, i = 0, \ldots, m' - 1\}$ and $\{\hat{u}^{FE}_{nm'+i}, i = 0, \ldots, m' - 1\}$, respectively, are generated using m' interleaved near- and far-end echo cancellers. The cancellation error is given by

$$z_\ell = x_\ell - \hat{u}^{NE}_\ell - \hat{u}^{FE}_\ell \qquad (7.18)$$

A different model is obtained if echo cancellation is accomplished before demodulation. In this case, two equivalent configurations for the echo canceller may be considered. In one configuration, the modulated symbols are input to the transversal filter, which approximates the passband echo response. Alternatively, the modulator can be placed after the transversal filter, which is then called a baseband transversal filter [Weinstein, 1977].

In the considered realization, the estimates of the echo signals after demodulation are given by

$$\hat{u}_{nm'+i}^{\text{NE}} = \sum_{k=0}^{N_{\text{NE}}-1} c_{n,km'+i}^{\text{NE}} a_{n-k}, \qquad i = 0, \ldots, m' - 1 \tag{7.19}$$

and

$$\hat{u}_{nm'+i}^{\text{FE}} = \left[\sum_{k=0}^{N_{\text{FE}}-1} c_{n,km'+i}^{\text{FE}} a_{n-k-D_{\text{FE}}} \right] e^{j\hat{\phi}_{nm'+i}}, \qquad i = 0, \ldots, m' - 1 \tag{7.20}$$

where $(c_{n,0}^{\text{NE}}, \ldots, c_{n,m'N_{\text{NE}}-1}^{\text{NE}})$ and $(c_{n,0}^{\text{FE}}, \ldots, c_{n,m'N_{\text{FE}}-1}^{\text{FE}})$ are the coefficients of the m' interleaved near- and far-end echo cancellers, respectively, $\{\hat{\phi}_{nm'+i}, i = 0, \ldots, m' - 1\}$ is the sequence of far-end echo phase estimates, and D_{FE} denotes the bulk delay accounting for the round-trip delay from the transmitter to the point of echo generation. To prevent overlap of the time span of the near-end echo canceller with the time span of the far-end echo canceller, the condition $D_{\text{FE}} > N_{\text{NE}}$ must be satisfied. We also note that, because of the different nature of near- and far-end echo generation, the time span of the far-end echo canceller needs to be larger than the time span of the near-end echo canceller, i.e, $N_{\text{FE}} > N_{\text{NE}}$.

Adaptation of the filter coefficients in the near- and far-end echo cancellers by the stochastic gradient algorithm leads to

$$c_{n+1,km'+i}^{\text{NE}} = c_{n,km'+i}^{\text{NE}} + \alpha z_{nm'+i} (a_{n-k})^*$$

$$k = 0, \ldots, N_{\text{NE}} - 1, \qquad i = 0, \ldots, m' - 1 \tag{7.21}$$

and

$$c_{n+1,km'+i}^{\text{FE}} = c_{n,km'+i}^{\text{FE}} + \alpha z_{nm'+i} (a_{n-k-D_{\text{FE}}})^* e^{-j\hat{\phi}_{nm'+i}}$$

$$k = 0, \ldots, N_{\text{FE}} - 1, \qquad i = 0, \ldots, m' - 1 \tag{7.22}$$

respectively, where the asterisk denotes complex conjugation.

The far-end echo phase estimate is computed by a second-order phase-lock loop algorithm, where the following stochastic gradient approach is adopted:

$$\begin{cases} \hat{\phi}_{\ell+1} = \hat{\phi}_{\ell} - \frac{1}{2}\gamma_{\text{FE}} \nabla_{\hat{\phi}} |z_{\ell}|^2 + \Delta\phi_{\ell} & (\text{mod } 2\pi) \\ \Delta\phi_{\ell+1} = \Delta\phi_{\ell} - \frac{1}{2}\zeta_{\text{FE}} \nabla_{\hat{\phi}} |z_{\ell}|^2 \end{cases} \tag{7.23}$$

where $\ell = nm' + i, \ i = 0, \ldots, m' - 1, \ \gamma_{\text{FE}}$ and ζ_{FE} are step-size parameters, and

$$\nabla_{\hat{\phi}} |z_{\ell}|^2 = \frac{\partial |z_{\ell}|^2}{\partial \hat{\phi}_{\ell}} = -2 \, \text{Im}\left\{ z_{\ell} (\hat{u}_{\ell}^{\text{FE}})^* \right\} \tag{7.24}$$

We note that algorithm (7.23) requires m' iterations per modulation interval, i.e., we cannot resort to interleaving to reduce the complexity of the computation of the far-end echo phase estimate.

7.4 Summary and Conclusions

Digital signal processing techniques for echo cancellation provide large echo attenuation and eliminate the need for additional line interfaces and digital-to-analog and analog-to-digital converters that are required by echo cancellation in the analog signal domain.

The realization of digital echo cancellers in transceivers for high-speed full-duplex data transmission is today possible at a low cost thanks to the advances in VLSI technology. Digital techniques for echo cancellation are also appropriate for near-end crosstalk cancellation in transceivers for transmission over voice-grade cables at rates of several megabit per second for local-area network applications.

In voiceband modems for data transmission over the telephone network, digital techniques for echo cancellation also permit precise tracking of the carrier phase and frequency shift of far-end echos.

References

Cherubini, G. 1993. Analysis of the convergence behavior of adaptive distributed-arithmetic echo cancellers. *IEEE Trans. Commun.* 41(11):1703–1714.

Duttweiler, D.L. 1982. Adaptive filter performance with nonlinearities in the correlation multiplier. *IEEE Trans. Acoust., Speech, Signal Processing,* 30(8):578–586.

Falconer, D.D. 1982. Adaptive reference echo-cancellation. *IEEE Trans. Commun.* 30(9):2083–2094.

Lee, E.A. and Messerschmitt, D.G. 1994. *Digital Communication,* 2nd ed. Kluwer Academic Publishers, Boston MA.

Messerschmitt, D.G. 1984. Echo cancellation in speech and data transmission. *IEEE J. Sel. Areas Commun.* 2(2):283–297.

Messerschmitt, D.G. 1986. Design issues for the ISDN U-Interface transceiver. *IEEE J. Sel. Areas Commun.* 4(8):1281–1293.

Smith, M.J., Cowan, C.F.N., and Adams, P.F. 1988. Adaptive nonlinear digital filters using distributed arithmetic. *IEEE Trans. Circuits and Systems* 35(1):6–18.

Ungerboeck, G. 1972. Theory on the speed of convergence in adaptive equalizers for digital communication. *IBM J. Res. Develop.* 16(6):546–555.

Weinstein, S.B. 1977. A passband data-driven echo-canceller for full-duplex transmission on two-wire circuits. *IEEE Trans. Commun.* 25(7):654–666.

Further Information

For further information on adaptive transversal filters with application to echo cancellation, see *Adaptive Filters: Structures, Algorithms, and Applications,* M.L. Honig and D.G. Messerschmitt, Kluwer, 1984.

8

Pseudonoise Sequences

Tor Helleseth
University of Bergen

P. Vijay Kumar
University of Southern California

8.1 Introduction

Pseudonoise sequences (PN sequences), also referred to as pseudorandom sequences, are sequences that are deterministically generated and yet possess some properties that one would expect to find in randomly generated sequences. Applications of PN sequences include signal synchronization, navigation, radar ranging, random number generation, spread-spectrum communications, multipath resolution, cryptography, and signal identification in multiple-access communication systems. The *correlation* between two sequences $\{x(t)\}$ and $\{y(t)\}$ is the complex inner product of the first sequence with a shifted version of the second sequence. The correlation is called 1) an autocorrelation if the two sequences are the same, 2) a crosscorrelation if they are distinct, 3) a periodic correlation if the shift is a cyclic shift, 4) an aperiodic correlation if the shift is not cyclic, and 5) a partial-period correlation if the inner product involves only a partial segment of the two sequences. More precise definitions are given subsequently.

Binary **m sequences**, defined in the next section, are perhaps the best-known family of PN sequences. The balance, run-distribution, and autocorrelation properties of these sequences mimic those of random sequences. It is perhaps the random-like correlation properties of PN sequences that makes them most attractive in a communications system, and it is common to refer to any collection of low-correlation sequences as a family of PN sequences.

Section 8.2 begins by discussing *m* sequences. Thereafter, the discussion continues with a description of sequences satisfying various correlation constraints along the lines of the accompanying self-explanatory figure, Fig. 8.1. Expanded tutorial discussions on pseudorandom sequences may be found in Sarwate and Pursley, 1980, in Chapter 5 of Simon et al., 1994, and in Helleseth and Kumar, 1996.

0-8493-8573-3/96/$0.00+$.50
© 1996 by CRC Press, Inc.

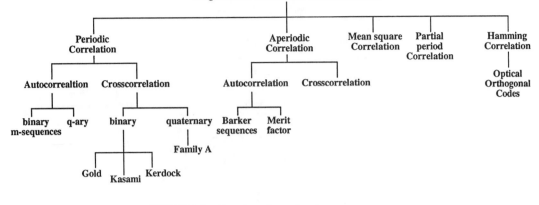

FIGURE 8.1   Overview of pseudonoise sequences.

8.2 *m* Sequences

A binary $\{0, 1\}$ **shift-register sequence** $\{s(t)\}$ is a sequence that satisfies a linear recurrence relation of the form

$$\sum_{i=0}^{r} f_i s(t + i) = 0, \qquad \text{for all } t \geq 0 \tag{8.1}$$

where $r \geq 1$ is the *degree* of the recursion; the coefficients f_i belong to the finite field $GF(2) = \{0, 1\}$ where the leading coefficient $f_r = 1$. Thus, both sequences $\{a(t)\}$ and $\{b(t)\}$ appearing in Fig. 8.2 are shift-register sequences. A sequence satisfying a recursion of the form in Eq. (8.1) is said to have *characteristic polynomial* $f(x) = \sum_{i=0}^{r} f_i x^i$. Thus, $\{a(t)\}$ and $\{b(t)\}$ have characteristic polynomials given by $f(x) = x^3 + x + 1$ and $f(x) = x^3 + x^2 + 1$, respectively.

Since an r-bit binary shift register can assume a maximum of 2^r different states, it follows that every shift-register sequence $\{s(t)\}$ is eventually periodic with period $n \leq 2^r$, i.e.,

$$s(t) = s(t + n), \qquad \text{for all } t \geq N$$

for some integer N. In fact, the maximum period of a shift-register sequence is $2^r - 1$, since a shift register that enters the all-zero state will remain forever in that state. The upper shift register in Fig. 8.2 when initialized with starting state 0 0 1 generates the periodic sequence $\{a(t)\}$ given by

$$0010111 \quad 0010111 \quad 0010111 \quad \cdots \tag{8.2}$$

of period $n = 7$. It follows then that this shift register generates sequences of maximal period starting from any nonzero initial state.

An m sequence is simply a binary shift-register sequence having maximal period. For every $r \geq 1$, m sequences are known to exist. The periodic **autocorrelation** function θ_s of a binary $\{0, 1\}$ sequence $\{s(t)\}$ of period n is defined by

$$\theta_s(\tau) = \sum_{t=0}^{n-1} (-1)^{s(t+\tau) - s(t)}, \qquad 0 \leq \tau \leq n - 1$$

An m sequence of length $2^r - 1$ has the following attributes. 1) *Balance property*: in each period of the m sequence there are 2^{r-1} ones and $2^{r-1} - 1$ zeros. 2) *Run property*: every nonzero binary s-tuple,

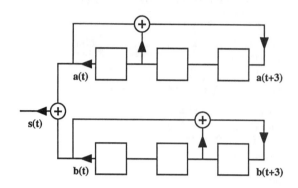

FIGURE 8.2   An example Gold sequence generator.   Here $\{a(t)\}$ and $\{b(t)\}$ are m sequences of length 7.

$s \le r$ occurs 2^{r-s} times, the all-zero s-tuple occurs $2^{r-s} - 1$ times. 3) *Two-level autocorrelation function:*

$$\theta_s(\tau) = \begin{cases} n & \text{if } \tau = 0 \\ -1 & \text{if } \tau \ne 0 \end{cases} \tag{8.3}$$

The first two properties follow immediately from the observation that every nonzero r-tuple occurs precisely once in each period of the m sequence. For the third property, consider the difference sequence $\{s(t + \tau) - s(t)\}$ for $\tau \ne 0$. This sequence satisfies the same recursion as the m sequence $\{s(t)\}$ and is clearly not the all-zero sequence. It follows, therefore, that $\{s(t+\tau)-s(t)\} \equiv \{s(t+\tau')\}$ for some τ', $0 \le \tau' \le n - 1$, i.e., is a different cyclic shift of the m sequence $\{s(t)\}$. The balance property of the sequence $\{s(t + \tau')\}$ then gives us attribute 3. The m sequence $\{a(t)\}$ in Eq. (8.2) can be seen to have the three listed properties.

If $\{s(t)\}$ is any sequence of period n and d is an integer, $1 \le d \le n$, then the mapping $\{s(t)\} \to \{s(dt)\}$ is referred to as a *decimation* of $\{s(t)\}$ by the integer d. If $\{s(t)\}$ is an m sequence of period $n = 2^r - 1$ and d is an integer relatively prime to $2^r - 1$, then the decimated sequence $\{s(dt)\}$ clearly also has period n. Interestingly, it turns out that the sequence $\{s(dt)\}$ is always also an m sequence of the same period. For example, when $\{a(t)\}$ is the sequence in Eq. (8.2), then

$$a(3t) = 0011101 \quad 0011101 \quad 0011101 \quad \cdots \tag{8.4}$$

and

$$a(2t) = 0111001 \quad 0111001 \quad 0111001 \quad \cdots \tag{8.5}$$

The sequence $\{a(3t)\}$ is also an m sequence of period 7, since it satisfies the recursion

$$s(t + 3) + s(t + 2) + s(t) = 0 \qquad \text{for all } t$$

of degree $r = 3$. In fact $\{a(3t)\}$ is precisely the sequence labeled $\{b(t)\}$ in Fig. 8.2. The sequence $\{a(2t)\}$ is simply a cyclically shifted version of $\{a(t)\}$ itself; this property holds in general. If $\{s(t)\}$ is any m sequence of period $2^r - 1$, then $\{s(2t)\}$ will always be a shifted version of the same m sequence. Clearly, the same is true for decimations by any power of 2.

Starting from an m sequence of period $2^r - 1$, it turns out that one can generate all m sequences of the same period through decimations by integers d relatively prime to $2^r - 1$. The set of integers d,

$1 \leq d \leq 2^r - 1$ satisfying $(d, 2^r - 1) = 1$ forms a group under multiplication modulo $2^r - 1$, with the powers $\{2^i \mid 0 \leq i \leq r - 1\}$ of 2 forming a subgroup of order r. Since decimation by a power of 2 yields a shifted version of the same m sequence, it follows that the number of distinct m sequences of period $2^r - 1$ is $[\phi(2^r - 1)/r]$ where $\phi(n)$ denotes the number of integers d, $1 \leq d \leq n$, relatively prime to n. For example, when $r = 3$, there are just two cyclically distinct m sequences of period 7, and these are precisely the sequences $\{a(t)\}$ and $\{b(t)\}$ discussed in the preceding paragraph. Tables provided in Peterson and Weldon, 1972 can be used to determine the characteristic polynomial of the various m sequences obtainable through the decimation of a single given m sequence. The classical reference on m sequences is Golomb, 1982.

If one obtains a sequence of some large length n by repeatedly tossing an unbiased coin, then such a sequence will very likely satisfy the balance, run, and autocorrelation properties of an m sequence of comparable length. For this reason, it is customary to regard the extent to which a given sequence possesses these properties as a measure of randomness of the sequence. Quite apart from this, in many applications such as signal synchronization and radar ranging, it is desirable to have sequences $\{s(t)\}$ with low autocorrelation sidelobes i.e., $|\theta_s(\tau)|$ is small for $\tau \neq 0$. Whereas m sequences are a prime example, there exist other methods of constructing binary sequences with low out-of-phase autocorrelation.

Sequences $\{s(t)\}$ of period n having an autocorrelation function identical to that of an m sequence, i.e., having θ_s satisfying Eq. (8.3) correspond to well-studied combinatorial objects known as *cyclic Hadamard difference sets*. Known infinite families fall into three classes 1) Singer and Gordon, Mills and Welch, 2) quadratic residue, and 3) twin-prime difference sets. These correspond, respectively, to sequences of period n of the form $n = 2^r - 1, r \geq 1$; n prime; and $n = p(p + 2)$ with both p and $p + 2$ being prime in the last case. For a detailed treatment of cyclic difference sets, see Baumert, 1971.

8.3 The q-ary Sequences with Low Autocorrelation

As defined earlier, the autocorrelation of a binary $\{0, 1\}$ sequence $\{s(t)\}$ leads to the computation of the inner product of an $\{-1, +1\}$ sequence $\{(-1)^{s(t)}\}$ with a cyclically shifted version $\{(-1)^{s(t+\tau)}\}$ of itself. The $\{-1, +1\}$ sequence is transmitted as a phase shift by either $0°$ and $180°$ of a radio-frequency carrier, i.e., using binary phase-shift keying (PSK) modulation. If the modulation is q-ary PSK, then one is led to consider sequences $\{s(t)\}$ with symbols in the set Z_q, i.e., the set of integers modulo q. The relevant autocorrelation function $\theta_s(\tau)$ is now defined by

$$\theta_s(\tau) = \sum_{t=0}^{n-1} \omega^{s(t+\tau) - s(t)}$$

where n is the period of $\{s(t)\}$ and ω is a complex primitive qth root of unity. It is possible to construct sequences $\{s(t)\}$ over Z_q whose autocorrelation function satisfies

$$\theta_s(\tau) = \begin{cases} n & \text{if } \tau = 0 \\ 0 & \text{if } \tau \neq 0 \end{cases}$$

For obvious reasons, such sequences are said to have an *ideal autocorrelation function*.

We provide without proof two sample constructions. The sequences in the first construction are given by

$$s(t) = \begin{cases} t^2/2 \pmod{n} & \text{when } n \text{ is even} \\ t(t + 1)/2 \pmod{n} & \text{when } n \text{ is odd} \end{cases}$$

Thus, this construction provides sequences with ideal autocorrelation for any period n. Note that the size q of the sequence symbol alphabet equals n when n is odd and $2n$ when n is even.

The second construction also provides sequences over Z_q of period n but requires that n be a perfect square. Let $n = r^2$ and let π be an arbitrary permutation of the elements in the subset $\{0, 1, 2, \ldots, (r-1)\}$ of Z_n: Let g be an arbitrary function defined on the subset $\{0, 1, 2, \ldots, r-1\}$ of Z_n. Then any sequence of the form

$$s(t) = rt_1\pi(t_2) + g(t_2) \quad (\text{mod } n)$$

where $t = rt_1 + t_2$ with $0 \leq t_1, t_2 \leq r - 1$ is the base-r decomposition of t, has an ideal autocorrelation function. When the alphabet size q equals or divides the period n of the sequence, ideal-autocorrelation sequences also go by the name *generalized bent functions*. For details, see Helleseth and Kumar, 1996.

8.4 Families of Sequences with Low Crosscorrelation

Given two sequences $\{s_1(t)\}$ and $\{s_2(t)\}$ over Z_q of period n, their **crosscorrelation** function $\theta_{1,2}(\tau)$ is defined by

$$\theta_{1,2}(\tau) = \sum_{t=0}^{n-1} \omega^{s_1(t+\tau)-s_2(t)}$$

where ω is a primitive qth root of unity. The crosscorrelation function is important in code-division multiple-access (CDMA) communication systems. Here, each user is assigned a distinct signature sequence and to minimize interference due to the other users, it is desirable that the signature sequences have pairwise, low values of crosscorrelation function. To provide the system in addition with a self-synchronizing capability, it is desirable that the signature sequences have low values of the autocorrelation function as well.

Let $\mathcal{F} = \{\{s_i(t)\} \mid 1 \leq i \leq M\}$ be a family of M sequences $\{s_i(t)\}$ over Z_q each of period n. Let $\theta_{i,j}(\tau)$ denote the crosscorrelation between the ith and jth sequence at shift τ, i.e.

$$\theta_{i,j}(\tau) = \sum_{t=0}^{n-1} \omega^{s_i(t+\tau)-s_j(t)}, \qquad 0 \leq \tau \leq n - 1$$

The classical goal in sequence design for CDMA systems has been minimization of the parameter

$$\theta_{\max} = \max\{|\theta_{i,j}(\tau)| \mid \text{either } i \neq j \text{ or } \tau \neq 0\}$$

for fixed n and M. It should be noted though that, in practice, because of data modulation the correlations that one runs into are typically of an aperiodic rather than a periodic nature (see Sec. 8.5). The problem of designing for low aperiodic correlation, however, is a more difficult one. A typical approach, therefore, has been to design based on periodic correlation, and then to analyze the resulting design for its aperiodic correlation properties. Again, in many practical systems, the mean square correlation properties are of greater interest than the worst-case correlation represented by a parameter such as $\theta_{\max}$. The mean square correlation is discussed in Sec. 8.6.

Bounds on the minimum possible value of $\theta_{\max}$ for given period n, family size M, and alphabet size q are available that can be used to judge the merits of a particular sequence design. The most efficient bounds are those due to Welch, Sidelnikov, and Levenshtein, see Helleseth and Kumar, 1996. In CDMA systems, there is greatest interest in designs in which the parameter $\theta_{\max}$ is in the range $\sqrt{n} \leq \theta_{\max} \leq 2\sqrt{n}$. Accordingly, Table 8.1 uses the Welch, Sidelnikov, and Levenshtein

TABLE 8.1 Bounds on Family Size M for Given n, $\theta_{\max}$

$\theta_{\max}$	Upper bound on M $q = 2$	Upper Bound on M $q > 2$
$\sqrt{n}$	$n/2$	n
$\sqrt{2n}$	n	$n^2/2$
$2\sqrt{n}$	$3n^2/10$	$n^3/2$

bounds to provide an order-of-magnitude upper bound on the family size M for certain $\theta_{\max}$ in the cited range.

Practical considerations dictate that q be small. The bit-oriented nature of electronic hardware makes it preferable to have q a power of 2. With this in mind, a description of some efficient sequence families having low auto- and crosscorrelation values and alphabet sizes $q = 2$ and $q = 4$ are described next.

Gold and Kasami Sequences

Given the low autocorrelation sidelobes of an m sequence, it is natural to attempt to construct families of low correlation sequences starting from m sequences. Two of the better known constructions of this type are the families of Gold and Kasami sequences.

Let r be odd and $d = 2^k + 1$ where k, $1 \leq k \leq r - 1$, is an integer satisfying $(k, r) = 1$. Let $\{s(t)\}$ be a cyclic shift of an m sequence of period $n = 2^r - 1$ that signifies $S(dt) \not\equiv 0$ and let $\mathcal{G}$ be the *Gold* family of $2^r + 1$ sequences given by

$$\mathcal{G} = \{s(t)\} \cup \{s(dt)\} \cup \{\{s(t) + s(d[t + \tau])\} \mid 0 \leq \tau \leq n - 1\}$$

Then each sequence in $\mathcal{G}$ has period $2^r - 1$ and the maximum-correlation parameter $\theta_{\max}$ of $\mathcal{G}$ satisfies

$$\theta_{\max} \leq \sqrt{2^{r+1}} + 1$$

An application of the Sidelnikov bound coupled with the information that $\theta_{\max}$ must be an odd integer yields that for the family $\mathcal{G}$, $\theta_{\max}$ is as small as it can possibly be. In this sense the family $\mathcal{G}$ is an optimal family. We remark that these comments remain true even when d is replaced by the integer $d = 2^{2k} - 2^k + 1$ with the conditions on k remaining unchanged.

The Gold family remains the best-known family of m sequences having low crosscorrelation. Applications include the Navstar Global Positioning System whose signals are based on Gold sequences.

The family of Kasami sequences has a similar description. Let $r = 2v$ and $d = 2^v + 1$. Let $\{s(t)\}$ be a cyclic shift of an m sequence of period $n = 2^r - 1$ that satisfies $s(dt) \not\equiv 0$, and consider the family of Kasami sequences given by

$$\mathcal{K} = \{s(t)\} \cup \{\{s(t) + s(d[t + \tau])\} \mid 0 \leq \tau \leq 2^v - 2\}$$

Then the Kasami family $\mathcal{K}$ contains 2^v sequences of period $2^r - 1$. It can be shown that in this case

$$\theta_{\max} = 1 + 2^v$$

This time an application of the Welch bound and the fact that $\theta_{\max}$ is an integer shows that the Kasami family is optimal in terms of having the smallest possible value of $\theta_{\max}$ for given n and M.

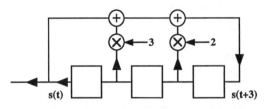

FIGURE 8.3   Shift register that generates family A quaternary sequences $\{s(t)\}$ of period 7.

Quaternary Sequences with Low Crosscorrelation

The entries in Table 8.1 suggest that nonbinary (i.e., $q > 2$) designs may be used for improved performance. A family of quaternary sequences that outperform the Gold and Kasami sequences is now discussed below.

Let $f(x)$ be the characteristic polynomial of a binary m sequence of length $2^r - 1$ for some integer r. The coefficients of $f(x)$ are either 0 or 1. Now, regard $f(x)$ as a polynomial over Z_4 and form the product $(-1)^r f(x) f(-x)$. This can be seen to be a polynomial in x^2. Define the polynomial $g(x)$ of degree r by setting $g(x^2) = (-1)^r f(x) f(-x)$. Let $g(x) = \sum_{i=0}^{r} g_i x^i$ and consider the set of all quaternary sequences $\{a(t)\}$ satisfying the recursion $\sum_{i=0}^{r} g_i a(t+i) = 0$ for all t.

It turns out that with the exception of the all-zero sequence, all of the sequences generated in this way have period $2^r - 1$. Thus, the recursion generates a family A of $2^r + 1$ cyclically distinct quaternary sequences. Closer study reveals that the maximum correlation parameter θ_{max} of this family satisfies $\theta_{max} \leq 1 + \sqrt{2^r}$. Thus, in comparison to the family of Gold sequences, the family A offers a lower value of θ_{max} (by a factor of $\sqrt{2}$) for the same family size. In comparison to the set of Kasami sequences, it offers a much larger family size for the same bound on θ_{max}.

We illustrate with an example. Let $f(x) = x^3 + x + 1$ be the characteristic polynomial of the m sequence $\{a(t)\}$ in Eq. (8.1). Then over Z_4

$$g(x^2) = (-1)^3 f(x) f(-x) = x^6 + 2x^4 + x^2 + 3$$

so that $g(x) = x^3 + 2x^2 + x + 3$. Thus, the sequences in family A are generated by the recursion $s(t+3) + 2s(t+2) + s(t+1) + 3s(t) = 0 \bmod 4$. The corresponding shift register is shown in Fig. 8.3. By varying initial conditions, this shift register can be made to generate nine cyclically distinct sequences, each of length 7. In this case $\theta_{max} \leq 1 + \sqrt{8}$.

Kerdock Sequences

The Gold and Kasami families of sequences are closely related to binary linear cyclic codes. It is well known in coding theory that there exists nonlinear binary codes whose performance exceeds that of the best possible linear code. Surprisingly, some of these examples come from binary codes, which are images of *linear quaternary* ($q = 4$) codes under the Gray map: $0 \rightarrow 00, 1 \rightarrow 01, 2 \rightarrow 11$, $3 \rightarrow 10$. A prime example of this is the Kerdock code, which recently has been shown to be the Gray image of a quaternary linear code. Thus, it is not surprising that the Kerdock code yields binary sequences that significantly outperform the family of Kasami sequences.

The Kerdock sequences may be constructed as follows: let $f(x)$ be the characteristic polynomial of an m sequence of period $2^r - 1$, r odd. As before, regarding $f(x)$ as a polynomial over Z_4 (which happens to have $\{0, 1\}$ coefficients), let the polynomial $g(x)$ over Z_4 be defined via $g(x^2) = -f(x) f(-x)$. [Thus, $g(x)$ is the characteristic polynomial of a family A sequence set of period

$2^r - 1$.] Set $h(x) = -g(-x) = \sum_{i=0}^r h_i x^i$, and let S be the set of all Z_4 sequences satisfying the recursion $\sum_{i=0}^r h_i s(t+i) = 0$. Then S contain 4^r-distinct sequences corresponding to all possible distinct initializations of the shift register.

Let T denote the subset S of size 2^r-consisting of those sequences corresponding to initializations of the shift register only using the symbols 0 and 2 in Z_4. Then the set $S - T$ of size $4^r - 2^r$ contains a set $\mathcal{U}$ of 2^{r-1} cyclically distinct sequences each of period $2(2^r - 1)$. Given $x = a + 2b \in Z_4$ with $a, b \in \{0, 1\}$, let μ denote the most significant bit (MSB) map $\mu(x) = b$. Let $\mathcal{K}_E$ denote the family of 2^{r-1} binary sequences obtained by applying the map μ to each sequence in $\mathcal{U}$. It turns out that each sequence in $\mathcal{U}$ also has period $2(2^r - 1)$ and that, furthermore, for the family $\mathcal{K}_E$, $\theta_{\max} \leq 2 + \sqrt{2^{r+1}}$. Thus, $\mathcal{K}_E$ is a much larger family than the Kasami family, while having almost exactly the same value of $\theta_{\max}$.

For example, taking $r = 3$ and $f(x) = x^3 + x + 1$, we have from the previous family $\mathcal{A}$ example that $g(x) = x^3 + 2x^2 + x + 3$, so that $h(x) = -g(-x) = x^3 + 2x^2 + x + 1$. Applying the MSB map to the head of the shift register, and discarding initializations of the shift register involving only 0's and 2's yields a family of four cyclically distinct binary sequences of period 14.

8.5 Aperiodic Correlation

Let $\{x(t)\}$ and $\{y(t)\}$ be complex-valued sequences of length (or period) n, not necessarily distinct. Their *aperiodic correlation* values $\{\rho_{x,y}(\tau)| - (n-1) \leq \tau \leq n - 1\}$ are given by

$$\rho_{x,y}(\tau) = \sum_{t=\max\{0,-\tau\}}^{\min\{n-1,n-1-\tau\}} x(t+\tau)y^*(t)$$

where $y^*(t)$ denotes the complex conjugate of $y(t)$. When $x \equiv y$, we will abbreviate and write ρ_x in place of $\rho_{x,y}$. The sequences described next are perhaps the most famous example of sequences with low-aperiodic autocorrelation values.

Barker Sequences

A binary $\{-1, +1\}$ sequence $\{s(t)\}$ of length n is said to be a *Barker sequence* if the aperiodic autocorrelation values $\rho_s(\tau)$ satisfy $|\rho_s(\tau)| \leq 1$ for all τ, $-(n-1) \leq \tau \leq n - 1$. The Barker property is preserved under the following transformations:

$$s(t) \to -s(t), \qquad s(t) \to (-1)^t s(t) \qquad \text{and} \qquad s(t) \to s(n-1-t)$$

as well as under compositions of the preceding transformations. Only the following Barker sequences are known:

$$
\begin{aligned}
n &= 2 & &+\,+ \\
n &= 3 & &+\,+\,- \\
n &= 4 & &+\,+\,+\,- \\
n &= 5 & &+\,+\,+\,-\,+ \\
n &= 7 & &+\,+\,+\,-\,-\,+\,- \\
n &= 11 & &+\,+\,+\,-\,-\,-\,+\,-\,-\,+\,- \\
n &= 13 & &+\,+\,+\,+\,+\,-\,-\,+\,+\,-\,+\,-\,+
\end{aligned}
$$

where $+$ denotes $+1$ and $-$ denotes -1 and sequences are generated from these via the transformations already discussed. It is known that if any other Barker sequence exists, it must have length $n > 1,898,884$, that is a multiple of 4.

For an upper bound to the maximum out-of-phase aperiodic autocorrelation of an m sequence, see Sarwate, 1984.

Sequences with High Merit Factor

The *merit factor* F of a $\{-1, +1\}$ sequence $\{s(t)\}$ is defined by

$$F = \frac{n^2}{2 \sum_{\tau=1}^{n-1} \rho_s^2(\tau)}$$

Since $\rho_s(\tau) = \rho_s(-\tau)$ for $1 \leq |\tau| \leq n-1$ and $\rho_s(0) = n$, factor F may be regarded as the ratio of the square of the in-phase autocorrelation, to the sum of the squares of the out-of-phase aperiodic autocorrelation values. Thus, the merit factor is one measure of the aperiodic autocorrelation properties of a binary $\{-1, +1\}$ sequence. It is also closely connected with the signal to self-generated noise ratio of a communication system in which coded pulses are transmitted and received.

Let F_n denote the largest merit factor of any binary $\{-1, +1\}$ sequence of length n. For example, at length $n = 13$, the Barker sequence of length 13 has a merit factor $F = F_{13} = 14.08$. Assuming a certain ergodicity postulate it was established by Golay that $\lim_{n \to \infty} F_n = 12.32$. Exhaustive computer searches carried out for $n \leq 40$ have revealed the following.

1. For $1 \leq n \leq 40, n \neq 11, 13$,

$$3.3 \leq F_n \leq 9.85,$$

2. $F_{11} = 12.1, F_{13} = 14.08$.

The value F_{11} is also achieved by a Barker sequence. From partial searches, for lengths up to 117, the highest known merit factor is between 8 and 9.56; for lengths from 118 to 200, the best-known factor is close to 6. For lengths > 200, statistical search methods have failed to yield a sequence having merit factor exceeding 5.

An *offset sequence* is one in which a fraction θ of the elements of a sequence of length n are chopped off at one end and appended to the other end, i.e., an offset sequence is a cyclic shift of the original sequence by $n\theta$ symbols. It turns out that the asymptotic merit factor of m sequences is equal to 3 and is independent of the particular offset of the m sequence. There exist offsets of sequences associated with quadratic-residue and twin-prime difference sets that achieve a larger merit factor of 6. Details may be found in Jensen, Jenson, and Høholdt, 1991.

Sequences with Low Aperiodic Crosscorrelation

If $\{u(t)\}$ and $\{v(t)\}$ are sequences of length $2n - 1$ defined by

$$u(t) = \begin{cases} x(t) & \text{if } 0 \leq t \leq n-1 \\ 0 & \text{if } n \leq t \leq 2n-2 \end{cases}$$

and

$$v(t) = \begin{cases} y(t) & \text{if } 0 \leq t \leq n-1 \\ 0 & \text{if } n \leq t \leq 2n-2 \end{cases}$$

then

$$\{\rho_{x,y}(\tau) \mid -(n-1) \le \tau \le n-1\} = \{\theta_{u,v}(\tau) \mid 0 \le \tau \le 2n-2\} \quad (8.6)$$

Given a collection

$$U = \{\{x_i(t)\} \mid 1 \le i \le M\}$$

of sequences of length n over Z_q, let us define

$$\rho_{\max} = \max\{|\rho_{a,b}(\tau)| \mid a, b \in U, \text{ either } a \ne b \text{ or } \tau \ne 0\}$$

It is clear from Eq. (8.6) how bounds on the *periodic* correlation parameter $\theta_{\max}$ can be adapted to give bounds on $\rho_{\max}$. Translation of the Welch bound gives that for every integer $k \ge 1$,

$$\rho_{\max}^{2k} \ge \left(\frac{n^{2k}}{M(2n-1)-1}\right)\left\{\frac{M(2n-1)}{\binom{2n+k-2}{k}} - 1\right\}$$

Setting $k = 1$ in the preceding bound gives

$$\rho_{\max} \ge n\sqrt{\frac{M-1}{M(2n-1)-1}}$$

Thus, for fixed M and large n, Welch's bound gives

$$\rho_{\max} \ge \mathcal{O}(n^{1/2})$$

There exist sequence families which asymptotically achieve $\rho_{\max} \approx \mathcal{O}(n^{1/2})$, [Mow, 1994].

8.6 Other Correlation Measures

Partial-Period Correlation

The *partial-period (p-p) correlation* between the sequences $\{u(t)\}$ and $\{v(t)\}$ is the collection $\{\Delta_{u,v}(l, \tau, t_0) \mid 1 \le l \le n, 0 \le \tau \le n-1, 0 \le t_0 \le n-1\}$ of inner products

$$\Delta_{u,v}(l, \tau, t_0) = \sum_{t=t_0}^{t=t_0+l-1} u(t+\tau)v^*(t)$$

where l is the length of the partial period and the sum $t + \tau$ is again computed modulo n.

In direct-sequence CDMA systems, the pseudorandom signature sequences used by the various users are often very long for reasons of data security. In such situations, to minimize receiver hardware complexity, correlation over a partial period of the signature sequence is often used to demodulate data, as well as to achieve synchronization. For this reason, the p-p correlation properties of a sequence are of interest.

Researchers have attempted to determine the moments of the p-p correlation. Here the main tool is the application of the Pless power-moment identities of coding theory [MacWilliams and Sloane, 1977]. The identities often allow the first and second p-p correlation moments to be completely determined. For example, this is true in the case of m sequences (the remaining moments turn out to depend upon the specific characteristic polynomial of the m sequence). Further details may be found in Simon et al., 1994.

Mean Square Correlation

Frequently in practice, there is a greater interest in the mean-square correlation distribution of a sequence family than in the parameter θ_{max}. Quite often in sequence design, the sequence family is derived from a linear, binary cyclic code of length n by picking a set of cyclically distinct sequences of period n. The families of Gold and Kasami sequences are so constructed. In this case, as pointed out by Massey, the mean square correlation of the family can be shown to be either optimum or close to optimum, under certain easily satisfied conditions, imposed on the minimum distance of the dual code. A similar situation holds even when the sequence family does not come from a linear cyclic code. In this sense, mean square correlation is not a very discriminating measure of the correlation properties of a family of sequences. An expanded discussion of this issue may be found in Hammons and Kumar, 1993.

Optical Orthogonal Codes

Given a pair of $\{0, 1\}$ sequences $\{s_1(t)\}$ and $\{s_2(t)\}$ each having period n, we define the *Hamming correlation* function $\theta_{12}(\tau)$, $0 \leq \tau \leq n - 1$, by

$$\theta_{12}(\tau) = \sum_{t=0}^{n-1} s_1(t + \tau)s_2(t)$$

Such correlations are of interest, for instance, in optical communication systems where the 1's and 0's in a sequence correspond to the presence or absence of pulses of transmitted light.

An (n, w, λ) optical orthogonal code (OOC) is a family $\mathcal{F} = \{\{s_i(t)\} \mid i = 1, 2, \ldots, M\}$, of M $\{0, 1\}$ sequences of period n, constant Hamming weight w, where w is an integer lying between 1 and $n - 1$ satisfying $\theta_{ij}(\tau) \leq \lambda$ whenever either $i \neq j$ or $\tau \neq 0$.

Note that the Hamming distance $d_{a,b}$ between a period of the corresponding codewords $\{a(t)\}$, $\{b(t)\}$, $0 \leq t \leq n - 1$ in an (n, w, λ) OOC having Hamming correlation ρ, $0 \leq \rho \leq \lambda$, is given by $d_{a,b} = 2(w - \rho)$, and, thus, OOCs are closely related to constant-weight error correcting codes. Given an (n, w, λ) OOC, by enlarging the OOC to include every cyclic shift of each sequence in the code, one obtains a constant-weight, minimum distance $d_{min} \geq 2(w - \lambda)$ code. Conversely, given a constant-weight cyclic code of length n, weight w and minimum distance d_{min}, one can derive an (n, w, λ) OOC code with $\lambda \leq w - d_{min}/2$ by partitioning the code into cyclic equivalence classes and then picking precisely one representative from each equivalence class of size n.

By making use of this connection, one can derive bounds on the size of an OOC from known bounds on the size of constant-weight codes. The bound given next follows directly from the Johnson bound for constant weight codes [MacWilliams and Sloane, 1977]. The number $M(n, w, \lambda)$ of codewords in a (n, w, λ) OOC satisfies

$$M(n, w, \lambda) \leq \frac{1}{w} \left\lfloor \frac{n-1}{w-1} \cdots \left\lfloor \frac{n-\lambda+1}{w-\lambda+1} \left\lfloor \frac{n-\lambda}{w-\lambda} \right\rfloor \right\rfloor \cdots \right\rfloor$$

An OOC code that achieves the Johnson bound is said to be optimal. A family $\{\mathcal{F}_n\}$ of OOCs indexed by the parameter n and arising from a common construction is said to be asymptotically optimum if

$$\lim_{n \to \infty} \frac{|\mathcal{F}_n|}{M(n, w, \lambda)} = 1$$

Constructions for optical orthogonal codes are available for the cases when $\lambda = 1$ and $\lambda = 2$. For larger values of λ, there exist constructions which are asymptotically optimum. Further details may be found in Helleseth and Kumar, 1996.

Defining Terms

Autocorrelation of a sequence: The complex inner product of the sequence with a shifted version itself.

Crosscorrelation of two sequences: The complex inner product of the first sequence with a shifted version of the second sequence.

***m* Sequence:** A periodic binary $\{0, 1\}$ sequence that is generated by a shift register with linear feedback and which has maximal possible period given the number of stages in the shift register.

Pseudonoise sequences: Also referred to as pseudorandom sequences (PN), these are sequences that are deterministically generated and yet possess some properties that one would expect to find in randomly generated sequences.

Shift-register sequence: A sequence with symbols drawn from a field, which satisfies a linear-recurrence relation and which can be implemented using a shift register.

References

Baumert, L.D. 1971. *Cyclic Difference Sets*, Lecture Notes in Mathematics 182, Springer–Verlag, New York.

Golomb, S.W. 1982. *Shift Register Sequences*, Aegean Park Press, San Francisco, CA.

Hammons, A.R., Jr. and Kumar, P.V. 1993. On a recent 4-phase sequence design for CDMA. *IEICE Trans. Commun.* E76-B(8).

Helleseth, T. and Kumar, P.V. 1996. (planned). Sequences with low correlation. In *Handbook of Coding Theory*, ed. R. Brualdi, C. Huffman, and V. Pless, Elsevier Science Publishers, Amsterdam.

Jensen, J.M., Jensen, H.E., and Høholdt, T. 1991. The merit factor of binary sequences related to difference sets. *IEEE Trans. Inform. Theory.* IT-37(May):617–626.

MacWilliams, F.J. and Sloane, N.J.A. 1977. *The Theory of Error-Correcting Codes*, North-Holland, Amsterdam.

Mow, W.H. 1994. On McEliece's open problem on minimax aperiodic correlation. In *Proc. IEEE Intern. Symp. Inform. Theory*, p. 75.

Peterson, W.W. and Weldon, E.J., Jr. 1972. *Error-Correcting Codes*, 2nd ed. MIT Press, Cambridge, MA.

Sarwate, D.V. 1984. An upper bound on the aperiodic autocorrelation function for a maximal-length sequence. *IEEE Trans. Inform. Theory.* IT-30(July):685–687.

Sarwate, D.V. and Pursley, M.B. 1980. Crosscorrelation properties of pseudorandom and related sequences. *Proc. IEEE*, 68(May):593–619.

Simon, M.K., Omura, J.K., Scholtz, R.A., and Levitt, B.K. 1994. *Spread Spectrum Communications Handbook*, revised ed. McGraw Hill, New York.

Further Information

A more in-depth treatment of pseudonoise sequences, may be found in the following.

Golomb, S.W., 1982, *Shift Register Sequences*, Aegean Park Press, San Francisco.

Helleseth, T., and Kumar, P.V., 1996 (planned), "Sequences with Low Correlation," in *Handbook of Coding Theory*, edited by R. Brualdi, C. Huffman and V. Pless, Elsevier Science Publishers, Amsterdam.

Sarwate, D.V., and Pursley, M.B., 1980, "Crosscorrelation Properties of Pseudorandom and Related Sequences," *Proc. IEEE*, Vol. 68, May, pp. 593–619.

Simon, M.K., Omura, J.K., Scholtz, R.A., and Levitt, B.K., 1994, *Spread Spectrum Communications Handbook*, revised ed. McGraw Hill, New York.

9

Optimum Receivers

Geoffrey C. Orsak
George Mason University

9.1 Introduction

Every engineer strives for optimality in design. This is particularly true for communications engineers since in many cases implementing suboptimal receivers and sources can result in dramatic losses in performance. As such, this chapter focuses on design principles leading to the implementation of optimum receivers for the most common communication environments.

The main objective in digital communications is to transmit a sequence of bits to a remote location with the highest degree of accuracy. This is accomplished by first representing bits (or more generally short bit sequences) by distinct waveforms of finite time duration. These time-limited waveforms are then transmitted (broadcasted) to the remote sites in accordance with the data sequence.

Unfortunately, because of the nature of the **communication channel**, the remote location receives a corrupted version of the concatenated signal waveforms. The most widely accepted model for the communication channel is the so-called **additive white Gaussian noise[1] channel (AWGN channel)**. Mathematical arguments based upon the central limit theorem [Shiryayev, 1984], together with supporting empirical evidence, demonstrate that many common communication channels are accurately modeled by this abstraction. Moreover, from the design perspective, this is quite fortuitous since design and analysis with respect to this channel model is relatively straightforward.

---

[1]For those unfamiliar with AWGN, a random process (waveform) is formally said to be white Gaussian noise if all collections of instantaneous observations of the process are jointly Gaussian and mutually independent. An important consequence of this property is that the power spectral density of the process is a constant with respect to frequency variation (spectrally flat). For more on AWGN, see Papoulis, 1991.

9.2 Preliminaries

To better describe the digital communications process, we shall first elaborate on so-called binary communications. In this case, when the source wishes to transmit a bit value of 0, the transmitter broadcasts a specified waveform $s_0(t)$ over the **bit interval** $t \in [0, T]$. Conversely, if the source seeks to transmit the bit value of 1, the transmitter alternatively broadcasts the signal $s_1(t)$ over the same bit interval. The received waveform $R(t)$ corresponding to the first bit is then appropriately described by the following hypotheses testing problem:

$$
\begin{aligned}
H_0 &: R(t) = s_0(t) + \eta(t) \qquad 0 \le t \le T \\
H_1 &: R(t) = s_1(t) + \eta(t)
\end{aligned}
\tag{9.1}
$$

where, as stated previously, $\eta(t)$ corresponds to AWGN with spectral height nominally given by $N_0/2$. It is the objective of the receiver to determine the bit value, i.e., the most accurate hypothesis from the received waveform $R(t)$.

The optimality criterion of choice in digital communication applications is the **total probability of error** normally denoted as P_e. This scalar quantity is expressed as

$$
\begin{aligned}
P_e = \ &Pr(\text{declaring } 1 \mid 0 \text{ transmitted}) Pr(0 \text{ transmitted}) \\
&+ Pr(\text{declaring } 0 \mid 1 \text{ transmitted}) Pr(1 \text{ transmitted})
\end{aligned}
\tag{9.2}
$$

The problem of determining the optimal binary receiver with respect to the probability of error is solved by applying stochastic representation theory [Wong and Hajek, 1985] to detection theory [Poor, 1988; Van Trees, 1968]. The specific waveform representation of relevance in this application is the **Karhunen–Loève (KL) expansion**.

9.3 Karhunen–Loève Expansion

The Karhunen–Loève expansion is a generalization of the Fourier series designed to represent a random process in terms of deterministic basis functions and uncorrelated random variables derived from the process. Whereas the Fourier series allows one to model or represent deterministic time-limited energy signals in terms of linear combinations of complex exponential waveforms, the Karhunen–Loève expansion allows us to represent a second-order random process in terms of a set of **orthonormal** basis functions scaled by a sequence of random variables. The objective in this representation is to choose the basis of time functions so that the coefficients in the expansion are mutually uncorrelated random variables.

To be more precise, if $R(t)$ is a zero mean second-order random process defined over $[0, T]$ with covariance function $K_R(t, s)$, then so long as the basis of deterministic functions satisfy certain integral constraints [Van Trees, 1968], one may write $R(t)$ as

$$
R(t) = \sum_{i=1}^{\infty} R_i \phi_i(t) \qquad 0 \le t \le T
\tag{9.3}
$$

where

$$
R_i = \int_0^T R(t) \phi_i(t) \, dt
$$

In this case the R_i will be mutually uncorrelated random variables with the ϕ_i being deterministic basis functions that are complete in the space of square integrable time functions over $[0, T]$.

Importantly, in this case, equality is to be interpreted as **mean-square equivalence**, i.e.,

$$\lim_{N \to \infty} E\left[\left(R(t) - \sum_{i=1}^{N} R_i \phi_i(t) \right)^2 \right] = 0$$

for all $0 \le t \le T$.

Fact 9.1. If $R(t)$ is AWGN, then any basis of the vector space of square integrable signals over $[0, T]$ results in uncorrelated and therefore independent Gaussian random variables.

The use of fact 9.1 allows for a conversion of a continuous time detection problem into a finite-dimensional detection problem. Proceeding, to derive the optimal binary receiver, we first construct our set of basis functions as the set of functions defined over $t \in [0, T]$ beginning with the signals of interest $s_0(t)$ and $s_1(t)$. That is,

$\{s_0(t), s_1(t), \text{plus a countable number of functions which complete the basis}\}$

In order to insure that the basis is orthonormal, we must apply the Gramm–Schmidt procedure[2] [Proakis, 1989] to the full set of functions beginning with $s_0(t)$ and $s_1(t)$ to arrive at our final choice of basis $\{\phi_i(t)\}$.

Fact 9.2. Let $\{\phi_i(t)\}$ be the resultant set of basis functions.

Then for all $i > 2$, the $\phi_i(t)$ are orthogonal to $s_0(t)$ and $s_1(t)$. That is,

$$\int_0^T \phi_i(t) s_j(t)\, dt = 0$$

for all $i > 2$ and $j = 0, 1$.

Using this fact in conjunction with Eq. (9.3), one may recognize that only the coefficients R_1 and R_2 are functions of our signals of interest. Moreover, since the R_i are mutually independent, the optimal receiver will, therefore, only be a function of these two values.

Thus, through the application of the KL expansion, we arrive at an equivalent hypothesis testing problem to that given in Eq. (9.1),

$$H_0 : R = \begin{bmatrix} \int_0^T \phi_1(t) s_0(t)\, dt \\ \int_0^T \phi_2(t) s_0(t)\, dt \end{bmatrix} + \begin{bmatrix} \eta_1 \\ \eta_2 \end{bmatrix}$$

$$H_1 : R = \begin{bmatrix} \int_0^T \phi_1(t) s_1(t)\, dt \\ \int_0^T \phi_2(t) s_1(t)\, dt \end{bmatrix} + \begin{bmatrix} \eta_1 \\ \eta_2 \end{bmatrix} \tag{9.4}$$

where it is easily shown that η_1 and η_2 are mutually independent, zero-mean, Gaussian random variables with variance given by $N_0/2$, and where ϕ_1 and ϕ_2 are the first two functions from our orthonormal set of basis functions. Thus, the design of the optimal binary receiver reduces to a simple two-dimensional detection problem that is readily solved through the application of detection theory.

[2]The Gramm-Schmidt procedure is a deterministic algorithm that simply converts an arbitrary set of basis functions (vectors) into an equivalent set of orthonormal basis functions (vectors).

9.4 Detection Theory

It is well known from detection theory [Poor, 1988] that under the minimum P_e criterion, the optimal detector is given by the *maximum a posteriori rule (MAP)*,

$$\text{choose}_i \text{ largest } p_{H_i|R}(H_i \mid R = r) \tag{9.5}$$

i.e., determine the hypothesis that is most likely, given that our observation vector is r. By a simple application of Bayes theorem [Papoulis, 1991], we immediately arrive at the central result in detection theory: the optimal detector is given by the likelihood ratio test (LRT),

$$L(R) = \frac{p_{R|H_1}(R)}{p_{R|H_0}(R)} \underset{H_0}{\overset{H_1}{\underset{<}{>}}} \frac{\pi_0}{\pi_1} \tag{9.6}$$

where the π_i are the a priori probabilities of the hypotheses H_i being true. Since in this case we have assumed that the noise is white and Gaussian, the LRT can be written as

$$L(R) = \frac{\prod_1^2 \frac{1}{\sqrt{\pi N_0}} \exp\left(-\frac{1}{2}\frac{(R_i - s_{1,i})^2}{N_0/2}\right)}{\prod_1^2 \frac{1}{\sqrt{\pi N_0}} \exp\left(-\frac{1}{2}\frac{(R_i - s_{0,i})^2}{N_0/2}\right)} \underset{H_0}{\overset{H_1}{\underset{<}{>}}} \frac{\pi_0}{\pi_1} \tag{9.7}$$

where

$$s_{j,i} = \int_0^T \phi_i(t) s_j(t)\, dt$$

By taking the logarithm and cancelling common terms, it is easily shown that the optimum binary receiver can be written as

$$\frac{2}{N_0}\sum_1^2 R_i(s_{1,i} - s_{0,i}) - \frac{1}{N_0}\sum_1^2 (s_{1,i}^2 - s_{0,i}^2) \underset{H_0}{\overset{H_1}{\underset{<}{>}}} \ln\frac{\pi_0}{\pi_1} \tag{9.8}$$

This finite-dimensional version of the optimal receiver can be converted back into a continuous time receiver by the direct application of Parseval's theorem [Papoulis, 1991] where it is easily shown that

$$\begin{aligned} \sum_{i=1}^2 R_i s_{k,i} &= \int_0^T R(t) s_k(t)\, dt \\ \sum_{i=1}^2 s_{k,i}^2 &= \int_0^T s_k^2(t)\, dt \end{aligned} \tag{9.9}$$

By applying Eq. (9.9) to Eq. (9.8) the final receiver structure is then given by

$$\int_0^T R(t)[s_1(t) - s_0(t)]dt - \frac{1}{2}(E_1 - E_0) \underset{H_0}{\overset{H_1}{\underset{<}{>}}} \frac{N_0}{2}\ln\frac{\pi_0}{\pi_1} \tag{9.10}$$

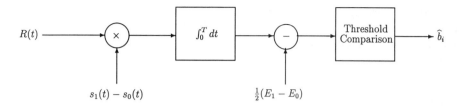

FIGURE 9.1   Optimal correlation receiver structure for binary communications.

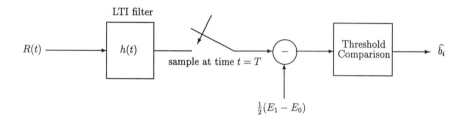

FIGURE 9.2   Optimal matched filter receiver structure for binary communications. In this case $h(t) = s_1(T - t) - s_0(t - t)$.

where E_1 and E_0 are the energies of signals $s_1(t)$ and $s_0(t)$, respectively. (See Fig. 9.1 for a block diagram.) Importantly, if the signals are equally likely ($\pi_0 = \pi_1$), the the optimal receiver is independent of the typically unknown spectral height of the background noise.

One can readily observe that the optimal binary communication receiver correlates the received waveform with the difference signal $s_1(t) - s_0(t)$ and then compares the statistic to a threshold. This operation can be interpreted as identifying the signal waveform $s_i(t)$ that best correlates with the received signal $R(t)$. Based on this interpretation, the receiver is often referred to as the **correlation receiver.**

As an alternate means of implementing the correlation receiver, we may reformulate the computation of the left-hand side of Eq. (9.10) in terms of standard concepts in filtering. Let $h(t)$ be the impulse response of a linear, time-invariant (LTI) system. By letting $h(t) = s_1(T - t) - s_0(T - t)$, then it is easily verified that the output of $R(t)$ to a LTI system with impulse response given by $h(t)$ and then sampled at time $t = T$ gives the desired result. (See Fig. 9.2 for a block diagram.) Since the impulse response is matched to the signal waveforms, this implementation is often referred to as the **matched filter receiver.**

9.5 Performance

Because of the nature of the statistics of the channel and the relative simplicity of the receiver, performance analysis of the optimal binary receiver in AWGN is a straightforward task. Since the conditional statistics of the log likelihood ratio are Gaussian random variables, the probability of error can be computed directly in terms of Marcum Q functions[3] as

$$P_e = Q\left(\frac{\|s_0 - s_1\|}{\sqrt{2N_0}}\right)$$

where the s_i are the two-dimensional signal vectors obtained from Eq. (9.4), and where $\|x\|$ denotes the Euclidean length of the vector x. Thus, $\|s_0 - s_1\|$ is best interpreted as the distance between

---

[3]The Q function is the probability that a standard normal random variable exceeds a specified constant, i.e., $Q(x) = \int_x^\infty 1/\sqrt{2\pi} \exp(-z^2/2)\, dz$.

the respective signal representations. Since the Q function is monotonically decreasing with an increasing argument, one may recognize that the probability of error for the optimal receiver decreases with an increasing separation between the signal representations, i.e., the more dissimilar the signals, the lower the P_e.

9.6 Signal Space

The concept of a **signal space** allows one to view the signal classification problem (receiver design) within a geometrical framework. This offers two primary benefits: first it supplies an often more intuitive perspective on the receiver characteristics (e.g., performance) and second it allows for a straightforward generalization to standard M-ary signalling schemes.

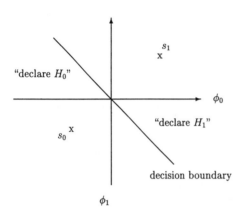

FIGURE 9.3   Signal space and decision boundary for optimal binary receiver.

To demonstrate this, in Fig. 9.3, we have plotted an arbitrary signal space for the binary signal classification problem. The axes are given in terms of the basis functions $\phi_1(t)$ and $\phi_2(t)$. Thus, every point in the signal space is a time function constructed as a linear combination of the two basis functions. By fact 9.2, we recall that both signals $s_0(t)$ and $s_1(t)$ can be constructed as a linear combination of $\phi_1(t)$ and $\phi_2(t)$ and as such we may identify these two signals in this figure as two points.

Since the decision statistic given in Eq. (9.8) is a linear function of the observed vector R which is also located in the signal space, it is easily shown that the set of vectors under which the receiver declares hypothesis H_i is bounded by a line in the signal space. This so-called **decision boundary** is obtained by solving the equation $\ln[L(R)] = 0$. (Here again we have assumed equally likely hypotheses.) In the case under current discussion, this decision boundary is simply the hyperplane separating the two signals in signal space. Because of the generality of this formulation, many problems in communication system design are best cast in terms of the signal space, that is, signal locations and decision boundaries.

9.7 Standard Binary Signalling Schemes

The framework just described allows us to readily analyze the most popular signalling schemes in binary communications: amplitude-shift keying (ASK), frequency-shift keying (FSK), and phase-shift keying (PSK). Each of these examples simply constitute a different selection for signals $s_0(t)$ and $s_1(t)$.

In the case of ASK, $s_0(t) = 0$, while $s_1(t) = \sqrt{2E/T} \sin(2\pi f_c t)$, where E denotes the energy of the waveform and f_c denotes the frequency of the carrier wave with $f_c T$ being an integer. Because $s_0(t)$ is the null signal, the signal space is a one-dimensional vector space with $\phi_1(t) = \sqrt{2/T} \sin(2\pi f_c t)$. This, in turn, implies that $\|s_0 - s_1\| = \sqrt{E}$. Thus, the corresponding probability of error for ASK is

$$P_e(\text{ASK}) = Q\left(\sqrt{\frac{E}{2N_0}}\right)$$

For FSK, the signals are given by equal amplitude sinusoids with distinct center frequencies, that is, $s_i(t) = \sqrt{2E/T} \sin(2\pi f_i t)$ with $f_i T$ being two distinct integers. In this case, it is easily verified that the signal space is a two-dimensional vector space with $\phi_i(t) = \sqrt{2/T} \sin(2\pi f_i t)$ resulting in

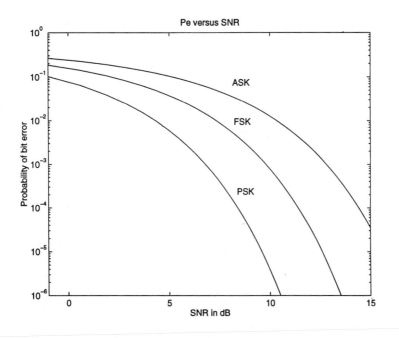

FIGURE 9.4 P_e vs the signal to noise ratio in decibels $[\text{dB} = 10\log(E/N_0)]$ for amplitude-shift keying, frequency-shift keying, and phase-shift keying; note that there is a 3-dB difference in performance from ASK to FSK to PSK.

$\| s_0 - s_1 \| = \sqrt{2E}$. The corresponding error rate is given to be

$$P_e(\text{FSK}) = Q\left(\sqrt{\frac{E}{N_0}}\right)$$

Finally, with regard to PSK signalling, the most frequently utilized binary PSK signal set is an example of an antipodal signal set. Specifically, the antipodal signal set results in the greatest separation between the signals in the signal space subject to an energy constraint on both signals. This, in turn, translates into the energy constrained signal set with the minimum P_e. In this case, the $s_i(t)$ are typically given by $\sqrt{2E/T}\sin[2\pi f_c t + \theta(i)]$, where $\theta(0) = 0$ and $\theta(1) = \pi$. As in the ASK case, this results in a one-dimensional signal space, however, in this case $\| s_0 - s_1 \| = 2\sqrt{E}$ resulting in probability of error given by

$$P_e(\text{PSK}) = Q\left(\sqrt{\frac{2E}{N_0}}\right)$$

In all three of the described cases, one can readily observe that the resulting performance is a function of only the signal-to-noise ratio E/N_0. In the more general case, the performance will be a function of the intersignal energy to noise ratio. To gauge the relative difference in performance of the three signalling schemes, in Fig. 9.4, we have plotted the P_e as a function of the SNR. Please note the large variation in performance between the three schemes for even moderate values of SNR.

9.8 *M*-ary Optimal Receivers

In binary signalling schemes, one seeks to transmit a single bit over the bit interval $[0, T]$. This is to be contrasted with M-ary signalling schemes where one transmits multiple bits simultaneously over the so-called symbol interval $[0, T]$. For example, using a signal set with 16 separate waveforms will allow one to transmit a length four-bit sequence per symbol (waveform). Examples of M-ary waveforms are quadrature phase-shift keying (QPSK) and quadrature amplitude modulation (QAM).

The derivation of the optimum receiver structure for M-ary signalling requires the straightforward application of fundamental results in detection theory. As with binary signalling, the Karhunen–Loève expansion is the mechanism utilized to convert a hypotheses testing problem based on continuous waveforms into a vector classification problem. Depending on the complexity of the M waveforms, the signal space can be as large as an M-dimensional vector space.

By extending results from the binary signalling case, it is easily shown that the optimum M-ary receiver computes

$$\xi_i[R(t)] = \int_0^T s_i(t)R(t)\,\mathrm{d}t - \frac{E_i}{2} + \frac{N_0}{2}\ln\pi_i \qquad i = 1, \dots, M$$

where, as before, the $s_i(t)$ constitute the signal set with the π_i being the corresponding a priori probabilities. After computing M separate values of ξ_i, the minimum probability of error receiver simply chooses the largest amongst this set. Thus, the M-ary receiver is implemented with a bank of correlation or matched filters followed by choose-largest decision logic.

In many cases of practical importance, the signal sets are selected so that the resulting signal space is a two-dimensional vector space irrespective of the number of signals. This simplifies the receiver structure in that the sufficient statistics are obtained by implementing only two matched filters. Both QPSK and QAM signal sets fit into this category. As an example, in Fig. 9.5, we have depicted

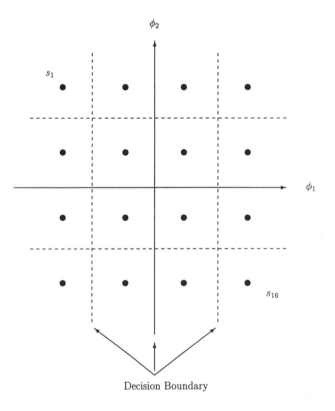

Decision Boundary

FIGURE 9.5 Signal space representation of 16-QAM signal set. Optimal decision regions for equally likely signals are also noted.

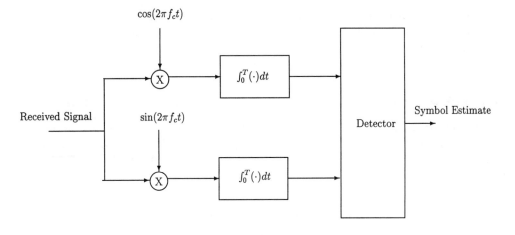

FIGURE 9.6 Optimum receiver structure for noncoherent (random or unknown phase) ASK demodulation.

the signal locations for standard 16-QAM signalling with the associated decision boundaries. In this case we have assumed an equally likely signal set. As can be seen, the optimal decision rule selects the signal representation that is closest to the received signal representation in this two-dimensional signal space.

9.9 More Realistic Channels

As is unfortunately often the case, many channels of practical interest are not accurately modeled as simply an AWGN channel. It is often that these channels impose nonlinear effects on the transmitted signals. The best example of this are channels that impose a random phase and random amplitude onto the signal. This typically occurs in applications such as in mobile communications, where one often experiences rapidly changing path lengths from source to receiver.

Fortunately, by the judicious choice of signal waveforms, it can be shown that the selection of the ϕ_i in the Karhunen–Loève transformation is often independent of these unwanted parameters. In these situations, the random amplitude serves only to scale the signals in signal space, whereas the random phase simply imposes a rotation on the signals in signal space.

Since the Karhunen–Loève basis functions typically do not depend on the unknown parameters, we may again convert the continuous time classification problem to a vector channel problem where the received vector R is computed as in Eq. (9.3). Since this vector is a function of both the unknown parameters (i.e., in this case amplitude A and phase ν), to obtain a likelihood ratio test independent of A and ν, we simply apply Bayes theorem to obtain the following form for the LRT:

$$L(R) = \frac{E\left[p_{R|H_1,A,\nu}(R \mid H_1, A, \nu)\right]}{E\left[p_{R|H_0,A,\nu}(R \mid H_0, A, \nu)\right]} \underset{H_0}{\overset{H_1}{\underset{<}{>}}} \frac{\pi_0}{\pi_1}$$

where the expectations are taken with respect to A and ν, and where $p_{R|H_i,A,\nu}$ are the conditional probability density functions of the signal representations. Assuming that the background noise is AWGN, it can be shown that the LRT simplifies to choosing the largest amongst

$$\xi_i[R(t)] = \pi_i \int_{A,\nu} \exp\left\{\frac{2}{N_0} \int_0^T R(t)s_i(t \mid A, \nu)\,dt - \frac{E_i(A,\nu)}{N_0}\right\} p_{A,\nu}(A,\nu)\,dA\,d\nu$$

$$i = 1, \ldots, M \quad (9.11)$$

It should be noted that in the Eq. (9.11) we have explicitly shown the dependence of the transmitted signals s_i on the parameters A and ν. The final receiver structures, together with their corresponding performance are, thus, a function of both the choice of signal sets and the probability density functions of the random amplitude and random phase.

Random Phase Channels

If we consider first the special case where the channel simply imposes a uniform random phase on the signal, then it can be easily shown that the so-called in-phase and quadrature statistics obtained from the received signal $R(t)$ (denoted by R_I and R_Q, respectively) are sufficient statistics for the signal classification problem. These quantities are computed as

$$R_I(i) = \int_0^T R(t) \cos[2\pi f_c(i)t] \, dt$$

and

$$R_Q(i) = \int_0^T R(t) \sin[2\pi f_c(i)t] \, dt$$

where in this case the index i corresponds to the center frequencies of hypotheses H_i, (e.g., FSK signalling). The optimum binary receiver selects the largest from amongst

$$\xi_i[R(t)] = \pi_i \exp\left(-\frac{E_i}{N_0}\right) I_0\left[\frac{2}{N_0}\sqrt{R_I^2(i) + R_Q^2(i)}\right] \qquad i = 1, \ldots, M$$

where I_0 is a zeroth-order, modified Bessel function of the first kind. If the signals have equal energy and are equally likely (e.g., FSK signalling), then the optimum receiver is given by

$$R_I^2(1) + R_Q^2(1) \underset{H_0}{\overset{H_1}{\gtrless}} R_I^2(0) + R_Q^2(0)$$

One may readily observe that the optimum receiver bases its decision on the values of the two envelopes of the received signal $\sqrt{R_I^2(i) + R_Q^2(i)}$ and, as a consequence, is often referred to as an envelope or square-law detector. Moreover, it should be observed that the computation of the envelope is independent of the underlying phase of the signal and is as such known as a noncoherent receiver.

The computation of the error rate for this detector is a relatively straightforward exercise resulting in

$$P_e(\text{noncoherent}) = \frac{1}{2} \exp\left(-\frac{E}{2N_0}\right)$$

As before, note that the error rate for the noncoherent receiver is simply a function of the SNR.

Rayleigh Channel

As an important generalization of the described random phase channel, many communication systems are designed under the assumption that the channel introduces both a random amplitude and a random phase on the signal. Specifically, if the original signal sets are of the form $s_i(t) = m_i(t)\cos(2\pi f_c t)$ where $m_i(t)$ is the baseband version of the message (i.e., what distinguishes one signal from another), then the so-called **Rayleigh channel** introduces random distortion in the received signal of the following form:

$$s_i(t) = A m_i(t) \cos(2\pi f_c t + \nu)$$

where the amplitude A is a Rayleigh random variable[4] and where the random phase ν is a uniformly distributed between zero and 2π.

To determine the optimal receiver under this distortion, we must first construct an alternate statistical model for $s_i(t)$. To begin, it can be shown from the theory of random variables [Papoulis, 1991] that if X_I and X_Q are statistically independent, zero mean, Gaussian random variables with variance given by σ^2, then

$$A m_i(t) \cos(2\pi f_c t + \nu) = m_i(t) X_I \cos(2\pi f_c t) + m_i(t) X_Q \sin(2\pi f_c t)$$

Equality here is to be interpreted as implying that both A and ν will be the appropriate random variables. From this, we deduce that the combined uncertainty in the amplitude and phase of the signal is incorporated into the Gaussian random variables X_I and X_Q. The in-phase and quadrature components of the signal $s_i(t)$ are given by $s_{Ii}(t) = m_i(t)\cos(2\pi f_c t)$ and $s_{Qi}(t) = m_i(t)\sin(2\pi f_c t)$, respectively. By appealing to Eq. (9.11), it can be shown that the optimum receiver selects the largest from

$$\xi_i[R(t)] = \frac{\pi_i}{1 + \dfrac{2E_i}{N_0}\sigma^2} \exp\left[\frac{\sigma^2}{\dfrac{1}{2} + \dfrac{E_i}{N_0}\sigma^2} \left(\langle R(t), s_{Ii}(t) \rangle^2 + \langle R(t), s_{Qi}(t) \rangle^2 \right) \right]$$

where the inner product

$$\langle R(t), S_i(t) \rangle = \int_0^T R(t) s_i(t)\, dt$$

Further, if we impose the conditions that the signals be equally likely with equal energy over the symbol interval, then optimum receiver selects the largest amongst

$$\xi_i[R(t)] = \sqrt{\langle R(t), s_{Ii}(t) \rangle^2 + \langle R(t), s_{Qi}(t) \rangle^2}$$

Thus, much like for the random phase channel, the optimum receiver for the Rayleigh channel computes the projection of the received waveform onto the in-phase and quadrature components of the hypothetical signals. From a signal space perspective, this is akin to computing the length of the received vector in the subspace spanned by the hypothetical signal. The optimum receiver then chooses the largest amongst these lengths.

---

[4]The density of a Rayleigh random variable is given by $p_A(a) = a/\sigma^2 \exp(-a^2/2\sigma^2)$ for $a \geq 0$.

As with the random phase channel, computing the performance is a straightforward task resulting in (for the equally likely, equal energy case)

$$P_e(\text{Rayleigh}) = \frac{\dfrac{1}{2}}{\left(1 + \dfrac{E\sigma^2}{N_0}\right)}$$

Interestingly, in this case the performance depends not only on the SNR, but also on the variance (spread) of the Rayleigh amplitude A. Thus, if the amplitude spread is large, we expect to often experience what is known as deep fades in the amplitude of the received waveform and as such expect a commensurate loss in performance.

9.10 Dispersive Channels

The **dispersive channel** model assumes that the channel not only introduces AWGN but also distorts the signal through a filtering process. This model incorporates physical realities such as multipath effects and frequency selective fading. In particular, the standard model adopted is depicted in the block diagram given in Fig. 9.7. As can be seen, the receiver observes a filtered version of the signal plus AWGN. If the impulse response of the channel is known, then we arrive at the optimum receiver design by applying the previously presented theory. Unfortunately, the duration of the filtered signal can be a complicating factor. More often than not, the channel will increase the duration of the transmitted signals, hence leading to the description, dispersive channel.

However, if the designers take this into account by shortening the duration of $s_i(t)$ so that the duration of $s_i^*(t)$ is less than T, then the optimum receiver chooses the largest amongst

$$\xi_i(R(t)) = \frac{N_0}{2} \ln \pi_i + \langle R(t), s_i^*(t) \rangle - \frac{1}{2} E_i^*$$

If we limit our consideration to equally likely binary signal sets, then the minimum P_e matches the received waveform to the filtered versions of the signal waveforms. The resulting error rate is given by

$$P_e(\text{dispersive}) = Q\left(\frac{\|s_0^* - s_0^*\|}{\sqrt{2N_0}}\right)$$

Thus, in this case the minimum P_e is a function of the separation of the filtered version of the signals in the signal space.

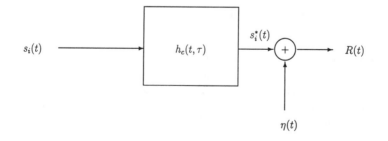

FIGURE 9.7   Standard model for dispersive channel. The time varying impulse response of the channel is denoted by $h_c(t, \tau)$.

The problem becomes substantially more complex if we cannot insure that the filtered signal durations are less than the symbol lengths. In this case we experience what is known as **intersymbol interference (ISI)**. That is, observations over one symbol interval contain not only the symbol information of interest but also information from previous symbols. In this case we must appeal to optimum sequence estimation [Poor, 1988] to take full advantage of the information in the waveform. The basis for this procedure is the maximization of the joint likelihood function conditioned on the sequence of symbols. This procedure not only defines the structure of the optimum receiver under ISI but also is critical in the decoding of convolutional codes and coded modulation. Alternate adaptive techniques to solve this problem involve the use of channel equalization.

Defining Terms

Communication channel: The medium over which communication signals are transmitted. Examples are fiber optic cables, free space, or telephone lines.

Additive white Gaussian noise (AWGN) channel: The channel whose model is that of corrupting a transmitted waveform by the addition of white (i.e., spectrally flat) Gaussian noise.

Bit (symbol) interval: The period of time over which a single symbol is transmitted.

Total probability of error: The probability of classifying the received waveform into any of the symbols that were not transmitted over a particular bit interval.

Karhunen–Loève expansion: A representation for second-order random processes. Allows one to express a random process in terms of a superposition of deterministic waveforms. The scale values are uncorrelated random variables obtained from the waveform.

Orthonormal: The property of two or more vectors or time-limited waveforms being mutually orthogonal and individually having unit length. Orthogonality and length are typically measured by the standard Euclidean inner product.

Mean-square equivalence: Two random vectors or time-limited waveforms are mean-square equivalent if and only if the expected value of their mean-square error is zero.

Correlation or matched filter receiver: The optimal receiver structure for digital communications in AWGN.

Signal space: An abstraction for representing a time limited waveform in a low-dimensional vector space. Usually arrived at through the application of the Karhunen–Loève transformation.

Decision boundary: The boundary in signal space between the various regions where the receiver declares H_i. Typically a hyperplane when dealing with AWGN channels.

Rayleigh channel: A channel that randomly scales the transmitted waveform by a Rayleigh random variable while adding an independent uniform phase to the carrier.

Dispersive channel: A channel that elongates and distorts the transmitted signal. Normally modeled as a time-varying linear system.

Intersymbol interference: The ill-effect of one symbol smearing into adjacent symbols thus interfering with the detection process. This is a consequence of the channel filtering the transmitted signals and therefore elongating their duration, see dispersive channel.

References

Gibson, J.D. 1993. *Principles of Digital and Analog Communications*, 2nd ed. MacMillan, New York.

Haykin, S. 1994. *Communication Systems*, 3rd ed. Wiley, New York.

Lee, E.A. and Messerschmitt, D.G. 1988. *Digital Communication*, Kluwer Academic Publishers, Norwell, MA.

Papoulis, A. 1991. *Probability, Random Variables, and Stochastic Processes*, 3rd ed. McGraw-Hill, New York.

Poor, H.V. 1988. *An Introduction to Signal Detection and Estimation*, Springer–Verlag, New York.

Proakis, J.G. 1989. *Digital Communications*, 2nd ed. McGraw-Hill, New York.

Shiryayev, A.N. 1984. *Probability*, Springer–Verlag, New York.

Sklar, B. 1988. *Digital Communications, Fundamentals and Applications*, Prentice Hall, Englewood Cliffs, NJ.

Van Trees, H.L. 1968. *Detection, Estimation, and Modulation Theory, Part I*, Wiley, New York.

Wong, E. and Hajek, B. 1985. *Stochastic Processes in Engineering Systems*, Springer–Verlag, New York.

Wozencraft, J.M. and Jacobs, I. 1990. *Principles of Communication Engineering*, reissue, Waveland Press, Inc, Prospect Heights, Illinois.

Ziemer, R.E. and Peterson, R.L. 1992. *Introduction to Digital Communication*, Macmillan, New York.

Further Information

Further Information The fundamentals of receiver design were put in place by Wozencraft and Jacobs in their seminal book. Since that time, there have been many outstanding textbooks in this area. For a sampling see Haykin, 1994; Sklar, 1988; Lee and Messerschmitt, 1988; Gibson, 1993; and Ziemer and Peterson, 1992. For a complete treatment on the use and application of detection theory in communications see van Trees, 1968 and Poor, 1988. For deeper insights into the Karhunen–Loève expansion and its use in communications and signal processing see Wong and Hajek, 1985.

10

Forward Error Correction Coding

V.K. Bhargava
University of Victoria

I.J. Fair
University of Victoria

10.1 Introduction

In 1948, Claude Shannon issued a challenge to communications engineers by proving that communication systems could be made arbitrarily reliable as long as a fixed percentage of the transmitted signal was redundant [Shannon, 1948]. He did not, however, indicate how this could be achieved. Subsequent research has led to a number of techniques that introduce redundancy to allow for correction of errors without retransmission. These techniques, collectively known as forward error correction (FEC) coding techniques, are used in systems where a reverse channel is not available for requesting retransmission, the delay with retransmission would be excessive, the expected number of errors would require a large number of retransmissions, or retransmission would be awkward to implement [Sklar, 1988].

A simplified model of a digital communication system which incorporates FEC coding is shown in Fig. 10.1. The FEC code acts on a **discrete data channel** comprising all system elements between the encoder output and decoder input. The encoder maps the source data to q-ary code symbols which are modulated and transmitted. During transmission, this signal can be corrupted, causing errors to arise in the demodulated symbol sequence. The FEC decoder attempts to correct these errors and restore the original source data.

A demodulator which outputs only a value for the q-ary symbol received during each symbol interval is said to make **hard decisions**. In the **binary symmetric channel** (BSC), hard decisions are made on binary symbols and the probability of error is independent of the value of the symbol. One example of a BSC is the coherently demodulated binary phase-shift-keyed (BPSK) signal corrupted by additive white Gaussian noise (AWGN). The conditional probability density functions which

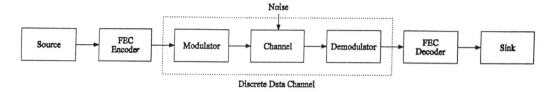

FIGURE 10.1   Block diagram of a digital communication system with forward error correction.

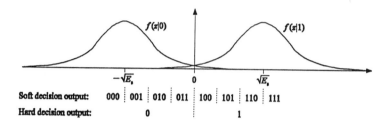

FIGURE 10.2   Hard and soft decision demodulation of a coherently demodulated BPSK signal corrupted by AWGN. $f(z \mid 1)$ and $f(z \mid 0)$ are the Gaussianly distributed conditional probability density functions at the threshold device.

result with this system are depicted in Fig. 10.2. The probability of error is given by the area under the density functions that lies across the decision threshold, and is a function of the symbol energy E_s and the one-sided noise power spectral density N_0.

Alternatively, the demodulator can make **soft decisions** or output an estimate of the symbol value along with an indication of its confidence in this estimate. For example, if the BPSK demodulator uses three-bit quantization, the two least significant bits can be taken as a confidence measure. Possible soft-decision thresholds for the BPSK signal are depicted in Fig. 10.2. In practice, there is little to be gained by using many soft-decision quantization levels.

Block and convolutional codes introduce redundancy by adding parity symbols to the message data. They map k source symbols to n code symbols and are said to have **code rate** $R = k/n$. With fixed information rates, this redundancy results in increased bandwidth and lower energy per transmitted symbol. At low signal-to-noise ratios, these codes cannot compensate for these impairments, and performance is degraded. At higher ratios of information symbol energy E_b to noise spectral density N_0, however, there is **coding gain** since the performance improvement offered by coding more than compensates for these impairments. Coding gain is usually defined as the reduction in required E_b/N_0 to achieve a specific error rate in an error-control coded system over one without coding. In contrast to block and convolutional codes, trellis-coded modulation introduces redundancy by expanding the size of the signal set rather than increasing the number of symbols transmitted, and so offers the advantages of coding to band-limited systems.

Each of these coding techniques is considered in turn. Following a discussion of interleaving and concatenated coding, this chapter concludes with a brief overview of FEC applications.

10.2 Fundamentals of Block Coding

In block codes there is a one-to-one mapping between k-symbol source words and n-symbol codewords. With q-ary signalling, q^k out of the q^n possible n-tuples are valid code vectors. The set of all n-tuples forms a **vector space** in which the q^k code vectors are distributed. The **Hamming**

distance between any two code vectors is the number of symbols in which they differ; the **minimum distance** d_{min} of the code is the smallest Hamming distance between any two codewords.

There are two contradictory objectives of block codes. The first is to distribute the code vectors in the vector space such that the distance between them is maximized. Then, if the decoder receives a corrupted vector, by evaluating the nearest valid code vector it will decode the correct word with high probability. The second is to pack the vector space with as many code vectors as possible to reduce the redundancy in transmission.

When code vectors differ in at least d_{min} positions, a decoder which evaluates the nearest code vector to each received word is guaranteed to correct up to t random symbol errors per word if

$$d_{min} \geq 2t + 1 \tag{10.1}$$

Alternatively, all $q^n - q^k$ illegal words can be detected, including all error patterns with $d_{min} - 1$ or fewer errors. In general, a block code can correct all patterns of t or fewer errors and detect all patterns of u or fewer errors provided that $u \geq t$ and

$$d_{min} \geq t + u + 1 \tag{10.2}$$

If $q = 2$, knowledge of the positions of the errors is sufficient for their correction; if $q > 2$, the decoder must determine both the positions and values of the errors. If the demodulator indicates positions in which the symbol values are unreliable, the decoder can assume their value unknown and has only to solve for the value of these symbols. These positions are called **erasures**. A block code can correct up to t errors and v erasures in each word if

$$d_{min} \geq 2t + v + 1 \tag{10.3}$$

10.3 Structure and Decoding of Block Codes

Shannon showed that the performance limit of codes with fixed code rate improves as the block length increases. As n and k increase, however, practical implementation requires that the mapping from message to code vector not be arbitrary but that an underlying structure to the code exist. The structures developed to date limit the error correcting capability of these codes to below what Shannon proved possible, on average, for a code with random codeword assignments. The search for good constructive codes continues.

A property which simplifies implementation of the coding operations is that of code linearity. A code is **linear** if the addition of any two code vectors forms another code vector, which implies that the code vectors form a subspace of the vector space of n-tuples. This subspace, which contains the all-zero vector, is spanned by any set of k linearly independent code vectors. Encoding can be described as the multiplication of the information k-tuple by a **generator matrix** G, of dimension $k \times n$, which contains these basis vectors as rows. That is, a message vector m_i is mapped to a code vector c_i according to

$$c_i = m_i\, G, \qquad i = 0, 1, \ldots, q^k - 1 \tag{10.4}$$

where elementwise arithmetic is defined in the **finite field** GF(q). In general, this encoding procedure results in code vectors with nonsystematic form in that the values of the message symbols cannot be determined by inspection of the code vector. However, if G has the form $[I_k, P]$ where I_k is the $k \times k$ identity matrix and P is a $k \times (n - k)$ matrix of parity checks, then the k most significant symbols of each code vector are identical to the message vector and the code has **systematic** form. This notation assumes that vectors are written with their most significant or first symbols in time on the left, a convention used throughout this chapter.

For each generator matrix there is an $(n-k) \times k$ **parity check matrix** H whose rows are orthogonal to the rows in G, i.e., $GH^T = 0$. If the code is systematic, $H = [P^T, I_{n-k}]$. Since all codewords are linear sums of the rows in G, it follows that $c_i H^T = 0$ for all $i, i = 0, 1, \ldots, q^k - 1$, and that the validity of the demodulated vectors can be checked by performing this multiplication. If a codeword c is corrupted during transmission so that the hard-decision demodulator outputs the vector $\hat{c} = c + e$, where e is a nonzero error pattern, the result of this multiplication is an $(n-k)$-tuple that is indicative of the validity of the sequence. This result, called the **syndrome** s, is dependent only on the error pattern since

$$s = \hat{c}H^T = (c + e)H^T = cH^T + eH^T = eH^T \tag{10.5}$$

If the error pattern is a code vector, the errors go undetected. For all other error patterns, however, the syndrome is nonzero. Since there are $q^{n-k} - 1$ nonzero syndromes, $q^{n-k} - 1$ error patterns can be corrected. When these patterns include all those with t or fewer errors and no others, the code is said to be a **perfect code**. Few codes are perfect; most codes are capable of correcting some patterns with more than t errors. **Standard array decoders** use lookup tables to associate each syndrome with an error pattern but become impractical as the block length and number of parity symbols increases. Algebraic decoding algorithms have been developed for codes with stronger structure. These algorithms are simplified with imperfect codes if the patterns corrected are limited to those with t or fewer errors, a simplification called **bounded distance decoding**.

Cyclic codes are a subclass of linear block codes with an algebraic structure that enables encoding to be implemented with a linear feedback shift register and decoding to be implemented without a lookup table. As a result, most block codes in use today are cyclic or are closely related to cyclic codes. These codes are best described if vectors are interpreted as polynomials and the arithmetic follows the rules for polynomials where the elementwise operations are defined in $GF(q)$. In a cyclic code, all codeword polynomials are multiples of a **generator polynomial** $g(x)$ of degree $n - k$. This polynomial is chosen to be a divisor of $x^n - 1$ so that a cyclic shift of a code vector yields another code vector, giving this class of codes its name. A message polynomial $m_i(x)$ can be mapped to a codeword polynomial $c_i(x)$ in nonsystematic form as

$$c_i(x) = m_i(x)g(x), \qquad i = 0, 1, \ldots, q^k - 1 \tag{10.6}$$

In systematic form, codeword polynomials have the form

$$c_i(x) = m_i(x)x^{n-k} - r_i(x), \qquad i = 0, 1, \ldots, q^k - 1 \tag{10.7}$$

where $r_i(x)$ is the remainder of $m_i(x)x^{n-k}$ divided by $g(x)$. Polynomial multiplication and division can be easily implemented with shift registers [Blahut, 1983].

The first step in decoding the demodulated word is to determine if the word is a multiple of $g(x)$. This is done by dividing it by $g(x)$ and examining the remainder. Since polynomial division is a linear operation, the resulting syndrome $s(x)$ depends only on the error pattern. If $s(x)$ is the all-zero polynomial, transmission is errorless or an undetectable error pattern has occurred. If $s(x)$ is nonzero, at least one error has occurred. This is the principle of the **cyclic redundancy check** (CRC). It remains to determine the most likely error pattern that could have generated this syndrome.

Single error correcting binary codes can use the syndrome to immediately locate the bit in error. More powerful codes use this information to determine the locations and values of multiple errors. The most prominent approach of doing so is with the iterative technique developed by Berlekamp. This technique, which involves computing an error-locator polynomial and solving for its roots, was subsequently interpreted by Massey in terms of the design of a minimum-length shift register. Once the location and values of the errors are known, Chien's search algorithm efficiently corrects them. The implementation complexity of these decoders increases only as the square of the number of errors to be corrected [Bhargava, 1983] but does not generalize easily to accomodate soft-decision information. Other decoding techniques, including Chase's algorithm and threshold decoding, are easier to implement with soft-decision input [Clark and Cain, 1981]. Berlekamp's algorithm can be used in conjunction with transform-domain decoding, which involves transforming the received block with a finite field Fourier-like transform and solving for errors in the transform domain. Since the implementation complexity of these decoders depends on the block length rather than the number of symbols corrected, this approach results in simpler circuitry for codes with high redundancy [Wu et al., 1987].

Other block codes have also been constructed, including product codes that extend the ideas to two dimensions, codes that are based on transform-domain spectral properties, codes that are designed specifically for correction of burst errors, and codes that are decodable with straightforward threshold or majority logic decoders [Blahut, 1983; Clark and Cain, 1981; Lin and Costello, 1983].

10.4 Important Classes of Block Codes

When errors occur independently, Bose–Chaudhuri–Hocquenghem (BCH) codes provide one of the best performances of known codes for a given block length and code rate. They are cyclic codes with $n = q^m - 1$, where m is any integer greater than 2. They are designed to correct up to t errors per word and so have **designed distance** $d = 2t + 1$; the minimum distance may be greater. Generator polynomials for these codes are listed in many texts, including [Clark and Cain, 1981]. These polynomials are of degree less than or equal to mt, and so $k \geq n - mt$. BCH codes can be shortened to accomodate system requirements by deleting positions for information symbols.

Some subclasses of these codes are of special interest. Hamming codes are perfect single error correcting binary BCH codes. Full length codes have $n = 2^m - 1$ and $k = n - m$ for any m greater than 2. The duals of these codes are maximal-length codes, with $n = 2^m - 1, k = m$, and $d_{min} = 2^{m-1}$. All $2^m - 1$ nonzero code vectors in these codes are cyclic shifts of a single nonzero code vector. Reed–Solomon (RS) codes are nonbinary BCH codes defined over GF(q), where q is often taken as a power of two so that symbols can be represented by a sequence of bits. In these cases, correction of even a single symbol allows for correction of a burst of bit errors. The block length is $n = q - 1$, and the minimum distance $d_{min} = 2t + 1$ is achieved using only $2t$ parity symbols. Since RS codes meet the Singleton bound of $d_{min} \leq n - k + 1$, they have the largest possible minimum distance for these values of n and k and are called **maximum distance separable** codes.

The Golay codes are the only nontrivial perfect codes that can correct more than one error. The (11, 6) ternary Golay code has minimum distance 5. The (23, 12) binary code is a triple error correcting BCH code with $d_{min} = 7$. To simplify implementation, it is often extended to a (24, 12) code through the addition of an extra parity bit. The extended code has $d_{min} = 8$.

The (23, 12) Golay code is also a binary quadratic residue code. These cyclic codes have prime length of the form $n = 8m \pm 1$, with $k = (n + 1)/2$ and $d_{min} \geq \sqrt{n}$. Some of these codes are as good as the best codes known with these values of n and k, but it is unknown if there are good quadratic residue codes with large n [Blahut, 1983].

Reed–Muller codes are equivalent to binary cyclic codes with an additional overall parity bit. For any m, the rth-order Reed-Muller code has $n = 2^m, k = \sum_{i=0}^{r} \binom{m}{i}$, and $d_{min} = 2^{m-r}$. The rth-order and $(m - r - 1)$th-order codes are duals, and the first-order codes are similar to maximal-length codes. These codes, and the closely related Euclidean geometry and projective geometry codes, can be decoded with threshold decoding.

The performance of several of these block codes is shown in Fig. 10.3 in terms of decoded bit error probability vs E_b/N_0 for systems using coherent, hard-decision demodulated BPSK signalling. Many other block codes have also been developed, including Goppa codes, quasicyclic codes, burst error correcting Fire codes, and other lesser known codes.

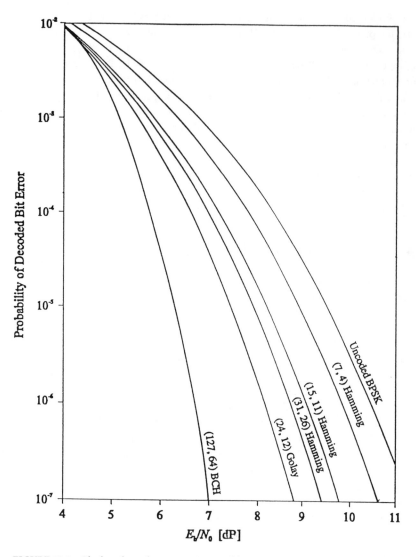

FIGURE 10.3   Block code performance. *Source*: Sklar, B., 1988, *Digital Communications: Fundamentals and Applications*, © 1988, p. 300. Reprinted by permission of Prentice-Hall, Inc., Englewood Cliffs, NJ.

10.5 Principles of Convolutional Coding

Convolutional codes map successive information k-tuples to a series of n-tuples such that the sequence of n-tuples has distance properties that allow for detection and correction of errors. Although these codes can be defined over any alphabet, their implementation has largely been restricted to binary signals, and only binary convolutional codes are considered here.

In addition to the code rate $R = k/n$, the **constraint length** K is an important parameter for these codes. Definitions vary; we will use the definition that K equals the number of k-tuples that affect formation of each n-tuple during encoding. That is, the value of an n-tuple depends on the k-tuple that arrives at the encoder during that encoding interval as well as the $K - 1$ previous information k-tuples.

Binary convolutional encoders can be implemented with kK-stage shift registers and n modulo-2 adders, an example of which is given in Fig. 10.4(a) for a rate 1/2, constraint length 3 code. The

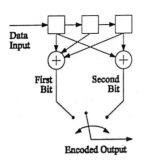

(a) Connection diagram

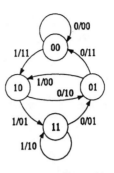

(b) State machine model

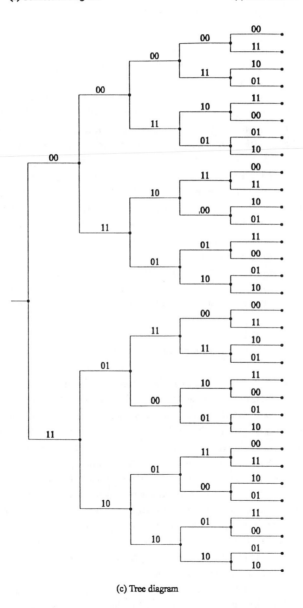

(c) Tree diagram

FIGURE 10.4 A rate 1/2, constraint length 3 convolutional code.

(Continues)

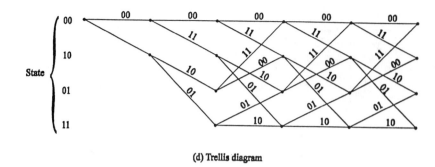

(d) Trellis diagram

FIGURE 10.4   (*Continued*)

encoder shifts in a new k-tuple during each encoding interval and samples the outputs of the adders sequentially to form the coded output.

Although connection diagrams similar to that of Fig. 10.4(a) completely describe the code, a more concise description can be given by stating the values of n, k, and K and giving the adder connections in the form of vectors or polynomials. For instance, the rate 1/2 code has the generator vectors $\mathbf{g}_1 = 111$ and $\mathbf{g}_2 = 101$, or equivalently, the generator polynomials $g_1(x) = x^2 + x + 1$ and $g_2(x) = x^2 + 1$. Alternatively, a convolutional code can be characterized by its impulse response, the coded sequence generated due to input of a single logic-1. It is straightforward to verify that the circuit in Fig. 10.4(a) has the impulse response 111011. Since modulo-2 addition is a linear operation, convolutional codes are linear, and the coded output can be viewed as the convolution of the input sequence with the impulse response, hence the name of this coding technique. Shifted versions of the impulse response or generator vectors can be combined to form an infinite-order generator matrix which also describes the code.

Shift register circuits can be modeled as finite state machines. A Mealy machine description of a convolutional encoder requires $2^{k(K-1)}$ states, each describing a different value of the $K-1$ k-tuples which have most recently entered the shift register. Each state has 2^k exit paths which correspond to the value of the incoming k-tuple. A state machine description for the rate 1/2 encoder depicted in Fig. 10.4(a) is given in Fig. 10.4(b). States are labeled with the contents of the two leftmost register stages; edges are labeled with information bit values and their corresponding coded output.

The dimension of time is added to the description of the encoder with tree and trellis diagrams. The tree diagram for the rate 1/2 convolutional code is given in Fig. 10.4(c), assuming the shift register is initially clear. Each node represents an encoding interval, from which the upper branch is taken if the input bit is a 0 and the lower branch is taken if the input bit is a 1. Each branch is labeled with the corresponding output bit sequence. A drawback of the tree representation is that it grows without bound as the length of the input sequence increases. This is overcome with the trellis diagram depicted in Fig. 10.4(d). Again, encoding results in left-to-right movement, where the upper of the two branches is taken whenever the input is a 0, the lower branch is taken when the input is a 1, and the output is the bit sequence which weights the branch taken. Each level of nodes corresponds to a state of the encoder as shown on the left-hand side of the diagram.

If the received sequence contains errors, it may no longer depict a valid path through the tree or trellis. It is the job of the decoder to determine the original path. In doing so, the decoder does not so much correct errors as find the closest valid path to the received sequence. As a result, the error correcting capability of a convolutional code is more difficult to quantify than that of a block code; it depends on how valid paths differ. One measure of this difference is the **column distance** $d_c(i)$, the minimum Hamming distance between all coded sequences generated over i encoding intervals which differ in the first interval. The nondecreasing sequence of column distance values is the **distance profile** of the code. The column distance after K intervals is the minimum distance of the code and is important for evaluating the performance of a code that uses threshold decoding. As i increases, $d_c(i)$ approaches the **free distance** of the code, d_{free}, which is the minimum Hamming distance in the set of arbitrarily long paths that diverge and then remerge in the trellis.

With maximum likelihood decoding, convolutional codes can generally correct up to t errors within three to five constraint lengths, depending on how the errors are distributed, where

$$d_{\text{free}} \geq 2t + 1 \tag{10.8}$$

The free distance can be calculated by exhaustively searching for the minimum-weight path that returns to the all-zero state, or evaluating the term of lowest degree in the generating function of the code.

The objective of a convolutional code is to maximize these distance properties. They generally improve as the constraint length of the code increases, and nonsystematic codes generally have better properties than systematic ones. Good codes have been found by computer search and are tabulated in many texts, including [Clark and Cain, 1981]. Convolutional codes with high code rate can be constructed by **puncturing** or periodically deleting coded symbols from a low rate code. A list of low rate codes and perforation matrices that result in good high rate codes can be found in many sources, including Wu et al., 1987. The performance of good punctured codes approaches that of the best convolutional codes known with similar rate, and decoder implementation is significantly less complex.

Convolutional codes can be **catastrophic**, having the potential to generate an unlimited number of decoded bit errors in response to a finite number of errors in the demodulated bit sequence. Catastrophic error propagation is avoided if the code has generator polynomials with a greatest common divisor of the form x^a for any a or, equivalently, if there are no closed-loop paths in the state diagram with all-zero output other than the one taken with all-zero input. Systematic codes are not catastrophic.

10.6 Decoding of Convolutional Codes

In 1967, Viterbi developed a maximum likelihood decoding algorithm that takes advantage of the trellis structure to reduce the complexity of the evaluation. This algorithm has become known as the **Viterbi algorithm**. With each received n-tuple, the decoder computes a **metric** or measure of likelihood for all paths that could have been taken during that interval and discards all but the most likely to terminate on each node. An arbitrary decision is made if path metrics are equal. The metrics can be formed using either hard or soft decision information with little difference in implementation complexity.

If the message has finite length and the encoder is subsequently flushed with zeros, a single decoded path remains. With a BSC, this path corresponds to the valid code sequence with minimum Hamming distance from the demodulated sequence. Full-length decoding becomes impractical as the length of the message sequence increases. The most likely paths tend to have a common stem, however, and selecting the trace value four or five times the constraint length prior to the present decoding depth results in near-optimum performance. Since the number of paths examined during each interval increases exponentially with the constraint length, the Viterbi algorithm also becomes impractical for codes with large constraint length. To date, Viterbi decoding has been implemented for codes with constraint lengths up to ten. Other decoding techniques, such as sequential and threshold decoding, can be used with larger constraint lengths.

Sequential decoding was proposed by Wozencraft, and the most widely used algorithm was developed by Fano. Rather than tracking multiple paths through the trellis, the sequential decoder operates on a single path while searching the code tree for a path with high probability. It makes tentative decisions regarding the transmitted sequence, computes a metric between its proposed path and the demodulated sequence, and moves forward through the tree as long as the metric indicates

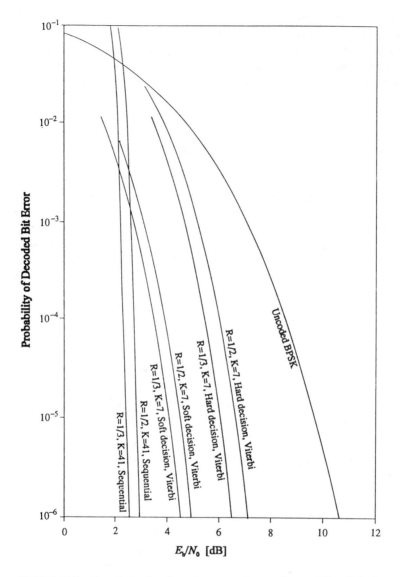

FIGURE 10.5 Convolutional code performance. *Source*: Omura, J.K. and Levitt, B.K., © 1982 IEEE, "Coded Error Probability Evaluation for Antijam Communication Systems," *IEEE Trans. Commun.*, vol. COM-30, no. 5, pp. 896–903. Reprinted by permission of IEEE.

that the path is likely. If the likelihood of the path becomes low, the decoder moves backward, searching other paths until it finds one with high probability. The number of computations involved in this procedure is almost independent of the constraint length and is typically quite small, but it can be highly variable, depending on the channel. Buffers must be provided to store incoming sequences as the decoder searches the tree. Their overflow is a significant limiting factor in the performance of these decoders.

Figure 10.5 compares the performance of the Viterbi and sequential decoding algorithms for several convolutional codes operating on coherently demodulated BPSK signals corrupted by AWGN. Other decoding algorithms have also been developed, including syndrome decoding methods such as table look-up feedback decoding and threshold decoding [Clark and Cain, 1981]. These algorithms are easily implemented but offer suboptimal performance.

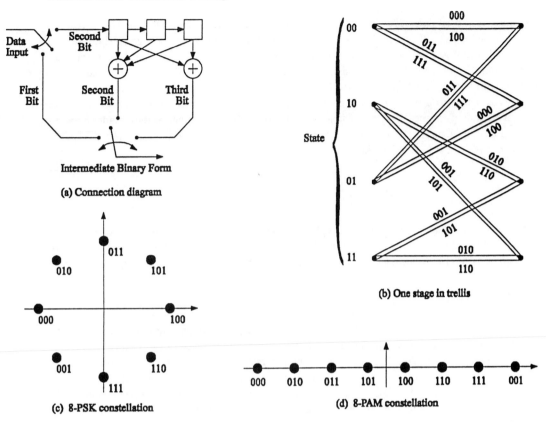

FIGURE 10.6   A rate 2/3 TCM code.

10.7 Trellis-Coded Modulation

Trellis-coded modulation (TCM) has received considerable attention since its development by Ungerboeck in the late 1970s [Ungerboeck, 1987]. Unlike block and convolutional codes, TCM schemes achieve coding gain by increasing the size of the signal alphabet and using multilevel/phase signalling. Like convolutional codes, sequences of coded symbols are restricted to certain valid patterns. In TCM, these patterns are chosen to have large Euclidean distance from one another so that a large number of corrupted sequences can be corrected. The Viterbi algorithm is often used to decode these sequences. Since the symbol transmission rate does not increase, coded and uncoded signals require the same transmission bandwidth. If transmission power is held constant, the signal constellation of the coded signal is denser. The loss in symbol separation, however, is more than overcome by the error correction capability of the code.

Ungerboeck investigated the increase in channel capacity that can be obtained by increasing the size of the signal set and restricting the pattern of transmitted symbols, and concluded that almost all of the additional capacity can be gained by doubling the number of points in the signal constellation. This is accomplished by encoding the binary data with a rate $R = k/(k + 1)$ code and mapping sequences of $k + 1$ coded bits to points in a constellation of 2^{k+1} symbols. For example, the rate 2/3 encoder of Fig. 10.6(a) encodes pairs of source bits to three coded bits. Figure 10.6(b) depicts one stage in the trellis of the coded output where, as with the convolutional code, the state of the encoder is defined by the values of the two most recent bits to enter the shift register. Note that unlike the trellis for the convolutional code, this trellis contains parallel paths between nodes.

The key to improving performance with TCM is to map the coded bits to points in the signal space such that the Euclidean distance between transmitted sequences is maximized. A method that ensures improved Euclidean distance is the method of **set partitioning**. This involves separating all parallel paths on the trellis with maximum distance and assigning the next greatest distance to paths that diverge from or merge onto the same node. Figures 10.6(c) and 10.6(d) give examples of mappings for the rate 2/3 code with 8-PSK and 8-PAM signal constellations respectively.

As with convolutional codes, the free distance of a TCM code is defined as the minimum distance between paths through the trellis, where the distance of concern is now Euclidean distance rather than Hamming distance. The free distance of an uncoded signal is defined as the distance between the closest signal points. When coded and uncoded signals have the same average power, the coding gain of the TCM system is defined as

$$\text{coding gain} = 20 \log_{10} \left(\frac{d_{\text{free, coded}}}{d_{\text{free, uncoded}}} \right) \tag{10.9}$$

It can be shown that the simple, rate 2/3 8 phase-shift keying (PSK) and 8 pulse-amplitude modulation (PAM) TCM systems provide gains of 3 dB and 3.3 dB, respectively [Clark and Cain, 1981]. More complex TCM systems yield gains up to 6 dB. Tables of good codes are given in Ungerboeck, 1987.

10.8 Additional Measures

When the demodulated sequence contains bursts of errors, the performance of codes designed to correct independent errors improves if coded sequences are **interleaved** prior to transmission and deinterleaved prior to decoding. Deinterleaving separates the burst errors, making them appear more random and increasing the likelihood of accurate decoding. It is generally sufficient to interleave several block lengths of a block coded signal or several constraint lengths of a convolutionally encoded signal. Block interleaving is the most straightforward approach, but delay and memory requirements are halved with convolutional and helical interleaving techniques. Periodicity in the way sequences are combined is avoided with pseudorandom interleaving.

Concatenated codes, first investigated by Forney, use two levels of coding to achieve a level of performance with less complexity than a single coding stage would require. The inner code interfaces with the modulator and demodulator and corrects the majority of the errors; the outer code corrects errors that appear at the output of the inner-code decoder. A convolutional code with Viterbi decoding is usually chosen as the inner code, and an RS code is often chosen as the outer code due to its ability to correct the bursts of bit errors, which can result with incorrect decoding of trellis-coded sequences. Interleaving and deinterleaving outer-code symbols between coding stages offers further protection against the burst error output of the inner code.

10.9 Applications

FEC coding remained of theoretical interest until advances in digital technology and improvements in decoding algorithms made their implementation possible. It has since become an attractive alternative to improving other system components or boosting transmission power. FEC codes are commonly used in digital storage systems, deep-space and satellite communication systems, terrestrial radio and band limited wireline systems, and have also been proposed for fiber optic transmission. Accordingly, the theory and practice of error correcting codes now occupies a prominent position in the field of communications engineering.

Deep-space systems began using forward error correction in the early 1970s to reduce transmission power requirements, and used multiple error correcting RS codes for the first time in 1977 to protect against corruption of compressed image data in the Voyager missions [Wicker and Bhargava, 1994]. The Consultative Committee for Space Data Systems (CCSDS) has since recommended use of a concatenated coding system which uses a rate 1/2, constraint length 7 convolutional inner code and a (255, 223) RS outer code.

Coding is now commonly used in satellite systems to reduce power requirements and overall hardware costs and to allow closer orbital spacing of geosynchronous satellites [Berlekamp et al., 1987]. FEC codes play integral roles in the VSAT, MSAT, INTELSAT, and INMARSAT systems [Wu et al., 1987]. Further, a (31, 15) RS code is used in the joint tactical information distribution system (JTIDS), a (7, 2) RS code is used in the air force satellite communication system (AFSATCOM), and a (204, 192) RS code has been designed specifically for satellite time division multiple access (TDMA) systems. Another code designed for military applications involves concatenation of a Golay and RS code with interleaving to ensure an imbalance of 1's and 0's in the transmitted symbol sequence and enhance signal recovery under severe noise and interference [Berlekamp et al., 1987].

TCM has become commonplace in transmission of data over voiceband telephone channels. Modems developed since 1984 use trellis coded QAM modulation to provide robust communication at rates above 9.6 kb/s.

FEC codes have also been widely used in digital recording systems, most prominently in the compact disc digital audio system. This system uses two levels of coding and interleaving in the cross-interleaved RS coding (CIRC) system to correct errors that result from disc imperfections and dirt and scratches, which accumulate during use. Steps are also taken to mute uncorrectable sequences [Wicker and Bhargava, 1994].

Defining Terms

Binary symmetric channel: A memoryless discrete data channel with binary signalling, hard-decision demodulation, and channel impairments that do not depend on the value of the symbol transmitted.

Bounded distance decoding: Limiting the error patterns, which are corrected in an imperfect code, to those with t or fewer errors.

Catastrophic code: A convolutional code in which a finite number of code symbol errors can cause an unlimited number of decoded bit errors.

Code rate: The ratio of source word length to codeword length, indicative of the amount of information transmitted per encoded symbol.

Coding gain: The reduction in signal-to-noise ratio required for specified error performance in a block or convolutional coded system over an uncoded system with the same information rate, channel impairments, and modulation and demodulation techniques. In TCM, the ratio of the squared free distance in the coded system to that of the uncoded system.

Column distance: The minimum Hamming distance between convolutionally encoded sequences of a specified length with different leading n-tuples.

Concatenated codes: Two levels of codes that achieve a level of performance with less complexity than a single coding stage would require. The inner code is often a convolutional code, and the outer code is usually an RS code.

Cyclic code: A block code in which cyclic shifts of code vectors are also code vectors.

Cyclic redundancy check: When the syndrome of a cyclic block code is used to detect errors.

Designed distance: The guaranteed minimum distance of a BCH code designed to correct up to t errors.

Discrete data channel: The concatenation of all system elements between FEC encoder output and decoder input.

Distance profile: The minimum Hamming distance after each encoding interval of convolutionally encoded sequences, which differ in the first interval.

Erasure: A position in the demodulated sequence where the symbol value is unknown.

Finite field: A finite set of elements and operations of addition and multiplication that satisfy specific properties. Often called Galois fields and denoted GF(q), where q is the number of elements in the field. Finite fields exist for all q, which are prime or the power of a prime.

Free distance: The minimum Hamming weight of convolutionally encoded sequences that diverge and remerge in the trellis. Equals the maximum column distance and the limiting value of the distance profile.

Generator matrix: A matrix used to describe a linear code. Code vectors equal the information vectors multiplied by this matrix.

Generator polynomial: The polynomial that is a divisor of all codeword polynomials in a cyclic block code; a polynomial that describes circuit connections in a convolutional encoder.

Hamming distance: The number of symbols in which codewords differ.

Hard decision: Demodulation that outputs only a value for each received symbol.

Interleaving: Shuffling the coded bit sequence prior to modulation and reversing this operation following demodulation. Used to separate and redistribute burst errors over several codewords (block codes) or constraint lengths (trellis codes) for higher probability of correct decoding by codes designed to correct random errors.

Linear code: A code whose code vectors form a vector space. Equivalently, a code where the addition of any two code vectors forms another code vector.

Maximum distance separable: A code with the largest possible minimum distance given the block length and code rate. These codes meet the Singleton bound of $d_{min} \leq n - k + 1$.

Metric: A measure of goodness against which items are judged. In the Viterbi algorithm, an indication of the probability of a path being taken given the demodulated symbol sequence.

Minimum distance: In a block code, the smallest Hamming distance between any two codewords. In a convolutional code, the column distance after K intervals.

Parity check matrix: A matrix whose rows are orthogonal to the rows in the generator matrix of a linear code. Errors can be detected by multiplying the received vector by this matrix.

Perfect code: A t error correcting (n, k) block code in which $q^{n-k} - 1 = \sum_{i=1}^{t} \binom{n}{i}$.

Puncturing: Periodic deletion of code symbols from the sequence generated by a convolutional encoder for purposes of constructing a higher rate code. Also, deletion of parity bits in a block code.

Set partitioning: Rules for mapping coded sequences to points in the signal constellation that always result in a larger Euclidean distance for a TCM system than an uncoded system, given appropriate construction of the trellis.

Soft decision: Demodulation that outputs an estimate of the received symbol value along with an indication of the reliability of this value. Usually implemented by quantizing the received signal to more levels than there are symbol values.

Standard array decoding: Association of an error pattern with each syndrome by way of a lookup table.

Syndrome: An indication of whether or not errors are present in the demodulated symbol sequence.

Systematic code: A code in which the values of the message symbols can be identified by inspection of the code vector.

Vector space: An algebraic structure comprised of a set of elements in which operations of vector addition and scalar multiplication are defined. For our purposes, a set of n-tuples consisting of symbols from GF(q) with addition and multiplication defined in terms of elementwise operations from this finite field.

Viterbi algorithm: A maximum-likelihood decoding algorithm for trellis codes that discards low-probability paths at each stage of the trellis, thereby reducing the total number of paths that must be considered.

References

Berlekamp, E.R., Peile, R.E., and Pope, S.P. 1987. The application of error control to communications. *IEEE Commun. Mag.* 25(4):44–57.

Bhargava, V.K. 1983. Forward error correction schemes for digital communications. *IEEE Commun. Mag.* 21(1):11–19.

Blahut, R.E. 1983. *Theory and Practice of Error Control Codes*, Addison-Wesley, Reading, MA.

Clark, G.C., Jr. and Cain, J.B. 1981. *Error Correction Coding for Digital Communications*, Plenum Press, New York.

Lin, S. and Costello, D.J., Jr. 1983. *Error Control Coding: Fundamentals and Applications*, Prentice-Hall, Englewood Cliffs, NJ.

Shannon, C.E. 1948. A mathematical theory of communication. *Bell Syst. Tech. J.* 27(3):379–423 and 623–656.

Sklar, B. 1988. *Digital Communications: Fundamentals and Applications*, Prentice-Hall, Englewood Cliffs, NJ.

Ungerboeck, G. 1987. Trellis-coded modulation with redundant signal sets. *IEEE Commun. Mag.* 25(2):5–11 and 12–21.

Wicker, S.B. and Bhargava, V.K. 1994. *Reed-Solomon Codes and Their Applications*, IEEE Press, NJ.

Wu, W.W., Haccoun, D., Peile, R., and Hirata, Y. 1987. Coding for satellite communication. *IEEE J. Selected Areas in Commun.* SAC-5(4):724–748.

Further Information

There is now a large amount of literature on the subject of FEC coding. An introduction to the philosophy and limitations of these codes can be found in the second chapter of Lucky's book *Silicon Dreams: Information, Man, and Machine*, St. Martin's Press, New York, 1989. More practical introductions can be found in overview chapters of many communications texts. The number of texts devoted entirely to this subject also continues to grow. Although these texts summarize the algebra underlying block codes, more in-depth treatments can be found in mathematical texts. Survey papers appear occasionally in the literature, but the interested reader is directed to the seminal papers by Shannon, Hamming, Reed and Solomon, Bose and Chaudhuri, Hocquenghem, Wozencraft, Fano, Forney, Berlekamp, Massey, Viterbi, and Ungerboeck, among others. The most recent advances in the theory and implementation of error control codes are published in *IEEE Transactions on Information Theory* and *IEEE Transactions on Communications*.

11

Spread Spectrum Communications

L.B. Milstein
University of California

M.K. Simon
Jet Propulsion Laboratory

11.1 A Brief History

Spread spectrum (SS) has its origin in the military arena where the friendly communicator is 1) susceptible to detection/interception by the enemy and 2) vulnerable to intentionally introduced unfriendly interference (jamming). Communication systems that employ spread spectrum to reduce the communicator's detectability and combat the enemy-introduced interference are respectively referred to as **low probability of intercept (LPI)** and **antijam** (AJ) **communication systems**. With the change in the current world political situation wherein the U.S. Department of Defense (DOD) has reduced its emphasis on the development and acquisition of new communication systems for the original purposes, a host of new commercial applications for SS has evolved, particularly in the area of cellular mobile communications. This shift from military to commercial applications of SS has demonstrated that the basic concepts that make SS techniques so useful in the military can also be put to practical peacetime use. In the next section, we give a simple description of these basic concepts using the original military application as the basis of explanation. The extension of these concepts to the mentioned commercial applications will be treated later on in the chapter.

11.2 Why Spread Spectrum?

Spread spectrum is a communication technique wherein the transmitted modulation is *spread* (increased) in bandwidth prior to transmission over the channel and then *despread* (decreased) in bandwidth by the same amount at the receiver. If it were not for the fact that the communication channel introduces some form of narrowband (relative to the spread bandwidth) interference, the receiver performance would be transparent to the spreading and despreading operations (assuming

0-8493-8573-3/96/$0.00+$.50

that they are identical inverses of each other). That is, after **despreading** the received signal would be identical to the transmitted signal prior to **spreading**. In the presence of narrowband interference, however, there is a significant advantage to employing the spreading/despreading procedure described. The reason for this is as follows. Since the interference is introduced after the transmitted signal is spread, then, whereas the despreading operation at the receiver shrinks the desired signal back to its original bandwidth, at the same time it spreads the undesired signal (interference) in bandwidth by the same amount, thus reducing its power spectral density. This, in turn, serves to diminish the effect of the interference on the receiver performance, which depends on the amount of interference power in the spread bandwidth. It is indeed this very simple explanation, which is at the heart of all spread spectrum techniques.

11.3 Basic Concepts and Terminology

To describe this process analytically and at the same time introduce some terminology that is common in spread spectrum parlance, we proceed as follows. Consider a communicator that desires to send a message using a transmitted power S Watts (W) at an information rate R_b bits/s (bps). By introducing a SS modulation, the bandwidth of the transmitted signal is increased from R_b Hz to W_{ss} Hz where $W_{ss} \gg R_b$ denotes the **spread spectrum bandwidth**. Assume that the channel introduces, in addition to the usual thermal noise (assumed to have a single-sided power spectral density (PSD) equal to N_0 W/Hz), an additive interference (jamming) having power J distributed over some bandwidth W_J. After despreading, the desired signal bandwidth is once again now equal to R_b Hz and the interference PSD is now $N_J = J/W_{ss}$. Note that since the thermal noise is assumed to be white, i.e., it is uniformly distributed over all frequencies, its PSD is unchanged by the despreading operation and, thus, remains equal to N_0. Regardless of the signal and interferer waveforms, the equivalent bit energy-to-total noise ratio is, in terms of the given parameters,

$$\frac{E_b}{N_t} = \frac{E_b}{N_0 + N_J} = \frac{S/R_b}{N_0 + J/W_{ss}} \qquad (11.1)$$

For most practical scenarios, the jammer limits performance and, thus, the effects of receiver noise in the channel can be ignored. Thus, assuming $N_J \gg N_0$, we can rewrite Eq. (11.1) as

$$\frac{E_b}{N_t} \cong \frac{E_b}{N_J} = \frac{S/R_b}{J/W_{ss}} = \frac{S}{J} \frac{W_{ss}}{R_b} \qquad (11.2)$$

where the ratio J/S is the *jammer-to-signal power ratio* and the ratio W_{ss}/R_b is the **spreading ratio** and is defined as the **processing gain** of the system. Since the ultimate error probability performance of the communication receiver depends on the ratio E_b/N_J, we see that from the communicator's viewpoint his goal should be to minimize J/S (by choice of S) and maximize the processing gain (by choice of W_{ss} for a given desired information rate). The possible strategies for the jammer will be discussed in the section on military applications dealing with AJ communications.

11.4 Spread Spectrum Techniques

By far the two most popular spreading techniques are **direct sequence (DS) modulation** and **frequency hopping (FH) modulation**. In the following subsections, we present a brief description of each.

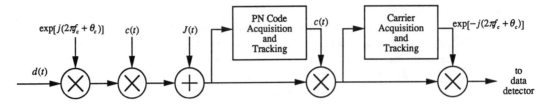

FIGURE 11.1   A DS-BPSK system (complex form).

Direct Sequence Modulation

A direct sequence modulation $c(t)$ is formed by linearly modulating the output sequence $\{c_n\}$ of a pseudorandom number generator onto a train of pulses, each having a duration T_c called the **chip time**. In mathematical form,

$$c(t) = \sum_{n=-\infty}^{\infty} c_n p(t - nT_c) \qquad (11.3)$$

where $p(t)$ is the basic pulse shape and is assumed to be of rectangular form. This type of modulation is usually used with binary phase-shift-keyed (BPSK) information signals, which have the complex form $d(t) \exp\{j(2\pi f_c t + \theta_c)\}$, where $d(t)$ is a binary-valued data waveform of rate $1/T_b$ bits/s and f_c and θ_c are the frequency and phase of the data-modulated carrier, respectively. As such, a DS/BPSK signal is formed by multiplying the BPSK signal by $c(t)$ (see Fig. 11.1), resulting in the real transmitted signal

$$x(t) = \mathrm{Re}\{c(t)d(t) \exp[j(2\pi f_c t + \theta_c)]\} \qquad (11.4)$$

Since T_c is chosen so that $T_b \gg T_c$, then relative to the bandwidth of the BPSK information signal, the bandwidth of the DS/BPSK signal[1] is effectively increased by the ratio $T_b/T_c = W_{ss}/2R_b$, which is one-half the spreading factor or processing gain of the system. At the receiver, the sum of the transmitted DS/BPSK signal and the channel interference $I(t)$ (as discussed before, we ignore the presence of the additive thermal noise) are ideally multiplied by the identical DS modulation (this operation is known as despreading), which returns the DS/BPSK signal to its original BPSK form whereas the real interference signal is now the real wideband signal $\mathrm{Re}\{I(t)c(t)\}$. In the previous sentence, we used the word ideally, which implies that the PN waveform used for despreading at the receiver is identical to that used for spreading at the transmitter. This simple implication covers up a multitude of tasks that a practical DS receiver must perform. In particular, the receiver must first acquire the PN waveform. That is, the local PN random generator that generates the PN waveform at the receiver used for despreading must be aligned (synchronized) to within one chip of the PN waveform of the received DS/BPSK signal. This is accomplished by employing some sort of **search algorithm** which typically steps the local PN waveform sequentially in time by a fraction of a chip (e.g., half a chip) and at each position searches for a high degree of correlation between the received and local PN reference waveforms. The search terminates when the correlation exceeds a given threshold, which is an indication that the alignment has been achieved. After bringing the two PN waveforms into **coarse alignment**, a **tracking algorithm** is employed to maintain **fine alignment**. The most popular forms of tracking loops are the continuous time **delay-locked loop** and its time-multiplexed version the **tau–dither loop**. It is the difficulty in synchronizing the receiver PN

---

[1]For the usual case of a rectangular spreading pulse $p(t)$, the PSD of the DS/BPSK modulation will have $(\sin x/x)^2$ form with first zero crossing at $1/T_c$, which is nominally taken as one-half the spread spectrum bandwidth W_{ss}.

generator to subnanosecond accuracy that limits PN chip rates to values on the order of hundreds of Mchips/s, which implies the same limitation on the DS spread spectrum bandwidth W_{ss}.

Frequency Hopping Modulation

A **frequency hopping (FH) modulation** $c(t)$ is formed by nonlinearly modulating a train of pulses with a sequence of pseudorandomly generated frequency shifts $\{f_n\}$. In mathematical terms, $c(t)$ has the complex form

$$c(t) = \sum_{n=-\infty}^{\infty} \exp\{j(2\pi f_n + \phi_n)\} p(t - nT_h) \qquad (11.5)$$

where $p(t)$ is again the basic pulse shape having a duration T_h, called the **hop time** and $\{\phi_n\}$ is a sequence of random phases associated with the generation of the hops. FH modulation is traditionally used with multiple-frequency-shift-keyed (MFSK) information signals, which have the complex form $\exp\{j[2\pi(f_c + d(t))t]\}$, where $d(t)$ is an M-level digital waveform (M denotes the symbol alphabet size) representing the information frequency modulation at a rate $1/T_s$ symbols/s (sps). As such, an FH/MFSK signal is formed by complex multiplying the MFSK signal by $c(t)$ resulting in the real transmitted signal

$$x(t) = \text{Re}\{c(t) \exp\{j[2\pi(f_c + d(t))t]\}\} \qquad (11.6)$$

In reality, $c(t)$ is never generated in the transmitter. Rather, $x(t)$ is obtained by applying the sequence of pseudorandom frequency shifts $\{f_n\}$ directly to the frequency synthesizer that generates the carrier frequency f_c (see Fig. 11.2). In terms of the actual implementation, successive (not necessarily disjoint) k-chip segments of a PN sequence drive a frequency synthesizer, which hops the carrier over 2^k frequencies. In view of the large bandwidths over which the frequency synthesizer must operate, it is difficult to maintain phase coherence from hop to hop, which explains the inclusion of the sequence $\{\phi_n\}$ in the Eq. (11.5) model for $c(t)$. On a short term basis, e.g., within a given hop, the signal bandwidth is identical to that of the MFSK information modulation, which is typically much smaller than W_{ss}. On the other hand, when averaged over many hops, the signal bandwidth is equal to W_{ss}, which can be on the order of several GHz, i.e., an order of magnitude larger than

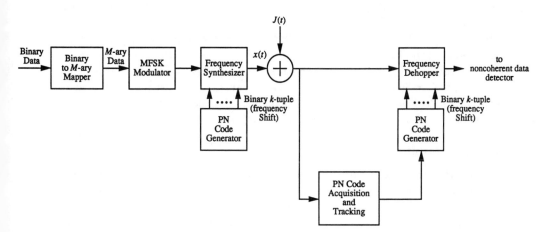

FIGURE 11.2 An FH-MFSK system.

that of implementable DS bandwidths. The exact relation between W_{ss}, T_h, T_s and the number of frequency shifts in the set $\{f_n\}$ will be discussed shortly.

At the receiver, the sum of the transmitted FH/MFSK signal and the channel interference $I(t)$ is ideally complex multiplied by the identical FH modulation (this operation is known as **dehopping**), which returns the FH/MFSK signal to its original MFSK form, whereas the real interference signal is now the wideband (in the average sense) signal $\text{Re}\{I(t)c(t)\}$. Analogous to the DS case, the receiver must acquire and track the FH signal so that the dehopping waveform is as close to the hopping waveform $c(t)$ as possible.

FH systems are traditionally classified in accordance with the relationship between T_h and T_s. **Fast frequency-hopped (FFH)** systems are ones in which there exists one or more hops per data symbol, that is, $T_s = NT_h$ (N an integer) whereas **slow frequency-hopped (SFH)** systems are ones in which there exists more than one symbol per hop, that is, $T_h = NT_s$. It is customary in SS parlance to refer to the FH/MFSK tone of shortest duration as a "chip", despite the same usage for the PN chips associated with the code generator that drives the frequency synthesizer. Keeping this distinction in mind, in an FFH system where, as already stated, there are multiple hops per data symbol, a chip is equal to a hop. For SFH, where there are multiple data symbols per hop, a chip is equal to an MFSK symbol. Combining these two statements, the chip rate R_c in an FH system is given by the larger of $R_h = 1/T_h$ and $R_s = 1/T_s$ and, as such, is the highest system clock rate.

The frequency spacing between the FH/MFSK tones is governed by the chip rate R_c and is, thus, dependent on whether the FH modulation is FFH or SFH. In particular, for SFH where $R_c = R_s$, the spacing between FH/MFSK tones is equal to the spacing between the MFSK tones themselves. For noncoherent detection (the most commonly encountered in FH/MFSK systems), the separation of the MFSK symbols necessary to provide orthogonality[2] is an integer multiple of R_s. Assuming the minimum spacing, i.e., R_s, the entire spread spectrum band is then partitioned into a total of $N_t = W_{ss}/R_s = W_{ss}/R_c$ equally spaced FH tones. One arrangement, which is by far the most common, is to group these N_t tones into $N_b = N_t/M$ contiguous, nonoverlapping bands, each with bandwidth $MR_s = MR_c$; see Fig. 11.3(a). Assuming symmetric MFSK modulation around the carrier frequency, then the center frequencies of the $N_b = 2^k$ bands represent the set of hop carriers, each of which is assigned to a given k-tuple of the PN code generator. In this fixed arrangement, each of the N_t FH/MFSK tones corresponds to the combination of a unique hop carrier (PN code k-tuple) and a unique MFSK symbol. Another arrangement, which provides more protection against the sophisticated interferer (jammer), is to overlap adjacent M-ary bands by an amount equal to R_c; see Fig. 11.3(b). Assuming again that the center frequency of each band corresponds to a possible hop carrier, then since all but $M-1$ of the N_t tones are available as center frequencies, the number of hop carriers has been increased from N_t/M to $N_t - (M-1)$, which for $N_t \gg M$ is approximately an increase in randomness by a factor of M.

For FFH, where $R_c = R_h$, the spacing between FH/MFSK tones is equal to the hop rate. Thus, the entire spread spectrum band is partitioned into a total of $N_t = W_{ss}/R_h = W_{ss}/R_c$ equally spaced FH tones, each of which is assigned to a unique k-tuple of the PN code generator that drives the frequency synthesizer. Since for FFH there are R_h/R_s hops per symbol, then the metric used to make a noncoherent decision on a particular symbol is obtained by summing up R_h/R_s detected chip (hop) energies, resulting in a so-called *noncoherent combining loss*.

Time Hopping Modulation

Time hopping (TH) is to spread spectrum modulation what pulse position modulation (PPM) is to information modulation. In particular, consider segmenting time into intervals of T_f seconds

---

[2]An optimum noncoherent MFSK detector consists of a bank of energy detectors each matched to one of the M frequencies in the MFSK set. In terms of this structure, the notion of *orthogonality* implies that for a given transmitted frequency there will be no crosstalk (energy spillover) in any of the other $M-1$ energy detectors.

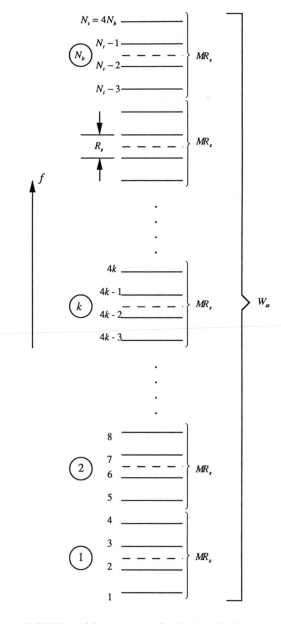

FIGURE 11.3(a) Frequency distribution for FH-4FSK
—nonoverlapping bands. Dashed lines indicate location
of hop frequencies.

and further segment each T_f interval into M_T increments of width T_f/M_T. Assuming a pulse of maximum duration equal to T_f/M_T, then a **time hopping spread spectrum** modulation would take the form

$$c(t) = \sum_{n=-\infty}^{\infty} p\left[t - \left(n + \frac{a_n}{M_T}\right)T_f\right] \tag{11.7}$$

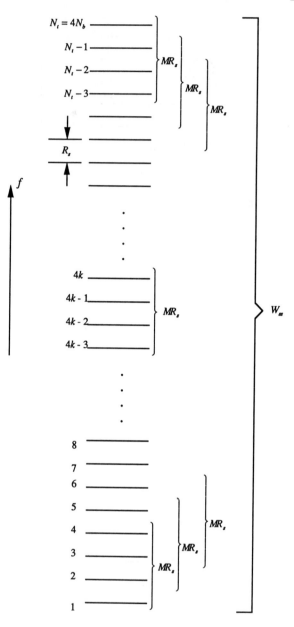

FIGURE 11.3(b)   Frequency distribution for FH-4FSK—over
lapping bands.

where a_n denotes the pseudorandom position (one of M_T uniformly spaced locations) of the pulse
within the T_r-second interval.

For DS and FH, we saw that *multiplicative* modulation, that is the transmitted signal is the product
of the SS and information signals, was the natural choice. For TH, *delay* modulation is the natural
choice. In particular, a TH-SS modulation takes the form

$$x(t) = \text{Re}\{c(t - d(t)) \exp[j(2\pi f_c + \phi_T)]\} \qquad (11.8)$$

where $d(t)$ is a digital information modulation at a rate $1/T_s$ sps. Finally, the dehopping procedure at the receiver consists of removing the sequence of delays introduced by $c(t)$, which restores the information signal back to its original form and spreads the interferer.

Hybrid Modulations

By blending together several of the previous types of SS modulation, one can form **hybrid** modulations that, depending on the system design objectives, can achieve a better performance against the interferer than can any of the SS modulations acting alone. One possibility is to multiply several of the $c(t)$ wideband waveforms [now denoted by $c^{(i)}(t)$ to distinguish them from one another] resulting in a SS modulation of the form

$$c(t) = \prod_i c^{(i)}(t) \tag{11.9}$$

Such a modulation may embrace the advantages of the various $c^{(i)}(t)$, while at the same time mitigating their individual disadvantages.

11.5 Applications of Spread Spectrum

Military

Antijam (AJ) Communications

As already noted, one of the key applications of spread spectrum is for antijam communications in a hostile environment. The basic mechanism by which a **direct sequence spread spectrum** receiver attenuates a noise jammer was illustrated in Sec. 11.3. Therefore, in this section, we will concentrate on tone jamming.

Assume the received signal, denoted $r(t)$, is given by

$$r(t) = Ax(t) + I(t) + n_w(t) \tag{11.10}$$

where $x(t)$ is given in Eq. (11.4), A is a constant amplitude,

$$I(t) = \alpha \cos(2\pi f_c t + \theta) \tag{11.11}$$

and $n_w(t)$ is additive white Gaussian noise (AWGN) having two-sided spectral density $N_0/2$. In Eq. (11.11), α is the amplitude of the tone jammer and θ is a random phase uniformly distributed in $[0, 2\pi]$.

If we employ the standard correlation receiver of Fig. 11.4, it is straightforward to show that the final test statistic out of the receiver is given by

$$g(T_b) = AT_b + \alpha \cos\theta \int_0^{T_b} c(t)\, dt + N(T_b) \tag{11.12}$$

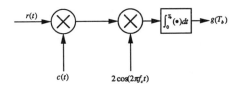

FIGURE 11.4

where $N(T_b)$ is the contribution to the test statistic due to the AWGN. Noting that, for rectangular chips, we can express

$$\int_0^{T_b} c(t)\, dt = T_c \sum_{i=1}^{M} c_i \tag{11.13}$$

where

$$M \triangleq \frac{T_b}{T_c} \tag{11.14}$$

is one-half of the processing gain, it is straightforward to show that, for a given value of θ, the signal-to-noise-plus-interference ratio, denoted by S/N_{total}, is given by

$$\frac{S}{N_{\text{total}}} = \frac{1}{\frac{N_0}{2E_b} + \left(\frac{J}{MS}\right)\cos^2\theta} \tag{11.15}$$

In Eq. (11.15), the jammer power is

$$J \triangleq \frac{\alpha^2}{2} \tag{11.16}$$

and the signal power is

$$S \triangleq \frac{A^2}{2} \tag{11.17}$$

If we look at the second term in the denominator of Eq. (11.15), we see that the ratio J/S is divided by M. Realizing that J/S is the ratio of the jammer power to the signal power before despreading, and J/MS is the ratio of the same quantity after despreading, we see that, as was the case for noise jamming, the benefit of employing direct sequence spread spectrum signalling in the presence of tone jamming is to reduce the effect of the jammer by an amount on the order of the processing gain.

Finally, one can show that an estimate of the average probability of error of a system of this type is given by

$$P_e = \frac{1}{2\pi} \int_0^{2\pi} \phi\left(-\sqrt{\frac{S}{N_{\text{total}}}}\right) d\theta \tag{11.18}$$

where

$$\phi(x) \triangleq \frac{1}{\sqrt{2\pi}} \int_{-\infty}^{x} e^{-y^2/2}\, dy \tag{11.19}$$

If Eq. (11.18) is evaluated numerically and plotted, the results are as shown in Fig. 11.5. It is clear from this figure that a large initial power advantage of the jammer can be overcome by a sufficiently large value of the processing gain.

Low-Probability of Intercept (LPI)

The opposite side of the AJ problem is that of LPI, that is, the desire to hide your signal from detection by an intelligent adversary so that your transmissions will remain unnoticed and, thus,

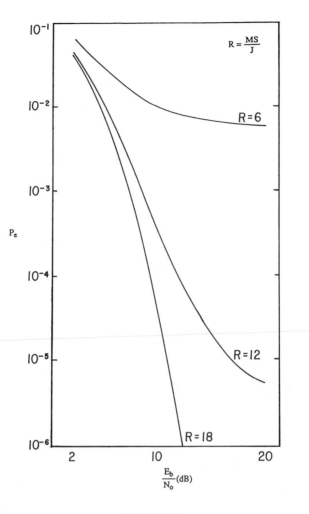

FIGURE 11.5

neither jammed nor exploited in any manner. This idea of designing an LPI system is achieved in a variety of ways, including transmitting at the smallest possible power level, and limiting the transmission time to as short an interval in time as is possible. The choice of signal design is also important, however, and it is here that spread spectrum techniques become relevant.

The basic mechanism is reasonably straightforward; if we start with a conventional narrowband signal, say a BPSK waveform having a spectrum as shown in Fig. 11.6(a), and then spread it so that its new spectrum is as shown in Fig. 11.6(b), the peak amplitude of the spectrum after spreading has been reduced by an amount on the order of the processing gain relative to what it was before spreading. Indeed, a sufficiently large processing gain will result in the spectrum of the signal after spreading falling below the ambient thermal noise level. Thus, there is no easy way for an unintended listener to determine that a transmission is taking place.

That is not to say the spread signal cannot be detected, however, merely that it is more difficult for an adversary to learn of the transmission. Indeed, there are many forms of so-called intercept receivers that are specifically designed to accomplish this very task. By way of example, probably the best known and simplest to implement is a **radiometer**, which is just a device that measures the total power present in the received signal. In the case of our intercept problem, even though we have lowered the power spectral density of the transmitted signal so that it falls below the noise

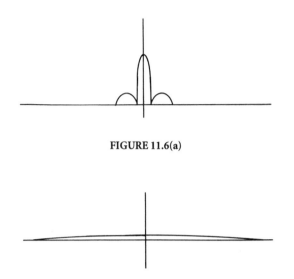

FIGURE 11.6(a)

FIGURE 11.6(b)

floor, we have not lowered its power (i.e., we have merely spread its power over a wider frequency range). Thus, if the radiometer integrates over a sufficiently long period of time, it will eventually determine the presence of the transmitted signal buried in the noise. The key point, of course, is that the use of the spreading makes the interceptor's task much more difficult, since he has no knowledge of the spreading code and, thus, cannot despread the signal.

Commercial

Multiple Access Communications

From the perspective of commercial applications, probably the most important use of spread spectrum communications is as a multiple accessing technique. When used in this manner, it becomes an alternative to either frequency division multiple access (FDMA) or time division multiple access (TDMA) and is typically referred to as either code division multiple access (CDMA) or spread spectrum multiple access (SSMA). When using CDMA, each signal in the set is given its own spreading sequence. As opposed to either FDMA, wherein all users occupy disjoint frequency bands but are transmitted simultaneously in time, or TDMA, whereby all users occupy the same bandwidth but transmit in disjoint intervals of time, in CDMA, all signals occupy the same bandwidth and are transmitted simultaneously in time; the different waveforms in CDMA are distinguished from one another at the receiver by the specific spreading codes they employ.

Since most CDMA detectors are correlation receivers, it is important when deploying such a system to have a set of spreading sequences that have relatively low-pairwise cross-correlation between any two sequences in the set. Further, there are two fundamental types of operation in CDMA, synchronous and asynchronous. In the former case, the symbol transition times of all of the users are aligned; this allows for orthogonal sequences to be used as the spreading sequences and, thus, eliminates interference from one user to another. Alternately, if no effort is made to align the sequences, the system operates asychronously; in this latter mode, multiple access interference limits the ultimate channel capacity, but the system design exhibits much more flexibility.

CDMA has been of particular interest recently for applications in wireless communications. These applications include cellular communications, personal communications services (PCS), and wireless local area networks. The reason for this popularity is primarily due to the performance that spread spectrum waveforms display when transmitted over a multipath fading channel.

To illustrate this idea, consider DS signalling. As long as the duration of a single chip of the spreading sequence is less than the multipath delay spread, the use of DS waveforms provides the system designer with one of two options. First, the multipath can be treated as a form of interference, which means the receiver should attempt to attenuate it as much as possible. Indeed, under this condition, all of the multipath returns that arrive at the receiver with a time delay greater than a chip duration from the multipath return to which the receiver is synchronized (usually the first return) will be attenuated because of the processing gain of the system.

Alternately, the multipath returns that are separated by more than a chip duration from the main path represent independent "looks" at the received signal and can be used constructively to enhance the overall performance of the receiver. That is, because all of the multipath returns contain information regarding the data that is being sent, that information can be extracted by an appropriately designed receiver. Such a receiver, typically referred to as a RAKE receiver, attempts to resolve as many individual multipath returns as possible and then to sum them coherently. This results in an *implicit* diversity gain, comparable to the use of *explicit* diversity, such as receiving the signal with multiple antennas.

The condition under which the two options are available can be stated in an alternate manner. If one envisions what is taking place in the frequency domain, it is straightforward to show that the condition of the chip duration being smaller than the multipath delay spread is equivalent to requiring that the spread bandwidth of the transmitted waveform exceed what is called the coherence bandwidth of the channel. This latter quantity is simply the inverse of the multipath delay spread and is a measure of the range of frequencies that fade in a highly correlated manner. Indeed, anytime the coherence bandwidth of the channel is less than the spread bandwidth of the signal, the channel is said to be *frequency selective* with respect to the signal. Thus, we see that to take advantage of DS signalling when used over a multipath fading channel, that signal should be designed such that it makes the channel appear frequency selective.

In addition to the desirable properties that spread spectrum signals display over multipath channels, there are two other reasons why such signals are of interest in cellular-type applications. The first has to do with a concept known as the reuse factor. In conventional cellular systems, either analog or digital, in order to avoid excessive interference from one cell to its neighbor cells, the frequencies used by a given cell are not used by its immediate neighbors (i.e., the system is designed so that there is a certain spatial separation between cells that use the same carrier frequencies). For CDMA, however, such spatial isolation is typically not needed, so that so-called *universal reuse* is possible.

Further, because CDMA systems tend to be interference limited, for those applications involving voice transmission, an additional gain in the capacity of the system can be achieved by the use of *voice activity detection*. That is, in any given two-way telephone conversation, each user is typically talking only about 50% of the time. During the time when a user is quiet, he is not contributing to the instantaneous interference. Thus, if a sufficiently large number of users can be supported by the system, statistically only about one-half of them will be active simultaneously, and the effective capacity can be doubled.

Interference Rejection

In addition to providing multiple accessing capability, spread spectrum techniques are of interest in the commercial sector for basically the same reasons they are in the military community, namely their AJ and LPI characteristics. However, the motivations for such interest differ. For example, whereas the military is interested in ensuring that systems they deploy are robust to interference generated by an intelligent adversary (i.e., exhibit jamming resistance), the interference of concern in commercial applications is unintentional. It is sometimes referred to as co-channel interference (CCI) and arises naturally as the result of many services using the same frequency band at the same time. And while such scenarios almost always allow for some type of spatial isolation between the interfering waveforms, such as the use of narrow-beam antenna patterns, at times the use of the inherent interference suppression property of a spread spectrum signal is also desired. Similarly, whereas the military is very much interested in the LPI property of a spread spectrum waveform, as indicated in Sec. 11.3, there are applications in the commercial segment where the same characteristic can be used to advantage.

To illustrate these two ideas, consider a scenario whereby a given band of frequencies is somewhat sparsely occupied by a set of conventional (i.e., nonspread) signals. To increase the overall spectral efficiency of the band, a set of spread spectrum waveforms can be overlaid on the same frequency band, thus forcing the two sets of users to share common spectrum. Clearly, this scheme is feasible only if the mutual interference that one set of users imposes on the other is within tolerable limits. Because of the interference suppression properties of spread spectrum waveforms, the despreading process at each spread spectrum receiver will attenuate the components of the final test statistic due to the overlaid narrowband signals. Similarly, because of the LPI characteristics of spread spectrum waveforms, the increase in the overall noise level as seen by any of the conventional signals, due to the overlay, can be kept relatively small.

Defining Terms

Antijam communication system: A communication system designed to resist intentional jamming by the enemy.

Chip time (interval):   The duration of a single pulse in a direct sequence modulation; typically much smaller than the information symbol interval.

Coarse alignment: The process whereby the received signal and the despreading signal are aligned to within a single chip interval.

Dehopping: Despreading using a frequency-hopping modulation.

Delay-locked loop: A particular implementation of a closed-loop technique for maintaining fine alignment.

Despreading: The notion of decreasing the bandwidth of the received (spread) signal back to its information bandwidth.

Direct sequence modulation: A signal formed by linearly modulating the output sequence of a pseudorandom number generator onto a train of pulses.

Direct sequence spread spectrum: A spreading technique achieved by multiplying the information signal by a direct sequence modulation.

Fast frequency-hopping: A spread spectrum technique wherein the hop time is less than or equal to the information symbol interval, i.e., there exist one or more hops per data symbol.

Fine alignment: The state of the system wherein the received signal and the despreading signal are aligned to within a small fraction of a single chip interval.

Frequency-hopping modulation: A signal formed by nonlinearly modulating a train of pulses with a sequence of pseudorandomly generated frequency shifts.

Hop time (interval):   The duration of a single pulse in a frequency-hopping modulation.

Hybrid spread spectrum: A spreading technique formed by blending together several spread spectrum techniques, e.g., direct sequence, frequency-hopping, etc.

Low-probability-of-intercept communication system: A communication system designed to operate in a hostile environment wherein the enemy tries to detect the presence and perhaps characteristics of the friendly communicator's transmission.

Processing gain (spreading ratio):   The ratio of the spread spectrum bandwidth to the information data rate.

Radiometer: A device used to measure the total energy in the received signal.

Slow frequency-hopping: A spread spectrum technique wherein the hop time is greater than the information symbol interval, i.e., there exists more than one data symbol per hop.

Spread spectrum bandwidth: The bandwidth of the transmitted signal after spreading.

Spreading: The notion of increasing the bandwidth of the transmitted signal by a factor far in excess of its information bandwidth.

Search algorithm: A means for coarse aligning (synchronizing) the despreading signal with the received spread spectrum signal.

Tau–dither loop: A particular implementation of a closed-loop technique for maintaining fine alignment.

Time-hopping spread spectrum: A spreading technique that is analogous to pulse position modulation.

Tracking algorithm: An algorithm (typically closed loop) for maintaining fine alignment.

Further Information

M.K. Simon, J.K. Omura, R.A. Scholtz, and B.K. Levitt, *Spread Spectrum Communications Handbook*, McGraw Hill, 1994 (previously published as *Spread Spectrum Communications*, Computer Science Press, 1985).

R.E. Ziemer and R.L. Peterson, *Digital Communications and Spread Spectrum Techniques*, Macmillan, 1985.

J.K. Holmes, *Coherent Spread Spectrum Systems*, John Wiley and Sons, Inc. 1982.

R.C. Dixon, *Spread Spectrum Systems*, 3rd ed., John Wiley and Sons, Inc. 1994.

C.F. Cook, F.W. Ellersick, L.B. Milstein, and D.L. Schilling, *Spread Spectrum Communications*, IEEE Press, 1983.

12

Diversity Techniques

A. Paulraj
Stanford University

12.1 Introduction

Diversity is a commonly used technique in mobile radio systems to combat signal fading. The basic principle of diversity is as follows. If several replicas of the same information carrying signal are received over multiple channels with comparable strengths and that exhibit independent fading, then there is a good likelihood that at least one or more of these of the received signals will not be in a fade at any given instant in time, thus making it possible to deliver adequate signal level to the receiver. Without diversity techniques, in noise limited conditions, the transmitter will have to deliver a much higher power level to protect the link during the short intervals when the channel is severely faded. In mobile radio, the power available on the reverse link is severely limited by the battery capacity in handheld subscriber units. Diversity methods play a crucial role in reducing transmit power needs. Also, cellular communication networks are mostly interference limited and once again mitigation of channel fading through use of diversity can translate into improved interference tolerance, which in turn means greater ability to support additional users and therefore higher system capacity.

The basic principles of diversity have been known since 1927 when the first experiments in space diversity were reported. There are many techniques for obtaining independently fading branches, and these can be subdivided into two main classes. The first called explicit techniques, users explicit redundant signal transmission to exploit diversity channels. Use of dual polarized signal transmission and reception in many point-to-point radios is an example of explicit diversity. Clearly, such redundant signal transmission involves a penalty in frequency spectrum or additional power. On the other hand, in the second main class called implicit techniques, the signal is transmitted only once, but the decorrelating effects in the propagation medium, such as multipaths, are exploited to receive signals over multiple diversity channels. A good example of implicit diversity is the **RAKE receiver** in code division multiple access (CDMA) systems that use independent fading of

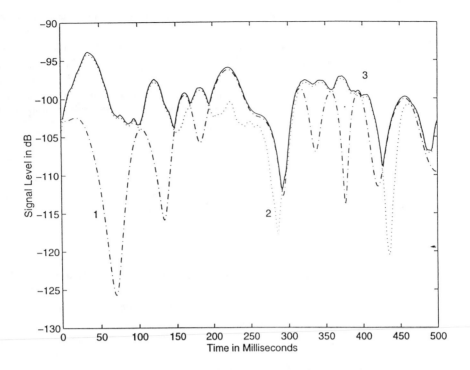

FIGURE 12.1 Example of diversity combining; two independently fading signals 1 and 2; the signal 3 is the result of selecting the strongest signal.

resolvable multipaths to achieve diversity gain. Figure 12.1 illustrates the principle of diversity where two independently fading signals are shown along with selection diversity output which selects the stronger signal. The fades in the resulting signal have been substantially smoothed out while also yielding higher average power.

Exploiting diversity needs careful design of the communication link. In explicit diversity, multiple copies of the same signal will have to be transmitted in channels using either frequency, or time, or polarization dimension. At the receiver end we need arrangements to receive the different diversity branches (this is true for both explicit and implicit diversity). The different diversity branches are then combined to reduce signal **outage probability** or bit-error rate.

In practice, the signals in the diversity branches may not show completely independent fading. The envelope cross correlation ρ between these signals is a measure of their independence.

$$\rho = \frac{E\left[[r_1 - \bar{r}_1][r_2 - \bar{r}_2]\right]}{\sqrt{E|r_1 - \bar{r}_1|^2 E|r_2 - \bar{r}_2|^2}}$$

where r_1 and r_2 represent the instantaneous envelope levels of the normalized signals at the two receivers and $\bar{r}_1$ and $\bar{r}_2$ are their respective means. It has been shown [Jakes, 1974] that a cross correlation of 0.7 between signal envelopes is sufficient to provide a reasonable degree of diversity gain. Depending on the type of the diversity employed, these diversity channels must be sufficiently *separated* along the appropriate diversity dimension. For spatial diversity, the antennas should be separated larger than the *coherence distance* to ensure a cross correlation less than 0.7. Likewise, in frequency diversity, the frequency separation must be larger than the *coherence bandwidth*, and in time diversity the separation between channel reuse in time should larger than the *coherence time*. These coherence factors, in turn, depend on the channel characteristics. The coherence distance, coherence bandwidth, and coherence time vary inversely as the angle spread, delay spread, and Doppler spread, respectively.

Once the receiver has a number of diversity branches, it has to combine these branches to maximize the signal level. Several techniques have been studied for diversity combining. We will describe three main techniques: selection combining, equal gain combining, and maximal ratio combining.

Finally, we should note that diversity is primarily used to combat fading, and if the signal does not have significant fading in the first place, as, for example, when there is a direct path component, diversity combining will not provide the gains normally expected.

12.2 Diversity Schemes

There are several techniques for obtaining diversity branches or, as they are sometimes also known, diversity dimensions. The most important of these are described in the following.

Space Diversity

This has historically been the most common form of diversity in mobile radio base stations. It is easy to implement and does not require additional frequency spectrum resources. Space diversity is exploited on the reverse link at the base station receiver by spacing antennas apart so as to obtain sufficient decorrelation. The key for obtaining uncorrelated fading of antenna outputs is adequate spacing of the antennas. The required spacing depends on the degree of multipath angle spread. For example, if the multipath signals arrive from all directions in the azimuth, as is usually the case at the mobile, antenna spacing of the order of 0.5λ–0.8λ is quite adequate [Lee, 1982]. On the other hand, if the multipath angle spread is small, as in the case of base stations, the coherence distance is much larger. Also, empirical measurements show a strong coupling between antenna height and spatial correlation. Larger antenna heights imply larger coherence distances. Typically, 10λ–20λ separation is adequate to achieve $\rho = 0.7$ at base stations in suburban settings when the signals arrive from the broadside direction. The coherence distance can be 3–4 times larger for endfire arrivals. The endfire problem is averted in base stations with trisectored antennas as each sector needs to handle only signals arriving ±60 deg off the broadside. The coherence distance depends strongly on the terrain. Also, base stations normally use space diversity in the horizontal plane only. Separation in the vertical plane can also be used, and the necessary spacing depends on vertical multipath angle spread. This can be small for distant mobiles, making vertical plane diversity less attractive in most applications.

Polarization Diversity

In mobile radio environments, signals transmitted on orthogonal polarizations exhibit decorrelated fading and, therefore, offer potential for diversity combining. Polarization diversity can be obtained either by explicit or implicit techniques. Note that with polarization only two diversity branches are available as opposed to space diversity where several branches can be obtained using multiple antennas. In explicit polarization diversity, the signal is both transmitted and received in two orthogonal polarizations. For a fixed total transmit power, the power in each branch will be 3 dB lower than if single polarization was used. In the implicit polarization technique, the signal is launched in a single polarization but is received with cross-polarized antennas. The propagation medium couples some energy into the cross-polarization plane. The observed cross-polarization coupling factor has been observed to be 10–12 dB in mobile ratio frequencies [Vaughan, 1990; Adachi et al. 1986]. Also, the cross-polarization envelope correlation has been found to be better than 0.7 and, therefore, provides significant diversity gain.

In recent years, with the increasing use of pocket telephones, the handset can be held at random orientations during a call. This results in energy being launched with varying polarization angles ranging from vertical to horizontal. This further increases the advantage of cross-polarized antennas at the base station since at least one of the two antennas will be well matched to the signal launch polarization. Recent work [Jefford, 1995] has shown that with variable launch polarization, a cross-polarized antenna can give comparable performance to a vertically polarized space diversity antenna.

Finally, we should note that cross-polarized antennas can be deployed in a compact antenna assembly and do not need the potentially large physical separation needed in space diversity antennas. This is an important advantage in the personal communication service (PCS) base stations, where low profile antennas are needed.

Angle Diversity

In situations where the angle spread is very high, as in the case of indoors or at the mobile unit in urban locations, signals collected from multiple nonoverlapping beams offer low-fade correlation with balanced power in the diversity branches. Clearly, since directional beams imply use of antenna aperture, angle diversity is closely related to the space diversity. Angle diversity has been utilized in indoor wireless local area networks (LANs), where its use allows substantial increase in LAN throughputs [Freeburg, 1991].

Frequency Diversity

Another technique to obtain decorrelated diversity branches is to transmit the same signal over different frequencies. The frequency separation between carriers should be larger than the coherence bandwidth. The coherence bandwidth, of course, depends on the multipath delay spread of the channel. The larger the delay spread is, the smaller the coherence bandwidth and the more closely the frequency diversity channels can be spaced. Clearly, frequency diversity is an explicit diversity technique and needs additional frequency spectrum.

A common form of frequency diversity is multicarrier (also known as multitone) modulation. This technique involves sending redundant data over a number of closely spaced carriers to benefit from frequency diversity, which is then exploited by applying **forward error correction (FEC)** across the carriers.

Path Diversity

A sophisticated form of diversity is based on using a signal bandwidth much larger than the channel coherence bandwidth, as is used in the so-called direct sequence spread spectrum modulation techniques. This modulation scheme is used in the CDMA mobile networks, one example of which is the IS-95 standard for 800-MHz cellular band. Spread spectrum signals can resolve multipath arrivals as long as the path delays are separated by at least one *chip* period. If the signal in each path shows low-fade correlation, as is usually the case, these paths offer a valuable source of diversity. A receiver that resolves the multipaths via code correlation and then combines them is referred to as a RAKE receiver [Viterbi, 1995]; see also Fig. 12.2.

In CDMA, diversity gain provided by the multiple paths (and other diversity branches, if any) not only reduces transmit power needs but also increases the number of users that can be supported per cell for a given bandwidth.

Time Diversity

In mobile communications channels, the mobile motion together with scattering in the vicinity of the mobile causes time selective fading of the signal with Rayleigh fading statistics for the signal envelope. Signal fade levels separated by the coherence time show low correlation and can be used as diversity branches if the same signal can be transmitted at multiple instants separated by the

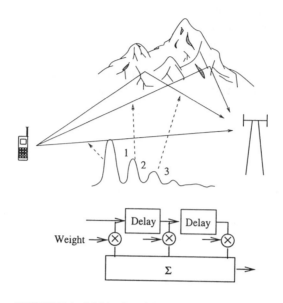

FIGURE 12.2 Multipath and RAKE receivers: three paths with independent fading are combined to provide path diversity.

coherence time. The coherence time depends on the Doppler spread of the signal, which in turn is a function of the mobile speed and the carrier frequency.

Time diversity is usually exploited via **interleaving**, forward-error correction coding, and **automatic request for repeat (ARQ)**. These are sophisticated techniques to exploit channel coding and time diversity. One fundamental drawback with time diversity approaches is the delay needed to collect the repeated or interleaved transmissions. If the coherence time is large, as, for example, when the vehicle is slow moving, the required delay becomes too large to be acceptable for interactive voice conversation.

The statistical properties of a fading signal depend on the field component used by the antenna, the vehicular speed, and the carrier frequency. For an idealized case of a mobile surrounded by scatterers in all directions, the auto-correlation function of the received signal $x(t)$ [note this is not the envelope $r(t)$] can be shown to be

$$E[x(t)x(t+\tau)] = J_0(2\pi\tau v/\lambda)$$

where J_0 is a Bessel function of zeroth order and v is the mobile velocity.

12.3 Diversity Combining Techniques

Several diversity combining methods are known. We describe three main techniques: selection, maximal ratio, and equal gain. They can be used with each of the diversity schemes just discussed.

Selection Combining

This is the simplest and perhaps the most frequently used form of diversity combining. In this technique, one of the two diversity branches with the highest carrier to noise ratio (C/N) is connected to the output; see Fig. 12.3(a).

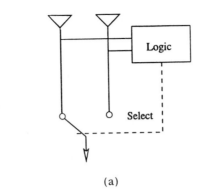

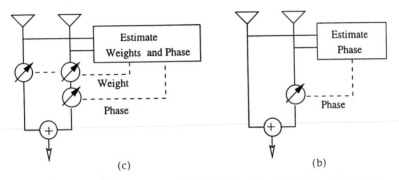

FIGURE 12.3 Diversity combining methods for two diversity branches. (a) Selection combining; (b) Maximal ratio combining; (c) Equal gain combining.

The performance improvement due to selection diversity can be seen as follows. Let the signal in each branch exhibit Rayleigh fading with mean power σ^2. The density function of the envelope is given by

$$p(r_i) = \frac{r_i}{\sigma^2} e^{\frac{-r_i^2}{2\sigma^2}} \tag{12.1}$$

where r_i is the signal envelope in each branch. If we define two new variables

$$\gamma_i = \frac{\text{instantaneous signal power in each branch}}{\text{mean noise power}}$$

$$\Gamma = \frac{\text{mean signal power in each branch}}{\text{mean noise power}}$$

then the probability that the C/N is less than or equal to some specified value γ_s is

$$\text{prob}[\gamma_i \leq \gamma_s) = 1 - e^{-\gamma_s/\Gamma} \tag{12.2}$$

The probability that γ_i in all branches with independent fading will be simultaneously less than or equal to γ_s is then

$$\text{prob}[\gamma_1, \gamma_2, \ldots, \gamma_M \leq \gamma_s] = (1 - e^{-\gamma_s/\Gamma})^M \tag{12.3}$$

This is the distribution of the best signal envelope from the two diversity branches. Figure 12.4 shows

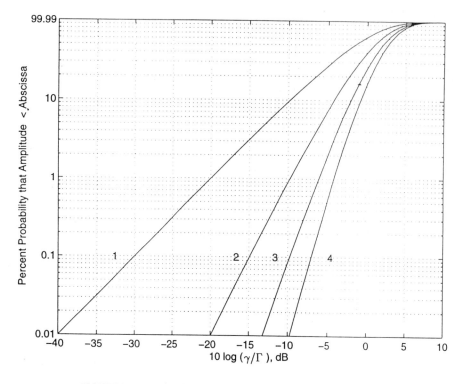

FIGURE 12.4 Probability distribution for signal envelope for selection combining.

the distribution of the combiner output C/N for $M = 1, 2, 3$, and 4 branches. The improvement in signal quality is significant. For example, at 99% reliability level, the improvement in C/N is 10 dB for 2 branches and 16 dB for 4 branches.

Selection combining also increases the mean C/N of the combiner output and can be shown [Jakes, 1974] to be

$$\text{mean}(\gamma_s) = \Gamma \sum_{k=1}^{M} \frac{1}{k} \qquad (12.4)$$

This indicates that with 4 branches, for example, the mean C/N of the selected branch is 2.03 better than the mean C/N in any one branch.

Maximal Ratio Combining

In this technique the M diversity branches are first cophased and then weighted proportional to their signal level before summing; see Fig. 12.3(b). The distribution of maximal ratio combining has been shown [Lee, 1982] to be

$$\text{prob}[\gamma \le \gamma_m] = 1 - e^{(-\gamma_m/\Gamma)} \sum_{k=1}^{M} \frac{(\gamma_m/\Gamma)^{k-1}}{(k-1)!} \qquad (12.5)$$

The distribution of the output of a maximal ratio combiner is shown in Fig. 12.5. Maximal ratio combining is known to be optimum in the sense that it yields the best statistical reduction of fading of any linear diversity combiner. In comparison to the selection combiner, at 99% reliability level,

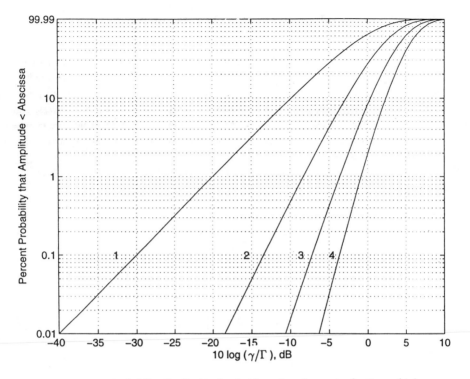

Percent Probability that Amplitude < Abscissa

$10 \log (\gamma / \Gamma)$, dB

FIGURE 12.5 Probability distribution for signal envelope for maximal ratio combining.

the maximal ratio combiner provides a 11.5-dB gain for 2 branches and a 19-dB gain for 4 branches. An improvement of 1.5 and 3 dB, respectively, over the selection diversity combiner.

The mean C/N of the combined signal may be easily shown to be

$$\text{mean}(\gamma_m) = M\Gamma \tag{12.6}$$

Therefore, combiner output mean varies linearly with M. This confirms the intuitive result that the output C/N averaged over fades should provide gain proportional to the number of diversity branches. This is a situation similar to conventional beam forming.

Equal Gain Combining

In some applications, it may be difficult to estimate the amplitude accurately, and the combining gains may all be set to unity, and the diversity branches merely summed after cophasing; see Fig. 12.3(c).

The distribution of the equal gain combiner does not have a neat expression and has been computed by numerical evaluation. Its performance has been shown to be very close (less than a decibel) to maximal ratio combining. The mean C/N can be shown [Jakes, 1974] to be

$$\text{mean}(\gamma_e) = \Gamma \left[1 + \frac{\pi}{4}(M - 1) \right] \tag{12.7}$$

Like maximal ratio combining, the mean C/N for equal gain combining grows almost linearly with M and is approximately only 1 dB poorer than the maximal ratio combiner even with infinite number of branches.

Loss of Diversity Gain Due to Branch Correlation and Unequal Branch Powers

The preceding analysis assumed that the fading signals in the diversity branches were all uncorrelated and of equal power. In practice, this may be difficult to achieve and, as we saw earlier, the branch cross-correlation coefficient $\rho = 0.7$ is considered to be acceptable. Also, equal mean power in diversity branches is rarely available. In such cases, we can expect a certain loss of diversity gain. Since most of the damage in fading is due to deep fades, however, and also since the chance of coincidental deep fades is small even for moderate branch correlation, one can expect reasonable tolerance to branch correlation.

The distribution of the output signal envelope of the maximal ratio combiner has been shown [Pahlavan and Levesque, 1995] to be

$$\text{prob}[\gamma_m] = \sum_{n=1}^{M} \frac{A_n}{2\lambda_n} e^{-\gamma_m/2\lambda_n} \tag{12.8}$$

where λ_n are the eigenvalues of the $M \times M$ branch envelope covariance matrix whose elements are defined by

$$\mathbf{R}_{ij} = E\left[r_i r_j^*\right] \tag{12.9}$$

and A_n is defined by

$$A_n = \prod_{\substack{k=1 \\ k \neq n}}^{M} \frac{1}{1 - \lambda_k/\lambda_n} \tag{12.10}$$

Recently extensive field measurements were carried out in the 1800-MHz band, and compact empirical results have been obtained for diversity gain at 90% signal reliability level as a function of branch correlation ρ and mean signal level difference Δ; see Jefford, 1995, for a full discussion.

Selection:

$$G = 5.71 \, e^{(-0.87\rho - 0.16\Delta)} \tag{12.11}$$

Equal gain:

$$G = -8.98 + 15.22 \, e^{(-0.20\rho - 0.04\Delta)} \tag{12.12}$$

Maximal ratio:

$$G = 7.14 \, e^{(-0.59\rho - 0.11\Delta)} \tag{12.13}$$

where G is in decibels.

12.4 Effect of Diversity Combining on Bit-Error Rate (BER)

So far we have studied the distribution of the instantaneous envelope or C/N after diversity combining. We will now briefly survey how diversity combining affects BER performance in digital radio links; we assume maximal ratio combining.

To begin let us first examine the effect of Rayleigh fading on BER performance of digital transmission links. This has been studied by several authors and is summarized in Proakis, 1989. Table 12.1

TABLE 12.1 Comparison of BER Performance for Unfaded and Rayleigh Faded Signals

Modulation	Unfaded BER	Faded BER
Coh BPSK	$\frac{1}{2}\,\mathrm{erfc}(\sqrt{E_b/N_0})$	$\frac{1}{4(\overline{E}_b/N_0)}$
Coh FSK	$\frac{1}{2}\,\mathrm{erfc}\left(\sqrt{\frac{1}{2}E_b/N_0}\right)$	$\frac{1}{2(\overline{E}_b/N_0)}$

TABLE 12.2 BER Performance for Coherent BPSK and FSK with Diversity

Modulation	Postdiversity BER
Coherent BPSK	$\left(\dfrac{1}{4\overline{E}_b/N_0}\right)^{L}\dbinom{2L-1}{L}$
Coherent FSK	$\left(\dfrac{1}{2\overline{E}_b/N_0}\right)^{L}\dbinom{2L-1}{L}$

gives the BER expressions in the large E_b/N_0 case for coherent binary phase-shift keying (PSK) and coherent binary orthogonal frequency-shift keying (FSK) for unfaded and Rayleigh faded additive white Gaussian noise channels (AWGN) channels. $\overline{E}_b/N_0$ represents the average E_b/N_0 for the fading channel.

Observe that error rates decrease only inversely with SNR as against exponential decrease for the unfaded channel. Also note that for fading channels, coherent binary PSK is 3 dB better than coherent binary FSK, exactly the same advantage as in the unfaded case. Even for modest target BER of 10^{-2} that is usually needed in mobile communications, the loss due to fading can be significant.

To obtain the BER with maximal ratio diversity combining we have to average the BER expression for the unfaded BER with the distribution obtained for the maximal ratio combiner given in Fig. 12.5. Analytical expressions have been derived for these in Proakis, 1989. For branch SNR greater than 10 dB, the BER after maximal ratio diversity combining is given in Table 12.2.

We observe that the probability of error varies as $1/(\overline{E}_b/N_0)$ raised to the Lth power. Thus, diversity reduces the error rate exponentially as the number of independent branches increases.

12.5 Conclusions

Diversity provides a powerful technique for combating fading in mobile communication systems. Diversity techniques seek to generate and exploit multiple branches over which the signal shows low-fade correlation. To obtain the best diversity performance, the multiple access, modulation, coding, and antenna design of the wireless link must all be carefully chosen so as to provide a rich and reliable level of well-balanced, low-correlation diversity branches in the target propagation environment. Successful diversity exploitation can impact a mobile network in several ways. Reduced power requirements can result in increased coverage or improved battery life. Low-signal outage improves voice quality and handoff performance. Finally, reduced fade margins directly translate to increased system capacity, particularly in CDMA networks.

Defining Terms

Automatic request for repeat: An error control mechanism in which received packets that cannot be corrected are retransmitted.
Fading: Fluctuation in the signal level due to shadowing and multipath effects.
Forward error correction (FEC): A technique that inserts redundant bits during transmission to help detect and correct bit errors during reception.
Interleaving: A form of data scrambling that spreads burst of bit errors evenly over the received data allowing efficient forward error correction.
Outage probability: The probability that the signal level falls below a specified minimum level.
RAKE receiver: A receiver used in direct sequence spread spectrum signals. The receiver extracts energy in each path and then adds them together with appropriate weighting and delay.

References

Adachi, F., Feeney, M.T., Williamson, A.G., and Parsons, J.D. 1986. Crosscorrelation between the envelopes of 900 MHz signals received at a mobile radio base station site. *Proc. IEE*, 133(6): 506–512.

Freeburg, T.A. 1991. Enabling technologies for in-building network communications—four technical challenges and four solutions. *IEEE Trans. Veh. Tech.* 29(4):58–64.

Jakes, W.C. 1974. *Microwave Mobile Communications*, John Wiley, New York.

Jefford, P.A., Turkmani, A.M.D., Arowojolu, A.A., and Kellett, C.J. 1995. An experimental evaluation of the performance of the two branch space and polarization schemes at 1800 MHz. *IEEE Trans. Veh. Tech.* VT-44(2):318–326.

Lee, W.C.Y. 1982. *Mobile Communications Engineering*, McGraw-Hill, New York.

Pahlavan, K. and Levesque, A. H. 1995. *Wireless Information Networks*, John Wiley, New York.

Proakis, J.G. 1989. *Digital Communications*, McGraw-Hill, New York.

Vaughan, R.G. 1990. Polarization diversity system in mobile communications. *IEEE Trans. Veh. Tech.* VT-39(3):177–86.

Viterbi, A.J. 1995. *CDMA: Principle of Spread Spectrum Communications*, Addison–Wesley, Reading, MA.

13

Digital Communication System Performance[1]

Bernard Sklar
Communications Engineering Services

13.1 Introduction

In this section we examine some fundamental tradeoffs among bandwidth, power, and error performance of digital communication systems. The criteria for choosing modulation and coding schemes, based on whether a system is bandwidth limited or power limited, are reviewed for several system examples. Emphasis is placed on the subtle but straightforward relationships we encounter when transforming from data-bits to channel-bits to symbols to chips.

The design or definition of any digital communication system begins with a description of the communication link. The *link* is the name given to the communication transmission path from the modulator and transmitter, through the channel, and up to and including the receiver and

---

[1] A version of this chapter has appeared as a paper in the *IEEE Communications Magazine*, November 1993, under the title "Defining, Designing, and Evaluating Digital Communication Systems."

demodulator. The *channel* is the name given to the propagating medium between the transmitter and receiver. A link description quantifies the average signal power that is received, the available bandwidth, the noise statistics, and other impairments, such as fading. Also needed to define the system are basic requirements, such as the data rate to be supported and the error performance.

The Channel

For radio communications, the concept of *free space* assumes a channel region free of all objects that might affect radio frequency (RF) propagation by absorption, reflection, or refraction. It further assumes that the atmosphere in the channel is perfectly uniform and nonabsorbing, and that the earth is infinitely far away or its reflection coefficient is negligible. The RF energy arriving at the receiver is assumed to be a function of distance from the transmitter (simply following the inverse-square law of optics). In practice, of course, propagation in the atmosphere and near the ground results in refraction, reflection, and absorption, which modify the free space transmission.

The Link

A radio transmitter is characterized by its average output signal power P_t and the gain of its transmitting antenna G_t. The name given to the product $P_t G_t$, with reference to an isotropic antenna is *effective radiated power* (*EIRP*) in watts (or dBW). The predetection average signal power S arriving at the output of the receiver antenna can be described as a function of the *EIRP*, the gain of the receiving antenna G_r, the path loss (or space loss) L_s, and other losses, L_o, as follows [Sklar, 1988; 1979]:

$$S = \frac{EIRP\, G_r}{L_s L_o} \tag{13.1}$$

The path loss L_s can be written as follows [Sklar, 1988]:

$$L_s = \left(\frac{4\pi d}{\lambda}\right)^2 \tag{13.2}$$

where d is the distance between the transmitter and receiver and λ is the wavelength.

We restrict our discussion to those links distorted by the mechanism of additive white Gaussian noise (AWGN) only. Such a noise assumption is a very useful model for a large class of communication systems. A valid approximation for average received noise power N that this model introduces is written as follows [Johnson, 1928; Nyquist, 1928]:

$$N \cong kT^\circ W \tag{13.3}$$

where k is Boltzmann's constant (1.38×10^{-23} joule/K), T° is effective temperature in kelvin, and W is bandwidth in hertz. Dividing Eq. (13.3) by bandwidth, enables us to write the received noise-power spectral density N_0 as follows:

$$N_0 = \frac{N}{W} = kT^\circ \tag{13.4}$$

Dividing Eq. (13.1) by N_0 yields the received average signal-power to noise-power spectral density S/N_0 as

$$\frac{S}{N_0} = \frac{EIRP\, G_r / T^\circ}{k L_s L_o} \tag{13.5}$$

where G_r/T° is often referred to as the receiver figure of merit. A link budget analysis is a compilation of the power gains and losses throughout the link; it is generally computed in decibels, and thus takes on the bookkeeping appearance of a business enterprise, highlighting the assets and liabilities of the link. Once the value of S/N_0 is specified or calculated from the link parameters, we then shift our attention to optimizing the choice of signalling types for meeting system bandwidth and error performance requirements.

Given the received S/N_0, we can write the received bit-energy to noise-power spectral density E_b/N_0, for any desired data rate R, as follows:

$$\frac{E_b}{N_0} = \frac{ST_b}{N_0} = \frac{S}{N_0}\left(\frac{1}{R}\right) \tag{13.6}$$

Equation (13.6) follows from the basic definitions that received bit energy is equal to received average signal power times the bit duration and that bit rate is the reciprocal of bit duration. Received E_b/N_0 is a key parameter in defining a digital communication system. Its value indicates the apportionment of the received waveform energy among the bits that the waveform represents. At first glance, one might think that a system specification should entail the symbol-energy to noise-power spectral density E_s/N_0 associated with the arriving waveforms. We will show, however, that for a given S/N_0 the value of E_s/N_0 is a function of the modulation and coding. The reason for defining systems in terms of E_b/N_0 stems from the fact that E_b/N_0 depends only on S/N_0 and R and is unaffected by any system design choices, such as modulation and coding.

13.2 Bandwidth and Power Considerations

Two primary communications resources are the received power and the available transmission bandwidth. In many communication systems, one of these resources may be more precious than the other and, hence, most systems can be classified as either bandwidth limited or power limited. In bandwidth-limited systems, spectrally efficient modulation techniques can be used to save bandwidth at the expense of power; in power-limited systems, power efficient modulation techniques can be used to save power at the expense of bandwidth. In both bandwidth- and power-limited systems, error-correction coding (often called channel coding) can be used to save power or to improve error performance at the expense of bandwidth. Recently, trellis-coded modulation (TCM) schemes have been used to improve the error performance of bandwidth-limited channels without any increase in bandwidth [Ungerboeck, 1987], but these methods are beyond the scope of this chapter.

The Bandwidth Efficiency Plane

Figure 13.1 shows the abscissa as the ratio of bit-energy to noise-power spectral density E_b/N_0 (in decibels) and the ordinate as the ratio of throughput, R (in bits per second), that can be transmitted per hertz in a given bandwidth W. The ratio R/W is called bandwidth efficiency, since it reflects how efficiently the bandwidth resource is utilized. The plot stems from the Shannon–Hartley capacity theorem [Sklar, 1988; Shannon, 1948; 1949], which can be stated as

$$C = W \log_2\left(1 + \frac{S}{N}\right) \tag{13.7}$$

where S/N is the ratio of received average signal power to noise power. When the logarithm is taken to the base 2, the capacity C, is given in bits per second. The capacity of a channel defines the maximum number of bits that can be reliably sent per second over the channel. For the case where the data (information) rate R is equal to C, the curve separates a region of practical communication systems from a region where such communication systems cannot operate reliably [Sklar, 1988; Shannon, 1948].

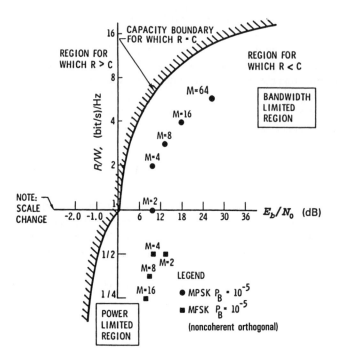

FIGURE 13.1   Bandwidth-efficiency plane.

M-ary Signalling

Each symbol in an M-ary alphabet can be related to a unique sequence of m bits, expressed as

$$M = 2^m \qquad \text{or} \qquad m = \log_2 M \tag{13.8}$$

where M is the size of the alphabet. In the case of digital transmission, the term symbol refers to the member of the M-ary alphabet that is transmitted during each symbol duration T_s. To transmit the symbol, it must be mapped onto an electrical voltage or current waveform. Because the waveform represents the symbol, the terms symbol and waveform are sometimes used interchangeably. Since one of M symbols or waveforms is transmitted during each symbol duration T_s, the data rate R in bits per second can be expressed as

$$R = \frac{m}{T_s} = \frac{\log_2 M}{T_s} \tag{13.9}$$

Data-bit-time duration is the reciprocal of data rate. Similarly, symbol-time duration is the recip-rocal of symbol rate. Therefore, from Eq. (13.9), we write that the effective time duration T_b of each bit in terms of the symbol duration T_s or the symbol rate R_s is

$$T_b = \frac{1}{R} = \frac{T_s}{m} = \frac{1}{m R_s} \tag{13.10}$$

Then, using Eqs. (13.8) and (13.10) we can express the symbol rate R_s in terms of the bit rate R as follows:

$$R_s = \frac{R}{\log_2 M} \tag{13.11}$$

From Eqs. (13.9) and (13.10), any digital scheme that transmits $m = \log_2 M$ bits in T_s seconds, using a bandwidth of W hertz, operates at a bandwidth efficiency of

$$\frac{R}{W} = \frac{\log_2 M}{WT_s} = \frac{1}{WT_b} \quad \text{(b/s)/Hz} \tag{13.12}$$

where T_b is the effective time duration of each data bit.

Bandwidth-Limited Systems

From Eq. (13.12), the smaller the WT_b product, the more bandwidth efficient will be any digital communication system. Thus, signals with small WT_b products are often used with bandwidth-limited systems. For example, the European digital mobile telephone system known as Global System for Mobile Communications (GSM) uses Gaussian minimum shift keying (GMSK) modulation having a WT_b product equal to 0.3 Hz/(b/s), where W is the 3-dB bandwidth of a Gaussian filter [Hodges, 1990].

For uncoded bandwidth-limited systems, the objective is to maximize the transmitted information rate within the allowable bandwidth, at the expense of E_b/N_0 (while maintaining a specified value of bit-error probability P_B). The operating points for coherent M-ary phase-shift keying (MPSK) at $P_B = 10^{-5}$ are plotted on the bandwidth-efficiency plane of Fig. 13.1. We assume Nyquist (ideal rectangular) filtering at baseband [Nyquist, April 1928]. Thus, for MPSK, the required double-sideband (DSB) bandwidth at an intermediate frequency (IF) is related to the symbol rate as follows:

$$W = \frac{1}{T_s} = R_s \tag{13.13}$$

where T_s is the symbol duration and R_s is the symbol rate. The use of Nyquist filtering results in the minimum required transmission bandwidth that yields zero intersymbol interference; such ideal filtering gives rise to the name Nyquist minimum bandwidth.

From Eqs. (13.12) and (13.13), the bandwidth efficiency of MPSK modulated signals using Nyquist filtering can be expressed as

$$R/W = \log_2 M \quad \text{(b/s)/Hz} \tag{13.14}$$

The MPSK points in Fig. 13.1 confirm the relationship shown in Eq. (13.14). Note that MPSK modulation is a bandwidth-efficient scheme. As M increases in value, R/W also increases. MPSK modulation can be used for realizing an improvement in bandwidth efficiency at the cost of increased E_b/N_0. Although beyond the scope of this chapter, many highly bandwidth-efficient modulation schemes are under investigation [Anderson and Sundberg, 1991].

Power-Limited Systems

Operating points for noncoherent orthogonal M-ary FSK (MFSK) modulation at $P_B = 10^{-5}$ are also plotted on Fig. 13.1. For MFSK, the IF minimum bandwidth is as follows [Sklar, 1988]

$$W = \frac{M}{T_s} = MR_s \tag{13.15}$$

where T_s is the symbol duration and R_s is the symbol rate. With MFSK, the required transmission bandwidth is expanded M-fold over binary FSK since there are M different orthogonal waveforms, each requiring a bandwidth of $1/T_s$. Thus, from Eqs. (13.12) and (13.15), the bandwidth efficiency of noncoherent orthogonal MFSK signals can be expressed as

$$\frac{R}{W} = \frac{\log_2 M}{M} \quad \text{(b/s)/Hz} \qquad (13.16)$$

The MFSK points plotted in Fig. 13.1 confirm the relationship shown in Eq. (13.16). Note that MFSK modulation is a bandwidth-expansive scheme. As M increases, R/W decreases. MFSK modulation can be used for realizing a reduction in required E_b/N_0 at the cost of increased bandwidth.

In Eqs. (13.13) and (13.14) for MPSK, and Eqs. (13.15) and (13.16) for MFSK, and for all the points plotted in Fig. 13.1, ideal filtering has been assumed. Such filters are not realizable! For realistic channels and waveforms, the required transmission bandwidth must be increased in order to account for realizable filters.

In the examples that follow, we will consider radio channels that are disturbed only by additive white Gaussian noise (AWGN) and have no other impairments, and for simplicity, we will limit the modulation choice to constant-envelope types, i.e., either MPSK or noncoherent orthogonal MFSK. For an uncoded system, MPSK is selected if the channel is bandwidth limited, and MFSK is selected if the channel is power limited. When error-correction coding is considered, modulation selection is not as simple, because coding techniques can provide power-bandwidth tradeoffs more effectively than would be possible through the use of any M-ary modulation scheme considered in this chapter [Clark and Cain, 1981].

In the most general sense, M-ary signalling can be regarded as a waveform-coding procedure, i.e., when we select an M-ary modulation technique instead of a binary one, we in effect have replaced the binary waveforms with better waveforms—either better for bandwidth performance (MPSK) or better for power performance (MFSK). Even though orthogonal MFSK signalling can be thought of as being a coded system, i.e., a first-order Reed-Muller code [Lindsey and Simon, 1973], we restrict our use of the term coded system to those traditional error-correction codes using redundancies, e.g., block codes or convolutional codes.

Minimum Bandwidth Requirements for MPSK and MFSK Signalling

The basic relationship between the symbol (or waveform) transmission rate R_s and the data rate R was shown in Eq. (13.11). Using this relationship together with Eqs. (13.13–13.16) and $R = 9600$ b/s, a summary of symbol rate, minimum bandwidth, and bandwidth efficiency for MPSK and noncoherent orthogonal MFSK was compiled for $M = 2, 4, 8, 16,$ and 32 (Table 13.1). Values of E_b/N_0 required to achieve a bit-error probability of 10^{-5} for MPSK and MFSK are also given for each value of M. These entries (which were computed using relationships that are presented later in this chapter) corroborate the tradeoffs shown in Fig. 13.1. As M increases, MPSK signalling provides more bandwidth efficiency at the cost of increased E_b/N_0, whereas MFSK signalling allows for a reduction in E_b/N_0 at the cost of increased bandwidth.

TABLE 13.1 Symbol Rate, Minimum Bandwidth, Bandwidth Efficiency, and Required E_b/N_0 for MPSK and Noncoherent Orthogonal MFSK Signalling at 9600 bit/s

M	m	R (b/s)	R_s (symb/s)	MPSK Minimum Bandwidth (Hz)	MPSK R/W	MPSK E_b/N_0 (dB) $P_B = 10^{-5}$	Noncoherent Orthog MFSK Min Bandwidth (Hz)	MFSK R/W	MFSK E_b/N_0 (dB) $P_B = 10^{-5}$
2	1	9600	9600	9600	1	9.6	19,200	1/2	13.4
4	2	9600	4800	4800	2	9.6	19,200	1/2	10.6
8	3	9600	3200	3200	3	13.0	25,600	3/8	9.1
16	4	9600	2400	2400	4	17.5	38,400	1/4	8.1
32	5	9600	1920	1920	5	22.4	61,440	5/32	7.4

13.3 Example 13.1: Bandwidth-Limited Uncoded System

Suppose we are given a bandwidth-limited AWGN radio channel with an available bandwidth of $W = 4000$ Hz. Also, suppose that the link constraints (transmitter power, antenna gains, path loss, etc.) result in the ratio of received average signal-power to noise-power spectral density S/N_0 being equal to 53 dB-Hz. Let the required data rate R be equal to 9600 b/s, and let the required bit-error performance P_B be at most 10^{-5}. The goal is to choose a modulation scheme that meets the required performance. In general, an error-correction coding scheme may be needed if none of the allowable modulation schemes can meet the requirements. In this example, however, we shall find that the use of error-correction coding is not necessary.

Solution to Example 13.1

For any digital communication system, the relationship between received S/N_0 and received bit-energy to noise-power spectral density, E_b/N_0 was given in Eq. (13.6) and is briefly rewritten as

$$\frac{S}{N_0} = \frac{E_b}{N_0} R \tag{13.17}$$

Solving for E_b/N_0 in decibels, we obtain

$$\begin{aligned}
\frac{E_b}{N_0} \text{ (dB)} &= \frac{S}{N_0} \text{ (dB-Hz)} - R \text{ (dB-b/s)} \\
&= 53 \text{ dB-Hz} - (10 \times \log_{10} 9600) \text{ dB-b/s} \\
&= 13.2 \text{ dB (or 20.89)}
\end{aligned} \tag{13.18}$$

Since the required data rate of 9600 b/s is much larger than the available bandwidth of 4000 Hz, the channel is bandwidth limited. We therefore select MPSK as our modulation scheme. We have confined the possible modulation choices to be constant-envelope types; without such a restriction, we would be able to select a modulation type with greater bandwidth efficiency. To conserve power, we compute the *smallest possible* value of M such that the MPSK minimum bandwidth does not exceed the available bandwidth of 4000 Hz. Table 13.1 shows that the smallest value of M meeting this requirement is $M = 8$. Next we determine whether the required bit-error performance of $P_B \leq 10^{-5}$ can be met by using 8-PSK modulation alone or whether it is necessary to use an error-correction coding scheme. Table 13.1 shows that 8-PSK alone will meet the requirements, since the required E_b/N_0 listed for 8-PSK is less than the received E_b/N_0 derived in Eq. (13.18). Let us imagine that we do not have Table 13.1, however, and evaluate whether or not error-correction coding is necessary.

Figure 13.2 shows the basic modulator/demodulator (MODEM) block diagram summarizing the functional details of this design. At the modulator, the transformation from data bits to symbols yields an output symbol rate R_s, that is, a factor $\log_2 M$ smaller than the input data-bit rate R, as is seen in Eq. (13.11). Similarly, at the input to the demodulator, the symbol-energy to noise-power spectral density E_S/N_0 is a factor $\log_2 M$ larger than E_b/N_0, since each symbol is made up of $\log_2 M$ bits. Because E_S/N_0 is larger than E_b/N_0 by the same factor that R_s is smaller than R, we can expand Eq. (13.17), as follows:

$$\frac{S}{N_0} = \frac{E_b}{N_0} R = \frac{E_s}{N_0} R_s \tag{13.19}$$

The demodulator receives a waveform (in this example, one of $M = 8$ possible phase shifts) during each time interval T_s. The probability that the demodulator makes a symbol error $P_E(M)$ is well

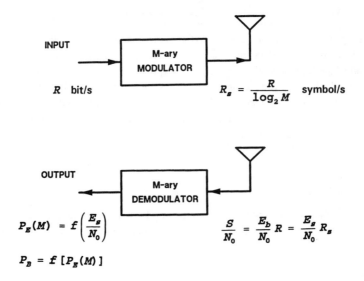

FIGURE 13.2   Basic modulator/demodulator (MODEM) without channel coding.

approximated by the following equation for $M > 2$ [Korn, 1985]:

$$P_E(M) \cong 2Q\left[\sqrt{\frac{2E_s}{N_0}}\sin\left(\frac{\pi}{M}\right)\right] \tag{13.20}$$

where $Q(x)$, sometimes called the complementary error function, represents the probability under the tail of a zero-mean unit-variance Gaussian density function. It is defined as follows [Van Trees, 1968]:

$$Q(x) = \frac{1}{\sqrt{2\pi}}\int_x^\infty \exp\left(-\frac{u^2}{2}\right)du \tag{13.21}$$

A good approximation for $Q(x)$, valid for $x > 3$, is given by the following equation [Borjesson and Sundberg, 1979]:

$$Q(x) \cong \frac{1}{x\sqrt{2\pi}}\exp\left(-\frac{x^2}{2}\right) \tag{13.22}$$

In Fig. 13.2 and all of the figures that follow, rather than show explicit probability relationships, the generalized notation $f(x)$ has been used to indicate some functional dependence on x.

A traditional way of characterizing communication efficiency in digital systems is in terms of the received E_b/N_0 in decibels. This E_b/N_0 description has become standard practice, but recall that there are no bits at the input to the demodulator; there are only waveforms that have been assigned bit meanings. The received E_b/N_0 represents a bit-apportionment of the arriving waveform energy.

To solve for $P_E(M)$ in Eq. (13.20), we first need to compute the ratio of received symbol-energy to noise-power spectral density E_s/N_0. Since from Eq. (13.18)

$$\frac{E_b}{N_0} = 13.2 \text{ dB (or } 20.89)$$

and because each symbol is made up of $\log_2 M$ bits, we compute the following using $M = 8$.

$$\frac{E_s}{N_0} = (\log_2 M)\frac{E_b}{N_0} = 3 \times 20.89 = 62.67 \tag{13.23}$$

Using the results of Eq. (13.23) in Eq. (13.20), yields the symbol-error probability $P_E = 2.2 \times 10^{-5}$. To transform this to bit-error probability, we use the relationship between bit-error probability P_B and symbol-error probability P_E, for multiple-phase signalling [Lindsey and Simon, 1973] for $P_E \ll 1$ as follows:

$$P_B \cong \frac{P_E}{\log_2 M} = \frac{P_E}{m} \tag{13.24}$$

which is a good approximation when Gray coding is used for the bit-to-symbol assignment [Korn, 1985]. This last computation yields $P_B = 7.3 \times 10^{-6}$, which meets the required bit-error performance. No error-correction coding is necessary, and 8-PSK modulation represents the design choice to meet the requirements of the bandwidth-limited channel, which we had predicted by examining the required E_b/N_0 values in Table 13.1.

13.4 Example 13.2: Power-Limited Uncoded System

Now, suppose that we have exactly the same data rate and bit-error probability requirements as in Example 13.1, but let the available bandwidth W be equal to 45 kHz, and the available S/N_0 be equal to 48 dB-Hz. The goal is to choose a modulation or modulation/coding scheme that yields the required performance. We shall again find that error-correction coding is not required.

Solution to Example 13.2

The channel is clearly not bandwidth limited since the available bandwidth of 45 kHz is more than adequate for supporting the required data rate of 9600 bit/s. We find the received E_b/N_0 from Eq. (13.18), as follows:

$$\frac{E_b}{N_0} \text{ (dB)} = 48 \text{ dB-Hz} - (10 \times \log_{10} 9600) \text{ dB-b/s} = 8.2 \text{ dB (or 6.61)} \tag{13.25}$$

Since there is abundant bandwidth but a relatively small E_b/N_0 for the required bit-error probability, we consider that this channel is power limited and choose MFSK as the modulation scheme. To conserve power, we search for the *largest possible M* such that the MFSK minimum bandwidth is not expanded beyond our available bandwidth of 45 kHz. A search results in the choice of $M = 16$ (Table 13.1). Next, we determine whether the required error performance of $P_B \le 10^{-5}$ can be met by using 16-FSK alone, i.e., without error-correction coding. Table 13.1 shows that 16-FSK alone meets the requirements, since the required E_b/N_0 listed for 16-FSK is less than the received E_b/N_0 derived in Eq. (13.25). Let us imagine again that we do not have Table 13.1, and evaluate whether or not error-correction coding is necessary.

The block diagram in Fig. 13.2 summarizes the relationships between symbol rate R_s, and bit rate R, and between E_s/N_0 and E_b/N_0, which is identical to each of the respective relationships in Example 13.1. The 16-FSK demodulator receives a waveform (one of 16 possible frequencies) during each symbol time interval T_s. For noncoherent orthogonal MFSK, the probability that the demodulator makes a symbol error $P_E(M)$ is approximated by the following upper bound [Viterbi, 1979]:

$$P_E(M) \le \frac{M-1}{2}\exp\left(-\frac{E_s}{2N_0}\right) \tag{13.26}$$

To solve for $P_E(M)$ in Eq. (13.26), we compute E_S/N_0, as in Example 13.1. Using the results of Eq. (13.25) in Eq. (13.23), with $M = 16$, we get

$$\frac{E_s}{N_0} = (\log_2 M)\frac{E_b}{N_0} = 4 \times 6.61 = 26.44 \tag{13.27}$$

Next, using the results of Eq. (13.27) in Eq. (13.26), yields the symbol-error probability $P_E = 1.4 \times 10^{-5}$. To transform this to bit-error probability, P_B, we use the relationship between P_B and P_E for orthogonal signalling [Viterbi, 1979], given by

$$P_B = \frac{2^{m-1}}{(2^m - 1)}P_E \tag{13.28}$$

This last computation yields $P_B = 7.3 \times 10^{-6}$, which meets the required bit-error performance. Thus, we can meet the given specifications for this power-limited channel by using 16-FSK modulation, without any need for error-correction coding, as we had predicted by examining the required E_b/N_0 values in Table 13.1.

13.5 Example 13.3: Bandwidth-Limited and Power-Limited Coded System

We start with the same channel parameters as in Example 13.1 ($W = 4000$ Hz, $S/N_0 = 53$ dB-Hz, and $R = 9600$ b/s), with one exception. In this example, we specify that P_B must be at most 10^{-9}. Table 13.1 shows that the system is both bandwidth limited and power limited, based on the available bandwidth of 4000 Hz and the available E_b/N_0 of 13.2 dB, from Eq. (13.18); 8-PSK is the only possible choice to meet the bandwidth constraint; however, the available E_b/N_0 of 13.2 dB is certainly insufficient to meet the required P_B of 10^{-9}. For this small value of P_B, we need to consider the performance improvement that error-correction coding can provide within the available bandwidth. In general, one can use convolutional codes or block codes.

The Bose–Chaudhuri–Hocquenghem (BCH) codes form a large class of powerful error-correcting cyclic (block) codes [Lin and Costello, 1983]. To simplify the explanation, we shall choose a block code from the BCH family. Table 13.2 presents a partial catalog of the available BCH codes in terms of n, k, and t, where k represents the number of information (or data) bits that the code transforms into a longer block of n coded bits (or channel bits), and t represents the largest number of incorrect channel bits that the code can correct within each n-sized block. The rate of a code is defined as the ratio k/n; its inverse represents a measure of the code's redundancy [Lin and Costello, 1983].

TABLE 13.2 BCH Codes (Partial Catalog)

n	k	t
7	4	1
15	11	1
	7	2
	5	3
31	26	1
	21	2
	16	3
	11	5
63	57	1
	51	2
	45	3
	39	4
	36	5
	30	6
127	120	1
	113	2
	106	3
	99	4
	92	5
	85	6
	78	7
	71	9
	64	10

Solution to Example 13.3

Since this example has the same bandwidth-limited parameters given in Example 13.1, we start with the same 8-PSK modulation used to meet the stated bandwidth constraint. We now employ error-correction coding, however, so that the bit-error probability can be lowered to $P_B \le 10^{-9}$.

To make the optimum code selection from Table 13.2, we are guided by the following goals.

1. The output bit-error probability of the combined modulation/coding system must meet the system error requirement.
2. The rate of the code must not expand the required transmission bandwidth beyond the available channel bandwidth.
3. The code should be as simple as possible. Generally, the shorter the code, the simpler will be its implementation.

The uncoded 8-PSK minimum bandwidth requirement is 3200 Hz (Table 13.1) and the allowable channel bandwidth is 4000 Hz, and so the uncoded signal bandwidth can be increased by no more than a factor of 1.25 (i.e., an expansion of 25%). The very first step in this (simplified) code selection example is to eliminate the candidates in Table 13.2 that would expand the bandwidth by more than 25%. The remaining entries form a much reduced set of bandwidth-compatible codes (Table 13.3).

TABLE 13.3 Bandwidth-Compatible BCH Codes

n	k	t	Coding Gain, G (dB) MPSK, $P_B = 10^{-9}$
31	26	1	2.0
63	57	1	2.2
	51	2	3.1
127	120	1	2.2
	113	2	3.3
	106	3	3.9

In Table 13.3, a column designated Coding Gain G (for MPSK at $P_B = 10^{-9}$) has been added. Coding gain in decibels is defined as follows:

$$G = \left(\frac{E_b}{N_0}\right)_{\text{uncoded}} - \left(\frac{E_b}{N_0}\right)_{\text{coded}} \tag{13.29}$$

G can be described as the reduction in the required E_b/N_0 (in decibels) that is needed due to the error-performance properties of the channel coding. G is a function of the modulation type and bit-error probability, and it has been computed for MPSK at $P_B = 10^{-9}$ (Table 13.3). For MPSK modulation, G is relatively independent of the value of M. Thus, for a particular bit-error probability, a given code will provide about the same coding gain when used with any of the MPSK modulation schemes. Coding gains were calculated using a procedure outlined in the subsequent Calculating Coding Gain section.

A block diagram summarizes this system, which contains both modulation and coding (Fig. 13.3). The introduction of encoder/decoder blocks brings about additional transformations. The relationships that exist when transforming from R b/s to R_c channel-b/s to R_s symbol/s are shown at the encoder/modulator. Regarding the channel-bit rate R_c, some authors prefer to use the units of channel-symbol/s (or code-symbol/s). The benefit is that error-correction coding is often described more efficiently with nonbinary digits. We reserve the term symbol for that group of bits mapped onto an electrical waveform for transmission, and we designate the units of R_c to be channel-b/s (or coded-b/s).

We assume that our communication system cannot tolerate any message delay, so that the channel-bit rate R_c must exceed the data-bit rate R by the factor n/k. Further, each symbol is made up of $\log_2 M$ channel bits, and so the symbol rate R_s is less than R_c by the factor $\log_2 M$. For a system containing both modulation and coding, we summarize the rate transformations as follows:

$$R_c = \left(\frac{n}{k}\right) R \tag{13.30}$$

$$R_s = \frac{R_c}{\log_2 M} \tag{13.31}$$

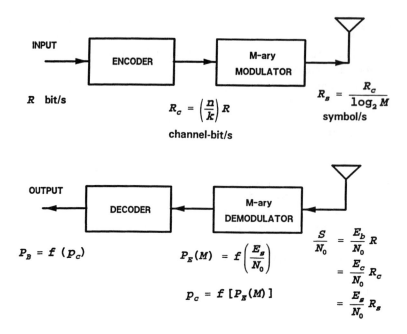

FIGURE 13.3   MODEM with channel coding.

At the demodulator/decoder in Fig. 13.3, the transformations among data-bit energy, channel-bit energy, and symbol energy are related (in a reciprocal fashion) by the same factors as shown among the rate transformations in Eqs. (13.30) and (13.31). Since the encoding transformation has replaced k data bits with n channel bits, then the ratio of channel-bit energy to noise-power spectral density E_c/N_0 is computed by decrementing the value of E_b/N_0 by the factor k/n. Also, since each transmission symbol is made up of $\log_2 M$ channel bits, then E_s/N_0, which is needed in Eq. (13.20) to solve for P_E, is computed by incrementing E_c/N_0 by the factor $\log_2 M$. For a system containing both modulation and coding, we summarize the energy to noise-power spectral density transformations as follows:

$$\frac{E_c}{N_0} = \left(\frac{k}{n}\right)\frac{E_b}{N_0} \tag{13.32}$$

$$\frac{E_s}{N_0} = (\log_2 M)\frac{E_c}{N_0} \tag{13.33}$$

Using Eqs. (13.30) and (13.31), we can now expand the expression for S/N_0 in Eq. (13.19), as follows(Appendix).

$$\frac{S}{N_0} = \frac{E_b}{N_0}R = \frac{E_c}{N_0}R_c = \frac{E_s}{N_0}R_s \tag{13.34}$$

As before, a standard way of describing the link is in terms of the received E_b/N_0 in decibels. However, there are no data bits at the input to the demodulator, and there are no channel bits; there are only waveforms that have bit meanings and, thus, the waveforms can be described in terms of bit-energy apportionments.

Since S/N_0 and R were given as 53 dB-Hz and 9600 b/s, respectively, we find as before, from Eq. (13.18), that the received $E_b/N_0 = 13.2$ dB. The received E_b/N_0 is fixed and independent of n, k, and t (Appendix). As we search, in Table 13.3 for the ideal code to meet the specifications, we can iteratively repeat the computations suggested in Fig. 13.3. It might be useful to program on a personal computer (or calculator) the following four steps as a function of n, k, and t. Step 1 starts by combining Eqs. (13.32) and (13.33), as follows.

Step 1:

$$\frac{E_s}{N_0} = (\log_2 M)\frac{E_c}{N_0} = (\log_2 M)\left(\frac{k}{n}\right)\frac{E_b}{N_0} \qquad (13.35)$$

Step 2:

$$P_E(M) \cong 2Q\left[\sqrt{\frac{2E_s}{N_0}}\,\sin\left(\frac{\pi}{M}\right)\right] \qquad (13.36)$$

which is the approximation for symbol-error probability P_E rewritten from Eq. (13.20). At each symbol-time interval, the demodulator makes a symbol decision, but it delivers a channel-bit sequence representing that symbol to the decoder. When the channel-bit output of the demodulator is quantized to two levels, 1 and 0, the demodulator is said to make hard decisions. When the output is quantized to more than two levels, the demodulator is said to make soft decisions [Sklar, 1988]. Throughout this paper, we shall assume hard-decision demodulation.

Now that we have a decoder block in the system, we designate the channel-bit-error probability out of the demodulator and into the decoder as p_c, and we reserve the notation P_B for the bit-error probability out of the decoder. We rewrite Eq. (13.24) in terms of p_c for $P_E \ll 1$ as follows.

Step 3:

$$p_c \cong \frac{P_E}{\log_2 M} = \frac{P_E}{m} \qquad (13.37)$$

relating the channel-bit-error probability to the symbol-error probability out of the demodulator, assuming Gray coding, as referenced in Eq. (13.24).

For traditional channel-coding schemes and a given value of received S/N_0, the value of E_s/N_0 with coding will always be less than the value of E_s/N_0 without coding. Since the demodulator with coding receives less E_s/N_0, it makes more errors! When coding is used, however, the system error-performance does not only depend on the performance of the demodulator, it also depends on the performance of the decoder. For error-performance improvement due to coding, the decoder must provide enough error correction to more than compensate for the poor performance of the demodulator.

The final output decoded bit-error probability P_B depends on the particular code, the decoder, and the channel-bit-error probability p_c. It can be expressed by the following approximation [Oldenwalder, 1976].

Step 4:

$$P_B \cong \frac{1}{n}\sum_{j=t+1}^{n} j\binom{n}{j}p_c^{j}(1-p_c)^{n-j} \qquad (13.38)$$

where t is the largest number of channel bits that the code can correct within each block of n bits. Using Eqs. (13.35–13.38) in the four steps, we can compute the decoded bit-error probability P_B as a function of n, k, and t for each of the codes listed in Table 13.3. The entry that meets the stated error requirement with the largest possible code rate and the smallest value of n is the double-error correcting (63, 51) code. The computations are as follows.

Step 1:

$$\frac{E_s}{N_0} = 3\left(\frac{51}{63}\right)20.89 = 50.73$$

where $M = 8$, and the received $E_b/N_0 = 13.2$ dB (or 20.89).

Step 2:

$$P_E \cong 2Q\left[\sqrt{101.5} \times \sin\left(\frac{\pi}{8}\right)\right] = 2Q(3.86) = 1.2 \times 10^{-4}$$

Step 3:

$$p_c \cong \frac{1.2 \times 10^{-4}}{3} = 4 \times 10^{-5}$$

Step 4:

$$P_B \cong \frac{3}{63}\binom{63}{3}(4 \times 10^{-5})^3(1 - 4 \times 10^{-5})^{60}$$

$$+ \frac{4}{63}\binom{63}{4}(4 \times 10^{-5})^4(1 - 4 \times 10^{-5})^{59} + \cdots$$

$$= 1.2 \times 10^{-10}$$

where the bit-error-correcting capability of the code is $t = 2$. For the computation of P_B in step 4, we need only consider the first two terms in the summation of Eq. (13.38) since the other terms have a vanishingly small effect on the result. Now that we have selected the (63, 51) code, we can compute the values of channel-bit rate R_c and symbol rate R_s using Eqs. (13.30) and (13.31), with $M = 8$,

$$R_c = \left(\frac{n}{k}\right)R = \left(\frac{63}{51}\right)9600 \approx 11{,}859 \text{ channel-b/s}$$

$$R_s = \frac{R_c}{\log_2 M} = \frac{11859}{3} = 3953 \text{ symbol/s}$$

Calculating Coding Gain

Perhaps a more direct way of finding the simplest code that meets the specified error performance is to first compute how much coding gain G is required in order to yield $P_B = 10^{-9}$ when using 8-PSK modulation alone; then, from Table 13.3, we can simply choose the code that provides this performance improvement. First, we find the uncoded E_s/N_0 that yields an error probability of $P_B = 10^{-9}$, by writing from Eqs. (13.24) and (13.36), the following:

$$P_B \cong \frac{P_E}{\log_2 M} \cong \frac{2Q\left[\sqrt{\frac{2E_s}{N_0}}\sin\left(\frac{\pi}{M}\right)\right]}{\log_2 M} = 10^{-9} \qquad (13.39)$$

At this low value of bit-error probability, it is valid to use Eq. (13.22) to approximate $Q(x)$ in Eq. (13.39). By trial and error (on a programmable calculator), we find that the uncoded $E_s/N_0 = 120.67 = 20.8$ dB, and since each symbol is made up of $\log_2 8 = 3$ bits, the required $(E_b/N_0)_{\text{uncoded}} = 120.67/3 = 40.22 = 16$ dB. From the given parameters and Eq. (13.18), we know that the received $(E_b/N_0)_{\text{coded}} = 13.2$ dB. Using Eq. (13.29), the required coding gain to meet the bit-error performance of $P_B = 10^{-9}$ in decibels is

$$G = \left(\frac{E_b}{N_0}\right)_{\text{uncoded}} - \left(\frac{E_b}{N_0}\right)_{\text{coded}} = 16 - 13.2 = 2.8$$

To be precise, each of the E_b/N_0 values in the preceding computation must correspond to exactly the same value of bit-error probability (which they do not). They correspond to $P_B = 10^{-9}$

and $P_B = 1.2 \times 10^{-10}$, respectively. At these low probability values, however, even with such a discrepancy, this computation still provides a good approximation of the required coding gain. In searching Table 13.3 for the simplest code that will yield a coding gain of at least 2.8 dB, we see that the choice is the (63, 51) code, which corresponds to the same code choice that we made earlier.

13.6 Example 13.4: Direct-Sequence (DS) Spread-Spectrum Coded System

Spread-spectrum systems are not usually classified as being bandwidth- or power-limited. They are generally perceived to be power-limited systems, however, because the bandwidth occupancy of the information is much larger than the bandwidth that is intrinsically needed for the information transmission. In a direct-sequence spread-spectrum (DS/SS) system, spreading the signal bandwidth by some factor permits lowering the signal-power spectral density by the same factor (the total average signal power is the same as before spreading). The bandwidth spreading is typically accomplished by multiplying a relatively narrowband data signal by a wideband spreading signal. The spreading signal or spreading code is often referred to as a pseudorandom code or PN code.

Processing Gain

A typical DS/SS radio system is often described as a two-step BPSK modulation process. In the first step, the carrier wave is modulated by a bipolar data waveform having a value $+1$ or -1 during each data-bit duration; in the second step, the output of the first step is multiplied (modulated) by a bipolar PN-code waveform having a value $+1$ or -1 during each PN-code-bit duration. In reality, DS/SS systems are usually implemented by first multiplying the data waveform by the PN-code waveform and then making a single pass through a BPSK modulator. For this example, however, it is useful to characterize the modulation process in two separate steps—the outer modulator/demodulator for the data, and the inner modulator/demodulator for the PN code (Fig. 13.4).

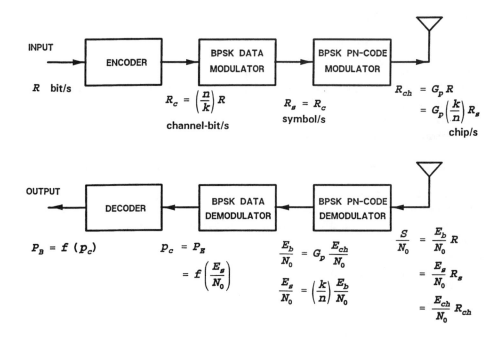

FIGURE 13.4 Direct-sequence spread-spectrum MODEM with channel coding.

A spread-spectrum system is characterized by a processing gain G_p, that is defined in terms of the spread-spectrum bandwidth W_{ss} and the data rate R as follows [Viterbi, 1979]:

$$G_p = \frac{W_{ss}}{R} \qquad (13.40)$$

For a DS/SS system, the PN-code bit has been given the name chip, and the spread-spectrum signal bandwidth can be shown to be about equal to the chip rate R_{ch} as follows:

$$G_p = \frac{R_{ch}}{R} \qquad (13.41)$$

Some authors define processing gain to be the ratio of the spread-spectrum bandwidth to the symbol rate. This definition separates the system performance that is due to bandwidth spreading from the performance that is due to error-correction coding. Since we ultimately want to relate all of the coding mechanisms relative to the information source, we shall conform to the most usually accepted definition for processing gain, as expressed in Eqs. (13.40) and (13.41).

A spread-spectrum system can be used for interference rejection and for multiple access (allowing multiple users to access a communications resource simultaneously). The benefits of DS/SS signals are best achieved when the processing gain is very large; in other words, the chip rate of the spreading (or PN) code is much larger than the data rate. In such systems, the large value of G_p allows the signalling chips to be transmitted at a power level well below that of the thermal noise. We will use a value of $G_p = 1000$. At the receiver, the despreading operation correlates the incoming signal with a synchronized copy of the PN code and, thus, accumulates the energy from multiple (G_p) chips to yield the energy per data bit. The value of G_p has a major influence on the performance of the spread-spectrum system application. We shall see, however, that the value of G_p has no effect on the received E_b/N_0. In other words, spread spectrum techniques offer no error-performance advantage over thermal noise. For DS/SS systems, there is no disadvantage either! Sometimes such spread-spectrum radio systems are employed only to enable the transmission of very small power-spectral densities and thus avoid the need for FCC licensing [Title, 47].

Channel Parameters for Example 13.4

Consider a DS/SS radio system that uses the same (63, 51) code as in the previous example. Instead of using MPSK for the data modulation, we shall use BPSK. Also, we shall use BPSK for modulating the PN-code chips. Let the received $S/N_0 = 48$ dB-Hz, the data rate $R = 9600$ b/s, and the required $P_B \leq 10^{-6}$. For simplicity, assume that there are no bandwidth constraints. Our task is simply to determine whether or not the required error performance can be achieved using the given system architecture and design parameters. In evaluating the system, we will use the same type of transformations used in the previous examples.

Solution to Example 13.4

A typical DS/SS system can be implemented more simply than the one shown in Fig. 13.4. The data and the PN code would be combined at baseband, followed by a single pass through a BPSK modulator. We will, however, assume the existence of the individual blocks in Fig. 13.4 because they enhance our understanding of the transformation process. The relationships in transforming from data bits, to channel bits, to symbols, and to chips (Fig. 13.4) have the same pattern of subtle but straightforward transformations in rates and energies as previous relationships (Figs. 13.2 and 13.3). The values of R_c, R_s, and R_{ch} can now be calculated immediately since the (63, 51) BCH code has already been selected. From Eq. (13.30) we write

$$R_c = \left(\frac{n}{k}\right)R = \left(\frac{63}{51}\right)9600 \approx 11,859 \text{ channel-b/s}$$

Since the data modulation considered here is BPSK, then from Eq. (13.31) we write

$$R_s = R_c \approx 11,859 \text{ symbol/s}$$

and from Eq. (13.41), with an assumed value of $G_p = 1000$

$$R_{\text{ch}} = G_p R = 1000 \times 9600 = 9.6 \times 10^6 \text{ chip/s}$$

Since we have been given the same S/N_0 and the same data rate as in Example 13.2, we find the value of received E_b/N_0 from Eq. (13.25) to be 8.2 dB (or 6.61). At the demodulator, we can now expand the expression for S/N_0 in Eq. (13.34) and the Appendix as follows:

$$\frac{S}{N_0} = \frac{E_b}{N_0} R = \frac{E_c}{N_0} R_c = \frac{E_s}{N_0} R_s = \frac{E_{\text{ch}}}{N_0} R_{\text{ch}} \tag{13.42}$$

Corresponding to each transformed entity (data bit, channel bit, symbol, or chip) there is a change in rate and, similarly, a reciprocal change in energy-to-noise spectral density for that received entity. Equation (13.42) is valid for any such transformation when the rate and energy are modified in a reciprocal way. There is a kind of *conservation of power* (or energy) phenomenon that exists in the transformations. The total received average power (or total received energy per symbol duration) is fixed regardless of how it is computed, on the basis of data bits, channel bits, symbols, or chips.

The ratio E_{ch}/N_0 is much lower in value than E_b/N_0. This can be seen from Eqs. (13.42) and (13.41), as follows:

$$\frac{E_{\text{ch}}}{N_0} = \frac{S}{N_0} \left(\frac{1}{R_{\text{ch}}} \right) = \frac{S}{N_0} \left(\frac{1}{G_p R} \right) = \left(\frac{1}{G_p} \right) \frac{E_b}{N_0} \tag{13.43}$$

But, even so, the despreading function (when properly synchronized) accumulates the energy contained in a quantity G_p of the chips, yielding the same value $E_b/N_0 = 8.2$ dB, as was computed earlier from Eq. (13.25). Thus, the DS spreading transformation has no effect on the error performance of an AWGN channel [Sklar, 1988], and the value of G_p has no bearing on the value of P_B in this example.

From Eq. (13.43), we can compute, in decibels,

$$\begin{aligned} \frac{E_{\text{ch}}}{N_0} &= E_b/N_0 - G_p \\ &= 8.2 - (10 \times \log_{10} 1000) \\ &= -21.8 \end{aligned} \tag{13.44}$$

The chosen value of processing gain ($G_p = 1000$) enables the DS/SS system to operate at a value of chip energy well below the thermal noise, with the same error performance as without spreading.

Since BPSK is the data modulation selected in this example, each message symbol therefore corresponds to a single channel bit, and we can write

$$\frac{E_s}{N_0} = \frac{E_c}{N_0} = \left(\frac{k}{n} \right) \frac{E_b}{N_0} = \left(\frac{51}{63} \right) \times 6.61 = 5.35 \tag{13.45}$$

where the received $E_b/N_0 = 8.2$ dB (or 6.61). Out of the BPSK data demodulator, the symbol-error probability P_E (and the channel-bit error probability p_c) is computed as follows [Sklar, 1988]:

$$p_c = P_E = Q\left(\sqrt{\frac{2E_c}{N_0}} \right) \tag{13.46}$$

Using the results of Eq. (13.45) in Eq. (13.46) yields

$$p_c = Q(3.27) = 5.8 \times 10^{-4}$$

Finally, using this value of p_c in Eq. (13.38) for the (63, 51) double-error correcting code yields the output bit-error probability of $P_B = 3.6 \times 10^{-7}$. We can, therefore, verify that for the given architecture and design parameters of this example the system does, in fact, achieve the required error performance.

13.7 Conclusion

The goal of this section has been to review fundamental relationships used in evaluating the performance of digital communication systems. First, we described the concept of a link and a channel and examined a radio system from its transmitting segment up through the output of the receiving antenna. We then examined the concept of bandwidth-limited and power-limited systems and how such conditions influence the system design when the choices are confined to MPSK and MFSK modulation. Most important, we focused on the definitions and computations involved in transforming from data bits to channel bits to symbols to chips. In general, most digital communication systems share these concepts; thus, understanding them should enable one to evaluate other such systems in a similar way.

Appendix: Received E_b/N_0 Is Independent of the Code Parameters

Starting with the basic concept that the received average signal power S is equal to the received symbol or waveform energy, E_s, divided by the symbol-time duration, T_s (or multiplied by the symbol rate, R_s), we write

$$\frac{S}{N_0} = \frac{E_s/T_s}{N_0} = \frac{E_s}{N_0}R_s \tag{A13.1}$$

where N_0 is noise-power spectral density.

Using Eqs. (13.27) and (13.25), rewritten as

$$\frac{E_s}{N_0} = (\log_2 M)\frac{E_c}{N_0} \quad \text{and} \quad R_s = \frac{R_c}{\log_2 M}$$

let us make substitutions into Eq. (A13.1), which yields

$$\frac{S}{N_0} = \frac{E_c}{N_0}R_c \tag{A13.2}$$

Next, using Eqs. (13.26) and (13.24), rewritten as

$$\frac{E_c}{N_0} = \left(\frac{k}{n}\right)\frac{E_b}{N_0} \quad \text{and} \quad R_c = \left(\frac{n}{k}\right)R$$

let us now make substitutions into Eq. (A13.2), which yields the relationship expressed in Eq. (13.11)

$$\frac{S}{N_0} = \frac{E_b}{N_0}R \tag{A13.3}$$

Hence, the received E_b/N_0 is only a function of the received S/N_0 and the data rate R. It is independent of the code parameters, n, k, and t. These results are summarized in Fig. 13.3.

References

Anderson, J.B. and Sundberg, C.-E.W. 1991. Advances in constant envelope coded modulation, *IEEE Commun., Mag.*, 29(12):36–45.

Borjesson, P.O. and Sundberg, C.E. 1979. Simple approximations of the error function $Q(x)$ for communications applications, *IEEE Trans. Comm.*, COM-27(March):639–642.

Clark, G.C., Jr. and Cain, J.B. 1981. *Error-Correction Coding for Digital Communications*, Plenum Press, New York.

Hodges, M.R.L. 1990. The GSM radio interface, *British Telecom Technol. J.* 8(1):31–43.

Johnson, J.B. 1928. Thermal agitation of electricity in conductors, *Phys. Rev.* 32(July):97–109.

Korn, I. 1985. *Digital Communications*, Van Nostrand Reinhold Co., New York.

Lin, S. and Costello, D.J., Jr. 1983. *Error Control Coding: Fundamentals and Applications*, Prentice-Hall, Englewood Cliffs, NJ.

Lindsey, W.C. and Simon, M.K. 1973. *Telecommunication Systems Engineering*, Prentice-Hall, Englewood Cliffs, NJ.

Nyquist, H. 1928. Thermal agitation of electric charge in conductors, *Phys. Rev.* 32(July):110–113.

Nyquist, H. 1928. Certain topics on telegraph transmission theory, *Trans. AIEE.* 47(April):617–644.

Odenwalder, J.P. 1976. *Error Control Coding Handbook*. Linkabit Corp., San Diego, CA, July 15.

Shannon, C.E. 1948. A mathematical theory of communication, *BSTJ.* 27:379–423, 623–657.

Shannon, C.E. 1949. Communication in the presence of noise, *Proc. IRE.* 37(1):10–21.

Sklar, B. 1988. *Digital Communications: Fundamentals and Applications*, Prentice-Hall Inc., Englewood Cliffs, N.J.

Sklar, B. 1979. What the system link budget tells the system engineer or how I learned to count in decibels, *Proc. of the Int'l. Telemetering Conf.*, San Diego, CA, Nov.

Title 47, *Code of Federal Regulations*, Part 15 Radio Frequency Devices.

Ungerboeck, G. 1987. Trellis-coded modulation with redundant signal sets, Pt. I and II, *IEEE Comm. Mag.*, 25(Feb.):5–21.

Van Trees, H.L. 1968. *Detection, Estimation, and Modulation Theory*, Pt. I, John Wiley and Sons, Inc., New York.

Viterbi, A.J. 1966. *Principles of Coherent Communication*, McGraw-Hill Book Co., New York.

Viterbi, A.J. 1979. Spread spectrum communications—myths and realities, *IEEE Comm. Mag.*, (May):11–18.

Further Information

A useful compilation of selected papers can be found in: *cellular Radio & Personal Communications– A Book of Selected Readings*, edited by Theodore S. Rappaport, Institute of Electrical and Electronics Engineers, Inc., Piscataway, New Jersey, 1995. Fundamental design issues, such as propagation, modulation, channel coding, speech coding, multiple-accessing and networking, are well represented in this volume.

Another useful sourcebook that covers the fundamentals of mobile communications in great detail is: *Mobile Radio Communications*, edited by Raymond Steele, Pentech Press, London 1992. This volume is also available through the Institute of Electrical and Electronics Engineers, Inc., Piscataway, New Jersey.

For spread spectrum systems, an excellent reference is: *Spread Spectrum Communications Handbook*, by Marvin K. Simon, Jim K Omura, Robert A. Scholtz, and Barry K. Levitt, McGraw-Hill Inc., New York, 1994.

14

Standards Setting Bodies

14.1 Introduction 196
14.2 Global Standardization 197
    ITU-T • ITU-R • BDT • ISO/IEC JTC 1
14.3 Regional Standardization 200
14.4 National Standardization 202
    ANSI T1 • TIA • TTC
14.5 Standards Coordination 204
14.6 Scientific 205
14.7 Standards Development Cycle 206

Spiros Dimolitsas
COMSAT Laboratories

14.1 Introduction

National economies are increasingly becoming information based, where networking and information transport provide a foundation for productivity and economic growth. Concurrently, many countries are rapidly adopting deregulation policies that are resulting in a telecommunications industry that is increasingly multicarrier and multivendor based, and where interconnectivity and compatibility between different networks is emerging as key to the success of this technological and regulatory transition. The communications industry has, consequently, become more interested in standardization; standards give manufacturers, service providers, and users freedom of choice at reasonable cost.

In this chapter, a review is provided of the primary telecommunications standards setting bodies. As will be seen, these bodies are often driven by slightly different underlying philosophies, but the output of their activities, i.e., the standards, possess essentially the same characteristics. An all-encompassing review of standardization bodies is not attempted here; this would clearly take many volumes to describe. Furthermore, as country after country increasingly deregulates its telecommunication industry, new standards setting bodies emerge to fill in the void of the de-facto (but no longer existing) standards setting bodies: the national telecommunications administration.

The principal communications standards bodies that will be covered are the following: the International Telecommunications Union (ITU); the United States ANSI Committee T1 on Telecommunications and the Telecommunications Industry Association (TIA); the European Telecommunications Standards Institute (ETSI); the Japanese Telecommunications Technology Committee (TTC); and the Institute of Electrical and Electronics Engineers (IEEE). Not addressed explicitly are other standards setting bodies that are either national and regional in character; even though it is recognized that sometimes there is overlap in scope with the bodies explicitly covered here.

196

0-8493-8573-3/96/$0.00+$.50
© 1996 by CRC Press, Inc.

Most notably, standards setting bodies that are not covered, but that are worth noting, include: the United States ANSI Committee X3, the Inter-American Telecommunications Commission (CITEL), the International Standards Organization (ISO), the International Electrotechnical Commission (IEC) [except ISO/IEC joint technical committee (JTC) 1], the Telecommunications Standards Advisory Council of Canada (TSACC), the Australian Telecommunications Standardization Committee (ATSC), the Telecommunication Technology Association (TTA) in Korea, and several forums (whose scope is, in principle, somewhat different) such as the asynchronous transfer mode (ATM) forum, the frame relay forum, the integrated digital services network (ISDN) users' forum, and telocator. As will be described later, many of these bodies operate in a coherent fashion through a mechanism developed by the interregional telecommunications standards conference (ITSC) and its successor, the global standards collaboration (GSC).

14.2 Global Standardization

When it comes to setting global communications standards, the ITU comes to the forefront. The ITU is an intergovernmental organization, whereby each sovereign state that is a member of the United Nations may become a member of the ITU. Member governments (in most cases represented by their telecommunications administrations) are constitutional members with a right to vote. Other organizations, such as network and service providers, manufacturers, and scientific and industrial organizations also participate in ITU activities but with a lower legal status.

ITU traces its history back to 1865 in the era of telegraphy. The supreme organ of the ITU is the plenipotentiary conference, which is held not less than every five years and plays a major role in the management of ITU. In 1993 the ITU as a U.N.-specialized agency was reorganized into three sectors (see Fig. 14.1): The *telecommunications standardization* sector (ITU-T), the *radiocommunications* sector (ITU-R), and the *development* sector (BDT). These sectors' activities are, respectively, standardization of telecommunications, including radio communications (although during a transition period considerable role in this field will be played by the ITU-R); regulation of telecommunications (mainly for radio communications); and promotion of telecommunications in developing countries.

It should be noted that, in general, the ITU-T is the successor of the international telephone and telegraph consultative committee (CCITT) of the ITU with additional responsibilities for standardization of network-related radio communications. Similarly, the ITU-R is the successor of the international radio consultative committee (CCIR) and the international frequency registration bureau (IFRB) of the ITU (after transferring some of its standardization activities to the ITU-T). The BDT is a new sector, which became operational in 1989.

ITU-T

Within the ITU structure, standardization work is undertaken by a number of study groups (SG) dealing with specific areas of communications. There are currently 15 study groups, as shown in Table 14.1.

Study groups develop standards for their respective work areas, which then have to be agreed upon by consensus—a process that for the time being is reserved to administrations only. The standards so developed are called recommendations to indicate their legal nonbinding nature. Technically, however, there is no distinction between recommendations developed by the ITU and standards developed by other standards setting bodies.

The study groups' work is undertaken by members, or delegates sent or sponsored by their national administrations. Because an ITU-T study group can have anywhere from 100 to more than 500 participating members and deal with 20–50 project standards, the work of each study group is often divided among working parties (WP). Such working parties are usually split further into experts' groups led by a chair or rapporteur with responsibility for a single (or part) of a project standards' topic.

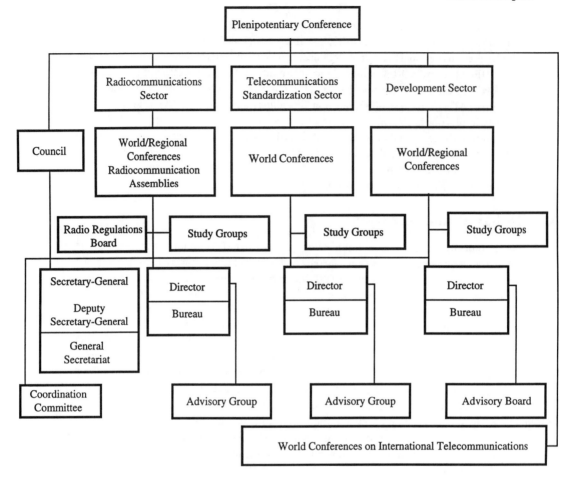

FIGURE 14.1   The (new) ITU structure.

TABLE 14.1   ITU-T Study Group Structure

SG 1	Service definition
SG 2	Network operation
SG 3	Tariff and accounting principles
SG 4	Network maintenance
SG 5	Protection against electromagnetic environmental effects
SG 6	Outside plant
SG 7	Data networks and open systems communications
SG 8	Terminals for telematic services
SG 9	Television and sound transmission (former CMTT)
SG 10	Languages for telecommunications applications
SG 11	Switching and signalling
SG 12	End-to-end transmission performance of networks and terminals
SG 13	General network aspects
SG 14	Modems and transmission techniques for data, telegraph, and telematic services
SG 15	Transmission systems and equipment

To coordinate standardization work that spans several study groups, a number of joint coordination groups (JCG) have also been established (not shown in Fig. 14.1). Presently, there are five such groups dealing with universal personal communications (UPT), transmission management network (TMN), audiovisual and multimedia services (AVMMS), quality of service and network performance (NP), and broadband ISDN (ATM/B-ISDN).

Such groups do not have executive powers but are merely there to coordinate work of pervasive interest within the ITU-T sector.

Also part of the ITU-T structure is the telecommunications standardization bureau (TSB) or, as it was formerly called, the CCITT secretariat. The TSB is responsible for the organization of numerous meetings held by the sector each year as well as all other support services required to ensure the smooth and efficient operation of the sector (including, but not limited to, document production and distribution). The TSB is headed by a director, who holds the executive power and, in collaboration with the study groups, bears full responsibility for the ITU-T activities. In this structure, unlike other U.N. organizations, the secretary general is the legal representative of the ITU, with the executive powers being vested in the director.

Finally, the ITU-T is supported by an advisory group, i.e. the telecommunications standardization advisory group (TSAG), which together with interested ITU members, the ITU-T Director, and ITU-T SG chairmen, guides standardization activities.

ITU-R

As noted earlier, standardization related work within the radiocommunications sector is gradually being transferred to the ITU-T, thus minimizing this sector's role in standardization and emphasizing its duties on regulatory and pure radio-interface aspects. The functional structure of the ITU-R includes a number (currently nine) of study groups, a radiocommunications bureau, and an advisory board. The role of the latter two elements is very similar to the ITU-T and, thus, need not be repeated here. The partitioning of the ITU-R sector into its nine study groups are shown in Table 14.2.

As within the ITU-T, there are areas of pervasive interest, and so areas of common interest can be found between the ITU-T and ITU-R where activities need to be coordinated. To achieve this objective, two intersector coordination groups (ICG) have been established (not shown in Fig. 14.1) dealing with future public land mobile telecommunications systems (FPLMTS) and ISDN and satellite matters.

BDT

Unlike the ITU-T (and to some extent ITU-R), which deals with standardization, the BDT deals with aspects that promote the integration and deployment of communications in developing countries. Typical outputs from this sector include implementation guides that expand the utility of ITU recommendations and ensure their expeditious implementation.

ISO/IEC JTC 1

Two global organizations are active in the information processing systems area, the ISO and the IEC, particularly through JTC 1.

TABLE 14.2 ITU-R Study Group Structure

SG 1	Spectrum management techniques
SG 2	Interservice sharing and compatibility
SG 3	Radio wave propagation
SG 4	Fixed satellite service
SG 7	Science services
SG 8	Mobile, radio determination, amateur and related services
SG 9	Fixed service
SG 10	Broadcasting services: sound
SG 11	Broadcasting services: television

TABLE 14.3   ISO/IEC/JTC1 Subcommittees

SC 1	Vocabulary
SC 2	Character sets and information coding
SC 6	Telecommunications information exchange between systems
SC 7	Software systems
SC 11	Flexible magnetic media for digital data interchange
SC 14	Representation of data elements
SC 15	Labeling and file structure
SC 17	Identification cards and related devices
SC 18	Document processing and related communication
SC 21	Information retrieval, transfer, and management for open systems interconnection
SC 22	Languages
SC 23	Optical disk cartridges for information interchange
SC 24	Computer graphics and image processing
SC 25	Interconnection information technology management
SC 26	Microprocessor systems
SC 27	Security techniques
SC 28	Office equipment
SC 29	Coded representation of picture, audio and multimedia/hypermedia information.

The ISO comprises national standards bodies, which have the responsibility for promoting and distributing ISO standards within their own countries. ISO technical work is carried out by some 170 technical committees (TC). Technical committees are established by the ISO council and their work program is approved by the technical board on behalf of the council.

The IEC comprises national committees (one from each country) and deals with almost all spheres of electrotechnology, including power, electronics, telecommunications, and nuclear energy. IEC technical work is performed by some 80 technical committees set up by its council. In 1987 a joint technical committee was established incorporating ISO TC97, IEC TC83, and subcommittee 47B to deal with generic information technology. The international standards developed by JTC1 are published under the ISO and IEC logos. The activities of ISO/IEC/JTC 1 are listed in Table 14.3 expressed in terms of its subcommittees (SC).

The ISO and IEC jointly issue directions for the work of the technical committees. The first step toward an international standard is the committee draft (CD). When agreement is reached within the relevant TC, the CD is sent to the central secretariat for registration as a draft international standard (DIS), which is subsequently circulated for voting.

14.3 Regional Standardization

Today the ETSI comes closest to being a true regional standards setting body, although CITEL, once fully developed, will most likely present a second regional (Latin-American) standardization body.

ETSI is the result of the Single act of the European community and the EC commission green paper in 1987 that analyzed the consequences of the Single act and recommended that a European telecommunications standards body be created to develop common standards for telecommunications equipment and networks. Out of this recommendation, the committee for harmonization (CCH) and the European conference for post and telecommunications (CEPT) evolved into ETSI, which formally came into being in March 1988. It should be noted, however, that even though ETSI attributes at least part of its existence to the European Community, its membership is wider than just the European Union Nations.

Because of the way ETSI came into being, ETSI is characterized by a unique aspect, namely, it is often called upon by the European commission to develop standards that are necessary to implement legislation. Such standards, which are referred to as technical basis reports (TBR) and whose application is usually mandatory, are often needed in public procurements, as well as in provisioning

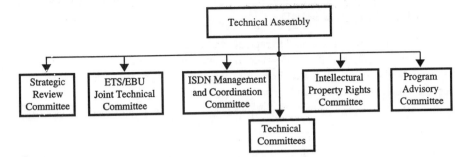

FIGURE 14.2 ETSI technical organization.

for open network interconnection as national telecommunications administrations are being deregulated. Like ITU, however, ETSI also develops voluntary standards in accordance with common international understanding against which industry is not obliged to produce conforming products. These standards fall into either the European technical standard (ETS) class when fully approved, or into the inter im-ETS class, when not fully stable or proven.

ETSI standards are typically sought when either the subject matter is not studied at the global level (such as when it may be required to support some piece of legislation), or the development of the standard is justified by market needs that exist in Europe and not in other parts of the world. In some cases, it may be necessary to adapt ITU standards for the European continent, although a simple endorsement of an ITU standard as a European standard is also possible. A more delicate case arises when both the ITU and ETSI are pursuing parallel standards activities, in which case close coordination with the ITU is sought either through member countries that may input ETSI standards to the ITU for consideration or through the global standards collaboration process.

TABLE 14.4 ETSI Technical Committees

NA	Network aspects
BT	Business telecommunications
SPS	Signaling protocols and switching
EE	Equipment engineering
RES	Radio equipment and systems
SES	Satellite earth stations
PS	Paging systems
TE	Terminal equipment
TM	Transmission & multiplexing
HF	Human factors
ATM	Advanced testing methods
SMG	Special mobile group

The highest authority of ETSI is the general assembly, which determines ETSI's policy, appoints its director and deputy, adopts the budget, and approves the audited accounts. The more technical issues are addressed by the technical assembly, which approves technical standards, advises on the work to be undertaken, and sets priorities. The structure of the technical part of ETSI is shown in Fig. 14.2, and the ETSI technical committees are listed in Table 14.4.

It can be seen that ETSI currently comprises 11 technical committees reporting to the technical assembly. These committees are responsible for the development of technical standards. In addition, these committees are responsible for prestandardization activities, that is, activities leading to ETSI technical reports (ETR) that eventually become the basis for future standards.

In addition to the technical assembly, a strategic review committee (SRC) is responsible for prospective examination of a single technical domain, whereas an intellectual property rights committee defines ETSI's policy in the area of intellectual property. Although by no means unique to ETSI, the rapid pace of technological progress has resulted in standards being adopted that embrace technologies that are still under patent protection. This creates a fundamental conflict between the private, exclusive nature of industrial property rights, and the open, public nature of standards. Harmonizing those conflicting claims has emerged as a thorny issue in all standards organizations; ETSI has established a formal function for this purpose. Finally, the ETS/EBU technical committee coordinates activities with the European broadcasting union (EBU), whereas the ISDN committee is in charge of managing and coordinating the standardization process for narrowband ISDN.

14.4 National Standardization

As standardization moves from global to regional and then to national levels, the number of actual participating entities rapidly grows. Here, the function of two national standards bodies are reviewed, primarily because these have been in existence the longest and secondarily because they also represent major markets for commercial communications.

ANSI T1

Unlike the ETSI, which came into being partly as a consequence of legislative recommendations, the ANSI Committee T1 on telecommunications came into being as a result of the realization that with the breakup of the Bell System, de-facto standards could no longer be expected. In fact, T1 came into being the very same year (1984) that the breakup of the Bell System came into effect.

The T1 membership comprises four types of interest groups: users and general interest groups, manufacturers, interexchange carriers, and exchange carriers. This rather broad membership is reflected, to some extent, by the scope to which T1 standards are being applied; this means that nontraditional telecommunications service providers are utilizing the technologies standardized by committee T1. This situation is the result of the rapid evolution and convergence of the telecommunications, computer, and cable television industries in the United States, and advances in wireless technology.

Committee T1 currently addresses approximately approved 150 projects, which led to the establishment of six, primarily functionally oriented, technical subcommittees (TSC), as shown in Table 14.5 and Fig. 14.3 [although not evident from Table 14.3, subcommittee T1P1 has primary responsibility for management of activities on personal communications systems (PCS)]. In-turn, each of these six subcommittees is divided into a number of subtending working groups, and subworking groups.

Committee T1 also has an advisory group (T1AG) made up of representatives from each of the four interest groups to carry out committee T1 directives and to develop proposals for consideration by the T1 membership.

In parallel to serving as the forum that establishes ANSI telecommunications network standards, committee T1 technical subcommittees draft candidate U.S. technical contributions to the ITU. These contributions are submitted to the U.S. Department of State National Committee for the ITU,

TABLE 14.5   T1 Subcommittee Structure

TSC: T1A1	Performance and signal processing
TSC: T1E1	Network interfaces and environmental considerations
TSC: T1M1	Interwork operations, administration, maintenance, and provisioning
TSC: T1P1	Systems engineering, standards planning, and program management
TSC: T1S1	Services, architecture, and signalling
TSC: T1X1	Digital hierarchy and synchronization

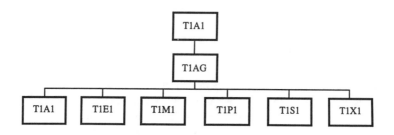

FIGURE 14.3   T1 committee structure.

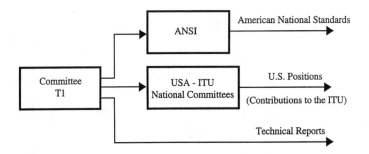

FIGURE 14.4   Committee T1 output.

which administers U.S. participation and contributions to the ITU (see Fig. 14.4). In this manner, activities within T1 are coordinated with those of the ITU. This coordination with other standards setting bodies is also reflected in T1's involvement with Latin-American standards, through the formation of an ad hoc group with CITEL's permanent technical committee 1 (PTC 1/T1). Further coordination with ETSI and other standards setting bodies is accomplished through the global standards collaboration process.

TIA

The TIA is a full-service trade organization that provides its members with numerous services including government relations, market support activities, educational programs, and standards setting activities.

TIA is a member-driven organization. Policy is formulated by 25 board members selected from member companies, and is carried out by a permanent professional staff located in Washington D.C. TIA comprises six issue-oriented standing committees, each of which is chaired by a board member. The six committees are membership scope and development, international, marketing and trade shows, public policy and government relations, and technical. It is this last committee that in 1992 was accredited by ANSI in the United States to standardize telecommunications products. Technology standardization activities are reflected by TIA's four product-oriented divisions, namely, user premises equipment, network equipment, mobile and personal communications equipment, and fiber optics.

In these divisions the legislative and regulatory concerns of product manufacturers and the preparation of standards dealing with performance testing and compatibility are addressed. For example, modem and telematic standards, as well as much of the cellular standards technology, has been standardized in the United States under the mandate of TIA.

TTC

The second national committee to be addressed is the TTC in Japan. TTC was established in October 1985 to develop and disseminate Japanese domestic standards for deregulated technical items and protocols. It is a nongovernmental, nonprofit standards setting organization established to ensure fair and transparent standardization procedures.

TTC's primary emphasis is to develop, conduct studies and research, and disseminate protocols and standards for the connection of telecommunications networks. TTC is organized along six technical subcommittees that report to a board of directors through a technical assembly (see Fig. 14.5).

The TTC organization comprises a general assembly, which is in charge of matters such as business plans and budgets. The councilors meeting examines standards development procedures in order to assure impartiality and clarity. The secretariat provides overall support to the organization;

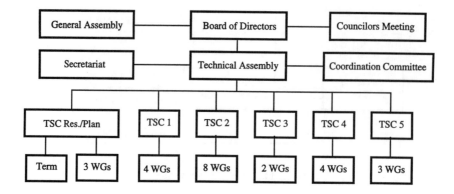

FIGURE 14.5   Organization of TTC.

the technical assembly develops standards and handles technical matters including surveys and research. Each technical subcommittee is partitioned into two or more working groups (WG). The coordination committee handles all issues in or between the TSCs and WGs, and it assures the smooth running of all technical committee meetings.

Under the coordination committee, a subcommittee examines users' requests and studies their applicability to the five-year standardization-project plan. This subcommittee also conducts user-request surveys.

TTC membership is divided into four categories. Type I telecommunications carriers, that is, those carriers that own telecommunications circuits and facilities; type II telecommunications carriers, that is, those with telecommunications circuits leased from type I carriers; related equipment manufacturers; and others, including users.

Underlying objectives that guide TTC's approach to standards development are 1) to conform to international recommendations or standards; 2) standardize items, where either international recommendations or standards are not clear, or where national standards need to be set, and where a consensus is achieved; and 3) to conduct further studies into any of the items just mentioned whenever the technical assembly is unable to arrive at a consensus.

These objectives, which give highest priority in developing standards that are compatible with international recommendations or standards, have often driven TTC to adapt international standards for national use through the use of supplements that:

- Give guidelines on users of TTC standards on how to apply them
- Help clarify the contents of standards
- Help with the implementation of standards in terminal equipment and adaptors
- Assure interconnection between terminal equipment and adaptors
- Provide background information regarding the content of standards
- Assure interconnection.

These supplements also include questions and answers that help in implementing the standards, including encoding examples of various parameters and explanation of the practical meaning of a standard.

14.5 Standards Coordination

Despite the growth of global telecommunications standards activities (between 1989 and 1992 the ITU has produced in three years 19,000 pages of standards, which is almost equal to the number produced in the prior 20 years), at the same time the industry is also witnessing an opposite trend towards regionalization.

In order to avoid duplication of work, waste of resources, and the possibility of arriving at conflicting standards, basic cooperation and coordination mechanisms were agreed upon between the ITU (then CCITT) and regional/national standardization organizations at the interregional telecommunications standardization conference (ITSC) hosted by committee T1 in Fredericksburg, Virginia in February 1990. At this conference there was a commitment made by all organizations represented to achieving and maintaining these objectives. Two working parties, the global standardization management and electronic information exchange, were set up to further pursue these objectives in detail. The second ITSC was held at ETSI's headquarters in Sophia/Antipolis (France), and the third conference was held in Tokyo in November 1992. The third conference adopted the Tokyo plan, a revised structure that streamlines the standards collaboration process while recognizing the rapidly changing telecommunications environment. Accordingly, it approved terms of reference that merged all necessary collaborative activities under a new global standards collaboration (GSC) group. The group will oversee the collaborative process, including work on electronic document handling (EDH) and five high-interest subjects:

- Broadband integrated services digital network (B-ISDN)
- Intelligent networks (IN)
- Transmission management network (TMN)
- Universal personal telecommunications (UPT)
- Synchronous digital hierarchy/synchronous optical network (SDH/SONET).

14.6 Scientific

Another global, scientifically based organization that has been particularly active in standards development (more recently emphasizing information processing) is the IEEE. Responsibility for standards adoption within the IEEE lies with the IEEE standards board. The board is supported by nine standing committees (see Fig. 14.6).

Proposed standards are normally developed in the technical committees of the IEEE Societies. There are occasions, however, when the scope of activity is too broad to be encompassed by a single society or where the societies are not able to do so for other reasons. In this case the standards board establishes its own standards developing committees, namely, the standards coordinating committees (SCC), to perform this function.

The adoption of IEEE standards is based on projects that have been approved by the IEEE standards board, while each project is the responsibility of a sponsor. Sponsors need not be an SCC but can also include technical committees of IEEE Societies; a standards, or standards coordinating committee

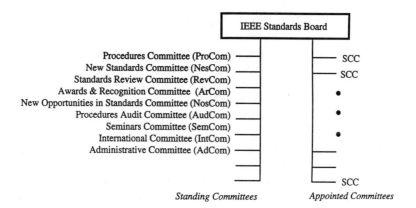

FIGURE 14.6 IEEE standards board organization.

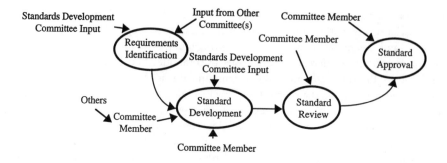

FIGURE 14.7   Typical standards development and approval process.

of an IEEE Society; an accredited standards committee; or another organization approved by the IEEE standards board.

14.7 Standards Development Cycle

Although the manner in which standards are developed and approved somewhat varies between standards organizations, there are common characteristics to be found.

For most standards, first a set of requirements is defined. This may be done either by the standards committee actually developing the standard or by another entity in collaboration with such a committee. Subsequently, the technical details of a standard are developed. The actual entity developing a standard may be a member of the standards committee, or the actual standards committee itself. Outsiders may also contribute to standards development but, typically, only if sponsored by a committee member. Membership in the standards committee and the right to contribute technical information towards the development of the standard differs among the various standards' organizations, as indicated. This process is illustrated in Fig. 14.7.

Finally, once the standard has been fully developed, it is placed under an approval cycle. Each standards setting body typically has precisely defined and often complex procedures for reviewing and then approving proposed standards, which although different in detail, are typically consensus driven.

Further Information

Irmer, T. 1994. Shaping future telecommunications: the challenge of global standardization, *IEEE Comm. Mag.* 32(1):20–28.

Matute, M.A. 1994. CITEL: formulating telecommunications in the Americas. *IEEE Comm. Mag.* 32(1):38–39.

Robin, G. 1994. The European perspective for telecommunications standards. *IEEE Comm. Mag.* 32(1):40–50.

Reilly, A.K. 1994. A US perspective on standards development. *IEEE Comm. Mag.* 32(1):30–36.

Iida, T. 1994. Domestic standards in a changing world. *IEEE Comm. Mag.* 32(1):46–50.

Habara, K. 1994. Cooperation in standardization. *IEEE Comm. Mag.* 32(1):78–84.

IEEE Standards Board Bylaws. 1993. Institute of Electrical and Electronic Engineers. Dec.

Chiarottino, W. and Pirani, G. 1993. International telecommunications standards organizations, *CSELT Tech. Repts.* XXI(2):207–236.

ITU, 1993. Book No. 1. Resolutions; Recommendations on the organization of the Work of ITU-T (series A); study groups and other groups; list of study questions (1993-1996). World Standardization Conf. Helsinki, 1–12, March.

Standards Committee T1. 1992. *Telecommunications.* Procedures Manual. 7th Iss. June.

II

Wireless

Wireless Personal Communications: A Perspective

Donald C. Cox
Stanford University

1 Introduction

Wireless personal communications has captured the attention of the media and with it, the imagi-
nation of the public. Hardly a week goes by without one seeing an article on the subject appearing in
a popular U.S. newspaper or magazine. Articles ranging from a short paragraph to many pages regularly appear in local newspapers, as well as in nationwide print media, e.g., *The Wall Street Journal,
The New York Times, Business Week*, and *U.S. News and World Report*. Countless marketing surveys
continue to project enormous demand, often projecting that at least half of the households, or half
of the people, want wireless personal communications. Trade magazines, newsletters, conferences,
and seminars on the subject by many different names have become too numerous to keep track
of, and technical journals, magazines, conferences, and symposia continue to proliferate and to

have ever increasing attendance and numbers of papers presented. It is clear that wireless personal communications is, by any measure, the fastest growing segment of telecommunications.

If you look carefully at the seemingly endless discussions of the topic, however, you cannot help but note that they are often describing different things, i.e., different versions of wireless personal communications [Cox, 1992; Padgett, Hattori, and Gunther, 1995]. Some discuss pagers, or messaging, or data systems, or access to the national information infrastructure, whereas others emphasize cellular radio, or cordless telephones, or dense systems of satellites. Many make reference to popular fiction entities such as Dick Tracy, Maxwell Smart, or *Star Trek*.

Thus, it appears that almost everyone wants wireless personal communications, but *What is it?* There are many different ways to segment the complex topic into different communications applications, modes, functions, extent of coverage, or mobility [Cox, 1992; Padgett, Hattori, and Gunther, 1995]. The complexity of the issues has resulted in considerable confusion in the industry, as evidenced by the many different wireless systems, technologies, and services being offered, planned, or proposed. Many different industry groups and regulatory entities are becoming involved. The confusion is a natural consequence of the massive dislocations that are occurring, and will continue to occur, as we progress along this large change in the paradigm of the way we communicate. Among the different changes that are occurring in our communications paradigm, perhaps the major constituent is the change from wired fixed place-to-place communications to wireless mobile person-to-person communications. Within this major change are also many other changes, e.g., an increase in the significance of data and message communications, a perception of possible changes in video applications, and changes in the regulatory and political climates.

This chapter attempts to identify different issues and to put many of the activities in wireless into a framework that can provide perspective on what is driving them, and perhaps even to yield some indication of where they appear to be going in the future. Like any attempt to categorize many complex interrelated issues, however, there are some that do not quite fit into neat categories, and so there will remain some dangling loose ends. Like any major paradigm shift, there will continue to be considerable confusion as many entities attempt to interpret the different needs and expectations associated with the new paradigm.

2 Background and Issues

Mobility and Freedom from Tethers

Perhaps the clearest constituents in all of the wireless personal communications activity are the desire for mobility in communications and the companion desire to be free from tethers, i.e., from physical connections to communications networks. These desires are clear from the very rapid growth of mobile technologies that provide primarily two-way voice services, even though economical wireline voice services are readily available. For example, cellular mobile radio has experienced rapid growth. Growth rates have been between 35 and 60% per year in the United States for a decade, with the total number of subscribers reaching 20 million by year-end 1994. The often neglected wireless companions to cellular radio, i.e., cordless telephones, have experienced even more rapid, but harder to quantify, growth with sales rates often exceeding 10 million sets a year in the United States, and with an estimated usage significantly exceeding 50 million in 1994. Telephones in airliners have also become commonplace. Similar or even greater growth in these wireless technologies has been experienced throughout the world.

Paging and associated messaging, although not providing two-way voice, do provide a form of tetherless mobile communications to many subscribers worldwide. These services have also experienced significant growth. There is even a glimmer of a market in the many different specialized wireless data applications evident in the many wireless local area network (WLAN) products on the market, the several wide area data services being offered, and the specialized satellite-based message services being provided to trucks on highways.

The topics discussed in the preceding two paragraphs indicate a dominant issue separating the different evolutions of wireless personal communications. That issue is the voice versus data communications issue that permeates all of communications today; this division also is very evident in fixed networks. The packet-oriented computer communications community and the circuit-oriented voice telecommunications (telephone) community hardly talk to each other and often speak different languages in addressing similar issues. Although they often converge to similar overall solutions at large scales (e.g., hierarchical routing with exceptions for embedded high-usage routes), the small-scale initial solutions are frequently quite different. Asynchronous transfer mode (ATM-) based networks are an attempt to integrate, at least partially, the needs of both the packet-data and circuit-oriented communities.

Superimposed on the voice-data issue is an issue of competing modes of communications that exist in both fixed and mobile forms. These different modes include the following.

Messaging is where the communication is not real time but is by way of message transmission, storage, and retrieval. This mode is represented by voice mail, electronic facsimile (fax), and electronic mail (e-mail), the latter of which appears to be a modern automated version of an evolution that includes telegraph and telex. Radio paging systems often provide limited one-way messaging, ranging from transmitting only the number of a calling party to longer alpha-numeric text messages.

Real-time two-way communications are represented by the telephone, cellular mobile radio telephone, and interactive text (and graphics) exchange over data networks. Two-way video phone always captures significant attention and fits into this mode; however, its benefit/cost ratio has yet to exceed a value that customers are willing to pay.

Paging, i.e., broadcast with no return channel, alerts a paged party that someone wants to communicate with him/her. Paging is like the ringer on a telephone without having the capability for completing the communications.

Agents are new high-level software applications or entities being incorporated into some computer networks. When launched into a data network, an agent is aimed at finding information by some title or characteristic and returning the information to the point from which the agent was launched.

There are still other ways in which wireless communications have been segmented in attempts to optimize a technology to satisfy the needs of some particular group. Examples include 1) user location, which can be differentiated by indoors or outdoors, or on an airplane or a train and 2) degree of mobility, which can be differentiated either by speed, e.g., vehicular, pedestrian, or stationary, or by size of area throughout which communications are provided.

At this point one should again ask: wireless personal communications—*What Is It?* The evidence suggests that what is being sought by users, and produced by providers, can be categorized according to the following two main characteristics.

Communications portability and mobility on many different scales:

- Within a house or building (cordless telephone, (WLANs))
- Within a campus, a town, or a city (cellular radio, WLANs, wide area wireless data, radio paging, extended cordless telephone)
- Throughout a state or region (cellular radio, wide area wireless data, radio paging, satellite-based wireless)
- Throughout a large country or continent (cellular radio, paging, satellite-based wireless)
- Throughout the world?

Communications by many different modes for many different applications:

- Two-way voice
- Data
- Messaging
- Video?

Thus, it is clear why wireless personal communications today is not one technology, not one system, and not one service but encompasses many technologies, systems, and services optimized for different applications.

3 Evolution of Technologies, Systems, and Services

Technologies and systems [Cox, 1992; Padgett, Hattori, and Gunther, 1995; Goodman, 1991; Steele, 1990; Schneideman, 1992; Cox, 1990; IEEE Comm. Mag. 1995] that are currently providing, or are proposed to provide, wireless communications services can be grouped into about seven relatively distinct groups, although there may be some disagreement on the group definitions, and in what group some particular technology or system belongs. All of the technologies and systems are evolving as technology advances and perceived needs change. Some trends are becoming evident in the evolutions. In this section, different groups and evolutionary trends are explored along with factors that influence the characteristics of members of the groups. The grouping is generally with respect to scale of mobility and communications applications or modes.

Cordless Telephones

Cordless telephones [Cox, 1992; Padgett, Hattori, and Gunther, 1995; Goodman, 1990] generally can be categorized as providing low-mobility, low-power, two-way tetherless voice communications, with low mobility applying both to the range and the user's speed. Cordless telephones using analog radio technologies appeared in the late 1970s, and have experienced spectacular growth. They have evolved to digital radio technologies in the forms of second-generation cordless telephone (CT-2), and digital European cordless telephone (DECT) standards in Europe, and several different industrial scientific medical (ISM) band technologies in the United States.[1]

Cordless telephones were originally aimed at providing economical, tetherless voice communications inside residences, i.e., at using a short wireless link to replace the cord between a telephone base unit and its handset. The most significant considerations in design compromises made for these technologies are to minimize total cost, while maximizing the talk time away from the battery charger. For digital cordless phones intended to be carried away from home in a pocket, e.g., CT-2 or DECT, handset weight and size are also major factors. These considerations drive designs toward minimizing complexity and minimizing the power used for signal processing and for transmitting.

Cordless telephones compete with wireline telephones. Therefore, high circuit quality has become a requirement. Early cordless sets had marginal quality. They were purchased by the millions, and discarded by the millions, until manufacturers produced higher-quality sets. Cordless telephones sales then exploded. Their usage has become commonplace, approaching, and perhaps exceeding, usage of corded telephones.

The compromises accepted in cordless telephone design in order to meet the cost, weight, and talk-time objectives are the following.

- Few users per megahertz
- Few users per base unit (many link together a particular handset and base unit)
- Large number of base units per unit area; one or more base units per wireline access line (in high-rise apartment buildings the density of base units is very large)
- Short transmission range

There is no added network complexity since a base unit looks to a telephone network like a wireline telephone. These issues are also discussed in Cox, 1992 and Padgett, Hattori, and Gunther, 1995.

---

[1]These ISM technologies either use spread spectrum techniques (direct sequence or frequency hopping) or very low-transmitter power ($< \sim 1$ mW) as required by the ISM band regulations.

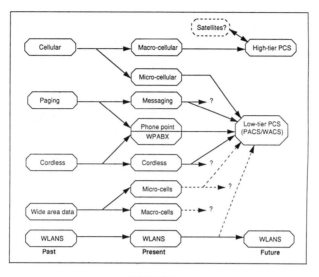

FIGURE 1

Digital cordless telephones in Europe have been evolving for a few years to extend their domain of use beyond the limits of inside residences. Cordless telephone, second generation, (CT-2) has evolved to provide telepoint or phone-point services. Base units are located in places where people congregate, e.g., along city streets and in shopping malls, train stations, etc. Handsets registered with the phone-point provider can place calls when within range of a telepoint. CT-2 does not provide capability for transferring (handing off) active wireless calls from one phone point to another if a user moves out of range of the one to which the call was initiated. A CT-2+ technology, evolved from CT-2 and providing limited handoff capability, is being deployed in Canada. Phone-point service was introduced in the United Kingdom twice, but failed to attract enough customers to become a viable service. In Singapore and Hong Kong, however, CT-2 phone point has grown rapidly, reaching over 150,000 subscribers in Hong Kong [Wong, 1994] in mid-1994. The reasons for success in some places and failure in others are still being debated, but it is clear that the compactness of the Hong Kong and Singapore populations make the service more widely available, using fewer base stations than in more spreadout cities. Complaints of CT-2 phone-point users in trials have been that the radio coverage was not complete enough, and/or they could not tell whether there was coverage at a particular place, and the lack of handoff was inconvenient. In order to provide the alerting or ringing function for phone-point service, conventional radio pagers have been built into some CT-2 handsets. (The telephone network to which a CT-2 phone point is attached has no way of knowing from which base units to send a ringing message, even though the CT-2 handsets can be rung from a home base unit).

Another European evolution of cordless telephones is DECT, which was optimized for use inside buildings. Base units are attached through a controller to private branch exchanges (PBXs), key telephone systems, or phone company CENTREX telephone lines. DECT controllers can hand off active calls from one base unit to another as users move, and can page or ring handsets as a user walks through areas covered by different base units.

These cordless telephone evolutions to more widespread usage outside and inside with telepoints and to usage inside large buildings are illustrated in Fig. 1, along with the integration of paging into handsets to provide alerting for phone-point services. They represent the first attempts to increase the service area of mobility for low-power cordless telephones.

Some of the characteristics of the digital cordless telephone technologies, CT-2 and DECT, are listed in Table 1. Additional information can be found in Padgett, Hattori, and Gunther, 1995 and Goodman, 1991. Even though there are significant differences between these technologies, e.g.,

TABLE 1   Wireless PCS Technologies

| | High-Power Systems | | | | Low-Power Systems | | | |
	Digital Cellular (High-Tier PCS)				Low-Tier PCS		Digital Cordless	
System	IS-54	IS-95 (DS)	GSM	DCS-1800	WACS/PACS	Handi-Phone	DECT	CT-2
Multiple access	TDMA/ FDMA	CDMA/ FDMA	TDMA/ FDMA	TDMA/ FDMA	TDMA/ FDMA	TDMA/ FDMA	TDMA/ FDMA	FDMA
Freq. band, MHz						1895–1907	1880–1990	864–868
Uplink, MHz	869–894	869–894	935–960	1710–1785	Emerg.			
Downlink, MHz	824–849	824–849	890–915	1805–1880	Tech.*			(Eur. and
	(USA)	(USA)	(Eur.)	(UK)	(USA)	(Japan)	(Eur.)	Asia)
RF ch. spacing						300	1728	100
Downlink, KHz	30	1250	200	200	300			
Uplink KHz	30	1250	200	200	300			
Modulation	$\pi/4$ DQPSK	BPSK/QPSK	GMSK	GMSK	$\pi/4$ QPSK	$\pi/4$ DQPSK	GFSK	GFSK
Portable txmit Power, max./avg.	600 mW/ 200 mW	600 mW	1 W/ 125 mW	1 W/ 125 mW	200 mW/ 25 mW	80 mW/ 10 mW	250 mW/ 10 mW	10 mW/ 5 mW
Speech coding	VSELP	QCELP	RPE-LTP	RPE-LTP	ADPCM	ADPCM	ADPCM	ADPCM
Speech rate, kb/s	7.95	8 (var.)	13	13	32/16/8	32	32	32
Speech ch./RF ch.	3	—	8	8	8/16/32	4	12	1
Ch. Bit rate, kb/s						384	1152	72
Uplink, kb/s	48.6		270.833	270.833	384			
Downlink, kb/s	48.6		270.833	270.833	384			
Ch. coding	1/2 rate conv.	1/2 rate fwd. 1/3 rate rev.	1/2 rate conv.	1/2 rate conv.	CRC	CRC	CRC (control)	None
Frame, ms	40	20	4.615	4.615	2.5	5	10	2

*Spectrum is 1.85–2.2 GHz allocated by the FCC for emerging technologies; DS is direct sequence.

multiple access technology [frequency division multiple access (FDMA) or time division multiple access (TDMA)/FDMA], and channel bit rate, there are many similarities that are fundamental to the design objectives discussed earlier and to a user's perception of them. These similarities and their implications are as follows.

32-kb/s Adaptive Differential Pulse Code Modulation (ADPCM) Digital Speech Encoding: This is a low-complexity (low-signal processing power) speech encoding process that provides wireline speech quality and is an international standard.

Average Transmitter Power ≤ 10 mW: This permits many hours of talk time with small, low-cost, lightweight batteries, but provides limited radio range.

Low-Complexity Radio Signal Processing: There is no forward error correction and no complex multipath mitigation (i.e., no equalization or spread spectrum).

Low Transmission Delay, e.g., < 50 ms, and for CT-2 < 10-ms Round Trip: This is a speech-quality and network-complexity issue. A maximum of 10 ms should be allowed, taking into account additional inevitable delay in long-distance networks. Echo cancellation is generally required for delays > 10 ms.

Simple Frequency-Shift Modulation and Noncoherent Detection: Although still being low in complexity, the slightly more complex 4QPSK modulation with coherent detection provides significantly more spectrum efficiency, range, and interference immunity.

Dynamic Channel Allocation: Although this technique has potential for improved system capacity, the cordless-telephone implementations do not take full advantage of this feature for handoff and, thus, cannot reap the full benefit for moving users [Chuang, 1993; Chuang, Sollenberger, and Cox, 1993].

Time Division Duplex (TDD): This technique permits the use of a single contiguous frequency band and implementation of diversity from one end of a radio link. Unless all base station transmissions are synchronized in time, however, it can incur severe cochannel interference penalties in outside environments [Chuang, 1993; 1992]. Of course, for cordless telephones used inside with base stations not having a propagation advantage, this is not a problem. Also, for small indoor PBX networks, synchronization of base station transmission is easier than is synchronization throughout a widespread outdoor network, which can have many adjacent base stations connected to different geographic locations for central control and switching.

Cellular Mobile Radio Systems

Cellular mobile radio systems are becoming known in the United States as high-tier personal communications service (PCS), particularly when implemented in the new 1.9-GHz PCS bands [Cook, 1994]. These systems generally can be categorized as providing high-mobility, wide-ranging, two-way tetherless voice communications. In these systems, high mobility refers to vehicular speeds, and also to widespread regional to nationwide coverage [Cox, 1992; Padgett, Hattori, and Gunther, 1995; Cox, 1990]. Mobile radio has been evolving for over 50 years. Cellular radio integrates wireless access with large-scale networks having sophisticated intelligence to manage mobility of users.

Cellular radio was designed to provide voice service to wide-ranging vehicles on streets and highways [Cox, 1992; Padgett, Hattori, and Gunther, 1995; Goodman, 1991; BSTJ, 1979], and generally uses transmitter power on the order of 100 times that of cordless telephones ($\approx$2 W for cellular). Thus, cellular systems can only provide reduced service to handheld sets that are disadvantaged by using somewhat lower transmitter power (<0.5 W) and less efficient antennas than vehicular sets. Handheld sets used inside buildings have the further disadvantage of attenuation through walls that is not taken into account in system design.

Cellular radio or high-tier PCS has experienced large growth as noted earlier. In spite of the limitations on usage of handheld sets already noted, handheld cellular sets have become very popular, with their sales becoming comparable to the sales of vehicular sets. Frequent complaints from handheld cellular users are that batteries are too large and heavy, and both talk time and standby time are inadequate.

Cellular radio at 800 MHz has evolved to digital radio technologies [Cox, 1992; Padgett, Hattori, and Gunther, 1995; Goodman, 1991] in the forms of the deployed systems standards

- Global Standard for Mobile (GSM) in Europe
- Japanese or personal digital cellular (JDC or PDC) in Japan
- U.S. TDMA digital cellular known as USDC or IS-54.

and in the form of the code division multiple access (CDMA) standard, IS-95, which is under development but not yet deployed.

The most significant consideration in the design compromises made for the U.S. digital cellular or high-tier PCS systems was the high cost of cell sites (base stations). A figure often quoted is U.S. $1 million for a cell site. This consideration drove digital system designs to maximize users per megahertz and to maximize the users per cell site.

Because of the need to cover highways running through low-population-density regions between cities, the relatively high transmitter power requirement was retained to provide maximum range from high antenna locations.

Compromises that were accepted while maximizing the two just cited parameters are as follows.

- High transmitter power consumption.
- High user-set complexity, and thus high signal-processing power consumption.
- Low circuit quality.

- High network complexity, e.g., the new IS-95 technology will require complex new switching and control equipment in the network, as well as high-complexity wireless- access technology.

Cellular radio or high-tier PCS has also been evolving for a few years in a different direction, toward very small coverage areas or microcells. This evolution provides increased capacity in areas having high user density, as well as improved coverage of shadowed areas. Some microcell base stations are being installed inside, in conference center lobbies and similar places of high user concentrations. Of course, microcells also permit lower transmitter power that conserves battery power when power control is implemented, and base stations inside buildings circumvent the outside wall attenuation. Low-complexity microcell base stations also are considerably less expensive than conventional cell sites, perhaps two orders of magnitude less expensive. Thus, the use of microcell base stations provides large increases in overall system capacity, while also reducing the cost per available radio channel and the battery drain on portable subscriber equipment. This microcell evolution, illustrated in Fig. 1, moves handheld cellular sets in a direction similar to that of the expanded-coverage evolution of cordless telephones to phone points and wireless PBX.

Some of the characteristics of digital-cellular or high-tier PCS technologies are listed in Table 1 for IS-54, IS-95, and GSM at 900 MHz, and DCS-1800, which is GSM at 1800 MHz. Additional information can be found in Cox, 1992; Padgett, Hattori, and Gunther, 1995; Goodman, 1991. The JDC or PDC technology, not listed, is similar to IS-54. As with the digital cordless technologies, there are significant differences among these cellular technologies, e.g., modulation type, multiple access technology, and channel bit rate. There are also many similarities, however, that are fundamental to the design objectives discussed earlier. These similarities and their implications are as follows.

Low Bit-Rate Speech Coding; $\leq 13\,kb/s$ with Some $\leq 8\,kb/s$: Low bit-rate speech coding obviously increases the number of users per megahertz and per cell site. However, it also significantly reduces speech quality [Cox, 1992], and does not permit speech encodings in tandem while traversing a network; see also the section on Other Issues later in this chapter.

Some Implementations Make Use of Speech Inactivity: This further increases the number of users per cell site i.e., the cell-site capacity. It also further reduces speech quality [Cox, 1992], however, because of the difficulty of detecting the onset of speech. This problem is even worse in an acoustically noisy environment like an automobile.

High Transmission Delay; ≈ 200-ms Round Trip: This is another important circuit-quality issue. Such large delay is about the same as one-way transmission through a synchronous-orbit communications satellite. A voice circuit with digital cellular technology on both ends will experience the delay of a full satellite circuit. It should be recalled that one reason long-distance circuits have been removed from satellites and put onto fiber-optic cable is because customers find the delay to be objectionable. This delay in digital cellular technology results from both computation for speech bit-rate reduction and from complex signal processing, e.g., bit interleaving, error correction decoding, and multipath mitigation [equalization or spread spectrum code division multiple access (CDMA)].

High-Complexity Signal Processing, Both for Speech Encoding and for Demodulation: Signal processing has been allowed to grow without bound and is about a factor of 10 greater than that used in the low-complexity digital cordless telephones [Cox, 1992]. Since several watts are required from a battery to produce the high transmitter power in a cellular or high-tier PCS set, signal-processing power is not as significant as it is in the low-power cordless telephones; see also the section on Complexity/Coverage Area Comparisons later in this chapter.

Fixed Channel Allocation: The difficulties associated with implementing capacity-increasing dynamic channel allocation to work with handoff [Chuang, 1993; Chuang, Sollenberger, and Cox, 1993] have impeded its adoption in systems requiring reliable and frequent handoff.

Frequency Division Duplex (FDD): Cellular systems have already been allocated paired-frequency bands suitable for FDD. Thus, the network or system complexity required for providing synchronized transmissions [Chuang, 1993; 1992] from all cell sites for TDD has not been embraced in these digital cellular systems. Note that TDD has not been employed in IS-95 even though such synchronization is required for other reasons.

Mobile/Portable Set Power Control: The benefits of increased capacity from lower overall cochannel interference and reduced battery drain have been sought by incorporating power control in the digital cellular technologies.

Wide-Area Wireless Data Systems

Existing wide area data systems generally can be categorized as providing high mobility, wide-ranging, low-data-rate digital data communications to both vehicles and pedestrians [Cox, 1992; Padgett, Hattori, and Gunther, 1995]. These systems have not experienced the rapid growth that the two-way voice technologies have, even though they have been deployed in many cities for a few years and have established a base of customers in several countries. Examples of these packet data systems are shown in Table 2.

The earliest and best known of these systems in the United States are the ARDIS network developed and run by Motorola, and the RAM mobile data network based on Ericsson Mobitex Technology. These technologies were designed to make use of standard, two-way voice, land mobile-radio channels, with 12.5- or 25-kHz channel spacing. In the United States these are specialized mobile radio services (SMRS) allocations around 450 MHz and 900 MHz. Initially, the data rates were low: 4.8 kb/s for ARDIS and 8 kb/s for RAM. The systems use high transmitter power (several tens of watts) to cover large regions from a few base stations having high antennas. The relatively low data capacity of a relatively expensive base station has resulted in economics that have not favored rapid growth.

The wide-area mobile data systems also are evolving in several different directions in an attempt to improve base station capacity, economics, and the attractiveness of the service. The technologies used in both the ARDIS and RAM networks are evolving to higher channel bit rates of 19.2 kb/s.

The cellular carriers and several manufacturers in the United States are developing and deploying a new wide area packet data network as an overlay to the cellular radio networks. This cellular digital packet data (CDPD) technology shares the 30-kHz spaced 800-MHz voice channels used by the analog FM advanced mobile phone service (AMPS) systems. Data rate is 19.2 kb/s. The CDPD base station equipment also shares cell sites with the voice cellular radio system. The aim is to reduce the cost of providing packet data service by sharing the costs of base stations with the better established and higher cell-site capacity cellular systems. This is a strategy similar to that used by nationwide fixed wireline packet data networks that could not provide an economically viable data service if they did not share costs by leasing a small amount of the capacity of the interexchange networks that are paid for largely by voice traffic.

TABLE 2 Wide-Area Wireless Packet Data Systems

	CDPD[1]	RAM Mobile (Mobitex)	ARDIS[2] (KDT)	Metricom (MDN)[3]
Data rate, kb/s	19.2	8 (19.2)	4.8 (19.2)	76
Modulation	GMSK BT = 0.5	GMSK	GMSK	GMSK
Frequency, MHz	800	900	800	915
Chan. spacing, kHz	30	12.5	25	160
Status	1994 service	Full service	Full service	In service
Access means	Unused AMPS channels	Slotted Aloha CSMA		FH SS (ISM)
Transmit power, W			40	1

Note: Data in parentheses ( ) indicates proposed.
[1] Cellular Digital Packet Data
[2] Advanced Radio Data Information Service
[3] Microcellular Data Network

Another evolutionary path in wide-area wireless packet data networks is toward smaller coverage areas or microcells. This evolutionary path also is indicated on Fig. 1. The microcell data networks are aimed at stationary or low-speed users. The design compromises are aimed at reducing service costs by making very small and inexpensive base stations that can be attached to utility poles, the sides of buildings and inside buildings and can be widely distributed throughout a region. Base-station-to-base-station wireless links are used to reduce the cost of the interconnecting data network. In one network this decreases the overall capacity to serve users, since it uses the same radio channels that are used to provide service. Capacity is expected to be made up by increasing the number of base stations that have connections to a fixed-distribution network as service demand increases. Another such network uses other dedicated radio channels to interconnect base stations. In the high-capacity limit, these networks will look more like a conventional cellular network architecture, with closely spaced, small, inexpensive base stations, i.e., microcells, connected to a fixed infrastructure. Specialized wireless data networks have been built to provide metering and control of electric power distribution, e.g., Celldata and Metricom in California.

A large microcell network of small inexpensive base stations has been installed in the lower San Francisco Bay Area by Metricom, and public packet-data service was offered during early 1994. Most of the small (shoe-box size) base stations are mounted on street light poles. Reliable data rates are about 75 kb/s. The technology is based on slow frequency-hopped spread spectrum in the 902–928 MHz U.S. ISM band. Transmitter power is 1 W maximum, and power control is used to minimize interference and maximize battery life time.

High-Speed Wireless Local-Area Networks (WLANs)

Wireless local-area data networks can be categorized as providing low-mobility high-data-rate data communications within a confined region, e.g., a campus or a large building. Coverage range from a wireless data terminal is short, tens to hundreds of feet, like cordless telephones. Coverage is limited to within a room or to several rooms in a building. WLANs have been evolving for a few years, but overall the situation is chaotic, with many different products being offered by many different vendors [Cox, 1992; Schneideman, 1992]. There is no stable definition of the needs or design objectives for WLANs, with data rates ranging from hundreds of kb/s to more than 10 Mb/s, and with several products providing one or two Mb/s wireless link rates. The best description of the WLAN evolutionary process is: having severe birth pains. An IEEE standards committee, 802.11, has been attempting to put some order into this topic, but their success has been somewhat limited. A partial list of some advertised products is given in Table 3. Users of WLANs are not nearly as numerous as the users of more voice-oriented wireless systems. Part of the difficulty stems from these systems being driven by the computer industry that views the wireless system as just another plug-in interface card, without giving sufficient consideration to the vagaries and needs of a reliable radio system.

There are two overall network architectures pursued by WLAN designers. One is a centrally coordinated and controlled network that resembles other wireless systems. There are base stations in these networks that exercise overall control over channel access [Katz, 1994].

The other type of network architecture is the self-organizing and distributed controlled network where every terminal has the same function as every other terminal, and networks are formed ad hoc by communications exchanges among terminals. Such adhoc networks are more like citizen band (CB) radio networks, with similar expected limitations if they were ever to become very widespread.

Nearly all WLANs in the United States have attempted to use one of the ISM frequency bands for unlicensed operation under part 15 of the FCC rules. These bands are 902–928 MHz, 2400–2483.5 MHz, and 5725–5850 MHz, and they require users to accept interference from any interfering source that may also be using the frequency. The use of ISM bands has further handicapped WLAN development because of the requirement for use of either frequency hopping or direct sequence spread spectrum as an access technology, if transmitter power is to be adequate to cover more

TABLE 3   Partial List of WLAN Products

Product Company Location	Freq., MHz	Link Rate, Mb/s	User Rate	Protocol(s)	Access	No. of chan. or Spread Factor	Mod./ Coding	Power,	Network Topology
Altair Plus Motorola Arlington Hts, IL	18–19 GHz	15	5.7 Mb/s	Ethernet			4-level FSK	25 peak	Eight devices/ radio; radio to base to ethernet
WaveLAN NCR/AT&T Dayton, OH	902–928	2	1.6 Mb/s	Ethernet-like	DS SS		DQPSK	250	Peer-to-peer
AirLan Solectek San Diego, CA	902–928		2 Mb/s	Ethernet	DS SS		DQPSK	250	PCMCIA w/ant.; radio to hub
Freeport Windata Inc. Northboro, MA	902–928	16	5.7 Mb/s	Ethernet	DS SS	32 chips/bit	16 PSK trellis coding	650	Hub
Intersect Persoft Inc. Madison, WI	902–928		2 Mb/s	Ethernet, token ring	DS SS		DQPSK	250	Hub
LAWN O'Neill Comm. Horsham, PA	902–928		38.4 kb/s	AX.25	SS	20 users/chan.; max. 4 chan.		20	Peer-to-peer
WILAN Wi-LAN Inc. Calgary, Alberta	902–928	20	1.5 Mb/s/ chan.	Ethernet, token ring	CDMA/ TDMA	3 chan. 10–15 links each	unconven- tional	30	Peer-to-peer
RadioPort ALPS Electric USA	902–928		242 kb/s	Ethernet	SS	7/3 channels		100	Peer-to-peer
ArLAN 600 Telesys. SLW Don Mills, Ont.	902–928; 2.4 GHz		1.35 Mb/s	Ethernet	SS			1 W max	PCs with ant.; radio to hub
RadioLink Cal. Microwave Sunnyvale, CA.	902–928; 2.4 GHz	250 kb/s	64 kb/s		FH SS	250 ms/hop 500 kHz space			Hub
Range LAN Proxim, Inc. Mountain View, CA	902–928		242 kb/s	Ethernet, token ring	DS SS	3 chan.		100	
RangeLAN2 Proxim, Inc. MountainView, CA	2.4 GHz	1.6	50 kb/s max.	Ethernet, token ring	FH SS	10 chan. at 5 kb/s; 15 sub-ch. each		100	Peer-to-peer bridge
Netwave Xircom Calabasas, CA	2.4 GHz	1/adaptor		Ethernet, token ring	FH SS	82 1-MHz chn. or "hops"			Hub
Freelink Cabletron Sys. Rochester, NH	2.4 and 5.8 GHz		5.7 Mb/s	Ethernet	DS SS	32 chips/bit	16 PSK trellis coding	100	Hub

than a few feet. One exception to the ISM band implementations is the Motorola ALTAIR, which operates in a licensed band at 18 GHz. The technical and economic challenges of operation at 18 GHz have hampered the adoption of this 10–15 Mb/s technology. The frequency-spectrum constraints have been improved in the United States with the recent FCC allocation of spectrum from 1910–1930 MHz for unlicensed data PCS applications. Use of this new spectrum requires implementation of an access etiquette incorporating listen before transmit in an attempt to provide some coordination of an otherwise potentially chaotic, uncontrolled environment [Steer, 1994]. Also, since spread spectrum is not a requirement, access technologies and multipath mitigation techniques more compatible with the needs of packet-data transmission [Schneideman, 1992], e.g., multipath equalization or multicarrier transmission can be incorporated into new WLAN designs.

Three other widely different WLAN activities also need mentioning. One is a large European Telecommunications Standards Institute (ETSI) activity to produce a standard for high performance radio local area network (HIPERLAN), a 20-Mb/s WLAN technology to operate near 5 GHz. Other activities are large U.S. Advance Research Projects Agency- (ARPA-) sponsored, WLAN research projects at the Universities of California at Berkeley (UCB), and at Los Angeles (UCLA). The UCB Infopad project is based on a coordinated network architecture with fixed coordinating nodes and direct-sequence spread spectrum (CDMA), whereas, the UCLA project is aimed at peer-to-peer networks and uses frequency hopping. Both ARPA sponsored projects are concentrated on the 900-MHz ISM band.

As computers shrink in size from desktop to laptop to palmtop, mobility in data network access is becoming more important to the user. This fact, coupled with the availability of more usable frequency spectrum, and perhaps some progress on standards, may speed the evolution and adoption of wireless mobile access to WLANs. From the large number of companies making products, it is obvious that many believe in the future of this market.

Paging/Messaging Systems

Radio paging began many years ago as a one-bit messaging system. The one bit was: some one wants to communicate with you. More generally, paging can be categorized as one-way messaging over wide areas. The one-way radio link is optimized to take advantage of the asymmetry. High transmitter power (hundreds of watts to kilowatts), and high antennas at the fixed base stations permit low-complexity, very low-power-consumption, pocket paging receivers that provide long usage time from small batteries. This combination provides the large radio-link margins needed to penetrate walls of buildings without burdening the user set battery. Paging has experienced steady rapid growth for many years and serves about 15 million subscribers in the United States.

Paging also has evolved in several different directions. It has changed from analog tone coding for user identification to digitally encoded messages. It has evolved from the 1-b message, someone wants you, to multibit messages from, first, the calling party's telephone number to, now, short e-mail text messages. This evolution is noted in Fig. 1.

The region over which a page is transmitted has also increased from 1) local, around one transmitting antenna; to 2) regional, from multiple widely-dispersed antennas; to 3) nationwide, from large networks of interconnected paging transmitters. The integration of paging with CT-2 user sets for phone-point call alerting was noted previously.

Another evolutionary paging route sometimes proposed is two-way paging. This is an ambiguous and unrealizable concept, however, since the requirement for two-way communications destroys the asymmetrical link advantage so well exploited by paging. Two-way paging puts a transmitter in the user's set and brings along with it all of the design compromises that must be faced in such a two-way radio system. Thus, the word paging is not appropriate to describe a system that provides two-way communications.

Satellite-Based Mobile Systems

Satellite-based systems are the epitome of wide-area coverage, expensive base station systems. They generally can be categorized as providing two-way (or one-way) limited quality voice and/or very limited data or messaging to very wide-ranging vehicles (or fixed locations). These systems can provide very widespread, often global, coverage, e.g., to ships at sea by INMARSAT. There are a few messaging systems in operation, e.g., to trucks on highways in the United States by Qualcomm's Omnitracs system.

A few large-scale mobile satellite systems have been proposed and are being pursued: perhaps the best known is Motorola's Iridium; others include Odyssey, Globalstar, and Teledesic. The strength of satellite systems is their ability to provide large regional or global coverage to users outside buildings. However, it is very difficult to provide adequate link margin to cover inside buildings, or even to cover locations shadowed by buildings, trees, or mountains. A satellite system's weakness is also its large coverage area. It is very difficult to provide from Earth orbit the small coverage cells that are necessary for providing high overall systems capacity from frequency reuse. This fact, coupled with the high cost of the orbital base stations, results in low capacity along with the wide overall coverage but also in expensive service. Thus, satellite systems are not likely to compete favorably with terrestrial systems in populated areas or even along well-traveled highways. They can complement terrestrial cellular or PCS systems in low-population-density areas. It remains to be seen whether there will be enough users with enough money in low-population-density regions of the world to make satellite mobile systems economically viable.

Proposed satellite systems range from 1) low-Earth-orbit systems (LEOS) having tens to hundreds of satellites through 2) intermediate- or medium-height systems (MEOS) to 3) geostationary or geosynchronous orbit systems (GEOS) having fewer than ten satellites. LEOS require more, but less expensive, satellites to cover the Earth, but they can more easily produce smaller coverage areas and, thus, provide higher capacity within a given spectrum allocation. Also, their transmission delay is significantly less (perhaps two orders of magnitude!), providing higher quality voice links, as discussed previously. On the other hand, GEOS require only a few, somewhat more expensive, satellites (perhaps only three) and are likely to provide lower capacity within a given spectrum allocation and suffer severe transmission-delay impairment on the order of 0.5 s. Of course, MEOS fall in between these extremes. The possible evolution of satellite systems to complement high-tier PCS is indicated in Fig. 1.

Reality Check

Before we go on to consider other applications and compromises, perhaps it would be helpful to see if there is any indication that the previous discussion is valid. For this check, we could look at cordless telephones for telepoint use (i.e., pocketphones) and at pocket cellular telephones that existed in the 1993 time frame.

Two products from one United States manufacturer are good for this comparison. One is a third-generation hand-portable analog FM cellular phone from this manufacturer that represents their second generation of pocketphones. The other is a first-generation digital cordless phone built to the United Kingdom CT-2 common air interface (CAI) standard. Both units are of flip phone type with the earpiece on the main handset body and the mouthpiece formed by or on the flip-down part. Both operate near 900 MHz and have 1/4 wavelength pull-out antennas. Both are fully functional within their class of operation (i.e., full number of U.S. cellular channels, full number of CT-2 channels, automatic channel setup, etc.). Table 4 compares characteristics of these two wireless access pocketphones from the same manufacturer.

The following are the most important items to note in the Table 4 comparison.

1. The talk time of the low-power pocketphone is four times that of the high-power pocketphone.
2. The battery inside the low-power pocketphone is about one-half the weight and size of the battery attached to the high-power pocketphone.
3. The battery-usage ratio, talk time/weight of battery, is eight times greater, almost an order of magnitude, for the low-power pocketphone compared to the high-power pocketphone!
4. Additionally, the lower power (5 mW) digital cordless pocketphone is slightly smaller and lighter than the high-power (500 mW) analog FM cellular mobile pocketphone.

TABLE 4 Comparison of CT-2 and Cellular Pocket Size Flip-Phones from the Same Manufacturer

Characteristics/Parameter	CT-2	Cellular
Weight, oz		
Flip phone only	5.2	4.2
Battery[1] only	1.9	3.6
Total unit	7.1	7.8
Size (max. dimensions), in		
Flip phone only	$5.9 \times 2.2 \times 0.95$ 8.5 in^3	$5.5 \times 2.4 \times 0.9$ —
Battery[1] only	$1.9 \times 1.3 \times 0.5$ internal	$4.7 \times 2.3 \times 0.4$ external
Total unit	$5.9 \times 2.2 \times 0.95$ 8.5 in^3	$5.9 \times 2.4 \times 1.1$ 11.6 in^3
Talk-time, min(h)		
Rechargeable battery[2]	180 (3)	45
Nonrechargeable battery	600 (10)	N/A
Standby time, h		
Rechargeable battery	30	8
Nonrechargeable battery	100	N/A
Speech quality	32 kb/s telephone quality	30 kHz FM depends on channel quality
Transmit power avg., W	0.005	0.5

[1] Rechargeable battery.
[2] Ni-cad battery.

The following should also be noted.

1. The room for technology improvement of the CT-2 cordless phone is greater since it is first generation and the cellular phone is second/third generation.
2. A digital cellular phone built to the IS-54, GSM, or JDC standard, or in the proposed United States CDMA technology, would either have less talk time or be heavier and larger than the analog FM phone, because: a) the low-bit-rate digital speech coder is more complex and will consume more power than the analog speech processing circuits; b) the digital units have complex digital signal-processing circuits for forward error correction—either for delay dispersion equalizing or for spread-spectrum processing—that will consume significant amounts of power and that have no equivalents in the analog FM unit; and c) power amplifiers for the shaped-pulse nonconstant-envelope digital signals will be less efficient than the amplifiers for constant-envelope analog FM. Although it may be suggested that transmitter power control will reduce the weight and size of a CDMA handset and battery, if that handset is to be capable of operating at full power in fringe areas, it will have to have capabilities similar to other cellular sets. Similar power control applied to a CT-2-like low-maximum-power set would also reduce its power consumption and thus also its weight and size.

The major difference in size, weight, and talk time between the two pocketphones is directly attributable to the two orders of magnitude difference in average transmitter power. The generation of transmitter power dominates power consumption in the analog cellular phone. Power consumption in the digital CT-2 phone is more evenly divided between transmitter-power generation and digital signal processing. Therefore, power consumption in complex digital signal processing would have more impact on talk time in small low-power personal communicators than in cellular handsets where the transmitter-power generation is so large. Other than reducing power consumption for both functions, the only alternative for increasing talk time and reducing battery weight is to invent new battery technology having greater energy density; see section on Other Issues later in this chapter.

In contrast, lowering the transmitter power requirement, modestly applying digital signal processing, and shifting some of the radio coverage burden to a higher density of small, low-power, low-complexity, low-cost fixed radio ports has the effect of shifting some of the talk time, weight, and cost constraints from battery technology to solid state electronics technology, which continues to experience orders-of-magnitude improvements in the span of several years. Digital signal-processing complexity, however, cannot be permitted to overwhelm power consumption in low-power handsets; whereas small differences in complexity will not matter much, orders-of-magnitude differences in complexity will continue to be significant.

Thus, it can be seen from Table 4 that the size, weight, and quality arguments in the preceding sections generally hold for these examples. It also is evident from the preceding paragraphs that they will be even more notable when comparing digital cordless pocketphones with digital cellular pocketphones of the same development generations.

Evolution Toward the Future and to Low-Tier Personal Communications Services

After looking at the evolution of several wireless technologies and systems in the preceding sections it appears appropriate to ask again: wireless personal communications, What is it? All of the technologies in the preceding sections claim to provide wireless personal communications, and all do to some extent. All have significant limitations, however, and all are evolving in attempts to overcome the limitations. It seems appropriate to ask, what are the likely endpoints? Perhaps some hint of the endpoints can be found by exploring what users see as limitations of existing technologies and systems and by looking at the evolutionary trends.

In order to do so, we summarize some important clues from the preceding sections and project them, along with some U.S. standards activity, toward the future.

Digital Cordless Telephones

- Strengths: good circuit quality; long talk time; small lightweight battery; low-cost sets and service.
- Limitations: limited range; limited usage regions.
- Evolutionary trends: phone points in public places; wireless PBX in business.
- Remaining limitations and issues: limited usage regions and coverage holes; limited or no handoff; limited range.

Digital Cellular Pocket Handsets

- Strength: widespread service availability.
- Limitations: limited talk time; large heavy batteries; high-cost sets and service; marginal circuit quality; holes in coverage and poor in-building coverage; limited data capabilities; complex technologies.
- Evolutionary trends: microcells to increase capacity and in-building coverage and to reduce battery drain; satellite systems to extend coverage.
- Remaining limitations and issues: limited talk time and large battery; marginal circuit quality; complex technologies.

Wide Area Data

- Strength: digital messages.
- Limitations: no voice; limited data rate; high cost.

- Evolutionary trends: microcells to increase capacity and reduce cost; share facilities with voice systems to reduce cost.
- Remaining limitations and issues: no voice; limited capacity.

Wireless Local Area Networks (WLANs)

- Strength: high data rate.
- Limitations: insufficient capacity for voice; limited coverage; no standards; chaos.
- Evolutionary trends: hard to discern from all of the churning.

Paging/messaging

- Strengths: widespread coverage; long battery life; small lightweight sets and batteries; economical.
- Limitations: one-way message only; limited capacity.
- Evolutionary desire: two-way messaging and/or voice; capacity.
- Limitations and issues: two-way link cannot exploit the advantages of one-way link asymmetry.

There is a strong trajectory evident in these systems and technologies aimed at providing the following features.

High Quality Voice and Data

- To small, lightweight, pocket carried communicators.
- Having small lightweight batteries.
- Having long talk time and long standby battery life.
- Providing service over large coverage regions.
- For pedestrians in populated areas (but not requiring high population density).
- Including low to moderate speed mobility with handoff.

Economical Service

- Low subscriber-set cost.
- Low network-service cost.

Privacy and Security of Communications

- Encrypted radio links.

This trajectory is evident in all of the evolving technologies but can only be partially satisfied by any of the existing and evolving systems and technologies! Trajectories from all of the evolving technologies and systems are illustrated in Fig. 1 as being aimed at low-tier personal communications systems or services, i.e., low-tier PCS. Taking characteristics from cordless, cellular, wide-area data and, at least moderate-rate, WLANs, suggests the following attributes for this low-tier PCS.

1. 32 kb/s ADPCM speech encoding in the near future to take advantage of the low complexity and low power consumption, and to provide low-delay high-quality speech.
2. Flexible radio link architecture that will support multiple data rates from several kilobits per second to several hundred kilobits per second. This is needed to permit evolution in the future to lower bit rate speech as technology improvements permit high quality without excessive power consumption or transmission delay and to provide multiple data rates for data transmission and messaging.
3. Low transmitter power (≤ 25 mW average) with adaptive power control to maximize talk time and data transmission time. This incurs short radio range that requires many base

stations to cover a large region. Thus, base stations must be small and inexpensive, like cordless telephone phone points or the Metricom wireless data base stations.

4. Low-complexity signal processing to minimize power consumption. Complexity one-tenth that of digital cellular or high-tier PCS technologies is required [Cox, 1992]. With only several tens of milliwatts (or less under power control) required for transmitter power, signal processing power becomes significant.

5. Low cochannel interference and high coverage area design criteria. In order to provide high-quality service over a large region, at least 99% of any covered area must receive good or better coverage and be below acceptable cochannel interference limits. This implies less than 1% of a region will receive marginal service. This is an order-of-magnitude higher service requirement than the 10% of a region permitted to receive marginal service in vehicular cellular system (high-tier PCS) design criteria.

6. Four-level phase modulation with coherent detection to maximize radio link performance and capacity with low complexity.

7. Frequency division duplexing to relax the requirement for synchronizing base station transmissions over a large region.

Such technologies and systems have been designed, prototyped, and laboratory and field tested and evaluated for several years [Cox, 1992; Padgett, Hattori, and Gunther, 1995; Cox, 1990; 1991; Bellcore, 1990; Cox, 1985; Cox, Arnold, and Portor, 1987; Cox, 1986; 1987; 1989; Cox, Gifford, and Sherry, 1991]. The viewpoint expressed here is consistent with the progress in the Joint Technical Committee (JTC) of the U.S. standards bodies, Telecommunications Industry Association (TIA) and Committee T1 of the Alliance for Telecommunications Industry Solutions (ATIS). Many technologies and systems were submitted to the JTC for consideration for wireless PCS in the new 1.9-GHz frequency bands for use in the United States [Cook, 1994]. Essentially all of the technologies and systems listed in Table 1, and some others, were submitted in late 1993. It was evident that there were at least two and perhaps three distinct different classes of submissions. No systems optimized for packet data were submitted, but some of the technologies are optimized for voice.

One class of submissions was the group labeled high-power systems, digital cellular (high-tier PCS) in Table 1. These are the technologies discussed previously in this chapter. They are highly optimized for low-bit-rate voice and, therefore, have somewhat limited capability for serving packet-data applications. Since it is clear that wireless services to wide ranging high-speed mobiles will continue to be needed, and that the technology already described for low-tier PCS may not be optimum for such services, Fig. 1 shows a continuing evolution and need in the future for high-tier PCS systems that are the equivalent of today's cellular radio. There are more than 100 million vehicles in the United States alone. In the future, most, if not all, of these will be equipped with high-tier cellular mobile phones. Therefore, there will be a continuing and rapidly expanding market for high-tier systems.

Another class of submissions to the JTC [Cook, 1994] included the Japanese personal handiphone system (PHS) and a technology and system originally developed at Bellcore but carried forward to prototypes and submitted to the JTC by Motorola and Hughes Network Systems. This system was known as wireless access communications system (WACS).[2] These two submissions were so similar in their design objectives and system characteristics that, with the agreement of the delegations from Japan and the United States, the PHS and WACS submissions were combined under a new name, personal access communication systems (PACS), that was to incorporate the best features of both. This advanced, low-power wireless access system, PACS, was to be known as low-tier PCS. Both WACS/PACS and Handiphone (PHS) are shown in Table 1 as low-tier PCS and represent the evolution to low-tier PCS, in Fig. 1. The WACS/PACS/ UDPC system and technology are discussed in Cox, 1992; Padgett, Hattori, and Gunther, 1995; Cox, 1991; Bellcore, 1990; Cox, 1985; Cox, Arnold, and Portor, 1987; Cox, 1986; 1987; 1989; and Cox, Gifford, and Sherry, 1991.

[2]WACS was known previously as Universal Digital Portable Communications (UDPC).

In the JTC, submissions for PCS of DECT and CT-2 and their variations were also lumped under the class of low-tier PCS, even though these advanced digital cordless telephone technologies were somewhat more limited in their ability to serve all of the low-tier PCS needs. They are included under digital cordless technologies in Table 1. Other technologies and systems were also submitted to the JTC for high-tier and low-tier applications, but they have not received widespread industry support.

One wireless access application discussed earlier that is not addressed by either high-tier or low-tier PCS is the high-speed WLAN application. Specialized high-speed WLANs also are likely to find a place in the future. Therefore, their evolution is also continued in Fig. 1. The figure also recognizes that widespread low-tier PCS can support data at several hundred kilobits per second and, thus, can satisfy many of the needs of WLAN users.

It is not clear what the future roles are for paging/messaging, cordless telephone appliances, or wide-area packet-data networks in an environment with widespread contiguous coverage by low-tier and high-tier PCS. Thus, their extensions into the future are indicated with a question mark in Fig. 1.

Those who may object to the separation of wireless PCS into high tier and low tier should review this section again, and note that we have two tiers of PCS now. On the voice side there is cellular radio, i.e., high-tier PCS, and cordless telephone, i.e., an early form of low-tier PCS. On the data side there is wide-area data, i.e., high-tier data PCS, and WLANs, i.e., perhaps a form of low-tier data PCS. In their evolutions, these all have the trajectories discussed and shown in Fig. 1 that point surely toward low-tier PCS. It is this low-tier PCS that marketing studies continue to project is wanted by more than half of the U.S. households or by half of the people, a potential market of over 100 million subscribers in the United States alone. Similar projections have been made worldwide.

4 Comparisons with Other Technologies

Complexity/Coverage Area Comparisons

Experimental research prototypes of radio ports and subscriber sets [Sollenberger et al., 1991; Sollenberger and Afrashteh, 1991] have been constructed to demonstrate the technical feasibility of the radio link requirements in Bellcore, 1990. These WACS prototypes generally have the characteristics and parameters previously noted, with the exceptions that 1) the portable transmitter power is lower (10 mW average, 100 mW peak), 2) dynamic power control and automatic time slot transfer are not implemented, and 3) a rudimentary automatic link-transfer implementation is based only on received power. The experimental base stations transmit near 2.17 GHz; the experimental subscriber sets transmit near 2.12 GHz. Both operated under a Bellcore experimental license. The experimental prototypes incorporate application-specific, very large-scale integrated circuits[3] fabricated to demonstrate the feasibility of the low-complexity high-performance digital signal-processing techniques [Sollenberger and Afrashteh, 1991; Sollenberger, 1991] for symbol timing and coherent bit detection. These techniques permit the efficient short TDMA bursts having only 100 b that are necessary for low-delay TDMA implementations. Other digital signal-processing functions in the prototypes are implemented in programmable logic devices. All of the digital signal-processing functions combined require about 1/10 of the logic gates that are required for digital signal processing in vehicular digital cellular mobile implementations [Sollenberger, 1990; 1991; Hinderling et al., 1992]; that is, this low-complexity PCS implementation having no delay-dispersion-compensating circuits and no forward error-correction decoding is about 1/10 as complex as the digital cellular implementations that include these functions.[4] The 32 kb/s ADPCM speech-encoding in the

---

[3]Application specific integrated circuits (ASIC), very large-scale integration (VLSI).

[4]Some indication of VLSI complexity can be seen by the number of people required to design the circuits. For the low-complexity TDMA ASIC set, only one person part time plus a student part time were required; the complex CDMA ASIC has six authors on the paper alone.

low-complexity PCS implementation is also about 1/10 as complex as the less than 10-kb/s speech encoding used in digital cellular implementations. This significantly lower complexity will continue to translate into lower power consumption and cost. It is particularly important for low-power pocket personal communicators with power control in which the DC power expended for radio frequency transmitting can be only tens of milliwatts for significant lengths of time.

The experimental radio links have been tested in the laboratory for detection sensitivity [bit error rate (BER) vs SNR] [Sollenberger et al., 1991; Sollenberger and Chuang, 1988; Chuang and Sollenberger, 1988] and for performance against cochannel interference [Afrashteh, Sollenberger, and Chukurov, 1991] and intersymbol interference caused by multipath delay spread [Sollenberger et al., 1991]. These laboratory tests confirm the performance of the radio-link techniques. In addition to the laboratory tests, qualitative tests have been made in several PCS environments to compare these experimental prototypes with several United States CT-1 cordless telephones at 50 MHz, with CT-2 cordless telephones at 900 MHz, and with DCT-900 cordless telephones at 900 MHz. Some of these comparisons have been reported [Tuthill, Granger, and Wurtz, 1992; AWC, 1992; MN, 1992; Bell South, 1991]. In general, depending on the criteria, e.g., either no degradation or limited degradation of circuit quality, these WACS experimental prototypes covered areas inside buildings that ranged from 1.4 to 4 times the areas covered by the other technologies. The coverage areas for the experimental prototypes were always substantially limited in two or three directions by the outside walls of the buildings. These area factors could be expected to be even larger if the coverage were not limited by walls, i.e., once all of a building is covered in one direction, no more area can be covered no matter what the radio link margin. The earlier comparisons [AWC, 1992; MN, 1992; Bell South, 1991] were made with only two-branch uplink diversity before subscriber-set transmitting antenna switching was implemented and, with only one radio port before automatic radio-link transfer was implemented. The later tests [Tuthill, Granger, and Wurtz, 1992] included these implementations. These reported comparisons agree with similar unreported comparisons made in a Bellcore laboratory building. Similar coverage comparison results have been noted for a 900-MHz ISM-band cordless telephone compared to the 2-GHz experimental prototype. The area coverage factors (e.g., ×1.4 to ×4) could be expected to be even greater if the cordless technologies had also been operated at 2 GHz since attenuation inside buildings between similar small antennas is about 7 dB greater at 2 GHz than at 900 MHz [Devaservatham et al., 1990a; 1990b] and the 900-MHz handsets transmitted only 3 dB less average power than the 2-GHz experimental prototypes. The greater area coverage demonstrated for this technology is expected because of the different compromises noted earlier; the following, in particular.

1. Coherent detection of QAM provides more detection sensitivity than noncoherent detection of frequency-shift modulations [Chuang, 1990].
2. Antenna diversity mitigates bursts of errors from multipath fading [Sollenberger et al., 1991].
3. Error detection and blanking of TDMA bursts having errors significantly improves perceived speech quality [Varma et al., 1987]. (Undetected errors in the most significant bits cause sharp audio pops that seriously degrade perceived speech quality.)
4. Robust symbol timing and burst and frame synchronization reduce the number of frames in error due to imperfect timing and synchronization [Sollenberger et al., 1991].
5. Transmitting more power from the radio port compared to the subscriber set offsets the less sensitive subscriber set receiver compared to the port receiver that results from power and complexity compromises made in a portable set.

Of course, as expected, the low-power (10-mW) radio links cover less area than high-power (0.5-W) cellular mobile pocketphone radio links because of the 17-dB transmitter power difference resulting from the compromises discussed previously. In the case of vehicular mounted sets, even more radio-link advantage accrues to the mobile set because of the higher gain of vehicle-mounted antennas and higher transmitter power (3 W).

5 Quality, Capacity, and Economic Issues

Although the several trajectories toward low-tier PCS discussed in the preceding section are clear, it does not fit the existing wireless communications paradigms. Thus, low-tier PCS has attracted less attention than the systems and technologies that are compatible with the existing paradigms. Some examples are cited in the following paragraphs.

The need for intense interaction with an intelligent network infrastructure in order to manage mobility is not compatible with the cordless telephone appliance paradigm. In that paradigm, independence of network intelligence and base units that mimic wireline telephones are paramount.

Wireless data systems often do not admit to the dominance of wireless voice communications and, thus, do not take advantage of the economics of sharing network infrastructure and base station equipment. Also, wireless voice systems often do not recognize the importance of data and messaging and, thus, only add them in as bandaids to systems.

The need for a dense collection of many low-complexity, low-cost, low-tier PCS base stations interconnected with inexpensive fixed-network facilities (copper or fiber based) does not fit the cellular high-tier paradigm that expects sparsely distributed $1 million cell sites. Also, the need for high transmission quality to compete with wireline telephones is not compatible with the drive toward maximizing users-per-cell-site and per megahertz to minimize the number of expensive cell sites. These concerns, of course, ignore the hallmark of frequency-reusing cellular systems. That hallmark is the production of almost unlimited overall system capacity by reducing the separation between base stations. The cellular paradigm does not recognize the fact that almost all houses in the U.S. have inexpensive copper wires connecting telephones to the telephone network. The use of low-tier PCS base stations that concentrate individual user services before backhauling in the network will result in less fixed interconnecting facilities than exist now for wireline telephones. Thus, inexpensive techniques for interconnecting many low-tier base stations are already deployed to provide wireline telephones to almost all houses.

This list could be extended, but the preceding examples are sufficient, along with the earlier sections of the paper, to indicate the many complex interactions among circuit quality, spectrum utilization, complexity (circuit and network), system capacity, and economics that are involved in the design compromises for a large, high-capacity wireless-access system. Unfortunately, the tendency has been to ignore many of the issues and focus on only one, e.g., the focus on cell site capacity that drove the development of digital-cellular high-tier systems in the United States. Interactions among circuit quality, complexity, capacity, and economics are considered in the following sections.

Capacity, Quality, and Complexity

Although capacity comparisons frequently are made without regard to circuit quality, complexity, or cost per base station, such comparisons are not meaningful. An example in Table 5 compares capacity factors for U.S. cellular or high-tier PCS technologies with the low-tier PCS technology, PACS/WACS. The mean opinion scores (MOS) (noted in Table 5) for speech coding are discussed later. Detection of speech activity and turning off the transmitter during times of no activity is implemented in IS-95. Its impact on MOS also is noted later. A similar technique has been proposed as E-TDMA for use with IS-54 and is discussed with respect to TDMA systems in Cox, 1992. Note that the use of low-bit-rate speech coding combined with speech activity degrades the high-tier system's quality by nearly one full MOS point on the five-point MOS scale when compared to 32 kb/s ADPCM. Tandem encoding is discussed in a later section. These speech quality degrading factors alone provide a base station capacity increasing factor of $\times 4 \times 2.5 = \times 10$ over the high-speech-quality low-tier system! Speech coding, of course, directly affects base station capacity and, thus, overall system capacity by its effect on the number of speech channels that can fit into a given bandwidth.

The allowance of extra system margin to provide coverage of 99% of an area for low-tier PCS versus 90% coverage for high-tier is discussed in the previous section and Cox, 1992. This additional

TABLE 5 Comparison of Cellular (IS-54/IS-95) and Low-Tier PCS (WACS/PACS). Capacity comparisons made without regard to quality factors, complexity, and cost per base station are not meaningful

Parameter	Cellular (High Tier)	Low-Tier PCS	Capacity Factor
Speech Coding, kb/s	8 (MOS 3.4) No tandem coding	32 (MOS 4.1) 3 or 4 tandem	×4
Speech activity	Yes (MOS 3.2)	No (MOS 4.1)	×2.5
Percentage of good areas, %	90	99	×2
Propagation σ, dB	8	10	×1.5
Total: trading quality for capacity			×30

quality factor costs a capacity factor of ×2. The last item in Table 5 does not change the actual system, but only changes the way that frequency reuse is calculated. The additional 2-dB margin in standard deviation σ, allowed for coverage into houses and small buildings for low-tier PCS, costs yet another factor of ×1.5 in calculation only. Frequency reuse factors affect the number of sets of frequencies required and, thus, the bandwidth available for use at each base station. Thus, these factors also affect the base station capacity and the overall system capacity.

For the example in Table 5, significant speech and coverage quality has been traded for a factor of ×30 in base station capacity! Whereas base station capacity affects overall system capacity directly, it should be remembered that overall system capacity can be increased arbitrarily by decreasing the spacing between base stations. Thus, if the PACS low-tier PCS technology were to start with a base station capacity of ×0.5 of AMPS cellular[5] (a much lower figure than the ×0.8 sometimes quoted [Cook, 1994]), and then were degraded in quality as described above to yield the ×30 capacity factor, it would have a resulting capacity of ×15 of AMPS! Thus, it is obvious that making such a base station capacity comparison without including quality is not meaningful.

Economics, System Capacity, and Coverage Area Size

Claims are sometimes made that low-tier PCS cannot be provided economically, even though IT is what the user wants. These claims are often made based on economic estimates from the cellular paradigm. These include the following.

- Very low estimates of market penetration, much less than cordless telephones, and often even less than cellular.
- High estimates of base station costs more appropriate to high-complexity, high-cost cellular technology than to low-complexity, low-cost, low-tier technology.
- Very low estimates of circuit useage time more appropriate to cellular than to cordless/wireline telephone usage, which is more likely for low-tier PCS.

Such economic estimates are often done by making absolute economic calculations based on very uncertain input data. The resulting estimates for low tier and high tier are often closer together than the large uncertainties in the input data. A perhaps more realistic approach for comparing such systems is to vary only one or two parameters while holding all others fixed and then looking at relative economics between high-tier and low-tier systems. This is the approach used in the following examples.

---

[5]Note that the ×0.5 factor is an arbitrary factor taken for illustrating this example. The so-called ×AMPS factors are only with regard to base station capacity, although they are often misused as system capacity.

TABLE 6 System Capacity/Coverage Area Size/Economics

Example 1

Assume channels/MHz are the same for cellular and PCS
   Cell site: spacing = 20,000 ft   cost $ = 1 M
   PCS port: spacing = 1,000 ft
   PCS system capacity is $(20000/1000)^2 = 400 \times$ cellular capacity

Then, for the system costs to be the same
   Port cost = ($ 1 M/400) $2,500, a reasonable figure

If, cell site and port each have 180 channels
   Cellular cost/circuit = $1 M/180 = $5,555/circuit
   PCS cost/circuit = $2500/180 = $14/circuit

Example 2

Assume equal cellular and PCS system capacity
   Cell site: spacing = 20,000 ft
   PCS port: spacing = 1,000 ft

If, a cell site has 180 channels
Then, for equal system capacity, a PCS port needs $180/400 < 1$ channel/port

Example 3

Quality/cost trade
   Cell site: Spacing = 20,000 ft   cost = $1 M   channels = 180
   PCS port: Spacing = 1,000 ft   cost = $2,500

Cellular to PCS, base station spacing capacity factor = $\times 400$
   PCS to cellular quality reduction factors:

32 to 8 kb/s speech	×4
Voice activity (buying)	×2
99–90% good areas	×2
Both in same environment (same σ)	×1
Capacity factor traded	×16

180 ch/16 = 11.25 channels/port   then, $2500/11.25 = $222/circuit
and remaining is $\times 400/16 = \times 25$ system capacity of PCS over cellular

Example 1: In the first example (see Table 6), the number of channels per megahertz is held constant for cellular and for low-tier PCS. Only the spacing is varied between base stations, e.g., cell sites for cellular and radio ports for low-tier PCS, to account for the differences in transmitter power, antenna height, etc. In this example, overall system capacity varies directly as the square of base station spacing, but base station capacity is the same for both cellular and low-tier PCS. For the typical values in the example, the resulting low-tier system capacity is ×400 greater, only because of the closer base station spacing. If the two systems were to cost the same, the equivalent low-tier PCS base stations would have to cost less than $2,500.

This cost is well within the range of estimates for such base stations, including equivalent infrastructure. These low-tier PCS base stations are of comparable or lower complexity than cellular vehicular subscriber sets, and large-scale manufacture will be needed to produce the millions that will be required. Also, land, building, antenna tower and legal fees for zoning approval, or rental of expensive space on top of commercial buildings, represent large expenses for cellular cell sites. Low-tier PCS base stations that are mounted on utility poles and sides of buildings will not incur such large additional expenses. Therefore, costs of the order of magnitude indicated seem reasonable in large quantities. Note that, with these estimates, the per-wireless-circuit cost of the low-tier PCS circuits would be only $14/circuit compared to $5,555/circuit for the high-tier circuits. Even if there were a factor of 10 error in cost estimates, or a reduction of channels per radio port of a factor of 10, the per-circuit cost of low-tier PCS would still be only $140/circuit, which is still much less than the per-circuit cost of high tier.

Example 2: In the second example (see Table 6), the overall system capacity is held constant, and the number of channels/port, i.e., channels/(base station) is varied. In this example, less than 1/2 channel/port is needed, again indicating the tremendous capacity that can be produced with close-spaced low-complexity base stations.

Example 3: Since the first two examples are somewhat extreme, the third example (see Table 6) uses a more moderate, intermediate approach. In this example, some of the cellular high-tier channels/(base station) are traded to yield higher quality low-tier PCS as in the previous subsection. This reduces the channels/port to 11 +, with an accompanying increase in cost/circuit up to $222/circuit, which is still much less than the $5,555/circuit for the high-tier system. Note, also, that the low-tier system still has ×25 the capacity of the high-tier system!

Low-tier base station (Port) cost would have to exceed $62,500 for the low-tier per-circuit cost to exceed that of the high-tier cellular system. Such a high port cost far exceeds any existing realistic estimate of low-tier system costs.

It can be seen from these examples, and particularly example 3, that the circuit economics of low-tier PCS are significantly better than for high-tier PCS, if the user demand and density is sufficient to make use of the large system capacity. Considering the high penetration of cordless telephones, the rapid growth of cellular handsets, and the enormous market projections for wireless PCS noted earlier in this chapter, filling such high capacity in the future would appear to be certain. The major problem is providing rapidly the widespread coverage (buildout) required by the FCC in the United States. If this unrealistic regulatory demand can be overcome, low-tier wireless PCS promises to provide the wireless personal communications that everyone wants.

6 Other Issues

Several issues in addition to those addressed in the previous two sections continue to be raised with respect to low-tier PCS. These are treated in this section.

Improvement of Batteries

Frequently, the suggestion is made that battery technology will improve so that high-power handsets will be able to provide the desired 5 or 6 hours of talk time in addition to 10 or 12 hours of standby time, and still weigh less than one-fourth of the weight of today's smallest cellular handset batteries. This hope does not take into account the maturity of battery technology, and the long history (many decades) of concerted attempts to improve it. Increases in battery capacity have come in small increments, a few percent, and very slowly over many years, and the shortfall is well over a factor of 10. In contrast, integrated electronics and radio frequency devices needed for low-power low-tier PCS continue to improve and to decrease in cost by factors of greater than 2 in time spans on the order of a year or so. It also should be noted that, as the energy density of a battery is increased, the energy release rate per volume must also increase in order to supply the same amount of power. If energy storage density and release rate are increased significantly, the difference between a battery and a bomb become indistinguishable! The likelihood of a ×10 improvement in battery capacity appears to be essentially zero. If even a modest improvement in battery capacity were possible, many people would be driving electric vehicles.

People Only Want One Handset

This issue is often raised in support of high-tier cellular handsets over low-tier handsets. Whereas the statement is likely true, the assumption that the handset must work with high-tier cellular is not. Such a statement follows from the current large usage of cellular handsets; but such usage results because that is the only form of widespread wireless service currently available, not because it is what people want. The statement assumes inadequate coverage of a region by low-tier PCS, and that low-tier handsets will not work in vehicles. The only way that high-tier handsets could serve the desires of people discussed earlier would be for an unlikely breakthrough in battery technology to occur. A low-tier system, however, can cover economically any large region having some people in it. (It will not cover rural or isolated areas but, by definition, there is essentially no one there to want communications anyway.)

Low-tier handsets will work in vehicles on village and city streets at speeds up to 30 or 40 mi/h, and the required handoffs make use of computer technology that is rapidly becoming inexpensive. Highways between populated areas, and also streets within them, will need to be covered by high-tier cellular PCS, but users are likely to use vehicular sets in these cellular systems. Frequently the vehicular mobile user will want a different communications device anyway, e.g., a hands-free phone. The use of hands-free phones in vehicles is becoming a legal requirement in some places now and is likely to become a requirement in many more places in the future. Thus, handsets may not be legally usable in vehicles anyway. With widespread deployment of low-tier PCS systems, the one handset of choice will be the low-power, low-tier PCS pocket handset or voice/data communicator.

There are approaches for integrating low-tier pocket phones or pocket communicators with high-tier vehicular cellular mobile telephones. The user's identity could be contained either in memory in the low-tier set or in a small smart card inserted into the set, as is a feature of the European GSM system. When entering an automobile, the small low-tier communicator or card could be inserted into a receptacle in a high-tier vehicular cellular set installed in the automobile.[6] The user's identity would then be transferred to the mobile set. The mobile set could then initiate a data exchange with the high-tier system, indicating that the user could now receive calls at that mobile set. This information about the user's location would then be exchanged between the network intelligence so that calls to the user could be correctly routed.[7] In this approach the radio sets are optimized for their specific environments, high-power, high-tier vehicular or low-power, low-tier pedestrian, as discussed earlier, and the network access and call routing is coordinated by the interworking of network intelligence. This approach does not compromise the design of either radio set or radio system. It places the burden on network intelligence technology that benefits from the large and rapid advances in computer technology.

The approach of using different communications devices for pedestrians than for vehicles is consistent with what has actually happened in other applications of technology in similarly different environments. For example, consider the case of audio cassette tape players. Pedestrians often carry and listen to small portable tape players with lightweight headsets (e.g., a Walkman[8]). When one of these people enters an automobile, he or she often removes the tape from the Walkman and inserts it into a tape player installed in the automobile. The automobile player has speakers that fill the car with sound. The Walkman is optimized for a pedestrian, whereas the vehicular-mounted player is optimized for an automobile. Both use the same tape, but they have separate tape heads, tape transports, audio preamps, etc. They do not attempt to share electronics. In this example, the tape cassette is the information-carrying entity similar to the user identification in the personal communications example discussed earlier. The main points are that the information is shared among different devices but that the devices are optimized for their environments and do not share electronics.

Similarly, a high-tier vehicular-cellular set does not need to share oscillators, synthesizers, signal processing, or even frequency bands or protocols with a low-power, low-tier pocket-size communicator. Only the information identifying the user and where he or she can be reached needs to be shared among the intelligence elements, e.g., routing logic, databases, and common channel signalling [Cox, 1992; 1989] of the infrastructure networks. This information exchange between

---

[6]Inserting the small personal communicator in the vehicular set would also facilitate charging the personal communicator's battery.

[7]This is a feature proposed for FPLMTS in CCIR Rec. 687.

[8]Walkman is a registered trademark of Sony Corporation.

network intelligence functions can be standardized and coordinated among infrastructure sub-networks owned and operated by different business entities (e.g., vehicular cellular mobile radio networks and intelligent low-tier PCS networks). Such standardization and coordination are the same as are required today to pass intelligence among local exchange networks and interexchange carrier networks.

Other Environments

Low-tier personal communications can be provided to occupants of airplanes, trains, and buses by installing compatible low-tier radio access ports inside these vehicles. The ports can be connected to high-power, high-tier vehicular cellular mobile sets or to special air–ground or satellite-based mobile communications sets. Intelligence between the internal ports and mobile sets could interact with cellular mobile, air–ground, or satellite networks in one direction, using protocols and spectrum allocated for that purpose, and with low-tier personal communicators in the other direction to exchange user identification and route calls to and from users inside these large vehicles. Radio isolation between the low-power units inside the large metal vehicles and low-power systems outside the vehicles can be ensured by using windows that are opaque to the radio frequencies. Such an approach also has been considered for automobiles, i.e., a radio port for low-tier personal communications connected to a cellular mobile set in a vehicle so that the low-tier personal communicator can access a high-tier cellular network. (This could be done in the United States using unlicensed PCS frequencies within the vehicle.)

Speech Quality Issues

All of the PCS and cordless telephone technologies that use CCITT standardized 32-kb/s ADPCM speech encoding can provide similar error-free speech distortion quality. This quality often is rated on a five-point subjective mean opinion score (MOS) with 5 excellent, 4 good, 3 fair, 2 poor, and 1 very poor. The error-free MOS of 32-kb/s ADPCM is about 4.1 and degrades very slightly with tandem encodings. Tandem encodings could be expected in going from a digital-radio PCS access link, through a network using analog transmission or 64-kb/s PCM, and back to another digital-radio PCS access link on the other end of the circuit. In contrast, a low-bit-rate (<10-kb/s) vocoder proposed for a digital cellular system was recently reported [QTF, 1992] to yield an MOS of 3.4 on an error-free link without speech-activity detection. This score dropped to 3.2 when speech-activity detection was implemented to increase system capacity. This nearly one full point decrease on the five-point MOS score indicates significant degradation below accepted CCITT wireline speech distortion quality. Either almost half of the population must have rated it as poor or most of the population must have rated it as only fair. It should also be noted that these MOS scores may not reflect additional degradation that may occur in low-bit-rate speech encoding when the speech being encoded is combined with acoustical noise in a mobile environment, e.g., tire, wind, and engine noise in automobiles and street noise, background talking, etc., in handheld phone environments along streets and in buildings. Comments from actual users of low-bit-rate speech technology in accoustically noisy environments suggest that the MOS scores just quoted are significantly degraded in these real world environments. Waveform coders, e.g., ADPCM are not likely to experience degradation from such background noise. In addition, the low-bit-rate speech encoding is not at all tolerant of the tandem speech encodings that will inevitably occur for PCS for many years. That is, when low-bit-rate speech is transcoded to a different encoding format, e.g., to 64 kb/s as is used in many networks or from an IS-54 phone on one end to a GSM or IS-95 phone on the other end, the speech quality deteriorates precipitously. Although this may not be a serious issue for a vehicular mobile user who has no choice other than not to communicate at all, it is likely to be a serious issue in an environment where a wireline telephone is available as an alternative. It is also less serious when there are few mobile-to-mobile calls through the network, but as wireless usage increases and digital mobile-to-mobile calls become commonplace, the marginal transcoded speech quality is likely to become a serious issue. These comments in this paragraph are generally applicable to speech encoding at rates of 13 kb/s or less.

In the arena of transmission delay, the short-frame (2-ms) FDMA/TDD and TDMA technologies (e.g., CT-2 and WACS noted earlier) can readily provide single-radio-link round-trip delays of <10 ms and, perhaps, even <5 ms. The longer frame (10 ms and greater) cordless-phone TDMA technologies, e.g., DCT-900/CT-3/DECT and some ISM-band implementations, inherently have a single-link round-trip delay of at least 20 ms and can range 30–40 ms or more in some implementations. As mentioned earlier, the digital vehicular-cellular technologies with low-bit-rate speech encoding, bit interleaving, forward error-correction decoding, and relatively long frame time (~16–20 ms) result in single-link round-trip delays on the order of 200 ms, well over an order of magnitude greater than the short-frame technologies, and on the same order of magnitude as a single-direction synchronous satellite link. It should be noted that almost all United States domestic long-distance telephone circuits have been removed from such satellite links, and many international satellite links also are being replaced by undersea fiber links. These long-distance-circuit technology changes are made partially to reduce the perceptual impairment of long transmission delay.

New Technology

New technology, e.g., spread spectrum or CDMA, is sometimes offered as a solution to both the high-tier cell site capacity and transmitter power issues. As these new technologies are pursued vigorously, however, it becomes increasingly evident that the early projections were considerably overoptimistic, that the base station capacity will be about the same as other technologies [Cox, 1992], and that the high complexity will result in more, not less, power consumption.

With the continuing problems and delays in initial deployments, there is increasing concern throughout the industry as to whether CDMA is a viable technology for high-capacity cellular applications. With the passage of time, it is becoming more obvious that Viterbi was correct in his 1985 paper in which he questioned the use of spread spectrum for commercial communications [Viterbi, 1985].

The IS-95 proposal is considerably more technically sophisticated than earlier spread spectrum proposals. It includes fast feedback control of mobile transmitter power, heavy forward error correction, speech detection and speech-encoding rate adjustment to take advantage of speech inactivity, and multiple receiver correlators to latch onto and track resolvable multipath maxima [Salmasi and Gilhousen, 1991]. The spreading sequence rate is 1.23 MHz.

The near–far problem is addressed directly and elegantly on the uplink by a combination of the fast-feedback power control and a technique called soft handoff that permits the instantaneous selection of the best paths between a mobile and two cell sites. Path selection is done on a frame-by-frame basis when paths between a mobile and the two cell sites are within a specified average level (perhaps 6–10 dB) of each other. This soft handoff provides a form of macroscopic diversity [Bernhardt, 1987] between pairs of cell sites when it is advantageous. Increasing capacity by soft handoff requires precise time synchronization (on the order of a microsecond) among all cell sites in a system. An advantage of this proposal is that frequency coordination is not needed among cell sites since all sites can share a frequency channel. Coordination of the absolute time delays of spreading sequences among cell sites is required, however, since these sequence delays are used to distinguish different cell sites for initial access and for soft handoff. Also, handoff from one frequency to another is complicated.

Initially, the projected cell site capacity of this CDMA system, determined by mathematical analysis and computer simulation of simplified versions of the system, was ×20 to ×40 that of the analog AMPS, with a coverage criterion of 99% of the covered area [CTIA, 1989]. However, some other early estimates [AWC, 1991] suggested that the factors were more likely to be ×6 to ×8 of AMPS.

A limited experiment was run in San Diego, California, during the fourth quarter of 1991 under the observation of cellular equipment vendors and service providers. This experiment had 42–62 mobile units in fewer than that many vehicles,[9] and four or five cell sites, one with three sectors. Well over half of the mobiles needed to provide the interference environment for system capacity tests were simulated by hardware noise simulation by a method not yet revealed for technical assessment. Estimates of the cell site capacity from this CDMA experiment center around ×10 that of AMPS [QTF, 1992; CTIA, 1991][10] with coverage criteria <99%, perhaps 90–95%, and with other capacity estimates ranging between ×8 and ×15.

This experiment did not exercise several potential capacity-reducing factors; the following, for example.

1. Only four cells participated in capacity tests. The test mobiles were all located in a relatively limited area and had limited choices of cell sites with which to communicate for soft handoffs. This excludes the effects of selecting a strong cell-site downlink for soft handoff that does not have the lowest uplink attenuation because of uncorrelated uplink and downlink multipath fading at slow vehicle speeds [Ariyavisitakul, 1992].

2. The distribution of time-dispersed energy in hilly environments like San Diego usually is more concentrated around one or two delays than is the dispersed energy scattered about in heavily built-up urban areas like downtown Manhattan [Chang, 1992; Chang and Ariyavisitakul, 1991] or Chicago. Energy concentrated at one or two delays is more fully captured by the limited number of receiver correlators than is energy more evenly dispersed in time.

3. Network delay in setting up soft handoff channels can result in stronger paths to other cell sites than to the one controlling uplink transmitter power. This effect can be more pronounced when coming out of shadows of tall buildings at intersections in heavily built-up areas. The effect will not occur as frequently in a system with four or five cell sites as it will in a large, many cell site system.

All of these effects and others [Ariyavisitakul et al.] will increase the interference in a large system, similarly to the increase in interference that results from additional mobiles and, thus, will decrease cell site capacity significantly below that estimated in the San Diego trial. Factors like these have been shown to reduce the San Diego estimate of ×10 to an expected CDMA capacity of ×5 or ×6 of analog AMPS [Ariyavisitakul, 1992; Chang, 1992; Ariyavisitakul et al.]. This further reduction in going from a limited experiment to a large-scale system in a large metropolitan area is consistent with the reduction already experienced in going from simplified theoretical estimates to the limited experiment in a restricted environment.

The San Diego trial also indicated a higher rate of soft(er) handoffs [QTF, 1992] between antenna sectors at a single cell site than expected for sectors well isolated by antenna patterns. This result suggests a lower realizable sectorization gain because of reflected energy than would be expected from more idealized antennas and locations. This could further reduce the estimated cell site capacity of a large-scale system.

Even considering the aforementioned factors, capacity increases of ×5 or ×6 are significant. These estimates, however, are consistent with the factor of ×3 obtained from low-bit-rate (<10-kb/s) speech coding and the ×2 to ×2.5 obtained by taking advantage of speech pauses. These factors result in an expected increase of ×6–7.5, with none of these speech-processing-related contributions being directly attributable to the spread-spectrum processing in CDMA. These results are consistent with the factor of ×6–8 estimate made earlier [AWC, 1991] and are not far from the factor of ×8 quoted recently [AWC, 1992b].

Thus, it is clear that new high-complexity, high-tier technology will not be a substitute for low-complexity, low-power, low-tier PCS.

---

[9] Some vehicles contained more than one mobile unit.

[10] AT&T stated that the San Diego data supported a capacity improvement over analog cellular of at least ×10.

Statistical Multiplexing, Speech Activity, CDMA, and TDMA

Factors of ×2–2.5 have been projected for capacity increase possible by taking advantage of pauses in speech. It has been suggested that implementing statistical multiplexing is easier for CDMA systems because it is sometimes thought to be time consuming to negotiate channels for speech spurts for implementation in TDMA systems. The most negative quality-impacting factor in implementing statistical multiplexing for speech, however, is not in obtaining a channel when needed but is in the detection of the onset of speech, particularly in an acoustically noisy environment. The effect of clipping at the onset of speech is evident in the MOS scores noted for the speech-activity implementation in the United States cellular CDMA proposal discusses earlier (i.e., an MOS of 3.4 without statistical multiplexing and of 3.2 with it). The degradation in MOS can be expected to be even greater for encoding that starts with a higher MOS, e.g., 32-kb/s ADPCM.

It was noted earlier that the proposed cellular CDMA implementation was ×10 as complex as the proposed WACS wireless access for personal communications TDMA implementation. From earlier discussion, the CDMA round-trip delay approaches 200 ms, whereas the short 2-ms-frame TDMA delay is <10-ms round trip. It should be noted that the TDMA architecture could permit negotiation for time slots when speech activity is detected. Since the TDMA frames already have capability for exchange of signalling data, added complexity for statistical multiplexing of voice could readily be added within less than 200 ms of delay and less than ×10 in complexity. That TDMA implementation supports 8 circuits at 32 kb/s or 16 circuits at 16 kb/s for each frequency. These are enough circuits to gain benefit from statistical multiplexing. Even more gain could be obtained at radio ports that support two or three frequencies and, thus, have 16–48 circuits over which to multiplex.

A statistical multiplexing protocol for speech and data has been researched at Rutgers WINLAB [Goodman, 1991]. The Rutgers packet reservation multiple access (PRMA) protocol has been used to demonstrate the feasibility of increasing capacity on TDMA radio links.. These PRMA TDMA radio links are equivalent to slotted ALOHA packet-data networks. Transmission delays of less than 50 ms are realizable. The capacity increase achievable depends on the acceptable packet-dropping ratio. This increase is soft in that a small increase in users causes a small increase in packet-dropping ratio. This is analogous to the soft capacity claimed for CDMA.

Thus, for similar complexity and speech quality, there appears to be no inherent advantage of either CDMA or TDMA for the incorporation of statistical multiplexing. It is not included in the personal communications proposal but is included in cellular proposals because of the different speech-quality/complexity design compromises discussed throughout this paper, not because of any inherent ease of incorporating it in any particular access technology.

High-Tier to Low-Tier or Low-Tier to High-Tier Dual Mode

Industry and the FCC in the United States appear willing to embrace multimode handsets for operating in very different high-tier cellular systems, e.g., analog FM AMPS, TDMA IS-54, and CDMA IS-95. Such sets incur significant penalties for dual mode operation with dissimilar air interface standards and, of course, incur the high-tier complexity penalties.

It has been suggested that multimode high-tier and low-tier handsets could be built around one air-interface standard, for example, TDMA IS-54 or GSM. When closely spaced low-power base stations were available, the handset could turn off unneeded power-consuming circuitry, e.g., the multipath equalizer. The problem with this approach is that the handset is still encumbered with power-consuming and quality-reducing signal processing inherent in the high-tier technology, e.g., error correction decoding and low-bit-rate speech encoding and decoding.

An alternative dual-mode low-tier, high-tier system based on a common air-interface standard can be configured around the low-tier PACS/WACS system, if such a dual-mode system is deemed desirable in spite of the discussion in this chapter. The range of PACS can readily be extended by increasing transmitter power and/or the height and gain of base station antennas. With increased range, the multipath delay spread will be more severe in some locations [Devasirvatham, 1988; Cox,

1972; 1977]. Two different solutions to the increased delay spread can be employed, one for the downlink and another for the uplink. The PACS radio-link architecture has a specified bit sequence, i.e., a unique word, between each data word on the TDM downlink [Cox, 1991; Bellcore, 1990]. This unique word can be used as a training sequence for setting the tap weights of a conventional equalizer added to subscriber sets for use in a high-tier PACS mode. Since received data can be stored digitally [Sollenberger and Chuang, 1990; Sollenberger, 1990], tap weights can be trimmed, if necessary, by additional passes through an adaptive equalizer algorithm, e.g., a decision feedback equalizer algorithm.

The PACS TDMA uplink has no unique word. The high-tier uplink, however, will terminate on a base station that can support greater complexity but still be no more complex than the high-tier cellular technologies. Research at Stanford University has indicated that blind equalization, using constant-modulus algorithms (CMA), [Treichler and Agee, 1983; Sato, 1975], can be effective for equalizing the PACS uplink. Techniques have been developed for converging the CMA equalizer on the short TDMA data burst.

The advantages of building a dual-mode high-tier, low-tier PCS system around the low-tier PACS air-interface standard follow.

1. The interface can still support small low-complexity, low-power, high-speech-quality low-tier handsets.
2. Both data and voice can be supported in a PACS personal communicator.
3. In high-tier low-tier dual mode PACS sets, circuits used for low-tier operation will also be used for high-tier operation, with additional circuits being activated only for high-tier operation.
4. The flexibility built into the PACS radio link to handle different data rates from 8 kb/s to several hundred kb/s will be available to both modes of operation.

7 Infrastructure Networks

It is beyond the scope of this chapter to consider the details of PCS network infrastructures. There are, however, perhaps as many network issues as there are wireless access issues discussed herein [Cox, 1989; Cox, Gifford, and Sherry, 1991; Jabbari et al., 1995; Zaid, 1994]. With the possible exception of the self-organizing WLANS, wireless PCS technologies serve as access technologies to large integrated intelligent fixed communications infrastructure networks.

These infrastructure networks must incorporate intelligence, i.e., database storage, signalling, processing and protocols, to handle both small-scale mobility, i.e., handoff from base station to base station as users move, and large-scale mobility, i.e., providing service to users who roam over large distances, and perhaps from one network to another. The fixed infrastructure networks also must provide the interconnection among base stations and other network entities, e.g., switches, databases, and control processors. Of course, existing cellular mobile networks now contain or are incorporating these infrastructure network capabilities. Existing cellular networks, however, are small compared to the expected size of future high-tier and low-tier PCS networks, e.g., 20 million cellular users in the United States compared with perhaps 100 million users or more each in the future for high-tier and low-tier PCS.

Several other existing networks have some of the capabilities needed to serve as access networks for PCS. Existing networks that could provide fixed base station interconnection include:

- Local exchange networks that could provide interconnection using copper or glass-fiber distribution facilities
- Cable TV networks that could provide interconnection using new glass-fiber and coaxial-cable distribution facilities
- Metropolitan fiber digital networks that could provide interconnection in some cities in which they are being deployed

Networks that contain intelligence, e.g., databases, control processors, and signalling that is suitable or could be readily adapted to support PCS access include:

- Local exchange networks that are equipped with signalling system 7 common channel signalling (SS7 CCS), databases, and digital control processors
- Interexchange networks that are similarly equipped

Data networks, e.g., the internet, could perhaps be adapted to provide the needed intelligence for wireless data access, but they do not have the capacity needed to support large voice/data wireless low-tier PCS access.

Many entities and standards bodies worldwide are working on the access network aspects of wireless PCS. The signalling, control processing, and database interactions required for wireless access PCS are considerably greater than those required for fixed place-to-place networks, but that fact must be accepted when considering such networks.

Low-tier PCS, when viewed from a cellular high-tier paradigm, requires much greater fixed interconnection for the much closer spaced base stations. When viewed from a cordless telephone paradigm of a base unit for every handset and, perhaps, several base units per wireline, however, the requirement is much less fixed interconnection because of the concentration of users and trunking that occurs at the multiuser base stations. One should remember that there are economical fixed wireline connections to almost all houses and business offices in the United States now. If wireless access displaces some of the wireline connections, as expected, the overall need for fixed interconnection could decrease!

8 Conclusion

Wireless personal communications embraces about seven relatively distinct groups of tetherless voice and data applications or services having different degrees of mobility for operation in different environments. Many different technologies and systems are evolving to provide the different perceived needs of different groups. Different design compromises are evident in the different technologies and systems. The evidence suggests that the evolutionary trajectories are aimed toward at least three large groups of applications or services, namely, high-tier PCS (current cellular radio), high-speed wireless local-area networks (WLANS), and low-tier PCS (an evolution from several of the current groups). It is not clear to what extent several groups, e.g., cordless telephones, paging, and wide-area data, will remain after some merging with the three large groups. Major considerations that separate current cellular technologies from evolving low-tier low-power PCS technologies are speech quality, complexity, flexibility of radio-link architecture, economics for serving high-user-density or low-user-density areas, and power consumption in pocket carried handsets or communicators. High-tier technologies make use of large complex expensive cell sites and have attempted to increase capacity and reduce circuit costs by increasing the capacity of the expensive cell sites. Low-tier technologies increase capacity by reducing the spacing between base stations, and achieve low circuit cost by using low-complexity, low-cost base stations. The differences between these approaches result in significantly different compromises in circuit quality and power consumption in pocket-sized handsets or communicators. These kinds of differences also can be seen in evolving wireless systems optimized for data. Advantages of the low-tier PACS/WACS technology are reviewed in the chapter, along with techniques for using that technology in high-tier PCS systems.

References

Afrashteh, A., Sollenberger, N.R., and Chukurov, D.D. 1991. Signal to interference performance for a TDMA portable radio link. *IEEE VTC '91*, St. Louis, MO, May 19–22.
Ariyavisitakul, S. 1992. SIR-based power control in a CDMA system. *IEEE GLOBECOM '92*, Orlando, FL, Paper 26.3, Dec. 6–9, to be published in *IEEE Trans. Comm.*

Ariyavisitakul, S. et al. private communications.

Bellcore Technical Advisories, 1990. Generic framework criteria for universal digital personal communications systems (PCS). FA-TSY-001013(1) March and FA-NWT-001013(2) Dec., and Tech. Ref. 1993. Generic criteria for version 0.1 wireless access communications systems (WACS). (1) Oct. rev. 1, June 1994.

BellSouth Services Inc., 1991. Quarterly progress report number 3 for experimental licenses KF2XFO and KF2XFN. To the Federal Communications Commission, Nov. 25.

Bernhardt, R.C. 1987. Macroscopic diversity in frequency reuse radio systems. *IEEE J. Sel. Areas in Comm.* SAC-5(June):862–870.

Cellular Telecommunications Industries Association. 1989. CDMA digital cellular technology open forum. June 6.

Cellular Telecommunications Industries Association. 1991. CTIA Meeting, Washington, DC, Dec. pp. 5–6.

Chang, L.F. 1992. Dispersive fading effects in CDMA radio systems. *Elect. Lett.* 28(19):1801–1802.

Chang, L.F. and Ariyavisitakul, S. 1991. Performance of a CDMA radio communications system with feed-back power control and multipath dispersion. *IEEE GLOBECOM '91*, Phoenix, AZ, Dec. pp. 1017–1021.

Chuang, J.C.-I. 1993. Performance issues and algorithms for dynamic channel assignment. *IEEE JSAC*, Aug.

Chuang, J.C.-I. 1992. Performance limitations of TDD wireless personal communications with asynchronous radio ports. *Electron. Lett.* 28(March):532–533.

Chuang, J.C.-I. 1990. Comparison of coherent and differential detection of BPSK and QPSK in a quasistatic fading channel. *IEEE Trans. Comm.* May pp. 565–576.

Chuang, J.C.-I. and Sollenberger, N.R. 1988. Burst coherent detection with robust frequency and timing estimation for portable radio communications . *IEEE GLOBECOM '88*, Hollywood, FL, Nov. 28–30.

Chuang, J.C.-I., Sollenberger, N.R., and Cox, D.C. 1993. A pilot based dynamic channel assignment scheme for wireless access TDMA/FDMA systems. *Proc. IEEE ICUPC '93*, Ottawa, Canada, Oct. 12–15, pp. 706–712, 1994. *Int'l J. of Wireless Info. Networks*, 1(1):37–48.

Cook, C.I. 1994. Development of air interface standards for PCS. *IEEE Personal Comm.* 4th Quarter:30–34.

Cox, D.C. 1992. Wireless network access for personal communications. *IEEE Comm. Mag.* Dec. pp. 96–115.

Cox, D.C. 1990. Personal communications—A viewpoint. *IEEE Comm. Mag.* Nov. pp. 8–20.

Cox, D.C. 1991. A radio system proposal for widespread low-power tetherless communications. *IEEE Trans. on Comm.* Feb. pp. 324–335.

Cox, D.C. 1985. Universal portable radio communications. *IEEE Trans. on Veh. Tech.* Aug. pp. 117–121.

Cox, D.C. 1986. Research toward a wireless digital loop. *Bellcore Exchange*, 2(Nov./Dec.):2–7.

Cox, D.C. 1987. Universal digital portable radio communications. *Proc. IEEE*, 75(April):436–477.

Cox, D.C. 1989. Portable digital radio communications—An approach to tetherless access. *IEEE Comm. Mag.* July pp. 30–40.

Cox, D.C. 1972. Delay-Doppler characteristics of multipath propagation at 910 MHz in a suburban mobile radio environment. *IEEE Trans. on Antennas and Propagation*, Sept. 625–635.

Cox, D.C. 1977. Multipath delay spread and path loss correlation for 910 MHz urban mobile radio propagation. *IEEE Trans. on Veh. Tech.* Nov. pp. 340–344.

Cox, D.C., Arnold, H.W., and Porter, P.T. 1987. Universal digital portable communications—a system perspective. *IEEE JSAC*, JSAC-5(June):764–773.

Cox, D.C., Gifford, W.G., and Sherry, H. 1991. Low-power digital radio as a ubiquitous subscriber loop. *IEEE Comm. Mag.* March pp. 92–95.

Devasirvatham, D.M.J. 1988. Radio propagation studies in a small city for universal portable communications. *IEEE VTC '88*, Conference Record. Philadelphia, PA, June 15–17, pp. 100–104.

Devasirvatham, D.M.J. et al.1990a. Radio propagation measurements at 850 MHz, 1.7 GHz and 4 GHz inside two dissimilar office buildings. *Elect. Lett.* 26(7):445–447.

Devasirvatham, D.M.J. et al. 1990b. Multi-frequency radiowave progation measurements in the portable radio environment. *IEEE ICC '90.* April, pp. 1334–1340.

Goodman, D.J. 1991. Trends in cellular and cordless communications. *IEEE Comm. Mag.* June pp. 31–40.

Hinderling, J. et al. 1992. CDMA mobile station modem ASIC. *IEEE CICC 92*, Boston, MA, May.

Int'l J. of Wireless Info. Networks. Jan. 1994, 1(1):37–48.

Jabbari B. et al. 1995. Network issues for wireless personal communications. *IEEE Comm. Mag.* Jan. pp. 88–98.

Katz, R.H. 1994. Adaptation and mobility in wireless information systems. *IEEE Personal Comm.* 1st Quarter:6–17.

Padgett, J.E., Hattori T., and Gunther, C. 1995. Overview of wireless personal communications. *IEEE Comm. Mag.* Jan. pp. 28–41.

Qualcomm Technology Forum, 1992. Open meeting on status of CDMA technology and review of San Diego experiment. San Diego, CA, Jan. 16–17.

Salmasi, A. and Gilhousen, K.S. 1991. On the system design aspects of code division multiple access (CDMA) applied to digital cellular and personal communications networks. *IEEE VTC '91*, St. Louis, MO, May, pp. 57–62.

Sato, Y. 1975. A method of self-recovering equalization for multilevel amplitude modulation systems. *IEEE Trans. on Comm.* June pp. 679–682.

Schneideman, R. 1992. Spread spectrum gains wireless applications. *Microwaves and RF.* May pp. 31–42.

Sollenberger, N.R. 1990. An experimental VLSI implementation of low-overhead symbol timing and frequency offset estimation for TDMA portable radio applications. *IEEE Globecom '90.* San Diego, CA, Dec. pp. 1701–1711.

Sollenberger, N.R. 1991. An experimental TDMA modulation/demodulation CMOS VLSI chip-set. *IEEE CICC '91*, San Diego, CA, May 12–15.

Sollenberger, N.R. and Afrashteh, A. 1991. An experimental low-delay TDMA portable radio link. *Wireless '91*, Calgary, Canada, July 8–10.

Sollenberger, N.R. and Chuang, J.C.-I. 1990. Low-overhead symbol timing and carrier recovery for TDMA portable radio systems. *IEEE Trans. on Comm.* Oct. pp. 1886–1892,.

Sollenberger, N.R. and Chuang, J.C.-I. 1988. Low overhead symbol timing and carrier recovery for TDMA portable radio systems. Third Nordic Sem. Digital Land Mobile Radio Comm., Copenhagen, Denmark, Sept. pp. 13–15.

Sollenberger, N.R. et al. 1991. Architecture and implementation of an efficient and robust TDMA frame structure for digital portable communications. *IEEE Trans. Veh. Tech.* 40(Feb.):250–260.

Steele, R. 1990. Deploying personal communications networks. *IEEE Comm. Mag.* Sept. pp. 12–15.

Steer, D.G. 1994. Coexistence and access etiquette in the United States unlicensed PCS band. *IEEE Personal Comm. Mag.* 4th Quarter:36–43.

Treichler, J.R. and Agee, B.G. 1983. A new approach to multipath correction of constant modulus signals. *IEEE Trans. on Acoustics, Speech and Signal Processing.* April pp. 459–472.

Tuthill, J.P., Granger, B.S., and Wurtz, J.L. 1992. Request for a pioneer's preference. Before the Federal Communications Commission, Pacific Bell submission for FCC General Docket No. 90–314, RM-7140, and RM-7175, May 4.

Varma, V.K. et al. 1987. Performance of sub-band and RPE coders in the portable communication environement. *Fourth Int. Conf. on Land Mobile Radio*, Coventry, UK, Dec. 14–17, pp. 221–227.

Viterbi, A.J. 1985. When not to spread spectrum—A sequel. *IEEE Comm. Mag.* April pp. 12–17.

Wong, A. 1994. Regulating public wireless networks. Workshop on Lightwave, Wireless and Networking Technologies, Chinese Univ. of Hong Kong, Hong Kong, Aug. 24.

Zaid, M. 1994. Personal mobility in PCS. *IEEE Personal Comm.* 4th Quarter:12–16.

1979. Special issue on advanced mobile phone service (AMPS). *Bell System Tech. J.* (BSTS). 58(Jan.).

1991. CDMA capacity seen as less than advertised. *Adv. Wireless Comm.* (AWC). Feb. 6, pp. 1–2.

1992. Bellcore's Wireless Prototype. *Microcell News.* (MN). Jan. 25, pp. 5–6.

1992a. Bellcore PCS phone shines in Bellsouth test; others fall short. *Adv. Wireless Comm.* (AWC). Feb. 19, pp. 2–3.

1992b. TDMA accelerates, but CDMA could still be second standard. *Adv. Wireless Comm.* (AWC). March 4, p. 6.

1995. Special issue on wireless personal communications. *IEEE Comm. Mag.* Jan.

15

Mobile Radio:
An Overview

Andy D. Kucar
*4U Communications Research
Inc.*

15.1 Introduction

The focus of this section is on terrestrial and satellite mobile radio communications. This includes: *cellular radio systems* such as existing North American **advanced mobile phone service (AMPS)**, Japanese **mobile communication systems (MCS)**, Scandinavian **nordic mobile telephone (system) (NMT)**, British **total access communication system (TACS)**, **groupe spécial mobile (GSM)**, **digital AMPS**, and spread-spectrum **code division multiple access (CDMA)**; *cordless telephony systems* such as existing *CT1* and *CT2* and the proposed *CT2Plus*, *CT3*, and *digital European cordless telecommunications (DECT)*; *mobile radio data systems* such as *ARDIS* and *RAM*; projects known as **personal communications networks (PCN)**, **personal communications systems (PCS)**, and **future public land mobile communications systems (FPLMTS)**; *satellite mobile radio systems*, such as existing *INMARSAT* and *OmniTRACS* and the proposed *INMARSAT*, *MSAT*, *Iridium*, *Globalstar*, and *ORBCOMM*. After brief prologue and historical overview, technical issues, such as the repertoire of systems and services, the airwaves management, the operating environment, service quality, network issues and cell size, channel coding and modulation, speech coding, diversity, multiplex and multiple access (**frequency division multiple accesss, FDMA; time division multiple accesss,**

0-8493-8573-3/96/$0.00+$.50
© 1996 by CRC Press, Inc.

TDMA; CDMA) are discussed. Potential economical and sociological impacts of the mobile radio communications in the wake of the redistribution of airwaves at World Administrative Radio Conferences are also addressed.

Most existing mobile radio communications systems collect the information on network behavior, users' positions, etc., with the purpose of enhancing the performance of communications, improving handover procedures, and increasing the system capacity. Coarse positioning is usually achieved inherently, whereas more precise *navigation* can be achieved by employing *LORAN-C* and/or **global positioning system (GPS)** signals, or some other means, at the marginal expense in cost and complexity.

15.2 Prologue

Mobile radio systems provide their users with opportunities to travel freely within the service area being able to communicate with any telephone, fax, data modem, and electronic mail subscriber anywhere in the world; to determine their own positions; to track precious cargo; to improve the management of fleets of vehicles and the distribution of goods; to improve traffic safety; and to provide vital communication links during emergencies, search and rescue operations, etc. These *tieless (wireless, cordless)* communications, exchanges of information, determination of position, course, and distance traveled are made possible by the unique property of the radio to employ an *aerial (antenna)* for radiating and receiving electromagnetic waves. When the user's radio antenna is stationary over a prolonged period of time, the term *fixed radio* is used; a radio transceiver capable of being carried or moved around, but stationary when in operation, is called a *portable radio*; a radio transceiver capable of being carried and used, by a vehicle or by a person on the move, is called *mobile radio*. Individual radio users may communicate directly or via one or more intermediaries, which may be *passive radio repeater(s), base station(s)*, or *switch(es)*. When all intermediaries are located on the Earth, the terms *terrestrial radio system* and *radio system* have been used; when at least one intermediary is satellite borne, the terms *satellite radio system* and *satellite system* have been used. According to the location of a user, the terms *land, maritime, aeronautical, space*, and *deep-space radio systems* have been used. The second unique property of all terrestrial and satellite radio systems is that they all share the same natural resource—the *airwaves (frequency bands and the space)*.

Recent developments in **microwave monolithic integrated circuit (MMIC)**, **application specific integrated circuit (ASIC)**, analog/digital signal processing (A/DSP), and battery technology, supported by **computer aided design (CAD)** and robotics manufacturing allow a viable implementation of miniature radio transceivers. The continuous flux of market forces (excited by the possibilities of a myriad of new services and great profits), international and domestic standard forces (who manage the common natural resource—the airwaves), and technology forces (capable of creating viable products), acted harmoniously and created a broad choice of communications (voice and data), information, and navigation systems, which propelled an explosive growth of mobile radio services for travelers.

15.3 A Glimpse of History

Many things have an epoch, in which they are found at the same time in several places, just as the violets appear on every side in spring.

Farkas Wolfgang Bolyai, in 1823

Late in the 19th century Heinrich Rudolf Hertz, Nikola Tesla, Alexander Popov, Edouard Branly, Oliver Lodge, Guglielmo Marconi, Adolphus Slaby, and some other engineers and scientists experimented with the transmission and reception of electromagnetic waves. In 1898 Tesla demonstrated in Madison Square Garden a radio remotely controlled boat; the same year Marconi established the first wireless ship-to-shore telegraph link with the royal yacht Osborne; these events are now accepted as the birth of the mobile radio. Since that time, mobile radio communications have provided safe navigation for ships and airplanes, saved many lives, dispatched diverse fleets of vehicles, won many battles, generated many new businesses, etc. A summary of some of the key historical developments related to the commercial mobile radio communications is provided in Table 15.1.

Satellite mobile radio systems launched in the 1970s and early 1980s use ultra high frequency (UHF) bands around 400 MHz and around 1.5 GHz for communications and navigation services.

In the 1950s and 1960s, numerous private mobile radio networks, **citizen band (CB) mobile radio**, ham operator mobile radio, and portable home radio telephones used diverse types and brands of radio equipment and chunks of airwaves located anywhere in the frequency band from

TABLE 15.1   A Summary of Events Related to Mobile Radio Communications

1898	Nikola Tesla demonstrated a radio remotely controlled boat in New York. Guglielmo Marconi established the wireless ship-to-shore telegraph link in England.
1903	First International Radiotelegraphic Conference held in Berlin.
1908	Public radio telephone between ships and land in Japan was established.
1921	Police car radio dispatch service was introduced in Detroit, MH police department.
1945	During WW II, significant progress in design and widespread use of mobile radio.
1958	LORAN-C commercial operation started. The initial development was started during WW II.
1964	Railway telephone service on Japanese Tokaido bullet train was introduced.
1968	Carterphone decision. FCC allows non-Bell equipment to be connected to (Bell) network.
1971	Fully automatic radiotelephone system, the B network, was introduced in West Germany. Later extended to the corresponding networks in Austria, Luxemburg, and the Netherlands.
1974	U.S. FCC allocated 40-MHz frequency band, paving the way for establishing what is now known as advanced mobile phone service.
1976	MARISAT consortium initiated commercial service for mobile maritime users, providing full duplex voice, data, and teleprinter services worldwide.
1979	Mobile communications system MCS-L1 introduced by NTT Japan.
1982	The Conference of European Postal and Telecommunications Administrations established Groupe Spécial Mobile with the mandate to define future Pan-European cellular radio standard.
1982	INMARSAT began providing similar services as MARISAT.
1982	Cospas—1 inclined orbit satellite was launched, with a search and rescue package compatible with future global maritime distress and safety system (FGMDSS) onboard.
1983	SARSAT search and rescue instrument package was placed onboard of U.S. National Oceanic and Atmospheric Administration satellite NOAA-8 and launched.
1984	First interagency tests of global positioning system receivers conducted in California.
1985	Total access communications system was introduced in U.K.
1985	CD900 cellular mobile radio system was introduced in West Germany.
1987	Japan launched its own experimental satellite ETS-V.
1988	Geostar introduced its *link one* radio-determination services. The radio-determination information is obtained from a LORAN-C receiver and sent over an L-band satellite payload toward Earth.
1988	Qualcomm/Omninet started its two-way data communication and radio determination (using a LORAN-C receiver) OmniTRACS services.
1988	Second high-capacity land mobile communications system (MCS-L2) was introduced in Japan.
1990	Pegasus rocket launched from the wing of a B-52; the rocket injected its 423-lb payload into a 273×370 nmi $94°$ inclined orbit.
1993	GSM (now global system for mobile communications) in commercial use. After almost two decades of studies and experiments, sponsored by Canadian and U.S. tax payers, North American mobile satellite system MSAT is entering its realization stage, Field trials of CTx, DCT, CDMA, TDMA, FDMA mobile radio communications systems in progress worldwide.

near 30 MHz to 3 GHz. Then, in the 1970s, Ericsson introduced the NMT system, and AT&T Bell Laboratories introduced AMPS. The impact of these two public land mobile telecommunication systems on the standardization and prospects of mobile radio communications may be compared with the impact of Apple and IBM on the personal computer industry. In Europe, systems such as AMPS competed with NMT systems; in the rest of the world, AMPS, backed by Bell Laboratories' reputation for technical excellence and the clout of AT&T, became de facto and de jure the technical standard (British TACS and Japanese MCS-L1 are based on AMPS). In 1982, the **Conference of European Postal and Telecommunications Administrations (CEPT)** established GSM with the mandate to define future Pan-European cellular radio standards. On January 1, 1984, during the phase of explosive growth of AMPS and similar cellular mobile radio communications systems and services, came the divestiture (breakup) of AT&T.

15.4 Πάντα ρει (Panta Rhei)

Based on the solid foundation established in the 1970s, the buildup of mobile radio systems and services in 1990s is continuing at a 20–50% rate per year, worldwide. Terrestrial mobile radio systems offer analog voice and low-to-medium rate data services compatible with existing public switching telephone networks in scope but with poorer voice quality and lower data throughput. Satellite mobile radio systems currently offer analog voice, low-to-medium rate data, radio determination, and global distress safety services for travelers. By the end of 1988 (1994) there were about 2 (8) million cellular telephones in North America, and additional 2 (8) million in the rest of the world. There are about 40 million cordless phones and about 9 million pagers in North America and about the same number in the rest of the world. Considerable progress has been made in recent years [Davis et al. eds., 1984; Cox, Hirade, and Mahmoud eds., 1987; Mahmoud, Rappaport, and Öhrvik eds., 1989; Kucar ed., 1991; Rhee and Lee eds., 1991; Steele ed., 1992; Chuang et al. eds., 1993; Cox and Greenspan eds., 1995].

Equipment miniaturization and price are important constraints on the systems providing these services. In the early 1950s, mobile radio equipment used a considerable amount of a car's trunk space and challenged the capacity of a car's alternator/battery source, while in transmit mode; today, the pocket-size (7.7 oz ≈ 218 g) handheld cellular radio telephone provides 45 min of talk capacity. The average cost of the least expensive models of battery powered cellular mobile radio telephones has dropped proportionally and has broken the $500 U.S. barrier.

There is a rapidly expanding market of *portable* communications, primarily devoted to the *indoor* (in building, around building) environment. Today, these cordless (wireless, fiberless) radio systems offer telepoint services similar in scope to those provided by the public telephone booths; their objectives are to provide a broad range of services similar to ones currently offered by the **public switched telephone network (PSTN)** and the **integrated service digital network (ISDN)**.

Mobile satellite systems are expanding in many directions: large and powerful single unit geostationary systems; medium-sized, low-orbit multisatellite systems; and small-sized, low-orbit multisatellite systems, launched from a plane, [Kucar et al. eds., 1992; Del Re et al. eds., 1995].

The growth and profit potentials of the mobile radio communications market attracted the *big league* players (network, systems, and switching). This caused profound changes in research and development, standardization, and the decision-making processes in the mobile radio communications industry. In the search for El Dorado, the mobile radio communications industry is following two main paths: terrestrial and satellite. The terrestrial mobile radio pioneers, now accompanied by large marketing teams, favor existing cellular radio systems concepts; the newcomers with telephony, switching, and software backgrounds promote **cordless telephony (CT)**, PCN, and PCS; those with backgrounds in administration promote FPLMTS concepts. The satellite mobile radio pioneers build on existing and new *geostationary* satellite systems, whereas the newcomers promote *inclined orbit* concepts. The promoters of each concept may further be subdivided into analog and digital, FDMA, TDMA, and spread spectrum CDMA, etc.

15.5 Repertoire of Systems and Services

The variety of services offered to travelers essentially consists of information in analog and/or digital form. Although most of today's traffic consists of analog voice transmitted by analog frequency modulation FM (or phase modulation PM), digital signalling and a combination of analog and digital traffic might provide superior frequency reuse capacity, processing, and network interconnectivity. By using a powerful and affordable microprocessor and digital signal processing chips, a myriad of different services particularly well suited to the needs of people on the move could be realized economically. A brief description of a few elementary systems/services currently available to travelers will follow. Some of these elementary services can be combined within the mobile radio units for a marginal increase in the cost and complexity with the respect to the cost of a single service system; for example, a mobile radio communications system can include a positioning receiver, digital map, etc.

Terrestrial Systems

In a terrestrial mobile radio network, a repeater was usually located at the nearest summit, offering maximum service area coverage. As the number of users increased, the available frequency spectrum became unable to handle the increased traffic, and a need for frequency reuse arose. The service area was split into many small subareas called cells, and the term cellular radio was born. The frequency reuse offers an increased system capacity, whereas the smaller cell size can offer an increased service quality but at the expense of increased complexity of the user's terminal and network infrastructure. The tradeoffs between real estate availability (base stations) and cost, the price of equipment (base and mobile), network complexity, and implementation dynamics dictate the shape and the size of a cellular network.

Satellite Systems

These employ one or more satellites to serve as base station(s) and/or repeater(s) in a mobile radio network. The position of satellites relative to the service area is of crucial importance for the coverage, service quality, price, and complexity of the overall network. When a satellite encompasses the Earth in 24-h periods, the term *geosynchronous orbit* has been used. An orbit that is inclined with the respect to the equatorial plane is called an inclined orbit; an orbit with a 90° inclination is called a *polar orbit*. A circular geosynchronous 24-h orbit over the equatorial plane (0° inclination) is known as *geostationary orbit*, since from any point at the surface of the Earth the satellite appears to be stationary; this orbit is particularly suitable for the land mobile services at low latitudes and for maritime and aeronautical services at latitudes of <80°. Systems that use geostationary satellites include INMARSAT, MSAT, and AUSSAT. An elliptical geosynchronous orbit with the inclination angle of 63.4° is known as *tundra orbit*. An elliptical 12-h orbit with the inclination angle of 63.4° is known as *Molniya orbit*. Both tundra and Molniya orbits have been selected for the coverage of the northern latitudes and the area around the North Pole; for users at those latitudes, the satellites appear to wander around the zenith for a prolonged period of time. The coverage of a particular region (*regional coverage*) and the whole globe (*global coverage*) can be provided by different constellations of satellites including those in inclined and polar orbits. For example, inclined circular orbit constellations have been proposed for GPS (18–24 satellites, 55–63° inclination), Globalstar (48 satellites, 47° inclination), and Iridium (66 satellites, 90° inclination, polar orbits) systems; all three systems will provide global coverage. ORBCOM system employs Pegasus launchable low-orbit satellites to provide uninterrupted coverage of the Earth below ±60° latitudes and an intermittent but frequent coverage over the polar regions.

Satellite antenna systems can have one (*single-beam global system*) or more beams (*multibeam spot system*). The multibeam satellite systems similar to the terrestrial cellular system, employs antenna directivity to achieve better frequency reuse, at the expense of system complexity.

Radio Paging

This is a nonspeech, one-way (from base station toward travelers), personal selective calling system with alert and without message or with defined message, such as numeric or alphanumeric. A person wishing to send a message contacts a system operator by PSTN and delivers his message. After an acceptable time (queuing delay), a system operator forwards the message to the traveler by radio repeater (FM broadcasting transmitter, VHF or UHF dedicated transmitter, satellite, cellular radio system). After receiving the message, a traveler's small (roughly the size of a cigarette pack) receiver (pager) stores the message into its memory and on demand either emits alerting tones or displays the message.

Examples. The Swedish system uses a 57-kHz subcarrier on FM broadcasting transmitters. The United States systems employ 150-, 450-, and 800-MHz mobile radio frequencies. The RPC1 system used in the United Kingdom, United States, Australia, New Zealand, the People's Republic of China, and Finland employs 150-MHz mobile radio frequencies. The Japanese system operates around 250 MHz, etc.

Global Distress Safety System (GDSS)

Here geostationary and inclined orbit satellites transfer emergency calls sent by vehicles to the central Earth station. Examples are COSPAS, **search and rescue satellite aided tracking system (SARSAT)**, **geostationary operational environmental satellites (GOES)**, and **search and rescue satellite (SERES)**. The recommended frequency for this transmission is 406.025 MHz.

Global Positioning System (GPS)

The United States Department of Defense Navstar GPS 18–24 planned satellites in inclined orbits emit L band ($L1 = 1575.42$ MHz, $L2 = 1227.6$ MHz) spread spectrum signals from which an intelligent microprocessor-based receiver will be able to extract extremely precise time and frequency information and accurately determine its own three-dimensional position, velocity, and acceleration worldwide. The coarse accuracy of <100 m available to commercial users has been demonstrated by using a handheld receiver. An accuracy of meters or centimeters is possible by using the precise (military) codes and/or differential GPS (additional reference) principals. [ION, 1980, 1984, 1986, 1993]

Glonass

This is the Russian's counterpart of the United States's GPS. It uses frequencies between 1602.56 MHz and 1615.50 MHz to achieve goals similar to GPS. Other systems have been studied by the European Space Agency (Navsat) and by West Germany (Granas, Popsat, and Navcom).

LORAN-C

This is the 100-kHz frequency navigation system that provides a positional accuracy between 10–150 m. A user's receiver measures the time difference between the master station transmitter and secondary stations signals and defines his hyperbolic line of position. North American LORAN-C coverage includes the Great Lakes, Atlantic, and Pacific Coast, with decreasing signal strength and accuracy as the user approaches the Rocky Mountains from the east. Similar radionavigation systems are the 100-kHz Decca and 10-kHz Omega.

Inmarsat

This communications system consists of three operational geostationary payloads located at 26° W (Atlantic Ocean), 63° E (Indian Ocean), and 180° W (Pacific Ocean). The standard-A L band system, by employing a 0.79–1.95-m-diam pointing antenna and about 200 kg of above/below deck equipment, can provide analog voice telephony, telex, facsimile, up to 56 kb/s data, group call broadcasting, and emergency calls to maritime users. The Standard-B system will provide digital voice (about 9.6 kb/s), data, and telex services, by employing smaller equipment than standard A.

The standard-C system, which employs a small antenna (about the size of a half-liter can) and a small transceiver (roughly the size of a telephone book directory) can offer up to 600 b/s data. Standard-M system is planned for land mobile and maritime mobile users, while aeronautical systems will provide data and voice services to the air travelers.

Volna

This is a Soviet system of satellites, which in conjunction with satellites and with L-band transponders on Raduga and Gorizont satellites, will provide voice and data services to a fleet of ships and aircrafts, worldwide.

Airphone

This is a public, fully automatic, air-to-ground telephone system that operates in the 900-MHz band using 6-kHz single-sideband (SSB) transmission. Each ground transceiver, by emitting an effective isotropic radiated power of 3 dBW, serves a cell with a radius of about 400 km. An aircraft uses two blade antennas, four transceivers each radiating 7 dBW, a telephone set, and an airborne computer that directs all call logistics.

Dispatch

This two-way radio land mobile or satellite system, with or without connection to the PSTN, consists of an operating center controlling the operation of a fleet of vehicles such as aircrafts, taxis, police cars, tracks, rail cars, etc. A summary of some of existing and planned terrestrial systems, including MOBITEX RAM and ARDIS, is given in Table 15.2. The OmniTRACS dispatch system employs Ku-band geostationary satellite located at 103° W to provide two-way digital message and position reporting (derived from incorporated satellite-aided LORAN-C receiver), throughout the contiguous U.S. (CONUS).

Cellular Radio or Public Land Mobile Telephone System

This offers a full range of services to the traveler that are equivalent to those provided by PSTN. Some of the operating cellular radio systems are: the North American AMPS, the Japanese land MCS-L1 and MCS-L2, the Nordic NMT-450 and NMT-900, the German C450, the Italian public land mobile

TABLE 15.2 Comparison of Dispatch Systems

Parameter	US	Sweden	Japan	Australia
TX freq. band, MHz				
Base	935–940	76.0–77.5	850–860	865.00–870.00
	851–866			415.55–418.05
Mobile	896–901	81.0–82.5	905–915	820.00–825.00
	806–821			406.10–408.60
Duplexing method	FDD/semi, full	FDD/semi	FDD/semi	FDD/semi, full
RF channel bw, kHz	12.5	25.0	12.5	25.0
	25.0			12.5
RF channel rate, kb/s	≤ 9.6	1.2	1.2	≤ 9.6
Number of traffic ch.	400	60 ?	799	200
	600			
Modulation type				
Voice	FM	FM	FM	FM
Data	FSK	MSK-FM	MSK-FM	FSK

Similar systems are used in the Netherlands, U.K., former U.S.S.R. and France. ARDIS is a commercial system compatible with U.S. specifications. MOBITEX/RAM is a commercial system compatible with U.S. specifications. *Source*: 4U Communications Research Inc., 1995.02.23–22:39, updated: 1994.10.31.

TABLE 15.3   Comparison of Cellular Mobile Radio Systems

Parameter	AMPS	MCS–L1 MCS–L2	NMT	C450	TACS	GSM	PCN	IS–54
TX freq., MHz								
Base	869–894	870–885	935–960	461–466	935–960	890–915	1710–1785	869–894
Mobile	824–849	925–940	890–915	451–456	890–915	935–960	1805–1880	824–849
Multiple access	FDMA	FDMA	FDMA	FDMA	FDMA	TDMA	TDMA	TDMA
Duplex method	FDD	FDD	FDD	FDD	FDD	FDD	FDD	FDD
Channel bw, kHz	30.0	25.0 12.5	12.5	20.0 10.0	25.0	200.0	200.0	30.0
Traffic channels per RF channel	1	1	1	1	1	8	16	3
Total traffic ch.	832	600 1200	1999	222 444	1000	125×8	375×16	832×3
Voice	analog	analog	analog	analog	analog	RELP	RELP	VSELP
Sylabic comp.	2:1	2:1	2:1	2:1	2:1	—	—	—
Speech rate, kb/s	—	—	—	—	—	13.0	6.7	8.0
Modulation	PM	PM	PM	PM	PM	GMSK	GMSK	$\pi/4$[1]
Peak dev., kHz	±12	±5	±5	±4	±9.5	—	—	—
Ch. rate, kb/s	—	—	—	—	—	270.8	270.8	48.6
Control	digital	digital	digital	digital	digital	digital	digital	digital
Modulation	FSK	FSK	FFSK	FSK	FSK	GMSK	GMSK	$\pi/4$
BB waveform	Manch.	Manch.	NRZ	NRZ	Manch.	NRZ	NRZ	NRZ
Peak dev., kHz	±8	±4.5	±3.5	±2.5	±6.4	—	—	—
Ch. rate, kb/s	10.0	0.3	1.2	5.3	8.0	270.8	270.8	48.6
Channel coding	BCH	BCH	B1	BCH	BCH	RS	RS	Conv.
Base→mobile	(40, 28)	(43, 31)	burst	(15, 7)	(40, 28)	(12, 8)	(12, 8)	1/2
Mobile→base	(48, 36)	a.(43, 31) p.(11, 07)	burst	(15, 7)	(48, 36)	(12, 8)	(12, 8)	1/2

[1]$\pi/4$ corresponds to the $\pi/4$ shifted differentialy encoded QPSK with $\alpha = 0.35$ square root raised-cosine filter.
Source: 4U Communications Research Inc., 1995.02.23–22:39, updated: 1994.10.31.

radio communication system at 450 MHz, the French radiotelephone multiservice network at 200, 400-MHz RADIOCOM 2000, and the United Kingdom's TACS. The technical characteristics of some of existing and planned systems are summarized in Table 15.3.

Cordless Telephony

The first generation of the U.K.'s cordless telephones (coded CT1) was developed as the answer to the large quantities of imported, technically superior, yet unlicensed mobile radio equipment. The simplicity and cost effectiveness of CT1 analog radio and base station products using eight RF channels and FDMA scheme stem from their applications limited to incoming calls from a limited number of mobile users to the isolated telepoints. As the number of users grew, so did the cochannel interference levels, while the quality of the service deteriorated. Anticipating this situation, the second generation digital cordless telecommunications radio equipment and *common air interface* standards (CT2/*CAI*), incompatible with the CT1 equipment, have been developed. CT2 schemes employ digital voice but the same FDMA principles as the CT1 schemes. Network and frequency reuse issues necessary to accommodate anticipated residential, business, and telepoint traffic growth have not been addressed adequately. Recognizing these limitations and anticipating the market requirements, different FDMA, TDMA, CDMA, and hybrid schemes aimed at cellular mobile and digital cordless telecommunications (DCT) services have been developed. The technical characteristics of some schemes are given in Table 15.4.

TABLE 15.4   Comparison of Digital Cordless Telephone Systems

Parameter	CT2Plus	CT3	DECT	CDMA
Multiple access method	(F/T)DMA	TDMA	TDMA	CDMA
Duplexing method	TDD	TDD	TDD	FDD
RF channel bw, MHz	0.10	1.00	1.73	2×1.25
RF channel rate, kb/s	72	640	1152	1228.80
Number of traffic ch. per one RF channel	1	8	12	32
Burst/frame length, ms	1/2	1/16	1/10	n/a
Modulation type	GFSK	GMSK	GMSK	BPSK/QPSK
Coding	Cyclic, RS	CRC 16	CRC 16	Conv 1/2, 1/3
Transmit power, mW	≤ 10	≤ 80	≤ 100	≤ 10
Transmit power steps	2	1	1	many
TX power range, dB	16	0	0	≥ 80
Vocoder type	ADPCM	ADPCM	ADPCM	CELP
Vocoder rate, kb/s	fixed 32	fixed 32	fixed 32	up to 8
Max data rate, kb/s	32	ISDN 144	ISDN 144	9.6
Processing delay, ms	2	16	16	80
Reuse efficiency[3]				
Minimum	1/25	1/15	1/15	1/4
Average	1/15	1/07	1/07	2/3
Maximum	1/02[1]	1/02[1]	1/02[1]	3/4
Theor. number of vc. per cell and 10 MHz	100×1	10×8	6×12	4×32
Practical per 10 MHz				
Minimum	4	5–6	5–6	32 (08)[2]
Average	7	11–12	11–12	85 (21)
Maximum	50[1]	40[1]	40[1]	96 (24)

[1] The capacity (in the number of voice channels) for a single isolated cell.
[2] The capacity in parentheses may correspond to a 32 kb/s vocoder.
[3] Reuse efficiency and associate capacities reflect our own estimates.
Source: 4U Communications Research Inc., 1995.02.23–22:39 updated: 1994.10.31.

Future Public Land Mobile Telecommunications Systems

This is a huge international administrative project, for which tasks and objectives are presented in ITU-R Document 8–1/292. It discusses different terrestrial and satellite mobile radio communications and broadcasting systems, the transmission of data, voice, and images, at rates between 8–1920 kb/s, and a very broad range of services and technical and administrative issues.

Amateur Satellite Services

These started in 1965 when the OSCAR 3 satellite was launched; successive OSCAR/AMSAT satellites use 144-, 432-, 1270-, and 2400-MHz carrier frequencies. The Russia's Iskra satellites use 21/29 MHz and RS-3 satellites use 145/29-MHz carrier frequencies.

Vehicle Information System

This is a synonym for the variety of systems and services aimed toward traffic safety and location. This includes traffic management, vehicle identification, digitized map information and navigation, radio navigation, speed sensing and adaptive cruise control, collision warning and prevention, etc.

Some of the vehicle information systems can easily be incorporated in mobile radio communications transceivers to enhance the service quality and capacity of respective communications systems.

15.6 The Airwaves Management

The airwaves (frequency spectrum and the space surrounding us) are a limited natural resource shared among several different radio users (military, government, commercial, public, and amateur). Its sharing (among different users and services described in the preceding section; TV and sound broadcasting, etc.), coordination, and administration is an ongoing process exercised on national as well as on international levels. National administrations (**Federal Communications Commission, FCC**, in the U.S., **Department of Communications, DOC**, in Canada, etc.), in cooperation with users and industry, set the rules and procedures for planning and utilization of scarce frequency bands. These plans and utilizations have to be further coordinated internationally.

The **International Telecommunications Union (ITU)** is a specialized agency of the United Nations, stationed in Geneva, Switzerland, with more than 150 government members, responsible for all policies related to radio, telegraph, and telephone. According to the ITU, the world is divided into three regions: region 1—Europe including the former Soviet Union, Mongolia, Africa, and the Middle East west of Iran; region 2—the Americas, and Greenland; and region 3—Asia (excluding parts west of Iran and Mongolia), Australia, and Oceania. Historically, these three regions have developed, more or less independently, their own frequency plans, which best suit local purposes. With the advent of satellite services and globalization trends, the coordination between different regions becomes more urgent. Frequency spectrum planning and coordination is performed through ITU's bodies, such as Comité Consultatif de International Radio (CCIR), now ITU-R; International Frequency Registration Board (IFRB), now ITU-R; World Administrative Radio Conference (WARC); and Regional Administrative Radio Conference (RARC).

ITU-R, through its study groups, deals with technical and operational aspects of radio communications. Results of these activities have been summarized in the form of reports and recommendations published every four years or more [ITU, 1990]. The IFRB serves as a custodian of the common and scarce natural resource, the airwaves; in its capacity, the IFRB records radio frequencies, advises the members on technical issues, and contributes on other technical matters. Based on the work of ITU-R and the national administrations, ITU members convene at appropriate RARC and WARC meetings, where documents on frequency planning and utilization, the *radio regulations*, are updated. Actions on a national level follow; see ITU, 1986 and 1992.

The far-reaching impact of the mobile radio communications on economies and the well being of the three main trading blocks, other developing and third world countries, potential manufacturers and users makes the airways (frequency spectrum) even more important.

15.7 Operating Environment

While traveling, a customer i.e., user of cellular mobile radio system, may experience sudden changes in signal quality caused by his movements relative to the corresponding base station and surroundings, multipath propagation, and unintentional jamming, such as man–made noise, adjacent channel interference, and cochannel interference inherent to the cellular systems. Such an environment belongs to the class of nonstationary random fields, where experimental data is difficult to obtain and their behavior is hard to predict and model satisfactorily. When reflected signal components become comparable in level to the attenuated direct component and their delays comparable to the inverse of the channel bandwidth, *frequency selective fading* occurs. The reception is further degraded due to movements of a user, relative to reflection points and relay station, causing the Doppler frequency shifts. The simplified model of this environment is known as the *Doppler multipath Rayleigh channel.*

The existing and planned cellular mobile radio systems employ sophisticated narrowband and wideband filtering, interleaving, coding, modulation, equalization, decoding, carrier and timing recovery, and multiple access schemes. The cellular mobile radio channel involves a *dynamic interaction* of signals arriving via different paths, adjacent and cochannel interference, and noise. Most channels exhibit some degree of memory; the description of which requires higher order statistics of (*spatial* and *temporal*) multidimensional random vectors (amplitude, phase, multipath delay, Doppler frequency, etc.) to be employed. This may require the evaluation of the usefulness of existing radio channel models and the eventual development of more accurate ones.

Cell engineering, prediction of service area and service quality, in an ever changing mobile radio channel environment, is a very difficult task. The average path loss depends on terrain microstructure within a cell, with considerable variation between different types of cells (i.e., urban, suburban, and rural environments). A variety of models based on experimental and theoretical work have been developed to predict path radio propagation losses in a mobile channel. Unfortunately, none of them are universally applicable. In almost all cases, an excessive transmitting power is necessary to provide an adequate system performance.

The first generation mobile satellite systems employ geostationary satellites (or payloads piggy backed on a host satellite) with small 18-dBi antennas covering the whole globe. When the satellite is positioned directly above the traveler (at zenith), a near constant signal environment, known as the *Gaussian channel*, is experienced. The traveler's movement relative to the satellite is negligible (i.e., Doppler frequency is practically equal zero). As the traveler moves—north or south, east or west—the satellite appears lower on the horizon. In addition to the direct path, many significant strength reflected components are present, resulting in a degraded performance. Frequencies of these components fluctuate due to movement of the traveler relative to the reflection points and the satellite. This environment is known as the *Doppler Ricean channel*. An inclined orbit satellite located for a prolonged period of time above 45° latitude north and 106° longitude west could provide travelers all over the U.S. and Canada, including the far North, a service quality unsurpassed by either geostationary satellite or terrestrial cellular radio. Similarly, a satellite located at 45° latitude north and 15° longitude east could provide travelers in Europe with improved service quality.

Inclined orbit satellite systems can offer a low startup cost, a near Gaussian channel environment, and improved service quality. Low-orbit satellites, positioned closer to the service area, can provide high-signal levels and short (a few milliseconds long) delays, and offer compatibility with the cellular terrestrial systems. These advantages need to be weighted against network complexity, intersatellite links, tracking facilities, etc.

15.8 Service Quality

The primary and the most important measure of service quality should be customer satisfaction. The customer's needs, both current and future, should provide guidance to a service offerer and an equipment manufacturer for both the system concept and product design stages. Acknowledging the importance of every single step of the complex service process and architecture, attention is limited here to a few technical merits of quality.

1. *Guaranteed quality* level is usually related to a percentage of the service area coverage for an adequate percentage of time.
2. *Data service quality* can be described by the average bit error rate (e.g., **BER** $< 10^{-5}$), packet BER (PBER $< 10^{-2}$), signal processing delay (1–10 ms), multiple access collision probability ($<20\%$), the probability of a false call (false alarm), the probability of a missed call (miss), the probability of a lost call (synchronization loss), etc.
3. *Voice quality* is usually expressed in terms of the mean opinion score (MOS) of subjective evaluations by service users. **MOS** marks are: bad = 0, poor = 1, fair = 2, good = 3,

and excellent $= 4$. MOS for PSTN voice service, pooled by leading service providers, relates the poor MOS mark to a signal–to–noise ratio (S/N) in a voice channel of $S/N \approx 35$ dB, whereas an excellent score corresponds to $S/N > 45$ dB. Currently, the users of the mobile radio services are giving poor marks to the voice quality associated with a $S/N \approx 15$ dB and an excellent mark for $S/N > 25$ dB. It is evident that there is a significant difference (20 dB) between the PSTN and mobile services. If digital speech is employed, both the speech and the speaker recognition have to be assessed. For more objective evaluation of speech quality under real conditions (with no impairments, in the presence of burst errors during fading, in the presence of random bit errors at BER $= 10^{-2}$, in the presence of Doppler frequency offsets, in the presence of truck acoustic background noise, in the presence of ignition noise, etc.), additional tests, such as the diagnostic acceptability measure (DAM), diagnostic rhyme test (DRT), Youden square rank ordering, Sino–Graeco–Latin square tests, etc., can be performed.

15.9 Network Issues and Cell Size

To understand ideas and technical solutions offered in existing schemes, and in proposals such as *cordless telephony (CT)*, DCT, PCS, PCN, etc., one need also to analyze the reasons for their introduction and success. Cellular mobile services are flourishing at an annual rate of 20–40%, worldwide. These systems (such as AMPS, NMT, TACS, MCS, etc.), use FDMA and digital modulation schemes for access, and command and control purposes and analog phase/frequency modulation schemes for the transmission of an analog voice. Most of the network intelligence is concentrated at fixed elements of the network including base stations, which seem to be well suited to the networks with a modest number of medium- to large-sized cells. To satisfy the growing number of potential customers, more cells and base stations were created by the cell splitting and frequency reuse process. Technically, the shape and size of a particular cell is dictated by the base station antenna pattern and the topography of the service area. Current terrestrial cellular radio systems employ cells with 0.5–50 km radius. The maximum cell size is usually dictated by the link budget, in particular, the gain of a mobile antenna and available output power. This situation arises in a rural environment, where the demand on capacity is very low and cell splitting is not economical. The minimum cell size is usually dictated by the need for an increase in capacity, in particular, in downtown cores. Practical constraints, such as real estate availability and price, and construction dynamics limit the minimum cell size to 0.5–2 km. In such types of networks, however, the complexity of the network and the cost of service grow exponentially with the number of base stations, whereas the efficiency of present handover procedures becomes inadequate.

Antennas with an omnidirectional pattern in a horizontal direction but with about 10 dBi gain in the vertical direction provide the frequency reuse efficiency of $N_{\text{FDMA}} = 1/12$. Base station antennas with similar directivity in the vertical direction and 60° directivity in the horizontal direction (a cell is divided into six sectors) can provide the reuse efficiency $N_{\text{FDMA}} = 1/4$, this results in a threefold increase in the system capacity; if CDMA is employed instead of FDMA, an increase in reuse efficiency $N_{\text{FDMA}} = 1/4 \rightarrow N_{\text{CDMA}} = 2/3$ may be expected.

Recognizing some of the limitations of existing schemes and anticipating the market requirements, the research in TDMA schemes aimed at cellular mobile and DCT services, and in CDMA schemes aimed towards mobile satellite system, cellular and personal mobile applications, has been initiated. Although employing different access schemes, TDMA (CDMA) network concepts rely on a smart mobile/portable unit that scans time slots (codes) to gain information on network behavior, free slots (codes), etc., improving frequency reuse and handover efficiency while hopefully keeping the complexity and cost of the overall network at reasonable levels. Some of the proposed system concepts depend on low-gain (0-dBi) base station antennas deployed in a license-free, uncoordinated fashion; small size cells (10–1000 m in radius) and an emitted isotropic radiated power of about 10 mW [+10 dB(1 mW)] per 100 kHz have been anticipated. A frequency reuse efficiency of $N = 1/9$ to $N = 1/36$ has been projected for DCT systems. $N = 1/9$ corresponds to the highest user capacity with the lowest transmission quality, whereas $N = 1/36$ has the lowest user capacity with the highest transmission quality. This significantly reduced frequency reuse capability of proposed system concepts will result in significantly reduced system capacity, which need to be compensated by other means, including new spectra.

In practical networks, the need for a capacity (and frequency spectrum) is distributed unevenly in space and time. In such an environment, the capacity and frequency reuse efficiency of the network may be improved by dynamic channel allocation, where an increase in the capacity at a particular hot spot may be traded for the decrease in the capacity in cells surrounding the hot spot, the quality of the transmission, and network instability.

To cover the same area (space) with smaller and smaller cells, one needs to employ more and more base stations. A linear increase in the number of base stations in a network usually requires an exponential increase in the number of connections between base stations, switches, and network centers. These connections can be realized by fixed radio systems (providing more frequency spectra will be available for this purpose), or, more likely, by a cord (wire, cable, fiber, etc.).

The first generation geostationary satellite system antenna beam covers the entire Earth (i.e., the cell radius equals ≈6500 km). The second generation geostationary satellites will use larger multibeam antennas providing 10–20 beams (cells) with 800–1600 km radius. Low-orbit satellites such as Iridium will use up to 37 beams (cells) with 670 km radius. The third generation geostationary satellite systems will be able to use very large reflector antennas (roughly the size of a baseball stadium) and provide 80–100 beams (cells) with a cell radius of ≈200 km. If such a satellite is tethered to a position 400 km above the Earth, the cell size will decrease to ≈2 km in radius, which is comparable in size with today's small-size cell in terrestrial systems. Yet, such a satellite system may have the potential to offer an improved service quality due to its near optimal location with respect to the service area. Similarly to the terrestrial concepts, an increase in the number of satellites in a network will require an increase in the number of connections between satellites and/or Earth network management and satellite tracking centers, etc. Additional factors that need to be taken into consideration include price, availability, reliability, and timeliness of the launch procedures, a few large vs many small satellites, tracking stations, etc.

15.10 Coding and Modulation

The conceptual transmitter and receiver of a mobile system may be described as follows. The transmitter signal processor accepts analog voice and/or data and transforms (by analog and/or digital means) these signals into a form suitable for a double-sided suppressed carrier amplitude modulator (also called quadrature amplitude modulator, QAM). Both analog and digital input signals may be supported, and either analog or digital modulation may result at the transmitter output. Coding and interleaving can also be included. Very often, the processes of coding and modulation are performed jointly; we will call this joint process *codulation*. A list of typical modulation schemes suitable for transmission of voice and/or data over a Doppler affected Ricean channel, which can be generated by this transmitter, is given in Table 15.5.

Existing cellular radio systems such as AMPS, TACS, MCS, and NMT employ hybrid (analog and digital) schemes. For example, in access mode AMPS uses a digital codulation scheme (BCH coding and frequency-shift keying, FSK, modulation). While in the information exchange mode, the frequency modulated analog voice is merged with discrete SAT and/or ST signals and occasionally blanked to send a digital message. These hybrid codulation schemes exhibit a constant envelope and as such allow the use of dc power efficient nonlinear amplifiers. On the receiver side, these schemes can be demodulated by an inexpensive but efficient limiter/discriminator device. They require modest to high $C/N (= 10\text{–}20)$ dB, are very robust in adjacent (a spectrum is concentrated near the carrier) and cochannel interference (up to $C/I = 0$ dB, due to capture effect) cellular radio environment, and react quickly to the signal fade outages (no carrier, code, or frame syn-

TABLE 15.5 Modulation Schemes, Glossary of Terms

Abbreviation	Description	Remarks/Use
ACSSB	Amplitude companded single sideband	Satellite transmission
AM	Amplitude modulation	Broadcasting
APK	Amplitude phase keying modulation	
BLQAM	Blackman quadrature amplitude modulation	
BPSK	Binary phase shift keying	Spread spectrum systems
CPFSK	Continuous phase frequency shift keying	
CPM	Continuous phase modulation	
DEPSK	Differentially encoded PSK(with carrier recovery)	
DPM	Digital phase modulation	
DPSK	Differential phase shift keying (no carrier recovery)	
DSB-AM	Double sideband amplitude modulation	
DSB-SC-AM	Double sideband suppressed carrier AM	Includes digital schemes
FFSK	Fast frequency shift keying (MSK)	
FM	Frequency modulation	Broadcasting, AMPS voice
FSK	Frequency shift keying	AMPS data and control
FSOQ	Frequency shift offset quadrature modulation	
GMSK	Gaussian minimum shift keying	GSM voice, data and control
GTFM	Generalized tamed frequency modulation	
HMQAM	Hamming quadrature amplitude modulation	
IJF	Intersymbol jitter free (SQORC)	
LPAM	L-ary pulse amplitude modulation	
LRC	LT symbols long raised cosine pulse shape	
LREC	LT symbols long rectangularly encoded pulse shape	
LSRC	LT symbols long spectrally raised cosine scheme	
MMSK	Modified minimum shift keying	
MPSK	M-ary phase shift keying	
MQAM	M-ary quadrature amplitude modulation	Subclass of DSB-SC-AM
MQPR	M-ary quadrature partial response	Radio-relay transmission
MQPRS	M-ary quadrature partial response system	$\equiv$ MQPR
MSK	Minimum shift keying	
m-h	Multi-h CPM	
OQPSK	Offset (staggered) quadrature phase shift keying	
PM	Phase modulation	Low-capacity radio
PSK	Phase shift keying	
QAM	Quadrature amplitude modulation	
QAPSK	Quadrature amplitude phase shift keying	
QPSK	Quadrature phase shift keying	$\equiv$ 4 QAM, low-capacity radio
QORC	Quadrature overlapped raised cosine	
SQAM	Staggered quadrature amplitude modulation	
SQPSK	Staggered quadrature phase shift keying	
SQORC	Staggered quadrature overlapped raised cosine	
SSB	Single sideband	Low- and high-capacity radio
S3MQAM	Staggered class-3 quadrature amplitude modulation	
TFM	Tamed frequency modulation	
TSI QPSK	Two-symbol-interval QPSK	
VSB	Vestigial sideband	TV
WQAM	Weighted quadrature amplitude modulation	Includes most digital schemes
XPSK	Crosscorrelated PSK	
$\pi/4$ QPSK	$\pi/4$ shift QPSK	IS-54 TDMA voice and data
3MQAM	Class-3 quadrature amplitude modulation	
4MQAM	Class-4 quadrature amplitude modulation	
12PM3	12-state PM with 3 b correlation	

Source: 4U Communications Research Inc., 1995.02.23–22:39, updated: 1994.10.31.

chronization). Frequency selective and Doppler affected mobile radio channels will cause modest to significant degradations known as the random phase/frequency modulation.

Tightly filtered codulation schemes, such as $\pi/4$ QPSK additionally filtered by a square root raised-cosine filter, exhibit a nonconstant envelope, which demands (quasi) linear, less dc power efficient amplifiers to be employed. On the receiver side, these schemes require complex demodulation receivers, a linear path for signal detection, and a nonlinear one for reference detection—differential detection or carrier recovery. When such a transceiver operates in a selective fading multipath channel environment, additional countermeasures (inherently sluggish equalizers, etc.) are necessary to improve the performance i.e., reduce the bit-error-rate floor. These codulation schemes require modest $C/N (= 8-16)$ dB and perform modestly in adjacent and/or cochannel (up to $C/I = 8$ dB) interference environment.

Codulation schemes employed in spread spectrum systems use low-rate-coding schemes and mildly filtered modulation schemes. When equipped with sophisticated amplitude gain control on the transmit and receive side and robust rake receiver, these schemes can provide superior $C/N (= 4-10$ dB$)$ and $C/I (<0$ dB$)$ performance.

15.11 Speech Coding

Human vocal tract and voice receptors, in conjunction with language redundancy (coding), are well suited for face to face conversation. As the channel changes (e.g., from telephone channel to mobile radio channel), different coding strategies are necessary to protect the loss of information.

In (analog) companded phase modulation/frequency modulation (PM/FM) mobile radio systems, speech is limited to 4 kHz, compressed in amplitude (2:1), pre-emphasized, and phase/frequency modulated. At a receiver, inverse operations are performed. Degradation caused by these conversions and channel impairments results in lower voice quality. Finally, the human ear and brain have to perform the estimation and decision processes on the received signal.

In digital schemes sampling and digitizing of an analog speech (source) are performed first. Then, by using knowledge of the properties of the human vocal tract and of the language itself, a spectrally efficient source coding is performed. A high rate 64-kb/s, 56-kb/s, and adaptive differential pulse code modulation (ADPCM) 32-kb/s digitized voice complies with ITU-T recommendations for toll quality but may be less practical for the mobile environment. One is primarily interested in 8–16-kb/s rate speech coders, which might offer satisfactory quality, spectral efficiency, robustness, and acceptable processing delays in a mobile radio environment. A summary of the major speech coding schemes is provided in Table 15.6.

At this point, a partial comparison between analog and digital voice should be made. The quality of 64-kb/s digital voice, transmitted over a telephone line, is essentially the same as the original analog voice (they receive nearly equal MOS). What does this near equal MOS mean in a radio environment? A mobile radio conversation consists of one (mobile to home) or a maximum of two (mobile to mobile) mobile radio paths, which dictate the quality of the overall connection. The results of a comparison between analog and digital voice schemes in different artificial mobile radio environments have been widely published. Generally, systems that employ digital voice and digital codulation schemes seem to perform well under modest conditions, whereas analog voice and analog codulation systems outperform their digital counterparts in fair and difficult (near threshold, in the presence of strong cochannel interference) conditions. Fortunately, present technology can offer a viable implementation of both analog and digital systems within the same mobile/portable radio telephone unit. This would give every individual a choice of either an analog or digital scheme, better service quality, and higher customer satisfaction. Tradeoffs between the quality of digital speech, the complexity of speech and channel coding, as well as dc power consumption have to be assessed carefully and compared with analog voice systems.

TABLE 15.6   Digitized Voice, Glossary of Terms

ADM	Adaptive delta modulation
ADPCM	Adaptive differential pulse code modulation
ACIT	Adaptive code subband excited transform (GTE)
APC	Adaptive predictive coding
APC-AB	APC with adaptive bit allocation
APC-HQ	APC with hybrid quantization
APC-MQL	APC with maximum likelihood quantization
AQ	Adaptive quantization
ATC	Adaptive transform coding
BAR	Backward adaptive re-encoding
CELP	Code excited linear prediction
CVSDM	Continuous variable slope delta modulation
DAM	Diagnostic acceptability measure
DM	Delta modulation
DPCM	Differential pulse code modulation
DRT	Diagnostic rhyme test
DSI	Digital speech interpolation
DSP	Digital signal processing
HCDM	Hybrid companding delta modulation
LDM	Linear delta modulation
LPC	Linear predictive coding
MPLPC	Multi pulse LPC
MSQ	Multipath search coding
NIC	Nearly instantanous companding
PVXC	Pulse vector excitation coding
PWA	Predicted wordlength assignment
QMF	Quadrature mirror filter
RELP	Residual excited linear prediction
RPE	Regular pulse excitation
SBC	Subband coding
TASI	Time assigned speech interpolation
TDHS	Time domain harmonic scaling
VAPC	Vector adaptive predictive coding
VCELP	Vector code excited linear prediction
VEPC	Voice excited predictive coding
VQ	Vector quantization
VQL	Variable quantum level coding
VSELP	Vector–sum excited linear prediction
VXC	Vector excitation coding

Source: 4U Communications Research Inc., 1995.02.23–22:39, updated: 1994.10.31.

15.12 Macrodiversity and Microdiversity

Macrodiversity

In a cellular system, the base station is usually located in the barocenter of the service area (center of the cell). Typically, the base antenna is omnidirectional in azimuth but with about 6–10 dBi gain in elevation and serves most of the cell area (e.g., > 95%). Some parts within the cell may experience a lower quality of service because the direct path signal may be attenuated due to obstruction losses caused by buildings, hills, trees, etc. The closest neighboring (the first tier) base stations serve corresponding neighboring areas cells by using different sets of frequencies, eventually causing adjacent channel interference. The second closest neighboring (the second tier) base stations might use the same frequencies (frequency reuse) causing cochannel interference. If the same real estate (base stations) is used in conjunction with 120° directional (in azimuth) antennas, the designated area may be served by three base stations. In this configuration one base station serves three cells by using three 120° directional antennas. Therefore, the same number of existing base stations equipped with new directional antennas and additional combining circuitry is required to serve the same number of cells, yet in a different fashion. The mode of operation in which two or more base stations serve the same area is called *macrodiversity*. Statistically, three base stations are able to provide a better coverage of an area similar in size to the system with a centrally located base station. The directivity of a base station antenna (120° or even 60°) provides additional discrimination against signals from neighboring cells, therefore reducing adjacent and cochannel interference (i.e., improving reuse efficiency and capacity). Effective improvement depends on the terrain configuration and the combining strategy and efficiency. However, it requires more complex antenna systems and combining devices.

Microdiversity

Microdiversity is when two or more signals are received at one site (base or mobile).

Space diversity systems employ two or more antennas spaced a certain distance apart from one another. A separation of only $\lambda/2 = 15$ cm at $f = 1$ GHz, which is suitable for implementation on the mobile side, can provide a notable improvement in some mobile radio channel environments. Microspace diversity is routinely used on cellular base sites. Macrodiversity is also a form of space diversity.

Field-component diversity systems employ different types of antennas receiving either the electric or the magnetic component of an electromagnetic signal.

Frequency diversity systems employ two or more different carrier frequencies to transmit the same information. Statistically, the same information signal may or may not fade at the same time at different carrier frequencies. Frequency hopping and very wideband signalling can be viewed as frequency diversity techniques.

Time diversity systems are primarily used for the transmission of data. The same data is sent through the channel as many times as necessary, until the required quality of transmission is achieved (automatic repeat request, ARQ). *Would you please repeat your last sentence* is a form of time diversity used in a speech transmission.

The improvement of any diversity scheme is strongly dependent on the combining techniques employed, i.e., the selective (switched) combining, the maximal ratio combining, the equal gain combining, the feedforward combining, the feedback (Granlund) combining, majority vote, etc.

15.13 Multiplex and Multiple Access

Communications networks for travelers have two distinct directions: the *forward link* i.e., from the base station (via satellite) to the traveler, and the *return link* i.e., from a traveler (via satellite) to the base station. In the forward direction a base station distributes information to travelers according to the previously established protocol, i.e., no multiple access is involved. In the reverse direction many travelers make attempts to access one of the base stations. This occurs in so-called *control channels*, in a particular time slot, at a particular frequency, or by using a particular code. If collisions occur, customers have to wait in a queue and try again until success is achieved. If successful (i.e., no collision occurred), a particular customer will exchange (automatically) the necessary information for call setup. The network management center (NMC) will verify the customer's status, his credit rating, etc. Then, the NMC may assign a channel frequency, time slot, or code on which the customer will be able to exchange information with his correspondent. The optimization of the forward and reverse links may require different coding and modulation schemes and different bandwidths in each direction.

In forward link, there are three basic distribution (multiplex) schemes: one that uses discrimination in frequency between different users and is called *frequency division multiplex* (*FDM*), another that discriminates in time and is called *time division multiplex* (*TDM*), and the last having different codes based on spread spectrum signalling, that is known as *code division multiplex* (*CDM*). It should be noted that hybrid schemes using a combination of basic schemes can also be developed.

In the reverse link, there are three basic access schemes: one that uses discrimination in frequency between different users and is called FDMA, another that discriminates in time and is called TDMA, and the last having different codes based on spread spectrum signalling that is known as CDMA. It should be noted that hybrid schemes using a combination of basic schemes can also be developed.

A performance comparison of multiple access schemes is a very difficult task. The strengths of FDMA schemes seem to be fully exploited in narrowband channel environments. To avoid the use of equalizers, channel bandwidths as narrow as possible should be employed; yet, in such narrowband channels the quality of service is limited by the maximal expected Doppler frequency and practical stability of frequency sources. Current practical limits are about 5 kHz.

The strengths of both TDMA and CDMA schemes seem to be fully exploited in wideband channel environments. TDMA schemes need many slots (and bandwidth) to collect information on network behavior. Once the equalization is necessary (at bandwidths >20 kHz), the data rate should be made as high as possible to increase frame efficiency and freeze the frame to ease equalization; yet, high-data rates require high-RF peak powers and a lot of signal processing power, which may be difficult to achieve in handheld units. Current practical bandwidths are about 0.1–1.0 MHz.

CDMA schemes need large spreading (processing) gains (and bandwidth) to realize spread spectrum potentials; yet, high-data rates require a lot of signal processing power, which may be difficult to achieve in handheld units. Current practical bandwidths are about 1.2 MHz. Narrow frequency bands seem to favor FDMA schemes, since both TDMA and CDMA schemes require more spectrum to fully develop their potentials. Once the adequate power spectrum is available, however, the later two schemes may be better suited for a complex (micro)cellular network environment. Multiple access schemes are also message sensitive. The length and type of message and the kind of service will influence the choice of multiple access, ARQ, frame and coding, among others.

15.14 System Capacity

The recent surge in the popularity of cellular radio and mobile service in general has resulted in an overall increase in traffic and a shortage of available system capacity in large metropolitan areas. Current cellular systems exhibit a wide range of traffic densities, from low in rural areas to overloading in downtown areas, with large daily variations between peak hours and quiet night hours. It is a great system engineering challenge to design a system that will make optimal use of the available frequency spectrum, offering a maximal traffic throughput (e.g., erlangs/megahertz/service area) at an acceptable service quality, constrained by the price and size of the mobile equipment. In a cellular environment, the overall system capacity in a given service area is a product of many (complexly interrelated) factors including the available frequency spectra, service quality, traffic statistics, type of traffic, type of protocol, shape and size of service area, selected antennas, diversity, frequency reuse capability, spectral efficiency of coding and modulation schemes, efficiency of multiple access, etc.

In the 1970s, so-called analog cellular systems employed omnidirectional antennas and simple or no diversity schemes offering modest capacity, which satisfied a relatively low number of customers. Analog cellular systems of the 1990s employ up to 60° sectorial antennas and improved diversity schemes; this combination results in a three- to fivefold increase in capacity. A further (twofold) increase in capacity can be expected from narrowband analog systems (25 kHz → 12.5 kHz); however, slight degradation in service quality might be expected. These improvements spurred the current growth in capacity, the overall success and prolonged life of analog cellular radio.

There are also numerous marketing results, where a 10- to 20-fold increase in capacity has been claimed. In this kind of campaign new (our) digital systems of the 21st century, operating under nice conditions, are usually compared with the old (their) systems of 1970s, operating under the worst conditions. There are plenty of ways of increasing the capacity of cellular radio, acquiring new frequency spectra is perhaps the easiest but not necessary the most cost effective way.

15.15 Conclusion

In this section, a broad repertoire of terrestrial and satellite systems and services for travelers is briefly described. The technical characteristics of the dispatch, cellular, and cordless telephony systems are tabulated for ease of comparison. Issues such as operating environment, service quality, network complexity, cell size, channel coding and modulation (codulation), speech coding, macro- and microdiversity, multiplex and multiple access, and the mobile radio communications system capacity are discussed.

Presented data reveals significant differences between existing and planned terrestrial cellular mobile radio communications systems and between terrestrial and satellite systems. These systems use different frequency bands, different bandwidths, different codulation schemes, different protocols, etc. (i.e., they are not compatible).

What are the technical reasons for this incompatibility? In this section, performance dependence on multipath delay (related to the cell size and terrain configuration), Doppler frequency (related to the carrier frequency, data rate, and the speed of vehicles), and message length (may dictate the choice of multiple access) are briefly discussed. A system optimized to serve the travelers in the Great Plains may not perform very well in mountainous Switzerland; a system optimized for downtown cores may not be well suited to a rural environment; a system employing geostationary (above equator) satellites may not be able to serve travelers at high latitudes very well; a system appropriate for slow moving vehicles may fail to function properly in a high-Doppler shift environment; a system optimized for voice transmission may not be very good for data transmission, etc. A system designed to provide a broad range of services to everyone, everywhere may not be as good as a system designed to provide a particular service in a particular local environment, as a decathlete world champion may not be as successful in competitions with specialists in particular disciplines.

There are plenty of opportunities, however, where compatibility between systems, their integration, and frequency sharing may offer improvements in service quality, efficiency, cost, and capacity (and therefore availability). Terrestrial systems offer a low startup cost and a modest cost per user in densely populated areas. Satellite systems may offer a high quality of the service and may be the most viable solution to serve travelers in scarcely populated areas, on oceans, and in the air. Terrestrial systems are confined to two dimensions, and radio propagation occurs in the near horizontal sectors. Barostationary satellite systems use the narrow sectors in the user's zenith nearly perpendicular to the Earth's surface having the potential for frequency reuse and an increase in the capacity in downtown areas during peak hours. A call setup in a forward direction (from the PSTN via base station to the traveler) may be a very cumbersome process in a terrestrial system when a traveler to whom a call is intended is roaming within an unknown cell; however, this is very easily realized in a global beam satellite system.

Defining Terms[1]

AMPS: Advanced mobile phone service
ASIC: Application specific integrated circuits
BER: Bit error rate

[1] *Source:* 4U, Communications Research Inc., 1995.02.23–22:39, updated: 1995.02.18.

CAD: Computer aided design
CB: Citizen band (mobile radio)
CDMA: Spread spectrum code division multiple access
CEPT: Conference of European Postal and Telecommunications (Administrations)
CT: Cordless telephony
DOC: Department of Communications (Canada)
DSP: Digital signal processing
FCC: Federal Communications Commission (U.S.)
FDMA: Frequency division multiple access
FPLMTS: Future public land mobile telecommunications systems
GDSS: Global distress safety system
GOES: Geostationary operational environmental satellites
GPS: Global positioning system
GSM: Groupe Spécial Mobile (now global system for mobile communications)
ISDN: Integrated service digital network
ITU: International Telecommunications Union
MOS: Mean opinion score
MMIC: Microwave monolitic integrated circuits
NMC: Network management center
NMT: Nordic mobile telephone (system)
PCN: Personal communications networks
PCS: Personal communications systems
PSTN: Public switched telephone network
SARSAT: Search and rescue satellite aided tracking system
SERES: Search and rescue satellite
TACS: Total access communication system
TDMA: Time division multiple access
WARC: World administrative radio conference

References

Chuang, J.C.-I., Anderson, J.B., Hattori, T., and Nettleton, R.W. eds. 1993. Special Issue on wireless personal communications. *IEEE J. Selected Areas in Comm.* Pt. I. 11(6).

Chuang, J.C.-I., Anderson, J.B., Hattori, T., and Nettleton, R.W. eds. 1993. Special Issue on wireless personal communications. *IEEE J. Selected Areas in Comm.* Pt. II. 11(7).

Cox, D.C., Hirade, K., and Mahmoud, S.A. eds. 1987. Special Issue on Portable and mobile communications. *IEEE J. Selected Areas in Comm.* 5(4).

Cox, D.C. and Greenstein, L.J. eds. 1995. Special Issue on Wireless personal communications. *IEEE Comm. Mag.* 33(1).

Davis, J.H., Mikulski, J.J., Porter, P.T., and King, B.L. eds. 1984. Special Issue on Mobile radio communications. *IEEE J. Selected Areas in Comm.* 2(4).

Del Re, E., Devieux, C.L., Jr., Kato, S., Raghavan, S., Taylor, D., and Ziemer, R. eds. 1995. Special Issue on Mobile satellite communications for seamless PCS. *IEEE J. on Selected Areas in Comm.* 13(2).

ION Global Positioning System. 1980–1993. Reprinted by Inst. of Navigation. Vol. I. Washington, D.C, 1980; Vol. II. Alexandria, VA, 1984; Vol. III. Alexandria, VA, 1986; Vol. IV. Alexandria, VA, 1993.

International Telecommunication Union. 1986. *Radio Regulations,* 1982 ed. rev. 1985 and 1986.

International Telecommunications Union, 1990. Mobile, Radio determination, Amateur and Related Satellite Services. Recommendations of the CCIR, (also Resolutions and Opinions). Vols. VIII, XVIIth Plenary Assembly, Düsseldorf, 1990.

Reports of the CCIR, 1990 (also Decisions). Land Mobile Service, Amateur Service, Amateur Satellite Service, Annex 1 to Volumes VIII, XVIIth Plenary Assembly, Düsseldorf, 1990.

Reports of the CCIR, 1990 (also Decisions). Maritime Mobile Service. Annex 2 to Vol. VIII, XVIIth Plenary Assembly, Düsseldorf, 1990.

Kucar, A.D. ed. 1991. Special Issue on Satellite and terrestrial systems and services for travelers. *IEEE Comm. Mag.* 29(11).

Kucar, A.D., Kato, S., Hirata, Y., and Lundberg, O. eds. 1992. Special Issue on Satellite systems and services for travelers. *IEEE J. on Selected Areas in Comm.* 10(8).

Mahmoud, S.A., Rappaport, S.S., and Öhrvik, S.O. eds. 1989. Special Issue on Portable and mobile communications. *IEEE J. on Selected Areas in Comm.* 7(1).

International Telecommunication Union. 1992. Final Acts of the World Administrative Radio Conference for Dealing with Frequency Allocations in Certain Parts of the Spectrum (WARC-92) at Málaga-Torremolinos, Geneva.

Rhee, S.B. and Lee, W.C.Y. eds. 1991. Special Issue on Digital cellular technologies. *IEEE Trans. on Vehicular Tech.* 40(2).

Steele, R. ed. 1992. Special Issue on PCS: The second generation. *IEEE Comm. Mag.* 30(12).

Further Information

This trilogy written by participants in AT&T Bell Labs projects on research and development in mobile radio is the Bible of diverse cellular mobile radio topics.

Jakes, W.C., Jr. ed. 1974. *Microwave Mobile Communications*, John Wiley & Sons, Inc. New York, 1974.

AT&T Bell Labs Technical Personnel, 1979. Advanced mobile phone service (AMPS). *Bell System Technical Journal*, 58(1).

Lee, W.C.Y. 1982. *Mobile Communications Engineering*, McGraw-Hill Book Co. New York, 1982.

An in-depth understanding of design, engineering, and use of cellular mobile radio networks, including PCS and PCN, requires knowledge of diverse subjects, such as three-dimensional cartography, electromagnetic propagation and scattering, antennas, analog and digital communications, project engineering, etc. The following is a list of books relating to these topics.

Balanis, C.A. 1982. *Antenna Theory Analysis and Design*, Harper & Row, New York, 1982.

Bowman, J.J., Senior, T.B.A., and Uslenghi, P.L.E. 1987. *Electromagnetic and Acoustic Scattering by Simple Shapes*, revised printing. Hemisphere Pub. Corp. New York, 1987.

Kucar, A. D. 1995. *Satellite and Terrestrial Radio Systems: Fixed, Mobile, PCS and PCN, Radio vs. Cable. A Practical Approach*, Stridon Press Inc.

Proakis, J.G., 1983. *Digital Communications*, McGraw-Hill Book Co. New York.

Sklar, B. 1988. *Digital Communications, Fundamentals and Applications*, Prentice-Hall Inc., Englewood Cliffs, NJ.

Snyder, J.P. 1987. *Map Projection—A Working Manual*, U.S. Geological Survey Professional Paper 1395, United States Government Printing Office, Washington DC, 2nd printing. 1989.

Spilker, J.J., Jr. 1977. *Digital Communications by Satellite*, Prentice-Hall Inc., Englewood Cliffs, NJ.

Van Trees, H. L., 1968–1971. *Detection, Estimation, and Modulation Theory*, Pt. I 1968, Pt. II and III, 1971, John Wiley & Sons, Inc., New York.

16

Base Station Subsystems

Chong Kwan Un
Korea Advanced Institute of Science and Technology

Chong Ho Yoon
Hankook Aviation University

16.1 Introduction

In this chapter, a **base station subsystem** (BSS) providing the interface between a mobile switching center (MSC) and a mobile is described. To investigate the functions of a BSS, three system configurations of the BSS are presented according to the channel access mechanism, the most widely known being frequency division multiple access (FDMA), time division multiple access (TDMA), and code division multiple access (CDMA) [Ehrlich, Fisher, and Wingard, 1979; Uebayashi, Ohno, and Nojima, 1993; Qualcomm, 1992]. Also, a teletraffic performance model of BSS and numerical results are presented.

A subscriber in a cell originates a call via an idle channel available among a given number of radio channels to the BSS in a cell. When the subscriber enters an adjacent cell, the BSS tries to acquire an idle channel in the new cell. If there are one or more idle channels, the call is successfully handed off without a service breakdown. Otherwise, the call is forced to terminate before completion. Since the quality of the telephone service is enhanced if the rate of breakdown during a conversation is lower than the blocking of **originating calls** (OCs), a **handoff** call (HC) must have a priority over an OC. To reduce this probability, some schemes based on cutoff priority with a fixed number of **guard channels** (GCs) have been introduced by several researchers [Yoon and Un, 1993; Hong and Rappaport, 1986; El-Dolil, Wong, and Steele, 1989; Guerin, 1988]. In this chapter we also present three call handling schemes with and without guard channels.

16.2 System Architectures

A BSS consists of several **base transceiver subsystems** (BTS) and a **base station controller** (BSC). Several configurations of the BSS might be possible, as shown in Fig. 16.1. For suburb sites Figs. 16.1(a)–16.1(d) are preferred, and for downtown sites Fig. 16(c) is preferred.

To investigate the functions of BSS, the functional diagrams of three BSSs are shown in Fig. 16.2, according to the channel access mechanism. Figure 16.2(a) is ATT's BSS for advance mobile phone

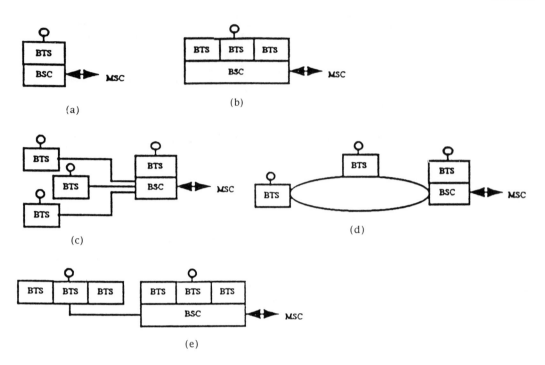

FIGURE 16.1   Configurations of BSS: (a) Combined omni, (b) combined star, (c) star, (d) ring, and (e) urban star configurations.

service (AMPS) using FDMA [Ehrlich, Fisher, and Wingard, 1979], Fig. 16.2(b) is NTT's BSS using TDMA [Uebayashi, Ohno, and Nojma, 1993], and Fig. 16.2(c) is Qualcomm's BSS using CDMA [1992].

Base Transceiver Subsystems (BTS)

A BTS provides RF radiation and reception with an appropriate channel access mechanism (e.g., FDMA, TDMA, or CDMA), and voice and data transmission interfaces between itself and the BSC. Typically, a BTS consists of several receive and transmit antennas, RF distributor, modulators and demodulators, and T1/E1 trunk line interfaces for voice and data traffic. For CDMA systems, a global positioning system (GPS) receive antenna is additionally included. Functions of each block are as follows.

1. *Antennas:* When each BTS functions in the omnidirectional mode, an omnidirectional transmit antenna and two-branch space-diversity receive antennas are used. If the BTS is configured in the directional mode with three 120° sectors, a directional transmit antenna and two receive antennas per each sector are used. The RF power is typically below 45 W. For the CDMA system, a GPS antenna is added which receives ticks of 1 pulse per second with a 10-MHz reference clock to generate system clocks.
2. *RF distributor:* It combines several carriers from amplifiers to transmit antennas with power amplifiers boosting the modulated signals to high-power signals. The transmit frequencies of FDMA, TDMA, and CDMA systems are in the range of 870–890 MHz, 940–956 MHz, and 869–894 MHz, respectively. The receive frequencies of FDMA, TDMA, and CDMA systems are in the range of 825–845 MHz, 810–826 MHz, and 824–849 MHz, respectively.
3. *Modulators and demodulators:* A modulator generates carrier signals of voice, supervisory audio tone (SAT), pilot, synch, and paging data. Each demodulator receives a two-diversity

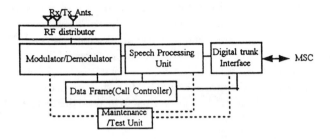

FIGURE 16.2(a) BSS structures with different channel access schemes: FDMA system.

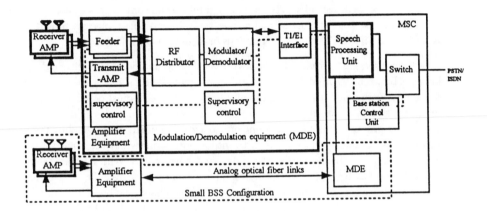

FIGURE 16.2(b) BSS structures with different channel access schemes: TDMA system.

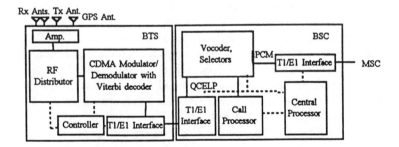

FIGURE 16.2(c) BSS structures with different channel access schemes: CDMA system.

input derived from the two receiving antennas. With these inputs and a local oscillator signal, it demodulates a baseband voice/SAT or data signals.

For the FDMA system, voice and data are modulated as phase modulation (PM) and frequency modulation (FM), respectively. The carrier spacing of the FDMA is 30 kHz, and the data transmission rate is 10 kb/s. The TDMA system uses a three-channel TDMA per carrier scheme with 42 kb/s $\pi/4$-shift quadrature differential phase shift keying (QDPSK) with a Nyquist filter whose rolloff factor is 0.5. The carrier spacing of the TDMA is 25 kHz. Thus, the bandwidth per channel is 8.3 kHz (=25 kHz/3). The CDMA system generates 19.2 Ksymbols per second and uses a convolutional encoder for either 9.6, 4.8, 2.4, or 1.2 kb/s voice and data, and a Viterbi decoder for providing a forward error correction mechanism over the multipath fading wireless channel.

4. *Trunk interface:* It performs voice and data communication between BTS and BSC over digital links, operating at T1 or E1 rate. Typically, each slot per T1/E1 frame is allocated for delivering a single voice channel traffic. A 64 kb/s slot of the TDMA system, however, can carry three 11.2 kb/s voice channels, and the CDMA system uses a compressed voice packet transmission scheme with an HDLC frame format to increase the channel efficiency.

Base Station Controller (BSC)

A BSC locates mobiles to the cell with highest signal strength (handoff), and performs call setup, call supervision, and call termination. Also, it performs remotely ordered equipment testing, updates the location information of mobile stations, and provides data transmission interfaces between itself and the MSC. The BSC consists of speech processing units, a call controller, a central processor, a maintenance and test unit, and digital trunk interfaces. In particular, selectors providing soft handoffs are included in the CDMA BSC. Functions of each block are as follows.

1. *Speech processing unit/vocoder:* It provides the per channel audio-level speech and data paths interface between MSC and RF. Certain orders and request data signals can be added on the transmitter path before the modulator, and these data signals must be removed from the receiver path after the demodulator.

    The speech processing unit employs one of two signalling methods for sending the orders and requests signalling over a voice channel without interfering with voice conversation. The two methods are the blank-and-burst and dim-and-burst modes. With the blank-and-burst mode, the signals are sent in the form of a binary data message over the voice channel by momentarily muting the voice and inserting a binary data sequence. The data sequence requires approximately one-tenth of a second. However, the blank-and-burst signal is sent and replaces voice traffic temporarily. For the CDMA system, the dim-and-burst as well as the blank-and-burst mode can be used. When a vocoder desires to transmit at its maximum rate under the dim-and-burst mode, it is permitted to supply data at half of this rate. The remaining rate is used for signalling and overhead.

    For the digital cellular system, both pulse code modulation (PCM) digitized voice and data suffer from the coding used to increase the whole system capacity. The TDMA system uses a 11.2 kb/s vector sum excited linear prediction (VSELP) transmission code, which consists of 6.7 kb/s source traffic and additional preamble bits, sync bits, and control bits. In the CDMA system, the vocoder handles variable rate vocoding [transforms the Qualcomm developed code excited linear prediction code (QCELP) to PCM, or vice versa]. The vocoder has a variable rate that supports 8, 4, 2, and 1 kb/s operation and corresponds to channel rates of 9.6, 4.8, 2.4, and 1.2 kb/s. For example, a low-rate vocoder, running at 4 kb/s would increase the system capacity by a factor of 1.7 times with some degraded voice quality.

    For the TDMA and CDMA systems, echo cancellors are added, which eliminate echo due to a 2-W/4-W hybrid transformer in the public switch telephone network.

2. *Data frame/call controller:* It maintains an independent setup channel for the shared use of the BTS in communicating with all mobiles within its zone. Only data traffic is transmitted on the setup channel. Any mobile wishing to initiate a call monitors the forward setup channel (land to mobile). If the channel is idle, the mobile can transmit a call request or a page response. Otherwise, the mobile must wait a short time interval and monitor the channel again, until the forward channel is idle.

    The call processor of the FDMA system initiates the hard handoff procedure with the MSC. To determine when and if a handoff is necessary, a signal-strength measurement or a phase range measurement in the speech processing unit is made once every few seconds on each active voice channel. The soft handoffs between BTSs of the CDMA system are handled in selectors.

The call processors must also detect and control the signalling tone, off and on hooks, and transmitter power control. In addition, the processor provides a paging procedure to find a mobile station when a land-originating call is initiated.

3. *Central controller:* It allocates or de-allocates voice and data channels, communicates with the MSC and BTS, and controls the maintenance and test unit.

4. *Digital trunk interface:* It performs voice and data communication between BTS and BSC or between BSC and MSC over digital links, operating at T1 or E1 rate.

5. *Selector for the CDMA system:* The selector handles soft handoffs between BTSs. The soft handoff allows both the original cell and a new cell to serve the call temporarily during the handoff transition. The transition is from the originating cell to both cells and then to the new cell. Thus, it provides the make-before-break switching function. After a cell is initiated, the mobile continues to scan neighboring cells to determine if the signal from another cell becomes comparable to that of the original cell. Then, it sends a control message to the MSC, which states that the new cell is now strong and identifies the new cell. The MSC indicates the handoff by establishing a link to the mobile through the new cell while maintaining the old link. While the mobile is located in the transition region between the two cells, the call is supported by both cells until the mobile notifies the MSC that one of the links is no longer useful. This decision is performed by the selector, which selects an appropriate BTS among the current three engaged BTSs.

16.3 Analysis of Call Handling Schemes

System Modeling

Here, three call handling schemes for a BSS are presented and analyzed. It is assumed that the BSS can handle two types of calls (originating and prioritized handoff calls) and store these calls in a finite storage buffer. One scheme is a prioritized handoff scheme with GCs to reduce the blocking probability of HCs with a penalty on OCs. Under this scheme, handoff calls can be stored exclusively and access all free channels without restriction, whereas OCs have access to idle channels, except for a fixed number of guard channels. The other two schemes are prioritized handoff schemes without GCs to increase the total grade of service by reducing the blocking probability of OCs without a severe penalty on the grade of service for HCs. Under these two schemes without guard channels, both types of calls are allowed to be stored, and prioritized handoff calls push out originating calls if the buffer is full.

Figure 16.3 shows a BSS which has a set of C duplex channels with a finite call storage buffer of size $K - C, C \leq K$. Prioritized HCs and ordinary OCs are handled in the BSS. We assume that these calls have the same exponential service time distribution with rate of μ, but different arrival rates of the Poisson process, λ_1 and λ_2, respectively. Here, λ_1 is assumed to be proportional to λ_2 such that $\lambda_1 = P_h \lambda_2$, where P_h is the probability of having an HC. Then, the total and arrival rate λ_T is $\lambda_1 + \lambda_2$, and the total traffic load per channel ρ_T is $\lambda_T / C\mu$. A hexagonal cell shape is assumed for the system. The cell radius R for a hexagonal cell is defined as the maximum distance from the BSS to the cell boundary. With the cell radius, the average new call origination rate per cell is given by $\lambda_2 \equiv 3\sqrt{3/2}\lambda_0 R^2$, where λ_0 denotes the arrival rate of originating calls per unit area [Hong and Rapport, 1986]. Under these assumptions, we can treat the BSS as an $M/M/C/K$ priority queueing

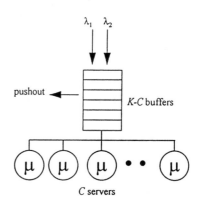

FIGURE 16.3 Base station model.

system which is formed by a set of C channels with a finite system (buffer plus channels) of size K, $C \leq K$.

Analysis of Call Handling Schemes [Yoon and Un, 1993]

Call Handling Scheme with Guard Channels (Scheme 1)

Under scheme 1, HCs can access C_g guard channels, $C_g \leq C$, exclusively among C channels. The remaining $(C - C_g)$ channels are shared by both types of calls. An OC is blocked if the number of available channels is less than or equal to C_g. If C channels have already been in use, $(K - C)$ HCs are buffered, but no buffering of OCs takes place. Hong and Rappaport [1986] analyzed this scheme with an infinite storage buffer for HCs. Here, we slightly modify their result to analyze our finite storage buffer model.

In Fig. 16.4, we show the state-transition-rate diagram for this scheme. Let P_j be the steady-state probability of a total of j calls being in the system. Using the state-transition-rate diagram, we obtain the probability P_j as

$$
P_j = \begin{cases}
\dfrac{(\lambda_1 + \lambda_2)^j}{j!\mu^j} P_0 & \text{for } 1 \leq j \leq C - C_g \\[3mm]
\dfrac{(\lambda_1 + \lambda_2)^{(C-C_g)}\lambda_1^{(j-C+C_g)}}{j!\mu^j} P_0 & \text{for } C - C_g + 1 \leq j \leq C \qquad (16.1) \\[3mm]
\dfrac{(\lambda_1 + \lambda_2)^{(C-C_g)}\lambda_1^{(j-C+C_g)}}{C!\mu^C (C\mu)^{j-C}} P_0 & \text{for } C + 1 \leq j \leq K
\end{cases}
$$

where

$$
P_0 = \left[\sum_{j=0}^{C-C_g} \frac{(\lambda_1 + \lambda_2)^j}{j!\mu^j} + \sum_{j=C-C_g+1}^{C} \frac{(\lambda_1 + \lambda_2)^{(C-C_g)}\lambda_1^{(j-C+C_g)}}{j!\mu^j} \right.
$$

$$
\left. + \sum_{j=C+1}^{K} \frac{(\lambda_1 + \lambda_2)^{(C-C_g)}\lambda_1^{(j-C+C_g)}}{C!\mu^C (C\mu)^{j-C}} \right]^{-1}
$$

Since an HC is blocked only when the buffer is full, its blocking probability P_{B1} is given by $P_{B1} = P_K$. Also, the blocking probability of an OC, P_{B2}, is given by the sum of the probabilities that the number of calls in the system is larger than or equal to $(C - C_g)$, that is

$$
P_{B2} = \sum_{j=C-C_g}^{K} P_j.
$$

Using Little's formula for a steady-state queueing system [Kleinrock, 1975], we can obtain the

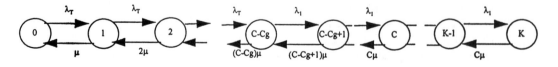

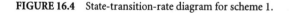

FIGURE 16.4   State-transition-rate diagram for scheme 1.

average waiting time of an HC, which is successfully served as

$$E[W_1] = \sum_{j=C}^{K} \frac{(j-C)P_j}{\lambda_1(1-P_{B1})}$$

Call Handling Schemes without Guard Channels

Here, we consider two prioritized call handling schemes without GCs as efficient call handling schemes for increasing the total grade of service (GOS).

Scheme 2. Under scheme 2, any type of calls has access to all channels as long as there is more than one free channel. When all channels are busy, these calls will be queued next to the newest HC in a buffer or will be queued at the first position of the buffer if there is no HC in the buffer. If an OC is in the position K and an arrival occurs, then it gets pushed out and the new arrival is queued next to the newest HC. When the system is full and the buffer has no more OCs, any type of arriving call is blocked. Accordingly, the HCs will be served by the first-in, first-out (FIFO) rule with the head-of-line priority basis, whereas the OCs will be served by the last-in, first-out (LIFO) rule with the pushout basis.

Let P_j, where $0 \le j \le K$, be the probability of having j calls in a basic $M/M/C/K$ system, and let $P_{j,k}$ be the probability of having j calls ($j = 0, 1, \ldots, K$) in the system and k OCs ($k = 0, 1, \ldots, j - C$) in the buffer. In Fig. 16.5, we show the state-transition-rate diagram for this

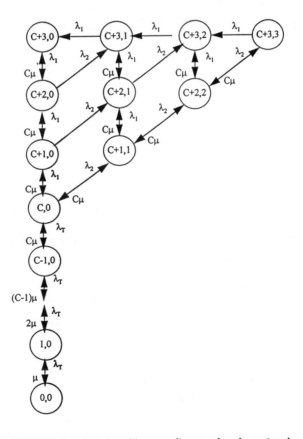

FIGURE 16.5 State-transition-rate diagrams for scheme 2 and scheme 3.

scheme. The states may be expressed by the following balanced equations:

$$(\lambda_T + C\mu)P_{j,k} = \lambda_2 P_{j-1,k-1} + \lambda_1 P_{j-1,k} + C\mu P_{j+1,k}$$
$$+ C\mu P_{j+1,k+1} \cdot \delta(j-k-C), \qquad \text{for } C \le j \le K, k \le j-C$$
$$C\mu P_{K,0} = \lambda_1 P_{K-1,0} + \lambda_2 P_{K,1}$$
$$(\lambda_1 + C\mu)P_{K,k} = \lambda_2 P_{K-1,k-1} + \lambda_1 P_{K-1,k} + \lambda_1 P_{K,k+1}, \qquad \text{for } 1 \le k \le K-C$$

$$(16.2)$$

where $\delta(\cdot)$ is a Dirac delta function. The solutions for the part $j-k<1$ or $k<0$ are all zero, and those for the part $0 \le j \le C$ and $k=0$ are $P_{j,0} = \rho_T^j/j! \cdot P_0$. Since an HC is blocked only when the system is full and the buffer has no more OCs, its blocking probability is given by $P_{B1} = P_{K,0}$.

Also, the blocking probability of an OC, P_{B2}, is given by the sum of the probability of blocking at its arrival and the probability of being pushed out caused by the arrival next to it. Since $\lambda_T P_K = \lambda_1 P_{B1} + \lambda_2 P_{B2}$, we can obtain P_{B2} as $P_{B2} = (\lambda_T P_K - \lambda_1 P_{B1})/\lambda_2$.

Next, we find the conditional waiting time probability distribution function (PDF) of each call getting served eventually. Since an HC gets served by the pure death process with rate of $C\mu$, its conditional waiting time PDF is given by

$$W_1(t) = W_1(0) + \left[\sum_{j=C}^{K-1} \sum_{k=0}^{j-C} P_{j,k} \int_{0+}^{t} \frac{C\mu(C\mu x)^{j-C-k}}{(j-C-k)!} e^{-C\mu x} dx \right.$$
$$\left. + \sum_{k=1}^{K-C} P_{K,k} \int_{0+}^{t} \frac{C\mu(C\mu x)^{K-C-k}}{(K-C-k)!} e^{-C\mu x} dx \right] \bigg/ (1 - P_{B1}) \qquad t > 0$$

$$(16.3)$$

where

$$W_1(0) = \sum_{j=0}^{C-1} P_j/(1 - P_{B1})$$

The second term of Eq. (16.3) corresponds to the case when an HC arrives at the system that is not full, and it joins at the position $j-C-k+1, C \le j < K$. As a result, k OCs change their positions after the HC. The third term corresponds to the case when an HC arrives at the full system, and it joins at the position $K-C-k+1, k > 0$. Thus, $(k-1)$ OCs change their position after the HC and the oldest OC in the buffer gets pushed out. Since

$$\int_{t}^{\infty} \frac{m(mx)^k}{k!} e^{-mx} dx = \sum_{i=0}^{k} \frac{(mt)^i e^{-mt}}{i!}$$

we can rewrite Eq. (16.3) as

$$W_1(t) = 1 - \left[\sum_{j=C}^{K-1} \sum_{k=0}^{j-C} P_{j,k} \sum_{i=0}^{j-C-k} \frac{(C\mu t)^i}{i!} e^{-C\mu t} \right.$$
$$\left. + \sum_{k=1}^{K-C} P_{K,k} \sum_{i=0}^{K-C-k} \frac{(C\mu t)^i}{i!} e^{-C\mu t} \right] \bigg/ (1 - P_{B1}), \qquad t > 0 \qquad (16.4)$$

To find the conditional waiting time PDF of an OC that is served successfully, we consider an OC arriving at the position i in the system. The OC may move into the new position $i - 1$ with probability $C\mu/(\lambda_T + C\mu)$ or the position $i + 1$ with probability $\lambda_T/(\lambda_T + C\mu)$. After a number of transitions in the buffer, the OC will be served eventually if it enters one of C servers, or it will get pushed out if it moves to the position $K + 1$. Let $u_{i,n}$ be the probability that an OC arrives at position i and gets served after n transitions. It can be obtained using the method of random walk with absorbing barriers at C and $K + 1$ as [Feller, 1968]

$$u_{i,n} = \sum_{k=0}^{\infty} \left(\frac{q}{r}\right)^{ka} \cdot w_{2ka+(i-C),n} - \sum_{k=0}^{\infty} \left(\frac{q}{r}\right)^{ka-(i-C)} \cdot w_{2ka-(i-C),n} \qquad (16.5)$$

where

$$q = \lambda_T/(\lambda_T + C\mu), \qquad r = 1 - q, \qquad a = K + 1 - C,$$

$$w_{i,n} = \frac{i}{n} \cdot \binom{n}{(n+i)/2} \cdot q^{(n-i)/2} \cdot r^{(n+i)/2}$$

and where $n \geq i$ and n of $w_{i,n}$ should have the same parity.

Also, let $g_{i,t}$ be the probability that an OC arrives at the position i and gets served after t. Since the time during which it completes after n transitions in the buffer has a gamma distribution of order $(n - 1)$, it is obtained as

$$g_{i,t} = \sum_{n=1}^{\infty} u_{i,n} \cdot \frac{(\lambda_T + C\mu)^n t^{n-1}}{(n-1)!} \cdot e^{-(\lambda_T + C\mu)t}, \qquad C < i \leq K \qquad (16.6)$$

When the system size is $j\,(j < K)$ with $k(k \geq 0)$ OCs in the buffer, an OC arrives at the position $j - k + 1$. If the system size is K with $k(k > 0)$ OCs in the buffer, the OC arrives at the position $K - k + 1$ by pushing out the oldest OC in the buffer. Thus, the conditional waiting time PDF of the OC which gets served is obtained with

$$W_2(0) = \sum_{j=0}^{C-1} P_{j,0}/(1 - P_{B2})$$

as

$$W_2(t) = W_2(0)$$

$$+ \frac{\displaystyle\sum_{j=C}^{K-1}\sum_{k=0}^{j-C} P_{j,k} \int_{0+}^{t} g_{j-k+1,x}\,dx + \sum_{k=1}^{K-C} P_{K,k} \int_{0+}^{t} g_{K-k+1,x}\,dx}{1 - P_{B2}}, \qquad t > 0$$

$$= 1 - \left[\sum_{j=C}^{K-1}\sum_{k=0}^{j-C} P_{j,k} \sum_{n=1}^{\infty} u_{j-k+1,n} \sum_{i=0}^{n-1} \frac{[(\lambda_T + C\mu)t]^i}{i!} \cdot e^{-(\lambda_T + C\mu)t} \right.$$

$$\left. + \sum_{k=1}^{K-C} P_{K,k} \sum_{n=1}^{\infty} u_{K-k+1,n} \sum_{i=0}^{n-1} \frac{[(\lambda_T + C\mu)t]^i}{i!} \cdot e^{-(\lambda_T + C\mu)t} \right] \Big/ (1 - P_{B2}),$$

$$t > 0 \qquad (16.7)$$

Scheme 3. Under scheme 3, any type of call has access to all channels as long as there is more than one free channel. When all channels are busy, HCs are served as under scheme 2. However, OCs will be queued next to the newest OC in a buffer or will be queued at the first position of the buffer, if the buffer is empty. When the system is full, an arriving OC is blocked. In addition, if an OC is in the position K and an HC arrives, then the OC gets pushed out and the new arrival is queued next to the newest HC. Thus, the HCs will be served by the FIFO rule with the head-of-line priority basis as done under scheme 2, whereas the OCs will be served by the FIFO rule with the pushout basis.

The state-transition-rate diagram for scheme 3 is the same as scheme 2. Thus, the blocking probabilities of HCs and OCs for scheme 3 are the same as scheme 2. Since the waiting time distribution of an HC is independent of the service rule of OCs, it is also obtained from Eq. (16.4). However, the waiting time distribution of an OC is different from the result of scheme 2 as follows.

Noting that an HC arriving only forces a move of the position of OC in i to the new position $i + 1$, we can obtain the probability of $u_{i,n}$ that an OC arrives at position i in the system and gets served after n transitions by using the result of Eq. (16.5) with $q = \lambda_1/(\lambda_1 + C\mu)$. Also, since the number of transitions in the buffer depends on the arrival rate of HCs and the service rate, the probability that an OC is in position i and the OC gets served after t is obtained from Eq. (16.6) as

$$g_{i,t} = \sum_{n=1}^{\infty} \mu_{i,n} \cdot \frac{(\lambda_1 + C\mu)^n t^{n-1}}{(n-1)!} \cdot e^{-(\lambda_1 + C\mu)t}, \qquad C < i \leq K$$

When the system size is j, $j < K$, an OC arriving can join at position $j + 1$ without considering the number of OCs in the buffer. It is blocked, however, when the system is full. Thus, the conditional waiting time PDF of the OC which gets served is obtained with

$$W_2(0) = \sum_{j=0}^{C-1} P_{j,0}$$

as

$$W_2(t) = W_2(0) + \left[\sum_{j=C}^{K-1} \sum_{k=0}^{j-C} P_{j,k} \int_{0+}^{t} g_{j+1,x} dx \right] \Big/ (1 - P_{B2})$$

$$= 1 - \frac{\displaystyle\sum_{j=C}^{K-1} \sum_{k=0}^{j-C} P_{j,k} \sum_{n=1}^{\infty} u_{j+1,n} \sum_{i=0}^{n-1} \frac{((\lambda_1 + C\mu)t)^i}{i!} \cdot e^{-(\lambda_1 + C\mu)t}}{(1 - P_{B2})}$$

$$t > 0 \quad (16.8)$$

Numerical Examples

In this section, some numerical examples are given to show the performance characteristics of the three call handling schemes for a personal portable radio telephone system. In these numerical examples, parameters are set as follows: $K = 30$, $C = 20$, $R = 0.8$ km, and time is normalized by the average holding time $(1/\mu)$.

In Fig. 16.6, we show the blocking probabilities of OCs and HCs vs the rate of OCs per unit area λ_0/km^2 and the total traffic load per channel ρ_T for the three schemes with the handoff probability $P_h = 0.1$. For P_{B2}, it is observed that the call handling schemes without GCs (schemes 2 and 3) are superior to the one with GCs (scheme 1), but the opposite holds for P_{B1}. This tradeoff becomes manifest as the number of guard channels C_g increases. The higher blocking probability of HCs under schemes without GCs, however, may be sufficiently small to be acceptable in the case where a typical requirement of P_{B2} is 0.5.

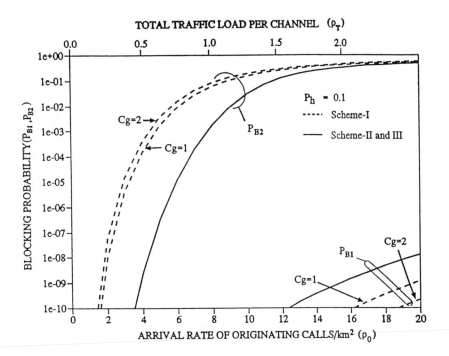

FIGURE 16.6 Comparison of blocking probabilities of three call handling schemes for $P_h = 0.1$.

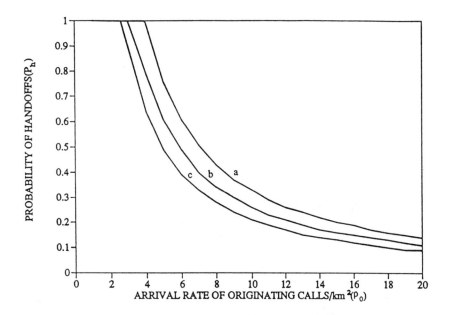

FIGURE 16.7 Boundaries between scheme 1 with a single GC and a scheme without GCs (scheme 2 or 3) for $P_h = 0.1$ and the value of $(1 - \alpha)$ equal to (a) 10^{-5}, (b) 10^{-6}, (c) 10^{-7}.

To make the choice of call handling schemes from the system provider's perspective by neglecting the waiting time of calls, we here define a cost function of P_{B1} and P_{B2}, which indicates the relative importance of blocking for HCs and OCs, as $CF = \alpha \cdot P_{B1} + (1 - \alpha) \cdot P_{B2}$, where α is in the interval $[0, 1]$. In Fig. 16.7, we show boundaries for the choice between the call handling schemes without GCs (schemes 2 and 3) and the one with single GC (scheme 1), which are obtained by

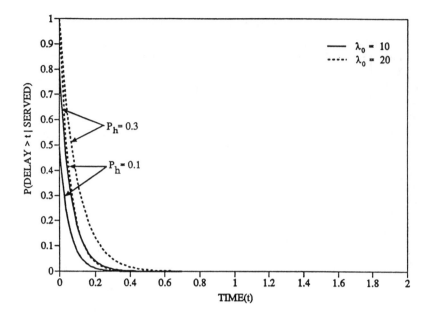

FIGURE 16.8 Comparison of waiting time distributions of HCs under schemes without guard channels, schemes 2 and 3.

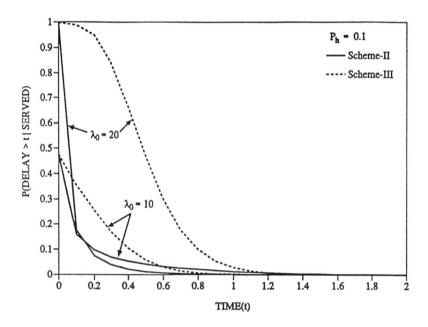

FIGURE 16.9 Comparison of waiting time distributions of OCs for $P_h = 0.1$ under schemes without guard channels, schemes 2 and 3.

comparing each CF. For given P_h, λ_0, and α, the region to the left of each curve is preferred to the call handling schemes without GCs, whereas that to the right of the curve is preferred to the one with a single GC. From the result, the call handling schemes without GCs (schemes 2 and 3) have an advantage over the one with GCs for a typical cell in the residential area, where the handoff probability is low.

We now show the waiting time distributions of OCs and HCs. Since the waiting time of OCs can be neglected in a typically lightly loaded cell, we here consider a heavy loaded environment (the total traffic load per channel is larger than 1). In Fig. 16.8, the waiting time distributions of HCs under schemes without guard channels (schemes 2 and 3) is shown. Since HCs are served with priority over OCs, the results of schemes 2 and 3 are the same. Although an HC suffers a longer delay as P_h or λ_0 increases, the delay is sufficiently short to be acceptable.

In Fig. 16.9, the waiting time distributions of OCs under schemes 2 and 3 with $P_h = 0.1$ are shown. From these results, one can find that scheme 2 shows a shorter average waiting time than scheme 3, although it has a longer tail of waiting time distribution. We also note that the average delay for $\lambda_0 = 20$ is shorter, relative to $\lambda_0 = 10$, under scheme 2. This effect is explained by the fact that as λ_0 increases, OCs that do not get served quickly are more likely to get pushed out. Considering the results of schemes without guard channels and the one with guard channels, one can find a tradeoff between the waiting time of an OC and its blocking probability. This tradeoff can be managed to satisfy both the system provider and subscribers by controlling the finite buffer length.

16.4 Summary

In this chapter we have described three system configurations of the base station subsystems for cellular mobile communication networks. We have observed that the prioritized pushout call handling schemes without guard channels reduce the call blocking probability of ordinary calls without a severe penalty on prioritized handoff calls. Also, with a cost function from the system provider's perspective, we have shown that two prioritized pushout call handling schemes without guard channels are better than the scheme with guard channels. By examining the waiting time distribution and average waiting time of originating calls under a heavy loaded environment, we have also shown that the scheme 2 performs better than the scheme 3. Thus, the scheme 2 can be a good candidate as a call handling scheme for a typical cell in a residential area where the handoff probability is low.

Defining Terms

Base transceiver subsystem: A system with RF transceivers and interfaces between BSC and itself.
Base station controller: A system for managing BTSs.
Base station subsystem: A system that consists of BTSs and a BSC.
Originating calls: Calls initiated in a cell.
Handoff: The procedure initiated by a base station when the mobile unit moves out of the coverage area of the base station without interrupting the call. Typically, the handoff calls have a priority over originating calls.
Guard channels: Several reserved channels used by higher priority calls exclusively.

References

Ehrlich, N., Fisher, R., and Wingard, T. 1979. Cell-site hardware, *Bell Sys. Tech. J.* 58(1):153–159.

El-Dolil, S.A., Wong, W., and Steele, R. 1989. Teletraffic performance of highway microcells with overlay macrocell. *IEEE J. Select. Areas Commun.* 7(1):71–78.

Feller, W. 1968. *An Introduction to Probability Theory and its Application*, Vol. 1, John Wiley and Sons, New York, pp. 349–370.

Guerin, R. 1988. Queueing-blocking system with two arrival streams and guard channels. *IEEE Trans. Commun.* COM-36(2):153–163.

Hong, D.H. and Rappaport, S.S. 1986. Traffic model and performance analysis for cellular mobile radio telephone systems with prioritized and nonprioritized handoff procedures. *IEEE Trans. Vech. Tech.* VT-28 (3):77–92.

Kleinrock, L. 1975. *Queueing System*, Vol. I, John Wiley and Sons, New York.

Qualcomm. 1992. *The CDMA Network Engineering Handbook*, Nov.

Uebayashi, S., Ohno, K. and Nojima, T. 1993. Development of TDMA cellular base station equipment, *43rd Vehicular Tech. Conf.* Secaucus, NJ, pp. 566–569.

Yoon, C.H. and Un, C.K. 1993. Performance of personal radio telephone system with and without guard channels. *IEEE J. Select. Areas Commun.* 11(6):911–917.

Further Information

For a more general description on microcell systems, see the paper by Sanrnecki, Vinodral, Javed, O'Kelly, and Dick in *IEEE Communications Magazine*, 1993, Vol. 31, No. 4.

The paper by Pavlidou presents the numerical method for finding the delay and blocking probabilities of the voice and data integrated cellular mobile system, in *IEEE Transactions on Communications*, 1994, Vol. 42, No. 2/3/4.

17

Access Methods

Bernd-Peter Paris
George Mason University

17.1 Introduction

The radio channel is fundamentally a broadcast communication medium. Therefore, signals transmitted by one user can potentially be received by all other users within range of the transmitter. Although this high connectivity is very useful in some applications, like broadcast radio or television, it requires stringent access control in wireless communication systems to avoid, or at least to limit, interference between transmissions. Throughout, the term wireless communication systems is taken to mean communication systems that facilitate two-way communication between a portable radio communication terminal and the fixed network infrastructure. Such systems range from mobile cellular systems through personal communication systems (PCS) to cordless telephones.

The objective of wireless communication systems is to provide communication channels on demand between a portable radio station and a radio port or base station that connects the user to the fixed network infrastructure. Design criteria for such systems include **capacity**, cost of implementation, and quality of service. All of these measures are influenced by the method used for providing multiple-access capabilities. However, the opposite is also true: the access method should be chosen carefully in light of the relative importance of design criteria as well as the system characteristics.

Multiple access in wireless radio systems is based on insulating signals used in different connections from each other. The support of parallel transmissions on the uplink and downlink, respectively, is called multiple access, whereas the exchange of information in both directions of a connection is referred to as **duplexing**. Hence, multiple access and duplexing are methods that facilitate the sharing of the broadcast communication medium. The necessary insulation is achieved by assigning to each transmission different components of the domains that contain the signals. The signal domains commonly used to provide multiple access capabilities include the following.

0-8493-8573-3/96/$0.00+$.50

Spatial domain: All wireless communication systems exploit the fact that radio signals experience rapid attenuation during propagation. The propagation exponent ρ on typical radio channels lies between $\rho = 2$ and $\rho = 6$ with $\rho = 4$ a typical value. As signal strength decays inversely proportional to the ρth power of the distance, far away transmitters introduce interference that is negligible compared to the strength of the desired signal. The cellular design principle is based on the ability to reuse signals safely if a minimum reuse distance is maintained. Directional antennas can be used to enhance the insulation between signals. We will not focus further on the spatial domain in this treatment of access methods.

Frequency domain: Signals which occupy nonoverlapping frequency bands can be easily separated using appropriate bandpass filters. Hence, signals can be transmitted simultaneously without interfering with each other. This method of providing multiple access capabilities is called **frequency-division multiple access (FDMA)**.

Time domain: Signals can be transmitted in nonoverlapping time slots in a round-robin fashion. Thus, signals occupy the same frequency band but are easily separated based on their time of arrival. This multiple access method is called **time-division multiple access (TDMA)**.

Code domain: In **code-division multiple access (CDMA)** different users employ signals that have very small cross-correlation. Thus, correlators can be used to extract individual signals from a mixture of signals even though they are transmitted simultaneously and in the same frequency band. The term code-division multiple-access is used to denote this form of channel sharing. Two forms of CDMA are most widely employed and will be described in detail subsequently, frequency hopping (FH) and direct sequence (DS).

System designers have to decide in favor of one, or a combination, of the latter three domains to facilitate multiple access. The three access methods are illustrated in Fig. 17.1. The principal idea in all three of these access methods is to employ signals that are orthogonal or nearly orthogonal.

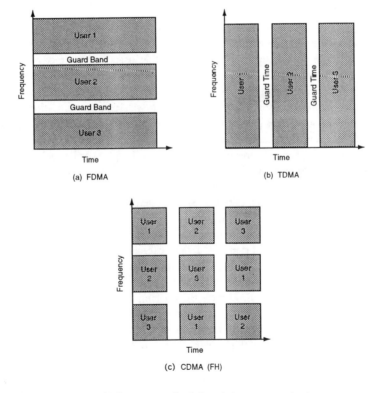

FIGURE 17.1 Multiple-access methods for wireless communication systems.

Then, correlators that project the received signal into the subspace of the desired signal can be employed to extract a signal without interference from other transmissions.

Preference for one access method over another depends largely on overall system characteristics, as we will see in the sequel. No single access method is universally preferable, and system considerations should be carefully weighed before the design decision is made. Before going into the detailed description of the different access methods, we will discuss briefly the salient features of some wireless communication systems. This will allow us later to assess the relative merits of the access methods in different scenarios.

17.2 Relevant Wireless Communication System Characteristics

Modern wireless radio systems range from relatively simple cordless telephones to mobile cellular systems and the emerging personal communication systems (PCS). It is useful to consider such diverse systems as cordless telephone and mobile cellular radio to illustrate some of the fundamental characteristics of wireless communication systems [Cox, 1992].

A summary of the relevant parameters and characteristics for cordless telephone and cellular radio is given in Table 17.1. As evident from that table, the fundamental differences between the two systems are speech quality and the area covered by a base station. The high speech quality requirement in the cordless application is the consequence of the availability of tethered access in the home and office and the resulting direct competition with wire-line telephone services. In the mobile cellular application, the user has no alternative to the wireless access and may be satisfied with lower, but still acceptable, quality of service.

In cordless telephone applications the transmission range is short because the base station can simply be moved to a conveniently located wire-line access point (wall jack) to provide wireless network access where desired. In contrast, the mobile cellular base station must provide access for users throughout a large geographical area of up to approximately 30 km (20 mi) around the base station. This large coverage area is necessary to economically meet the promise of of uninterrupted service to roaming users.

The different range requirements directly affect the transmit power and antenna height for the two systems. High-power transmitters used in mobile cellular user sets consume far more power than even complex signal processing hardware. Hence, sophisticated signal processing, including speech compression, voice activity detection, error correction and detection, and adaptive equalization, can be employed without substantial impact on the battery life in portable hand sets. Furthermore, such techniques are consistent with the goals of increased range and support of large numbers of users with a single, expensive base station. On the other hand, the high mobile cellular base station antennas introduce delay spreads that are one or two orders of magnitude larger than those commonly observed in cordless telephone applications.

TABLE 17.1 Summary of Relevant Characteristics of Cordless Telephone and Cellular Mobile Radio

Characteristic or Parameter	Cordless Telephone	Cellular Radio
Speech quality	Toll quality	Varying with channel quality; possibly decreased by speech pause exploitation
Transmission range	<100 m	100 m–30 km
Transmit power	Milliwatts	Approx. 1 W
Base station antenna height	Approx. 1 m	Tens of meters
Delay spread	Approx. 1 μs	Approx. 10 μs
Complexity of base station	Low	High
Complexity of user set	Low	High

Clearly, the two systems just considered are at extreme ends of the spectrum of wireless communications systems. Most notably, the emerging PCS systems fall somewhere between the two. However, the comparison above highlights some of the system characteristics that should be considered when discussing access methods for wireless communication systems.

17.3 Frequency Division Multiple Access

As mentioned in Sec. 17.1, in FDMA nonoverlapping frequency bands are allocated to different users on a continuous time basis. Hence, signals assigned to different users are clearly orthogonal, at least ideally. In practice, out-of-band spectral components can not be completely suppressed leaving signals not quite orthogonal. This necessitates the introduction of guard bands between frequency bands to reduce adjacent channel interference, i.e., inference from signals transmitted in adjacent frequency bands; see also Fig. 17.1(a).

It is advantageous to combine FDMA with time-division duplexing (TDD) to avoid simultaneous reception and transmission that would require insulation between receive and transmit antennas. In this scenario, the base station and portable take turns using the same frequency band for transmission. Nevertheless, combining FDMA and frequency division duplex is possible in principle, as is evident from the analog FM-based systems deployed throughout the world since the early 1980s.

Channel Considerations

In principle there exists the well-known duality between TDMA and FDMA; see Bertsekas and Gallager, 1987, p. 113 ff. In the wireless environment, however, propagation related factors have a strong influence on the comparison between FDMA and TDMA. Specifically, the duration of a transmitted symbol is much longer in FDMA than in TDMA. As an immediate consequence, an equalizer is typically not required in an FDMA-based system because the delay spread is small compared to the symbol duration.

To illustrate this point, consider a hypothetical system that transmits information at a constant rate of 50 kb/s. This rate would be sufficient to support 32-kb/s adaptive differential pulse code modulation (ADPCM) speech encoding, some coding for error protection, and control overhead. If we assume further that some form of QPSK modulation is employed, the resulting symbol duration is 40 μs. In relation to delay spreads of approximately 1 μs in the cordless application and 10 μs in cellular systems, this duration is large enough that only little intersymbol interference is introduced. In other words, the channel is frequency nonselective, i.e., all spectral components of the signal are affected equally by the channel. In the cordless application an equalizer is certainly not required; cellular receivers may require equalizers capable of removing intersymbol interference between adjacent bits. Furthermore, it is well known that intersymbol interference between adjacent bits can be removed without loss in SNR by using maximum-likelihood sequence estimation; e.g., Proakis, 1989, p. 622.

Hence, rather simple receivers can be employed in FDMA systems at these data rates. However, there is a flip side to the argument. Recall that the Doppler spread, which characterizes the rate at which the channel impulse response changes, is given approximately by $B_d = v/cf_c$, where v denotes the speed of the mobile user, c is the propagation speed of the electromagnetic waves carrying the signal, and f_c is the carrier frequency. Thus, for systems operating in the vicinity of 1 GHz, B_d will be less than 1 Hz in the cordless application and typically about 100 Hz for a mobile traveling on a highway. In either case, the signal bandwidth is much larger than the Doppler spread B_d, and the channel can be characterized as slowly fading. Whereas this allows tracking of the carrier phase and the use of coherent receivers, it also means that fade durations are long in comparison to the symbol duration and can cause long sequences of bits to be subject to poor channel conditions. The problem is compounded by the fact that the channel is frequency nonselective because it implies that the entire signal is affected by a fade.

To overcome these problems either time diversity, frequency diversity, or spatial diversity could be employed. Time diversity can be accomplished by a combination of coding and interleaving if the fading rate is sufficiently large. For very slowly fading channels, such as the cordless application, the necessary interleaving depth would introduce too much delay to be practical. Frequency diversity can be introduced simply by slow frequency hopping, a technique that prescribes users to change the carrier frequency periodically. Frequency hopping is a form of spectrum spreading because the bandwidth occupied by the resulting signal is much larger than the symbol rate. In contrast to direct sequence spread spectrum discussed subsequently, however, the instantaneous bandwidth is not increased. The jumps between different frequency bands effectively emulate the movement of the portable and, thus, should be combined with the just described time-diversity methods. Spatial diversity is provided by the use of several receive or transmit antennas. At carrier frequencies exceeding 1 GHz, antennas are small and two or more antennas can be accommodated even in the hand set. Furthermore, if FDMA is combined with time-division duplexing, multiple antennas at the base station can provide diversity on both uplink and downlink. This is possible because the channels for the two links are virtually identical, and the base station, using channel information gained from observing the portable's signal, can transmit signals at each antenna such that they combine coherently at the portable's antenna. Thus, signal processing complexity is moved to the base station extending the portable's battery life.

Influence of Antenna Height

In the cellular mobile environment base station antennas are raised considerably to increase the coverage area. Antennas mounted on towers and rooftops are a common sight, and antenna heights of 50 m above ground are no exceptions. Besides increasing the coverage area, this has the additional effect that frequently there exists a better propagation path between two base station antennas than between a mobile and the base station; see Fig. 17.2.

Assuming that FDMA is used in conjunction with TDD as specified at the beginning of this section, then base stations and mobiles transmit on the same frequency. Now, unless there is tight synchronization between all base stations, signals from other base stations will interfere with the reception of signals from portables at the base station. To keep the interference at acceptable levels, it is necessary to increase the reuse distance, i.e., the distance between cells using the same frequencies. In other words, sufficient insulation in the spatial domain must be provided to facilitate the separation of signals. Note that these comments apply equally to cochannel and adjacent channel interference.

This problem does not arise in cordless applications. Base station antennas are generally of the same height as user sets. Hence, interference created by base stations is subject to the same

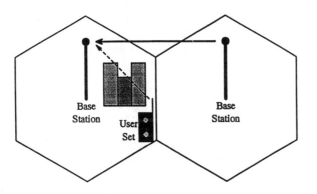

FIGURE 17.2 High base station antennas lead to stronger propagation paths between base stations than between a user set and its base stations.

propagation conditions as signals from user sets. Furthermore, in cordless telephone applications there are frequently attenuating obstacles, such as walls, between base stations that reduce intracell interference further. Note that this reduction is vital for the proper functioning of cordless telephones since there is typically no network planning associated with installing a cordless telephone. As a safety feature, to overcome intercell interference, adaptive channel management strategies based on sensing interference levels can be employed.

Example 17.1: CT2

The CT2 standard was originally adopted in 1987 in Great Britain and improved with a common air interface (CAI) in 1989. The CAI facilitates interoperability between equipment from different vendors whereas the original standard only guarantees noninterference. The CT2 standard is used in home and office cordless telephone equipment and has been used for telepoint applications [Goodman, 1991b].

CT2 operates in the frequency band 864–868 MHz and uses carriers spaced at 100 kHz. FDMA with time division duplexing is employed. The combined gross bit rate is 72 kb/s, transmitted in frames of 2-ms duration of which the first-half carries downlink and the second-half carries uplink information. This setup supports a net bit rate of 32 kb/s of user data (32-kb/s ADPCM encoded speech) and 2-kb/s control information in each direction. The CT2 modulation technique is binary frequency shift keying.

Further Remarks

From the preceding discussion it is obvious that FDMA is a good candidate for applications like cordless telephone. In particular, the simple signal processing makes it a good choice for inexpensive implementation in the benign cordless environment. The possibility of concentration of signal processing functions in the base station strengthens this aspect.

In the cellular application, on the other hand, FDMA is inappropriate because of the lack of built-in diversity and the potential for severe intercell interference between base stations. A further complication arises from the difficulty of performing handovers if base-stations are not tightly synchronized.

For PCS the decision is not as obvious. Depending on whether the envisioned PCS application resembles more a cordless private branch exchange (PBX) than a cellular system, FDMA may be an appropriate choice. We will see later that it is probably better to opt for a combined TDMA/FDMA or a CDMA-based system to avoid the pitfalls of pure FDMA systems and still achieve moderate equipment complexities.

Finally, there is the problem of channel assignment. Clearly, it is not reasonable to assign a unique frequency to each user as there are not sufficient frequencies and the spectral resource would be unused whenever the user is idle. Instead, methods that allocate channels on demand can make much more efficient use of the spectrum. Such methods will be discussed further during the description of TDMA systems.

17.4 Time Division Multiple Access

In TDMA systems users share the same frequency band by accessing the channel in non-overlapping time intervals in a round-robin fashion [Falconer, Adachi, and Gudmundson, 1995]. Since the signals do not overlap, they are clearly orthogonal, and the signal of interest is easily extracted by switching the receiver on only during the transmission of the desired signal. Hence, the receiver filters are simply windows instead of the bandpass filters required in FDMA. As a consequence, the guard time between transmissions can be made as small as the synchronization of the network permits. Guard times of 30–50 μs between time slots are commonly used in TDMA-based systems. As a consequence, all users must be synchronized with the base station to within a fraction of the guard time. This is achievable by distributing a master clock signal on one of the base station's broadcast channels.

TDMA can be combined with TDD or frequency-division duplexing (FDD). The former duplexing scheme is used, for example, in the Digital European Cordless Telephone (DECT) standard and is well suited for systems in which base-to-base and mobile-to-base propagation paths are similar, i.e., systems without extremely high base station antennas. Since both the portable and the base station transmit on the same frequency, some signal processing functions for the downlink can be implemented in the base station, as discussed earlier for FDMA/TDD systems.

In the cellular application, the high base station antennas make FDD the more appropriate choice. In these systems, separate frequency bands are provided for uplink and downlink communication. Note that it is still possible and advisable to stagger the uplink and downlink transmission intervals such that they do not overlap, to avoid the situation that the portable must transmit and receive at the same time. With FDD the uplink and downlink channel are not identical and, hence, signal processing functions can not be implemented in the base-station; antenna diversity and equalization have to be realized in the portable.

Propagation Considerations

In comparison to a FDMA system supporting the same user data rate, the transmitted data rate in a TDMA system is larger by a factor equal to the number of users sharing the frequency band. This factor is eight in the pan-European global system for mobile communications (GSM) and three in the advanced mobile phone service (D-AMPS) system. Thus, the symbol duration is reduced by the same factor and severe intersymbol interference results, at least in the cellular environment.

To illustrate, consider the earlier example where each user transmits 25 K symbols per second. Assuming eight users per frequency band leads to a symbol duration of 5 μs. Even in the cordless application with delay spreads of up to 1 μs, an equalizer may be useful to combat the resulting interference between adjacent symbols. In cellular systems, however, the delay spread of up to 20 μs introduces severe intersymbol interference spanning up to 5 symbol periods. As the delay spread often exceeds the symbol duration, the channel can be classified as frequency selective, emphasizing the observation that the channel affects different spectral components differently.

The intersymbol interference in cellular TDMA systems can be so severe that linear equalizers are insufficient to overcome its negative effects. Instead, more powerful, nonlinear decision feedback or maximum-likelihood sequence estimation equalizers must be employed [Proakis, 1991]. Furthermore, all of these equalizers require some information about the channel impulse response that must be estimated from the received signal by means of an embedded training sequence. Clearly, the training sequence carries no user data and, thus, wastes valuable bandwidth.

In general, receivers for cellular TDMA systems will be fairly complex. On the positive side of the argument, however, the frequency selective nature of the channel provides some built-in diversity that makes transmission more robust to channel fading. The diversity stems from the fact that the multipath components of the received signal can be resolved at a resolution roughly equal to the symbol duration, and the different multipath components can be combined by the equalizer during the demodulation of the signal. To further improve robustness to channel fading, coding and interleaving, slow frequency hopping and antenna diversity can be employed as discussed in connection with FDMA.

Initial Channel Assignment

In both FDMA and TDMA systems, channels should not be assigned to a mobile on a permanent basis. A fixed assignment strategy would either be extremely wasteful of precious bandwidth or highly susceptible to cochannel interference. Instead, channels must be assigned on demand. Clearly, this implies the existence of a separate uplink channel on which mobiles can notify the base

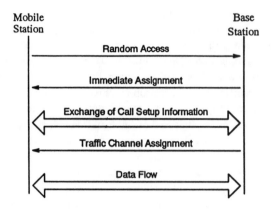

FIGURE 17.3   Mobile-originating call establishment.

station of their need for a traffic channel. This uplink channel is referred to as the **random-access channel** because of the type of strategy used to regulate access to it.

The successful procedure for establishing a call that originates from the mobile station is outlined in Fig. 17.3. The mobile initiates the procedure by transmitting a request on the random-access channel. Since this channel is shared by all users in range of the base station, a random access protocol, like the ALOHA protocol, has to be employed to resolve possible collisions. Once the base station has received the mobile's request, it responds with an immediate assignment message that directs the mobile to tune to a dedicated control channel for the ensuing call setup. Upon completion of the call setup negotiation, a traffic channel, i.e., a frequency in FDMA systems or a time slot in TDMA systems, is assigned by the base station and all future communication takes place on that channel. In the case of a mobile-terminating call request, the sequence of events is preceded by a paging message alerting the base station of the call request.

Example 17.2: GSM

Named after the organization that created the system standards (Groupe Speciale Mobile) this pan-European digital cellular system has been deployed in Europe since the early 1990s [Hodges, 1990]. GSM uses combined TDMA and FDMA with frequency-division duplex for access. Carriers are spaced at 200 kHz and support eight TDMA time slots each. For the uplink the frequency band 890–915 MHz is allocated, whereas the downlink uses the band 935–960 MHz. Each time slot is of duration 577 μs, which corresponds to 156.26-b periods, including a guard time of 8.25-b periods. Eight consecutive time slots form a GSM frame of duration 4.62 ms.

The GSM modulation is Gaussian minimum shift keying with time-bandwidth product of 0.3, i.e., the modulator bandpass has a cutoff frequency of 0.3 times the bit rate. At the bit rate of 270.8 kb/s, severe intersymbol interference arises in the cellular environment. To facilitate coherent detection, a 26-b training sequence is embedded into every time slot. Time diversity is achieved by interleaving over 8 frames for speech signals and 20 frames for data communication. Sophisticated error-correction coding with varying levels of protection for different outputs of the speech coder is provided. Note that the round-trip delay introduced by the interleaver is on the order of 80 ms for speech signals. GSM provides slow frequency hopping as a further mechanism to improve the efficiency of the interleaver.

Further Remarks

In cellular systems, such as GSM or the North-American D-AMPS, TDMA is combined with FDMA. Different frequencies are used in neighboring cells to provide orthogonal signalling without the need for tight synchronization of base stations. Furthermore, channel assignment can then be performed in each cell individually. Within a cell, one or more frequencies are shared by users in the time domain.

From an implementation standpoint TDMA systems have the advantage that common radio and signal processing equipment at the base station can be shared by users communicating on the same frequency. A somewhat more subtle advantage of TDMA systems arises from the possibility of monitoring surrounding base stations and frequencies for signal quality to support mobile assisted handovers.

17.5 Code Division Multiple Access

CDMA systems employ wideband signals with good cross-correlation properties [Kohno, Meidan, and Milstein, 1995]. That means the output of a filter matched to one user's signal is small when a different user's signal is input. A large body of work exists on spreading sequences that lead to signal sets with small cross correlations [Sarwate and Pursley, 1980]. Because of their noise-like appearance such sequences are often referred to as pseudonoise (PN) sequences, and because of their wideband nature CDMA systems are often called spread-spectrum systems.

Spectrum spreading can be achieved mainly in two ways: through frequency hopping as explained earlier or through direct sequence spreading. In direct sequence spread spectrum, a high-rate, antipodal pseudorandom spreading sequence modulates the transmitted signal such that the bandwidth of the resulting signal is roughly equal to the rate of the spreading sequence. The cross correlation of the signals is then largely determined by the cross-correlation properties of the spreading signals. Clearly, CDMA signals overlap in both time and frequency domains but are separable based on their spreading waveforms.

An immediate consequence of this observation is that CDMA systems do not require tight synchronization between users as do TDMA systems. By the same token, frequency planning and management are not required as frequencies are reused throughout the coverage area.

Propagation Considerations

Spread spectrum is well suited for wireless communication systems because of its built-in frequency diversity. As discussed before, in cellular systems the delay spread measures several microseconds and, hence, the coherence bandwidth of the channel is smaller than 1 MHz. Spreading rates can be chosen to exceed the coherence bandwidth such that the channel becomes frequency selective, i.e., different spectral components are affected unequally by the channel and only parts of the signal are affected by fades. Expressing the same observation in time domain terms, multipath components are resolvable at a resolution equal to the chip period and can be combined coherently, for example, by means of a RAKE receiver [Proakis, 1989]. An estimate of the channel impulse response is required for the coherent combination of multipath components. This estimate can be gained from a training sequence or by means of a so-called pilot signal.

Even for cordless telephone systems, operating in environments with submicrosecond delay spread and corresponding coherence bandwidths of a few megahertz, the spreading rate can be chosen large enough to facilitate multipath diversity. If the combination of multipath components already described is deemed too complex, a simpler, but less powerful, form of diversity can be used that decorrelates only the strongest received multipath component and relies on the suppression of other path components by the matched filter.

Multiple-Access Interference

If it is possible to control the relative timing of the transmitted signals, such as on the downlink, the transmitted signals can be made perfectly orthogonal, and if the channel only adds white

Gaussian noise, matched filter receivers are optimal for extracting a signal from the superposition of waveforms. If the channel is dispersive because of multipath, the signals arriving at the receiver will no longer be orthogonal and will introduce some multiple-access interference, i.e., signal components from other signals that are not rejected by the matched filter.

On the uplink, extremely tight synchronization between users to within a fraction of a chip period, which is defined as the inverse of the spreading rate, is generally not possible, and measures to control the impact of multiple-access interference must be taken. Otherwise, the near–far problem, i.e., the problem of very strong undesired users' signals overwhelming the weaker signal of the desired user, can severely decrease performance. Two approaches are proposed to overcome the near–far problem: power control with soft handovers and multiuser detection.

Power control attempts to ensure that signals from all mobiles in a cell arrive at the base station with approximately equal power levels. To be effective, power control must be accurate to within about 1 dB and fast enough to compensate for channel fading. For a mobile moving at 55 mph and transmitting at 1 GHz, the Doppler bandwidth is approximately 100 Hz. Hence, the channel changes its characteristic drastically about 100 times per second and on the order of 1000 b/s must be sent from base station to mobile for power control purposes. As different mobiles may be subject to vastly different fading and shadowing conditions, a large dynamic range of about 80 dB must be covered by power control. Notice, that power control on the downlink is really only necessary for mobiles that are about equidistant from two base stations, and even then neither the update rate nor the dynamic range of the uplink is required.

The interference problem that arises at the cell boundaries where mobiles are within range of two or more base stations can be turned into an advantage through the idea of soft handover. On the downlink, all base stations within range can transmit to the mobile, which in turn can combine the received signals to achieve some gain from the antenna diversity. On the uplink, a similar effect can be obtained by selecting the strongest received signal from all base stations that received a user's signal. The base station that receives the strongest signal will also issue power control commands to minimize the transmit power of the mobile. Note, however, that soft handover requires fairly tight synchronization between base stations, and one of the advantages of CDMA over TDMA is lost.

Multiuser detection is still an emerging technique. It is probably best used in conjunction with power control. The fundamental idea behind this technique is to model multiple-access interference explicitly and devise receivers that reject or cancel the undesired signals. A variety of techniques have been proposed ranging from optimum maximum-likelihood sequence estimation via multistage schemes, reminiscent of decision feedback algorithms, to linear decorrelating receivers. An excellent survey of the theory and practice of multiuser detection is given by Verdu, 1992.

Further Remarks

CDMA systems work well in conjunction with frequency division duplexing. This arrangement decouples the power control problem on the uplink and downlink, respectively.

Signal quality enhancing methods, such as time diversity through coding and interleaving, can be applied just as with the other access methods. In spread spectrum systems, however, coding can be built into the spreading process, avoiding the loss of bandwidth associated with error protection. Additionally, CDMA lends itself naturally to the exploitation of speech pauses that make up more than half the time of a connection. If no signals are transmitted during such pauses, then the instantaneous interference level is reduced and the total number of users supportable by the system can be approximately doubled.

17.6 Comparison and Outlook

The question of which of the access methods is best does not have a single answer. Based on the preceding discussion FDMA is only suited for applications such as cordless telephone with very small cells and submicrosecond delay spreads. In cellular systems and for most versions of personal communication systems, the choice reduces to TDMA vs CDMA.

In terms of complexity, TDMA receivers require adaptive, nonlinear equalizers when operating in environments with large delay spreads. CDMA systems, in turn, need RAKE receivers and sophisticated power control algorithms. In the future, some form of multiple-access interference rejection is likely to be implemented as well. Time synchronization is required in both systems, albeit for different reasons. The additional complexity for coding and interleaving is comparable for both access methods.

An often quoted advantage of CDMA systems is the fact that the performance will degrade gracefully as the load increases. In TDMA systems, in turn, requests will have to be blocked once all channels in a cell are in use. Hence, there is a hard limit on the number of channels per cell. There are proposals for extended TDMA systems, however, that incorporate reassignment of channels during speech pauses. Not only would such extended TDMA systems match the advantage of the exploitation of speech pauses of CDMA systems, they would also lead to a soft limit on the system capacity. The extended TDMA proposals would implement the statistical multiplexing of the user data, e.g., by means of the packet reservation multiple access protocol [Goodman, 1991a]. The increase in capacity depends on the acceptable packet loss rate; in other words, small increases in the load lead to small increases in the packet loss probability.

Many comparisons in terms of capacity between TDMA and CDMA can be found in the recent literature. Such comparisons, however, are often invalidated by making assumptions that favor one access method over the other. An important exception constitutes the recent paper by Wyner, 1994. Under a simplified model that nevertheless captures the essence of cellular systems, he computes the Shannon capacity. Highlights of his results include the following.

- TDMA is distinctly suboptimal in cellular systems.
- When the signal-to-noise-ratio is large, CDMA appears to achieve twice the capacity of TDMA.
- Multiuser detectors are essential to realize near-optimum performance in CDMA systems.
- Intercell interference in CDMA systems has a detrimental effect when the signal-to-noise ratio is large, but it can be exploited via diversity combining to increase capacity when the signal-to-noise ratio is small.

More research along this avenue is necessary to confirm the validity of the results. In particular, incorporation of realistic channel models into the analysis is required. However, this work represents a substantial step towards quantifying capacity increases achievable with CDMA.

Defining Terms

Capacity: Shannon originally defined capacity as the maximum data rate which permits error-free communication in a given environment. A looser interpretation is normally employed in wireless communication systems. Here capacity denotes the traffic density supported by the system under consideration normalized with respect to bandwidth and coverage area.

Multiple access: Denotes the support of simultaneous transmissions over a shared communication channel.

Duplexing: Refers to the exchange of messages in both directions of a connection.

Frequency-division multiple access (FDMA): Simultaneous access to the radio channel is facilitated by assigning nonoverlapping frequency bands to different users.

Time-division multiple access (TDMA): Systems assign nonoverlapping time slots to different users in a round-robin fashion.

Code-division multiple access (CDMA): Systems use signals with very small cross-correlations to facilitate sharing of the broadcast radio channel. Correlators are used to extract the desired user's signal while simultaneously suppressing interfering, parallel transmissions.

Random-access channel: This uplink control channel is used by mobiles to request assignment of a traffic channel. A random access protocol is employed to arbitrate access to this channel.

References

Bertsekas, D. and Gallager, R. 1987. *Data Networks*, Prentice Hall, Englewood Cliffs, NJ.

Cox, D.C. 1992. Wireless network access for personal communications, *IEEE Comm. Mag.*, pp. 96–115.

Falconer, D.D., Adachi, F., and Gudmundson, B. 1995. Time division multiple access methods for wireless personal communications. *IEEE Comm. Mag.* 33(1):50–57.

Goodman, D. 1991a. Trends in cellular and cordless communications. *IEEE Comm. Mag.* pp 31–40.

Goodman, D.J. 1991b. Second generation wireless information networks. *IEEE Trans. on Vehicular Tech.* 40(2):366–374.

Hodges, M.R.L. 1990. The GSM radio interface. *Br. Telecom Tech. J.* 8(1):31–43.

Kohno, R., Meidan, R., and Milstein, L.B. 1995. Spread spectrum access methods for wireless communications. *IEEE Comm. Mag.* 33(1):58.

Proakis, J.G. 1989. *Digital Communications.* 2nd ed. McGraw-Hill, New York.

Proakis, J.G . 1991. Adaptive equalization for TDMA digital mobile radio. *IEEE Trans. on Vehicular Tech.* 40(2):333–341.

Sarwate, D.V. and Pursley, M.B. 1980. Crosscorrelation properties of pseudorandom and related sequences. *Proceedings of the IEEE*, 68(5):593–619.

Verdu, S. 1992. Multi-user detection. In *Advances in Statistical Signal Processing—Vol. 2: Signal Detection*, JAI Press, Greenwich, CT.

Wyner, A.D. 1994. Shannon-theoretic approach to a Gaussian cellular multiple-access channel. *IEEE Trans. on Information Theory*, 40(6):1713–1727.

Further Information

Several of the IEEE publications, including the *Transactions on Communications, Journal on Selected Areas in Communications, Transactions on Vehicular Technology, Communications Magazine*, and *Personal Communications Magazine* contain articles the on subject of access methods on a regular basis.

18

Location Strategies for Personal Communications Services

Ravi Jain
Bell Communications Research

Yi-Bing Lin
Bell Communications Research

Seshadri Mohan[1]
Bell Communications Research

18.1 Introduction

The vision of nomadic personal communications is the ubiquitous availability of services to facilitate exchange of information (voice, data, video, image, etc.) between nomadic end users independent of time, location, or access arrangements. To realize this vision, it is necessary to locate

[1]Address correspondence to: Seshadri Mohan, MCC-1A216B, Bellcore, 445 South St, Morristown, NJ 07960; Phone: 201-829-5160, Fax: 201-829-5888, E-mail: **smohan@thumper.bellcore.com**.

users that move from place to place. The strategies commonly proposed are two-level hierarchical strategies, which maintain a system of mobility databases, home location registers (HLR) and visitor location resisters (VLR), to keep track of user locations. Two standards exist for carrying out two-level hierarchical strategies using HLRs and VLRs. The standard commonly used in North America is the EIA/TIA Interim Standard 41 (IS 41) [EIA/TIA, 1991] and in Europe the Global System for Mobile Communications (GSM) [Mouly and Pautet, 1992; Lycksell, 1991]. In this chapter, we refer to these two strategies as *basic* location strategies.

We introduce these two strategies for locating users and provide a tutorial on their usage. We then analyze and compare these basic location strategies with respect to load on mobility databases and signalling network. Next we propose an auxiliary strategy, called the *per-user caching* or, simply, the *caching* strategy, that augments the basic location strategies to reduce the signalling and database loads.

The outline of this chapter is as follows. In Sec. 18.2 we discuss different forms of mobility in the context of personal communications services (PCS) and describe a reference model for a PCS architecture. In Secs. 18.3 and 18.4, we describe the user location strategies specified in the IS-41 and GSM standards, respectively, and in Sec. 18.5, using a simple example, we present a simplified analysis of the database loads generated by each strategy. In Sec. 18.6, we briefly discuss possible modifications to these protocols that are likely to result in significant benefits by either reducing query and update rate to databases or reducing the signalling traffic or both. Section 18.7 introduces the caching strategy followed by an analysis in the next two sections. This idea attempts to exploit the spatial and temporal locality in calls received by users, similar to the idea of exploiting locality of file access in computer systems [Silberschatz and Peterson, 1988]. A feature of the caching location strategy is that it is useful only for certain classes of PCS users, those meeting certain call and mobility criteria. We encapsulate this notion in the definition of the user's call-to-mobility ratio (CMR), and local CMR (LCMR), in Sec. 18.8. We then use this definition and our PCS network reference architecture to quantify the costs and benefits of caching and the threshold LCMR for which caching is beneficial, thus characterizing the classes of users for which caching should be applied. In Sec. 18.9 we describe two methods for estimating users' LCMR and compare their effectiveness when call and mobility patterns are fairly stable, as well as when they may be variable. In Sec. 18.10, we briefly discuss alternative architectures and implementation issues of the strategy proposed and mention other auxiliary strategies that can be designed. Section 18.11 provides some conclusions and discussion of future work.

The choice of platforms on which to realize the two location strategies (IS-41 and GSM) may vary from one service provider to another. In this paper, we describe a possible realization of these protocols based on the advanced intelligent network (AIN) architecture (see Bellcore, 1991, and Berman and Brewster, 1992), and signalling system 7 (SS7). It is also worthwhile to point out that several strategies have been proposed in the literature for locating users, many of which attempt to reduce the signalling traffic and database loads imposed by the need to locate users in PCS.

18.2 An Overview of PCS

This section explains different aspects of mobility in PCS using an example of two nomadic users who wish to communicate with each other. It also describes a reference model for PCS.

Aspects of Mobility—Example 18.1

PCS can involve two possible types of mobility, terminal mobility and personal mobility, that are explained next.

Terminal Mobility: This type of mobility allows a terminal to be identified by a unique terminal identifier independent of the point of attachment to the network. Calls intended for that terminal can therefore be delivered to that terminal regardless of its network point of attachment. To facilitate terminal mobility, a network must provide several functions, which include those that locate, identify, and validate a terminal and provide services (e.g., deliver calls) to the terminal based on the location information. This implies that the network must store and maintain the location information of the terminal based on a unique identifier assigned to that terminal. An example of a terminal identifier is the IS-41 EIA/TIA cellular industry term mobile identification number (MIN), which is a North American Numbering Plan (NANP) number that is stored in the terminal at the time of manufacture and cannot be changed. A similar notion exists in GSM (see Sec. 18.4).

Personal Mobility: This type of mobility allows a PCS user to make and receive calls independent of both the network point of attachment and a specific PCS terminal. This implies that the services that a user has subscribed to (stored in that user's service profile) are available to the user even if the user moves or changes terminal equipment. Functions needed to provide personal mobility include those that identify (authenticate) the end user and provide services to an end user independent of both the terminal and the location of the user. An example of a functionality needed to provide personal mobility for voice calls is the need to maintain a user's location information based on a unique number, called the universal personal telecommunications (UPT) number, assigned to that user. UPT numbers are also NANP numbers. Another example is one that allows end users to define and manage their service profiles to enable users to tailor services to suit their needs. In Sec. 18.4, we describe how GSM caters to personal mobility via smart cards.

For the purposes of the example that follows, the terminal identifiers (TID) and UPT numbers are NANP numbers, the distinction being TIDs address terminal mobility and UPT numbers address personal mobility. Though we have assigned two different numbers to address personal and terminal mobility concerns, the same effect could be achieved by a single identifier assigned to the terminal that varies depending on the user that is currently utilizing the terminal. For simplicity we assume that two different numbers are assigned.

Figure 18.1 illustrates the terminal and personal mobility aspects of PCS, which will be explained via an example. Let us assume that users Kate and Al have, respectively, subscribed to PCS services from PCS service provider (PSP) A and PSP B. Kate receives the UPT number, say, 500 111 4711, from PSP A. She also owns a PCS terminal with TID 200 777 9760. Al too receives his UPT number 500 222 4712 from PSP B, and he owns a PCS terminal with TID 200 888 5760. Each has been provided a personal identification number (PIN) by their respective PSP when subscription began. We assume that the two PSPs have subscribed to PCS access services from a certain network provider such as, for example, a local exchange carrier (LEC). (Depending on the capabilities of the PSPs, the access services provided may vary. Examples of access services include translation of UPT number to a routing number, terminal and personal registration, and call delivery. Refer to Bellcore, 1993a, for further details). When Kate plugs in her terminal to the network, or when she activates it, the terminal registers itself with the network by providing its TID to the network. The network creates an entry for the terminal in an appropriate database, which, in this example, is entered in the terminal mobility database (TMDB) A. The entry provides a mapping of her terminal's TID, 200 777 9760, to a routing number (RN), RN1. All of these activities happen without Kate being aware of them. After activating her terminal, Kate registers herself at that terminal by entering her UPT number (500 111 4711) to inform the network that all calls to her UPT number are to be delivered to her at the terminal. For security reasons, the network may want to authenticate her and she may be prompted to enter her PIN number into her terminal. (Alternatively, if the terminal is equipped with a smart card reader, she may enter her smart card into the reader. Other techniques, such as, for example, voice recognition, may be employed). Assuming that she is authenticated, Kate has now registered herself. As a result of personal registration by Kate, the network creates an entry for her in the personal mobility database (PMDB) A that maps her UPT number to the TID of the terminal at which she registered. Similarly, when Al activates his terminal and registers

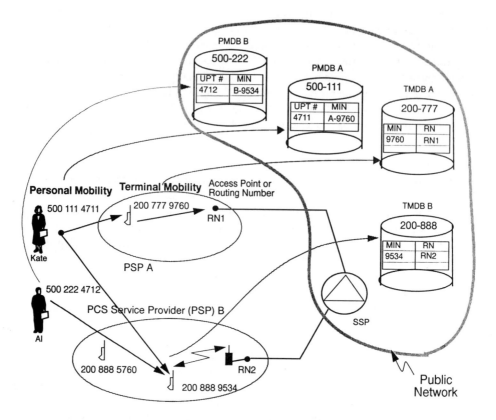

FIGURE 18.1 Illustrating terminal and personal mobility.

himself, appropriate entries are created in TMDB B and PMDB B. Now Al wishes to call Kate and, hence, he dials Kate's UPT number (500 111 4711). The network carries out the following tasks.

1. The switch analyzes the dialed digits and recognizes the need for AIN service, determines that the dialed UPT number needs to be translated to a RN by querying PMDB A and, hence, it queries PMDB A.
2. PMDB A searches its database and determines that the person with UPT number 500 111 4711 is currently registered at terminal with TID 200 777 9760.
3. PMDB A then queries TMDB A for the RN of the terminal with TID 200 777 9760. TMDB A returns the RN (RN1).
4. PMDB A returns the RN (RN1) to the originating switch.
5. The originating switch directs the call to the switch RN1, which then alerts Kate's terminal. The call is completed when Kate picks up her terminal.

Kate may take her terminal wherever she goes and perform registration at her new location. From then on, the network will deliver all calls for her UPT number to her terminal at the new location. In fact, she may actually register on someone else's terminal too. For example, suppose that Kate and Al agree to meet at Al's place to discuss a school project they are working on together. Kate may register herself on Al's terminal (TID 200 888 9534). The network will now modify the entry corresponding to 4711 in PMDB A to point to B 9534. Subsequent calls to Kate will be delivered to Al's terminal.

The scenario given here is used only to illustrate the key aspects of terminal and personal mobility; an actual deployment of these services may be implemented in ways different from those suggested here. We will not discuss personal registration further. The analyses that follow consider only terminal mobility but may easily be modified to include personal mobility.

A Model for PCS

Figure 18.2 illustrates the reference model used for the comparative analysis. The model assumes that the HLR resides in a service control point (SCP) connected to a regional signal transfer point (RSTP). The SCP is a storehouse of the AIN service logic, i.e., functionality used to perform the processing required to provide advanced services, such as speed calling, outgoing call screening, etc., in the AIN architecture (see Bellcore, 1991 and Berman and Brewster, 1992). The RSTP and the local STP (LSTP) are packet switches, connected together by various links such A links or D links, that perform the signalling functions of the SS7 network. Such functions include, for example, global title translation for routing messages between the AIN switching system, which is also referred to as the service switching point (SSP), and SCP and IS-41 messages [EIA/TIA, 1991]. Several SSPs may be connected to an LSTP.

The reference model in Fig. 18.2 introduces several terms which are explained next. We have tried to keep the terms and discussions fairly general. Wherever possible, however, we point to equivalent cellular terms from IS-41 or GSM.

For our purposes, the geographical area served by a PCS system is partitioned into a number of radio port coverage areas (or cells, in cellular terms) each of which is served by a radio port (or, equivalently, base station) that communicates with PCS terminals in that cell. A registration area (also known in the cellular world as location area) is composed of a number of cells. The base stations of all cells in a registration area are connected by wireline links to a mobile switching center (MSC). We assume that each registration area is served by a single VLR. The MSC of a registration area is responsible for maintaining and accessing the VLR and for switching between radio ports. The VLR associated with a registration area is responsible for maintaining a subset of the user information contained in the HLR.

Terminal registration process is initiated by terminals whenever they move into a new registration area. The base stations of a registration area periodically broadcast an identifier associated with that area. The terminals periodically compare an identifier they have stored with the identifier to the registration area being broadcast. If the two identifiers differ, the terminal recognizes that it has

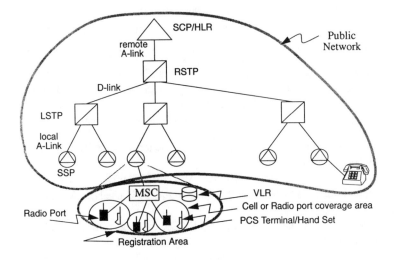

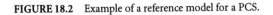

FIGURE 18.2 Example of a reference model for a PCS.

moved from one registration area to another and will, therefore, generate a registration message. It also replaces the previous registration area identifier with that of the new one. Movement of a terminal within the same registration area will not generate registration messages. Registration messages may also be generated when the terminals are switched on. Similarly, messages are generated to deregister them when they are switched off.

PCS services may be provided by different types of commercial service vendors. Bellcore, 1993a describes three different types of PSPs and the different access services that a public network may provide to them. For example, a PSP may have full network capabilities with its own switching, radio management, and radio port capabilities. Certain others may not have switching capabilities, and others may have only radio port capabilities. The model in Fig. 18.2 assumes full PSP capabilities. The analysis in Sec. 18.5 is based on this model and modifications may be necessary for other types of PSPs.

It is also quite possible that one or more registration areas may be served by a single PSP. The PSP may have one or more HLRs for serving its service area. In such a situation users that move within the PSP's serving area may generate traffic to the PSP's HLR (not shown in Fig. 18.2) but not to the network's HLR (shown in Fig. 18.2). In the interest of keeping the discussions simple, we have assumed that there is one-to-one correspondence between SSPs and MSCs and also between MSCs, registration areas, and VLRs. One impact of locating the SSP, MSC, and VLR in separate physical sites connected by SS7 signalling links would be to increase the required signalling message volume on the SS7 network. Our model assumes that the messages between the SSP and the associated MSC and VLR do not add to signalling load on the public network. Other configurations and assumptions could be studied for which the analysis may need to be suitably modified. The underlying analysis techniques will not, however, differ significantly.

18.3 IS-41 Preliminaries

We now describe the message flow for call origination, call delivery, and terminal registration, sometimes called location registration, based on the IS-41 protocol. This protocol is described in detail in EIA/TIA, 1991. Only an outline is provided here.

Terminal/Location Registration

During IS-41 registration, signalling is performed between the following pairs of network elements:

- New serving MSC and the associated database (or VLR)
- New database (VLR) in the visited area and the HLR in the public network
- HLR and the VLR in former visited registration area or the old MSC serving area.

Figure 18.3 shows the signalling message flow diagram for IS-41 registration activity, focusing only on the essential elements of the message flow relating to registration; for details of variations from the basic registration procedure, see Bellcore, 1993a.

The following steps describe the activities that take place during registration.

1. Once a terminal enters a new registration area, the terminal sends a registration request to the MSC of that area.
2. The MSC sends an authentication request (AUTHRQST) message to its VLR to authenticate the terminal, which in turn sends the request to the HLR. The HLR sends its response in the authrqst message.
3. Assuming the terminal is authenticated, the MSC sends a registration notification (REGNOT) message to its VLR.
4. The VLR in turn sends a REGNOT message to the HLR serving the terminal. The HLR updates the location entry corresponding to the terminal to point to the new serving MSC/VLR.

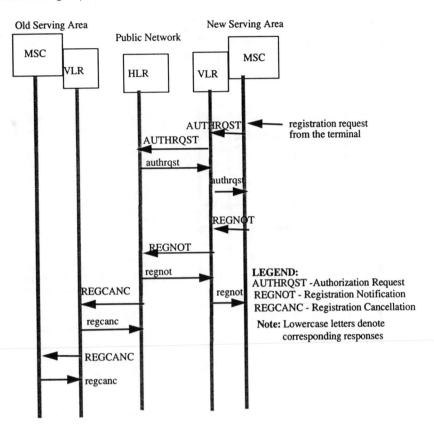

FIGURE 18.3 Signalling flow diagram for registration in IS-41.

The HLR sends a response back to the VLR, which may contain relevant parts of the user's service profile. The VLR stores the service profile in its database and also responds to the serving MSC.

5. If the user/terminal was registered previously in a different registration area, the HLR sends a registration cancellation (REGCANC) message to the previously visited VLR. On receiving this message, the VLR erases all entries for the terminal from the record and sends a REGCANC message to the previously visited MSC, which then erases all entries for the terminal from its memory.

The protocol shows authentication request and registration notification as separate messages. If the two messages can be packaged into one message, then the rate of queries to HLR may be cut in half. This does not necessarily mean that the total number of messages are cut in half.

Call Delivery

The signalling message flow diagram for IS-41 call delivery is shown in Fig. 18.4. The following steps describe the activities that take place during call delivery.

1. A call origination is detected and the number of the called terminal (for example, MIN) is received by the serving MSC. Observe that the call could have originated from within the public network from a wireline phone or from a wireless terminal in an MSC/VLR serving area. (If the call originated within the public network, the AIN SSP analyzes the dialed digits and sends a query to the SCP.)

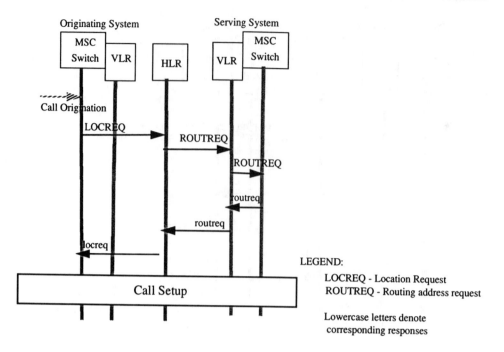

FIGURE 18.4   Signalling flow diagram for call delivery in IS-41.

2. The MSC determines the associated HLR serving the called terminal and sends a location request (LOCREQ) message to the HLR.
3. The HLR determines the serving VLR for that called terminal and sends a routing address request (ROUTEREQ) to the VLR, which forwards it to the MSC currently serving the terminal.
4. Assuming that the terminal is idle, the serving MSC allocates a temporary identifier, called a temporary local directory number (TLDN), to the terminal and returns a response to the HLR containing this information. The HLR forwards this information to the originating SSP/MSC in response to its LOCREQ message.
5. The originating SSP requests call setup to the serving MSC of the called terminal via the SS7 signalling network using the usual call setup protocols.

Similar to the considerations for reducing signalling traffic for location registration, the VLR and HLR functions could be united in a single logical database for a given serving area and collocated; further, the database and switch can be integrated into the same piece of physical equipment or be collocated. In this manner, a significant portion of the messages exchanged between the switch, HLR and VLR as shown in Fig. 18.4 will not contribute to signalling traffic.

18.4 Global System for Mobile Communications

In this section we describe the user location strategy proposed in the European Global System for Mobile Communications (GSM) standard and its offshoot, digital cellular system 1800 (DCS1800). There has recently been increased interest in GSM in North America, since it is possible that early deployment of PCS will be facilitated by using the communication equipment already available from European manufacturers who use the GSM standard. Since the GSM standard is relatively unfamiliar to North American readers, we first give some background and introduce the various abbreviations. The reader will find additional details in Mouley and Pautet, 1992. For an overview on GSM, refer to Lyckscll, 1991.

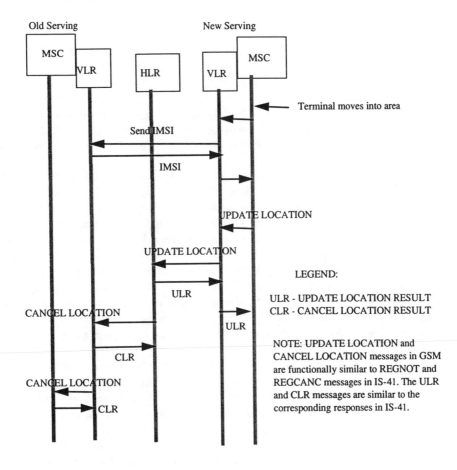

FIGURE 18.5 Flow diagram for registration in GSM.

The abbreviation GSM originally stood for Groupe Special Mobile, a committee created within the pan-European standardization body Conference Europeenne des Posts et Telecommunications (CEPT) in 1982. There were numerous national cellular communication systems and standards in Europe at the time, and the aim of GSM was to specify a uniform standard around the newly reserved 900-MHz frequency band with a bandwidth of twice 25 MHz. The phase 1 specifications of this standard were frozen in 1990. Also in 1990, at the request of the United Kingdom, specification of a version of GSM adapted to the 1800-MHz frequency, with bandwidth of twice 75 MHz, was begun. This variant is referred to as DCS1800; the abbreviation GSM900 is sometimes used to distinguish between the two variations, with the abbreviation GSM being used to encompass both GSM900 and DSC1800. The motivation for DCS1800 is to provide higher capacities in densely populated urban areas, particularly for PCS. The DCS1800 specifications were frozen in 1991, and by 1992 all major GSM900 European operators began operation.

At the end of 1991, activities concerning the post-GSM generation of mobile communications were begun by the standardization committee, using the name universal mobile telecommunications system (UMTS) for this effort. In 1992, the name of the standardization committee was changed from GSM to special mobile group (SMG) to distinguish it from the 900-MHz system itself, and the term GSM was chosen as the commercial trademark of the European 900-MHz system, where GSM now stands for global system for mobile communications.

The GSM standard has now been widely adopted in Europe and is under consideration in several other non-European countries, including the United Arab Emirates, Hong Kong, and New Zealand. In 1992, Australian operators officially adopted GSM.

Architecture

In this section we describe the GSM architecture, focusing on those aspects that differ from the architecture assumed in the IS-41 standard.

A major goal of the GSM standard was to enable users to move across national boundaries and still be able to communicate. It was considered desirable, however, that the operational network within each country be operated independently. Each of the operational networks is called a public land mobile network (PLMN) and its commercial coverage area is confined to the borders of one country (although some radio coverage overlap at national boundaries may occur), and each country may have several competing PLMNs.

A GSM customer subscribes to a single PLMN called the home PLMN, and subscription information includes the services the customer subscribes to. During normal operation, a user may elect to choose other PLMNs as their service becomes available (either as the user moves or as new operators enter the marketplace). The user's terminal [GSM calls the terminal a mobile station (MS)] assists the user in choosing a PLMN in this case, either presenting a list of possible PLMNs to the user using explicit names (e.g., DK Sonofon for the Danish PLMN) or choosing automatically based on a list of preferred PLMNs stored in the terminal's memory. This PLMN selection process allows users to choose between the services and tariffs of several competing PLMNs. Note that the PLMN selection process differs from the cell selection and handoff process that a terminal carries out automatically without any possibility of user intervention, typically based on received radio signal strengths and, thus, requires additional intelligence and functionality in the terminal.

The geographical area covered by a PLMN is partitioned into MSC serving areas, and a registration area is constrained to be a subset of a single MSC serving area. The PLMN operator has complete freedom to allocate cells to registration areas. Each PLMN has, logically speaking, a single HLR, although this may be implemented as several physically distributed databases, as for IS-41. Each MSC also has a VLR, and a VLR may serve one or several MSCs. As for IS-41, it is interesting to consider how the VLR should be viewed in this context. The VLR can be viewed as simply a database off loading the query and signalling load on the HLR and, hence, logically tightly coupled to the HLR or as an ancillary processor to the MSC. This distinction is not academic; in the first view, it would be natural to implement a VLR as serving several MSCs, whereas in the second each VLR would serve one MSC and be physically closely coupled to it. For GSM, the MSC implements most of the signalling protocols, and at present all switch manufacturers implement a combined MSC and VLR, with one VLR per MSC [Mouly and Pautet, 1992].

A GSM mobile station is split in two parts, one containing the hardware and software for the radio interface and the other containing subscribers-specific and location information, called the subscriber identity module (SIM), which can be removed from the terminal and is the size of a credit card or smaller. The SIM is assigned a unique identity within the GSM system, called the international mobile subscriber identity (IMSI), which is used by the user location strategy as described the next subsection. The SIM also stores authentication information, services lists, PLMN selection lists, etc., and can itself be protected by password or PIN.

The SIM can be used to implement a form of large-scale mobility called SIM roaming. The GSM specifications standardize the interface between the SIM and the terminal, so that a user carrying his or her SIM can move between different terminals and use the SIM to personalize the terminal. This capability is particularly useful for users who move between PLMNs which have different radio interfaces. The user can use the appropriate terminal for each PLMN coverage area while obtaining the personalized facilities specified in his or her SIM. Thus, SIMs address personal mobility. In the European context, the usage of two closely related standards at different frequencies, namely, GSM900 and DCS1800, makes this capability an especially important one and facilitates interworking between the two systems.

User Location Strategy

We present a synopsis of the user location strategy in GSM using call flow diagrams similar to those used to describe the strategy in IS-41.

In order to describe the registration procedure, it is first useful to clarify the different identifiers used in this procedure. The SIM of the terminal is assigned a unique identity, called the IMSI, as already mentioned. To increase confidentiality and make more efficient use of the radio bandwidth, however, the IMSI is not normally transmitted over the radio link. Instead, the terminal is assigned a temporary mobile subscriber identity (TMSI) by the VLR when it enters a new registration area. The TMSI is valid only within a given registration area and is shorter than the IMSI. The IMSI and TMSI are identifiers that are internal to the system and assigned to a terminal or SIM and should not be confused with the user's number that would be dialed by a calling party; the latter is a separate number called the mobile subscriber integrated service digital network (ISDN) number (MSISDN), and is similar to the usual telephone number in a fixed network.

We now describe the procedure during registration. The terminal can detect when it has moved into the cell of a new registration area from the system information broadcast by the base station in the new cell. The terminal initiates a registration update request to the new base station; this request includes the identity of the old registration area and the TMSI of the terminal in the old area. The request is forwarded to the MSC, which, in turn, forwards it to the new VLR. Since the new VLR cannot translate the TMSI to the IMSI of the terminal, it sends a request to the old VLR to send the IMSI of the terminal corresponding to that TMSI. In its response, the old VLR also provides the required authentication information. The new VLR then initiates procedures to authenticate the terminal. If the authentication succeeds, the VLR uses the IMSI to determine the address of the terminal's HLR.

The ensuing protocol is then very similar to that in IS-41, except for the following differences. When the new VLR receives the registration affirmation (similar to regnot in IS-41) from the HLR, it assigns a new TMSI to the terminal for the new registration area. The HLR also provides the new VLR with all relevant subscriber profile information required for call handling (e.g., call screening lists, etc.) as part of the affirmation message. Thus, in contrast with IS-41, authentication and subscriber profile information are obtained from both the HLR and old VLR and not just the HLR.

The procedure for delivering calls to mobile users in GSM is very similar to that in IS-41. The sequence of messages between the caller and called party's MSC/VLRs and the HLR is identical to that shown in the call flow diagrams for IS-41, although the names, contents and lengths of messages may be different and, hence, the details are left out. The interested reader is referred to Mouly and Pautet, 1992, or Lycksell, 1991, for further details.

18.5 Analysis of Database Traffic Rate for IS-41 and GSM

In the two subsections that follow, we state the common set of assumptions on which we base our comparison of the two strategies.

The Mobility Model for PCS Users

In the analysis that follows in the IS-41 analysis subsection, we assume a simple mobility model for the PCS users. The model, which is described in Thomas, Gilbert, and Mazziotto, 1988, assumes that PCS users carrying terminals are moving at an average velocity of v and their direction of movement is uniformly distributed over $[0, 2\pi]$. Assuming that the PCS users are uniformly populated with a density of ρ and the registration area boundary is of length L, it has been shown that the rate of registration area crossing R is given by

$$R = \frac{\rho v L}{\pi} \tag{18.1}$$

Using Eq. (18.1), we can calculate the signalling traffic due to registration, call origination, and delivery. We new need a set of assumptions so that we may proceed to derive the traffic rate to the databases using the model in Fig. 18.2.

Additional Assumptions

The following assumptions are made in performing the analysis.

- 128 total registration areas
- Square registration area size: $(7.575 \text{ km})^2 = 57.5 \text{ km}^2$, with border length $L = 30.3$ km
- Average call origination rate = average call termination (delivery) rate = 1.4/h/terminal
- Mean density of mobile terminals = $\rho = 390/\text{km}^2$
- Total number of mobile terminals = $128 \times 57.4 \times 390 = 2.87 \times 10^6$
- Average call origination rate = average call termination (delivery) rate = 1.4/h/terminal
- Average speed of a mobile, $v = 5.6$ km/h
- Fluid flow mobility model

The assumptions regarding the total number of terminals may also be obtained by assuming that a certain public network provider serves 19.15×10^6 users and that 15% (or 2.87×10^6) of the users also subscribe to PCS services from various PSPs.

Note that we have adopted a simplified model that ignores situations where PCS users may turn their handsets on and off that will generate additional registration and deregistration traffic. The model also ignores wireline registrations. These activities will increase the total number of queries and updates to HLR and VLRs.

Analysis of IS-41

Using Eq. (18.1) and the parameter values assumed in the preceding subsection, we can compute the traffic due to registration. The registration traffic is generated by mobile terminals moving into a new registration area, and this must equal the mobile terminals moving out of the registration area, which per second is

$$R_{\text{reg, VLR}} = \frac{390 \times 30.3 \times 5.6}{3600\pi} = 5.85$$

This must also be equal to the number of deregistrations (registration cancellations),

$$R_{\text{dereg, VLR}} = 5.85$$

The total number of registration messages per second arriving at the HLR will be

$$R_{\text{reg, HLR}} = R_{\text{reg, VLR}} \times \text{total No. of registration areas} = 749$$

The HLR should, therefore, be able to handle, roughly, 750 updates per second. We observe from Fig. 18.3 that authenticating terminals generate as many queries to VLR and HLR as the respective number of updates generated due to registration notification messages.

The number of queries that the HLR must handle during call origination and delivery can be similarly calculated. Queries to HLR are generated when a call is made to a PCS user. The SSP that receives the request for a call, generates a location request (LOCREQ) query to the SCP controlling the HLR. The rate per second of such queries must be equal to the rate of calls made to PCS users.

TABLE 18.1 IS-41 Query and Update Rates to HLR and VLR

Activity	HLR Updates/s	VLR Updates/s	HLR Queries/s	VLR queries/s
Mobility-related activities at registration	749	5.85	749	5.85
Mobility-related activities at deregistration		5.85		
Call origination				8.7
Call delivery			1116	8.7
Total (per RA)	5.85	11.7	14.57	23.25
Total (Network)	749	1497.6	1865	2976

This is calculated as

$$R_{\text{CallDeliv, HLR}} = \text{call rate per user} \times \text{total number of users}$$
$$= \frac{1.4 \times 2.87 \times 10^5}{3600}$$
$$= 1116$$

For calls originated from a mobile terminal by PCS users, the switch authenticates the terminal by querying the VLR. The rate per second of such queries is determined by the rate of calls originating in an SSP serving area, which is also a registration area (RA). This is given by

$$R_{\text{CallOrig, VLR}} = \frac{1116}{128} = 8.7$$

This is also the number of queries per second needed to authenticate terminals of PCS users to which calls are delivered:

$$R_{\text{CallDeliv, VLR}} = 8.7$$

Table 18.1 summarizes the calculations.

Analysis of GSM

Calculations for query and update rates for GSM may be performed in the same manner as for IS-41, and they are summarized in Table 18.2. The difference between this table and Table 18.1 is that in GSM the new serving VLR does not query the HLR separately in order to authenticate the terminal during registration and, hence, there are no HLR queries during registration. Instead, the entry (749 queries) under HLR queries in Table 18.1, corresponding to mobility-related authentication activity at registration, gets equally divided between the 128 VLRs. Observe that with either protocol the total database traffic rates are conserved, where the total database traffic for the entire network is given by the sum of all of the entries in the last row total (Network), i.e.,

$$\text{HLR updates} + \text{VLR updates} + \text{HLR queries} + \text{VLR queries}$$

From Tables 18.1 and 18.2 we see that this quantity equals 7087.

The conclusion is independent of any variations we may provide to the assumptions in earlier in the section . For example, if the PCS penetration (the percentage of the total users subscribing to PCS services) were to increase from 15 to 30%, all of the entries in the two tables will double and, hence, the total database traffic generated by the two protocols will still be equal.

TABLE 18.2   GSM Query and Update Rates to HLR and VLR

Activity	HLR Updates/s	VLR Updates/s	HLR Queries/s	VLR Queries/s
Mobility-related activities at registration	749	5.85		11.7
Mobility-related activities at deregistration		5.85		
Call origination				8.7
Call delivery			1116	8.7
Total (per VLR)	749	11.7	1116	29.1
Total (Network)	749	1497.6	1116	3724.8

18.6 Reducing Signalling During Call Delivery

In the preceding section, we provided a simplified analysis of some scenarios associated with user location strategies and the associated database queries and updates required. Previous studies [Meier-Hellstern and Alonso, 1992; Lo, Wolff, and Bernhardt, 1992] indicate that the signalling traffic and database queries associated with PCS due to user mobility are likely to grow to levels well in excess of that associated with a conventional call. It is, therefore, desirable to study modifications to the two protocols that would result in reduced signalling and database traffic. We now provide some suggestions.

For both GSM and IS-41, delivery of calls to a mobile user involves four messages: from the caller's VLR to the called party's HLR, from the HLR to the called party's VLR, from the called party's VLR to the HLR, and from the HLR to the caller's VLR. The last two of these messages involve the HLR, whose role is to simply relay the routing information provided by the called party's VLR to the caller's VLR. An obvious modification to the protocol would be to have the called VLR directly send the routing information to the calling VLR. This would reduce the total load on the HLR and on signalling network links substantially. Such a modification to the protocol may not be easy, of course, due to administrative, billing, legal, or security concerns. Besides, this would violate the query/response model adopted in IS-41, requiring further analysis.

A related question which arises is whether the routing information obtained from the called party's VLR could instead be stored in the HLR. This routing information could be provided to the HLR, for example, whenever a terminal registers in a new registration area. If this were possible, two of the four messages involved in call delivery could be eliminated. This point was discussed at length by the GSM standards body, and the present strategy was arrived at. The reason for this decision was to reduce the number of temporary routing numbers allocated by VLRs to terminals in their registration area. If a temporary routing number (TLDN in IS-41 or MSRN in GSM) is allocated to a terminal for the whole duration of its stay in a registration area, the quantity of numbers required is much greater than if a number is assigned on a per-call basis. Other strategies may be employed to reduce signalling and database traffic via intelligent paging or by storing user's mobility behavior in user profiles (see, for example, Tabbane, 1993). A discussion of these techniques is beyond the scope of the paper.

18.7 Per-User Location Caching

The basic idea behind per-user location caching is that the volume of SS7 message traffic and database accesses required in locating a called subscriber can be reduced by maintaining local storage, or cache, of user location information at a switch. At any switch, location caching for a given user should be employed only if a large number of calls originate for that user from that switch, relative to the user's mobility. Note that the cached information is kept at the switch from which calls originate, which may or may not be the switch where the user is currently registered.

Location caching involves the storage of location pointers at the originating switch; these point to the VLR (and the associated switch) where the user is currently registered. We refer to the procedure of locating a PCS user a *FIND* operation, borrowing the terminology from Awerbuch and Peleg, 1991. We define a basic *FIND*, or *BasicFIND*( ), as one where the following sequence of steps takes place.

1. The incoming call to a PCS user is directed to the nearest switch.
2. Assuming that the called party is not located within the immediate RA, the switch queries the HLR for routing information.
3. The HLR contains a pointer to the VLR in whose associated RA the subscriber is currently situated and launches a query to that VLR.
4. The VLR, in turn, queries the MSC to determine whether the user terminal is capable of receiving the call (i.e., is idle) and, if so, the MSC returns a routable address (TLDN in IS-41) to the VLR.
5. The VLR relays the routing address back to the originating switch via the HLR.

At this point, the originating switch can route the call to the destination switch. Alternately, *BasicFIND*( ) can be described by pseudocode as follows. (We observe that a more formal method of specifying PCS protocols may be desirable).

> *BasicFIND*( ){
>
>   Call to PCS user is detected at local switch;
>   *if* called party is in same RA *then* return;
>   Switch queries called party's HLR;
>   Called party's HLR queries called party's current VLR, *V* ;
>   *V* returns called party's location to HLR;
>   HLR returns location to calling switch;
> }

In the *FIND* procedure involving the use of location caching, or *CacheFIND*( ), each switch contains a local memory (cache) that stores location information for subscribers. When the switch receives a call origination (from either a wire-line or wireless caller) directed to a PCS subscriber, it first checks its cache to see if location information for the called party is maintained. If so, a query is launched to the pointed VLR; if not, *BasicFIND*( ), as just described, is followed. If a cache entry exists and the pointed VLR is queried, two situations are possible. If the user is still registered at the RA of the pointed VLR (i.e., we have a *cache hit*), the pointed VLR returns the user's routing address. Otherwise, the pointed VLR returns a *cache miss*.

> *CacheFIND*( ){
>
>   Call to PCS user is detected at local switch;
>   *if* called is in same RA *then* return;
>   *if* there is no cache entry for called user
>   *then* invoke *BasicFIND*( ) and return;
>   Switch queries the VLR, *V* , specified in the cache entry;
>   *if* called is at *V* , *then*
>   *V* returns called party's location to calling switch;
>   *else* {
>   *V* returns "miss" to calling switch;
>   Calling switch invokes *BasicFIND*( );
>   }
> }

When a cache hit occurs we save one query to the HLR [a VLR query is involved in both *CacheFIND*( ) and *BasicFIND*( )], and we also save traffic along some of the signalling links; instead of four message transmissions, as in *BasicFIND*( ), only two are needed. In steady-state operation, the cached pointer for any given user is updated only upon a miss.

Note that the *BasicFIND*( ) procedure differs from that specified for roaming subscribers in the IS-41 standard EIA/TIA, 1991. In the IS-41 standard, the second line in the *BasicFIND*( ) procedure is omitted, i.e., every call results in a query of the called user's HLR. Thus, in fact, the procedure specified in the standard will result in an even higher network load than the *BasicFIND*( ) procedure specified here. To make a fair assessment of the benefits of *CacheFIND*( ), however, we have compared it against *BasicFIND*( ). Thus, the benefits of *CacheFIND*( ) investigated here depend specifically on the use of caching and not simply on the availability of user location information at the local VLR.

18.8 Caching Threshold Analysis

In this section we investigate the classes of users for which the caching strategy yields net reductions in signalling traffic and database loads. We characterize classes of users by their CMR. The CMR of a user is the average number of calls to a user per unit time, divided by the average number of times the user changes registration areas per unit time. We also define a LCMR, which is the average number of calls to a user from a given originating switch per unit time, divided by the average number of times the user changes registration areas per unit time.

For each user, the amount of savings due to caching is a function of the probability that the cached pointer correctly points to the user's location and increases with the user's LCMR. In this section we quantify the minimum value of LCMR for caching to be worthwhile. This caching threshold is parameterized with respect to costs of traversing signalling network elements and network databases and can be used as a guide to select the subset of users to whom caching should be applied. The analysis in this section shows that estimating user's LCMRs, preferably dynamically, is very important in order to apply the caching strategy. The next section will discuss methods for obtaining this estimate.

From the pseudocode for *BasicFIND*( ), the signalling network cost incurred in locating a PCS user in the event of an incoming call is the sum of the cost of querying the HLR (and receiving the response), and the cost of querying the VLR which the HLR points to (and receiving the response). Let

$\alpha =$ cost of querying the HLR and receiving a response

$\beta =$ cost of querying the pointed VLR and receiving a response

Then, the cost of *BasicFIND*( ) operation is

$$C_B = \alpha + \beta \tag{18.2}$$

To quantify this further, assume costs for traversing various network elements as follows.

$A_l =$ cost of transmitting a location request or response message on A link between SSP and LSTP

$D =$ cost of transmitting a location request on response message or D link

$A_r =$ cost of transmitting a location request or response message on A link between RSTP and SCP

$L =$ cost of processing and routing a location request or response message by LSTP

$R =$ cost of processing and routing a location request or response message by RSTP

$H_Q =$ cost of a query to the HLR to obtain the current VLR location

$V_Q =$ cost of a query to the VLR to obtain the routing address

Then, using the PCS reference network architecture (Fig. 18.2),

$$\alpha = 2(A_l + D + A_r + L + R) + H_Q \tag{18.3}$$

$$\beta = 2(A_l + D + A_r + L + R) + V_Q \tag{18.4}$$

From Eqs. (18.2)–(18.4)

$$C_B = 4(A_l + D + A_r + L + R) + H_Q + V_Q \tag{18.5}$$

We now calculate the cost of *CacheFIND( )*. We define the *hit ratio* as the relative frequency with which the cached pointer correctly points to the user's location when it is consulted. Let

$p =$ cache hit ratio
$C_H =$ cost of the *CacheFIND( )* procedure when there is a hit
$C_M =$ cost of the *CacheFIND( )* procedure when there is a miss

Then the cost of *CacheFIND( )* is

$$C_C = p C_H + (1 - p)C_M \tag{18.6}$$

For *CacheFIND( )*, the signalling network costs incurred in locating a user in the event of an incoming call depend on the hit ratio as well as the cost of querying the VLR, which is stored in the cache; this VLR query may or may not involve traversing the RSTP. In the following, we say a VLR is a *local* VLR if it is served by the same LSTP as the originating switch, and a *remote* VLR otherwise. Let

$q =$ Prob (VLR in originating switch's cache is a local VLR)
$\delta =$ cost of querying a local VLR
$\epsilon =$ cost of querying a remote VLR
$\eta =$ cost of updating the cache upon a miss

Then,

$$\delta = 4A_l + 2L + V_Q \tag{18.7}$$

$$\epsilon = 4(A_l + D + L) + 2R + V_Q \tag{18.8}$$

$$C_H = q\delta + (1 - q)\epsilon \tag{18.9}$$

Since updating the cache involves an operation to a fast local memory rather than a database operation, we shall assume in the following that $\eta = 0$. Then,

$$C_M = C_H + C_B = q\delta + (1 - q)\epsilon + \alpha + \beta \tag{18.10}$$

From Eqs. (18.6), (18.9) and (18.10) we have

$$C_C = \alpha + \beta + \epsilon - p(\alpha + \beta) + q(\delta - \epsilon) \tag{18.11}$$

For net cost savings we require $C_C < C_B$, or that the hit ratio exceeds a *hit ratio threshold* p_T, derived using Eqs. (18.6), (18.9), and (18.2),

$$p > p_T = \frac{C_H}{C_B} = \frac{\epsilon + q(\delta - \epsilon)}{\alpha + \beta} \tag{18.12}$$

$$= \frac{4A_l + 4D + 4L + 2R + V_Q - q(4D + 2L + 2R)}{4A_l + 4D + 4A_r + 4L + 4R + H_Q + V_Q} \tag{18.13}$$

Equation (18.13) specifies the hit ratio threshold for a user, evaluated at a given switch, for which local maintenance of a cached location entry produces cost savings. As pointed out earlier, a given user's hit ratio may be location dependent, since the rates of calls destined for that user may vary widely across switches.

The hit ratio threshold in Eq. (18.13) is comprised of heterogeneous cost terms, i.e., transmission link utilization, packet switch processing, and database access costs. Therefore, numerical evaluation of the hit ratio threshold requires either detailed knowledge of these individual quantities or some form of simplifying assumptions. Based on the latter approach, the following two possible methods of evaluation may be employed.

1. Assume one or more cost terms dominate, and simplify Eq. (18.13) by setting the remaining terms to zero.
2. Establish a common unit of measure for all cost terms, for example, *time delay.* In this case, A_l, A_r, and D may represent transmission delays of fixed transmission speed (e.g., 56 kb/s) signalling links, L and R may constitute the sum of queueing and service delays of packet switches (i.e., STPs), and H_Q and V_Q the transaction delays for database queries.

In this section we adopt the first method and evaluate Eq. (18.13) assuming a single term dominates. (In Sec. 18.9 we present results using the second method). Table 18.3 shows the hit ratio threshold required to obtain net cost savings, for each case in which one of the cost terms is dominant.

In Table 18.3 we see that if the cost of querying a VLR or of traversing a local A link is the dominant cost, caching for users who may move is never worthwhile, regardless of users' call reception and mobility patterns. This is because the caching strategy essentially distributes the functionality of the HLR to the VLRs. Thus, the load on the VLR and the local A link is always increased, since any move by a user results in a cache miss. On the other hand, for a fixed user (or telephone), caching is always worthwhile. We also observe that if the remote A links or HLR querying are the bottlenecks, caching is worthwhile even for users with very low hit ratios.

As a simple average-case calculation, consider the net network benefit of caching when HLR access and update is the performance bottleneck. Consider a scenario where $u = 50\%$ of PCS users receive $c = 80\%$ of their calls from $s = 5$ RAs where their hit ratio $p > 0$, and $s' = 4$ of the SSPs at those RAs contain sufficiently large caches. Assume that caching is applied only to this subset of users and to no other users. Suppose that the average hit ratio for these users is $p = 80\%$, so that 80% of the HLR accesses for calls to these users from these RA are avoided. Then the net saving in the accesses to the system's HLR is $H = (u\,c\,s'\,p)/s = 25\%$.

We discuss other quantities in Table 18.3 next. It is first useful to relate the cache hit ratio to users' calling and mobility patterns directly via the LCMR. Doing so requires making assumptions about the distribution of the user's calls and moves. We consider the steady state where the incoming call stream from an SSP to a user is a Poisson process with arrival rate λ, and the time that the user

TABLE 18.3 Minimum Hit Ratios and LCMRs for Various Individual Dominant Signalling Network Cost Terms

Dominant Cost Term	Hit ratio Threshold, p_T	LCMR Threshold, $LCMR_T$	LCMR Threshold $(q = 0.043)$	LCMR Threshold $(q = 0.25)$
A_l	1	∞	∞	∞
A_r	0	0	0	0
D	$1 - q$	$1/q - 1$	22	3
L	$1 - q/2$	$2/q - 1$	45	7
R	$1 - q/2$	$2/q - 1$	45	7
H_Q	0	0	0	0
V_Q	1	∞	∞	∞

resides in an RA has a general distribution $F(t)$ with mean $1/\mu$. Thus,

$$LCMR = \frac{\lambda}{\mu} \qquad (18.14)$$

Let t be the time interval between two consecutive calls from the SSP to the user and t_1 be the time interval between the first call and the time when the user moves to a new RA. From the random observer property of the arrival call stream [Feller, 1966], the hit ratio is

$$p = Pr[t < t_1] = \int_{t=0}^{\infty} \lambda e^{-\lambda t} \int_{t_1=t}^{\infty} \mu[1 - F(t_1)] \, dt_1 \, dt$$

If $F(t)$ is an exponential distribution, then

$$p = \frac{\lambda}{\lambda + \mu} \qquad (18.15)$$

and we can derive the *LCMR threshold*, the minimum LCMR required for caching to be beneficial assuming incoming calls are a Poisson process and intermove times are exponentially distributed,

$$LCMR_T = \frac{p_T}{1 - p_T} \qquad (18.16)$$

Equation (18.16) is used to derive LCMR thresholds assuming various dominant costs terms, as shown in Table 18.3.

Several values for $LCMR_T$ in Table 18.3 involve the term q, i.e., the probability that the pointed VLR is a local VLR. These values may be numerically evaluated by simplifying assumptions. For example, assume that all of the SSPs in the network are uniformly distributed amongst l LSTPs. Also, assume that all of the PCS subscribers are uniformly distributed in location across all SSPs and that each subscriber exhibits the same incoming call rate at every SSP. Under those conditions, q is simply $1/l$. Consider the case of the public switched telephone network. Given that there are a total of 160 local access transport area (LATA) across the 7 Regional Bell Operating Company (RBOC) regions [Bellcore, 1992c], the average number of LATAs, or l, is 160/7 or 23. Table 18.3 shows the results with $q = 1/l$ in this case.

We observe that the assumption that all users receive calls uniformly from all switches in the network is extremely conservative. In practice, we expect that user call reception patterns would display significantly more locality, so that q would be larger and the LCMR thresholds required to make caching worthwhile would be smaller. It is also worthwhile to consider the case of a RBOC region with PCS deployed in a few LATA only, a likely initial scenario, say, 4 LATAs. In either case the value of q would be significantly higher; Table 18.3 shows the LCMR threshold when $q = 0.25$.

It is possible to quantify the net costs and benefits of caching in terms of signalling network impacts in this way and to determine the hit ratio and LCMR threshold above which users should have the caching strategy applied. Applying caching to users whose hit ratio and LCMR is below this threshold results in net increases in network impacts. It is, thus, important to estimate users' LCMRs accurately. The next section discusses how to do so.

18.9 Techniques for Estimating Users' LCMR

Here we sketch some methods of estimating users' LCMR. A simple and attractive policy is to not estimate these quantities on a per-user basis at all. For instance, if the average LCMR over all users in a PCS system is high enough (and from Table 18.3, it need not be high depending on which network elements are the dominant costs), then caching could be used at every SSP to yield net system-wide benefits. Alternatively, if it is known that at any given SSP the average LCMR over all users is high enough, a cache can be installed at that SSP. Other variations can be designed.

One possibility for deciding about caching on a per-user basis is to maintain information about a user's calling and mobility pattern at the HLR and to download it periodically to selected SSPs during off-peak hours. It is easy to envision numerous variations on this idea.

In this section we investigate two possible techniques for estimating LCMR on a per-user basis when caching is to be deployed. The first algorithm, called the *running average* algorithm, simply maintains a running average of the hit ratio for each user. The second algorithm, called the *reset-K* algorithm, attempts to obtain a measure of the hit ratio over the recent history of the user's movements. We describe the two algorithms next and evaluate their effectiveness using a stochastic analysis taking into account user calling and mobility patterns.

The Running Average Algorithm

The running average algorithm maintains, for every user that has a cache entry, the running average of the hit ratio. A running count is kept of the number of calls to a given user, and, regardless of the *FIND* procedure used to locate the user, a running count of the number of times that the user was at the same location for any two consecutive calls; the ratio of these numbers provides the measured running average of the hit ratio. We denote the measured running average of the hit ratio by p_M; in steady state, we expect that $p_M = p$. The user's previous location as stored in the cache entry is used only if the running average of the hit ratio p_M is greater than the cache hit threshold p_T. Recall that the cache scheme outperforms the basic scheme if $p > p_T = C_H/C_B$. Thus, in steady state, the running average algorithm will outperform the basic scheme when $p_M > p_T$.

We consider, as before, the steady state where the incoming call stream from an SSP to a user is a Poisson process with arrival rate λ, and the time that the user resides in an RA has an exponential distribution with mean $1/\mu$. Thus $LCMR = \lambda/\mu$ [Eq. (18.14)] and the location tracking cost at steady state is

$$C_C = \begin{cases} p_M C_H + (1 - p_M)C_B, & p_M > p_T \\ C_B, & \text{otherwise} \end{cases} \tag{18.17}$$

Figure 18.6 plots the cost ratio C_C/C_B from Eq. (18.17) against *LCMR*. (This corresponds to assigning uniform units to all cost terms in Eq. (18.13), i.e., the second evaluation method as discussed in Sec. 18.8. Thus, the ratio C_C/C_B may represent the percentage reduction in user location time with the caching strategy compared to the basic strategy.) The figure indicates that in the steady state, the caching strategy with the running average algorithm for estimating LCMR can significantly outperform the basic scheme if *LCMR* is sufficiently large. For instance with $LCMR \sim 5$, caching can lead to cost savings of 20–60% over the basic strategy.

Equation (18.17) (cf., solid curves in Fig. 18.6) is validated against a simple Monte Carlo simulation (cf., dashed curves in Fig. 18.6). In the simulation, the confidence interval for the 95% confidence level of the output measure C_C/C_B is within 3% of the mean value. This simulation model will later be used to study the running average algorithm when the mean of the movement distribution changes from time to time [which cannot be modeled by using Eq. (18.17)].

One problem with the running average algorithm is that the parameter p is measured from the entire past history of the user's movement, and the algorithm may not be sufficiently dynamic to adequately reflect the recent history of the user's mobility patterns.

The Reset-*K* Algorithm

We may modify the running average algorithm such that p is measured from the recent history. Define every K incoming calls as a *cycle*. The modified algorithm, which is referred to as the reset-

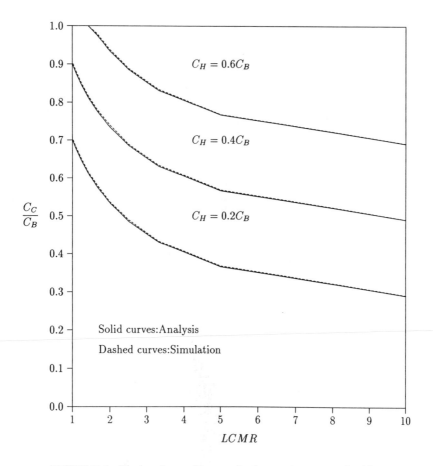

FIGURE 18.6 The location tracking cost for the running average algorithm.

K algorithm, counts the number of cache hits n in a cycle. If the measured hit ratio for a user, $p_M = n/K \geq p_T$, then the cache is enabled for that user, and the cached information is always used to locate the user in the next cycle. Otherwise, the cache is disabled for that user, and the basic scheme is used. At the beginning of a cycle, the cache hit count is reset, and a new p_M value is measured during the cycle.

To study the performance of the reset-K algorithm, we model the number of cache misses in a cycle by a Markov process. Assume as before that the call arrivals are a Poisson process with arrival rate λ and the time period the user resides in an RA has an exponential distribution withe mean $1/\mu$. A pair (i, j), where $i > j$, represents the state that there are j cache misses before the first i incoming phone calls in a cycle. A pair $(i, j)^*$, where $i \geq j \geq 1$, represents the state that there are $j - 1$ cache misses before the first i incoming phone calls in a cycle, and the user moves between the ith and the $i + 1$ phone calls. The difference between (i, j) and $(i, j)^*$ is that if the Markov process is in the state (i, j) and the user moves, then the process moves into the state $(i, j + 1)^*$. On the other hand, if the process is in state $(i, j)^*$ when the user moves, the process remains in $(i, j)^*$ because at most one cache miss occurs between two consecutive phone calls.

Figure 18.7(a) illustrates the transitions for state $(i, 0)$ where $2 < i < K + 1$. The Markov process moves from $(i - 1, 0)$ to $(i, 0)$ if a phone call arrives before the user moves out. The rate is λ. The process moves from $(i, 0)$ to $(i, 1)^*$ if the user moves to another RA before the $i + 1$ call arrival. Let $\pi(i, j)$ denote the probability of the process being in state (i, j). Then the transition

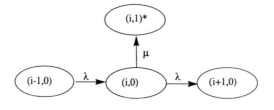

(a) Transitions for state (i,0) (2 < i < K+1)

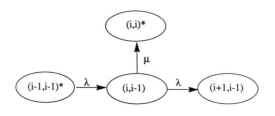

(b) Transitions for state (i,i-1)(1 < i < K+1)

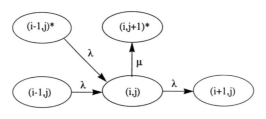

(c) Transitions for state (i,j) (2< i < K+1, 0 < j < i-1)

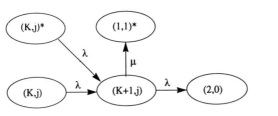

(d) Transitions for state (K+1,j) (0 < j < K+1)

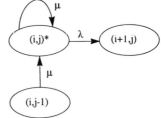

(e) Transitions for state (i,j)* (0 < j ≤ i, 1 < i < K

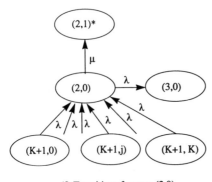

(f) Transitions for state (2,0)

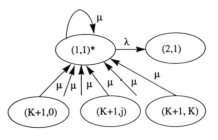

(g) Transitions for state (1,1)*

FIGURE 18.7 State transitions.

equation is

$$\pi(i, 0) = \frac{\lambda}{\lambda + \mu} \pi(i - 1, 0), \qquad 2 < i < K + 1 \qquad (18.18)$$

Figure 18.7(b) illustrates the transitions for state $(i, i - 1)$ where $1 < i < K + 1$. The only transition into the state $(i, i - 1)$ is from $(i - 1, i - 1)^*$, which means that the user always moves to another RA after a phone call. [Note that there can be no state $(i - 1, i - 1)$ by definition and, hence, no transition from such a state.] The transition rate is λ. The process moves from $(i, i - 1)$ to $(i, i)^*$ with rate μ, and moves to $(i + 1, i - 1)$ with rate λ. Let $\pi^*(i, j)$ denote the probability of the process being in state $(i, j)^*$. Then the transition equation is

$$\pi(i, i - 1) = \frac{\lambda}{\lambda + \mu} \pi^*(i - 1, i - 1), \qquad 1 < i < K + 1 \qquad (18.19)$$

Figure 18.7(c) illustrates the transitions for state (i, j) where $2 < i < K + 1, 0 < j < i - 1$. The process may move into state (i, j) from two states $(i - 1, j)$ and $(i - 1, j)^*$ with rate λ, respectively. The process moves from (i, j) to $(i, j + 1)^*$ or $(i + 1, j)$ with rates μ and λ, respectively. The transition equation is

$$\pi(i, j) = \frac{\lambda}{\lambda + \mu} [\pi(i - 1, j) + \pi^*(i - 1, j)],$$

$$2 < i < K + 1, \qquad 0 < j < i - 1 \qquad (18.20)$$

Figure 18.7(d) illustrates the transitions for state $(K + 1, j)$ where $0 < j < K + 1$. Note that if a phone call arrives when the process is in (K, j) or $(K, j)^*$, the system enters a new cycle (with rate λ), and we could represent the new state as $(1, 0)$. In our model, we introduce the state $(K + 1, j)$ instead of $(1, 0)$, where

$$\sum_{0 \leq j \leq K} \pi(K + 1, j) = \pi(1, 0)$$

so that the hit ratio, and thus the location tracking cost, can be derived [see Eq. (18.25)]. The process moves from $(K + 1, j)$ [i.e., $(1, 0)$] to $(1, 1)^*$ with rate μ if the user moves before the next call arrives. Otherwise, the process moves to $(2, 0)$ with rate λ. The transition equation is

$$\pi(K + 1, j) = \frac{\lambda}{\lambda + \mu} [\pi(K, j) + \pi^*(K, j)], \qquad 0 < j < K + 1 \qquad (18.21)$$

For $j = 0$, the transition from $(K, j)^*$ to $(K + 1, 0)$ should be removed in Fig. 18.7(d) because the state $(K, 0)^*$ does not exist. The transition equation for $(K + 1, 0)$ is given in Eq. (18.18). Figure 18.7(e) illustrates the transitions for state $(i, j)^*$ where $0 < j < i, 1 < i < K + 1$. The process can only move to $(i, j)^*$ from $(i, j - 1)$ (with rate μ). From the definition of $(i, j)^*$, if the user moves when the process is in $(i, j)^*$, the process remains in $(i, j)^*$ (with rate μ). Otherwise, the process moves to $(i + 1, j)$ with rate λ. The transition equation is

$$\pi^*(i, j) = \frac{\mu}{\lambda} \pi(i, j - 1), \qquad 0 < j \leq i, \qquad 1 < i < K + 1, \qquad i \geq 2 \quad (18.22)$$

The transitions for $(2, 0)$ are similar to the transitions for $(i, 0)$ except that the transition from $(1, 0)$

is replaced by $(K+1, 0), \ldots, (K+1, K)$ [cf., Fig. 18.7(f)]. The transition equation is

$$\pi(2, 0) = \frac{\lambda}{\lambda + \mu}\left[\sum_{0 \leq j \leq K} \pi(K+1, j)\right] \tag{18.23}$$

Finally, the transitions for $(1, 1)^*$ is similar to the transitions for $(i, j)^*$ except that the transition from $(1, 0)$ is replaced by $(K+1, 0), \ldots, (K+1, K)$ [cf., Fig. 18.7(g)]. The transition equation is

$$\pi^*(1, 1) = \frac{\mu}{\lambda}\left[\sum_{0 \leq j \leq K} \pi(K+1, j)\right] \tag{18.24}$$

Suppose that at the beginning of a cycle, the process is in state $(K+1, j)$, then it implies that there are j cache misses in the previous cycle. The cache is enabled if and only if

$$p_M \geq p_T = \frac{C_H}{C_B} \Rightarrow 1 - \frac{j}{K} \geq \frac{C_H}{C_B} \Rightarrow 0 \leq j \leq \left\lceil K\left(1 - \frac{C_H}{C_B}\right)\right\rceil$$

Thus, the probability that the measured hit ratio $p_M < p_T$ in the previous cycle is

$$Pr[p_M < p_T] = \frac{\displaystyle\sum_{\lceil k[1-(C_H/C_B)]\rceil < j \leq K} \pi(K+1, j)}{\displaystyle\sum_{0 \leq j \leq K} \pi(K+1, j)}$$

and the location tracking cost for the reset-K algorithm is

$$C_C = C_B Pr[p_M < p_T] + (1 - Pr[p_M < p_T])$$
$$\times \left\{\sum_{0 \leq j \leq K} \left(\frac{(K-j)C_H}{K} + \frac{j(C_H + C_B)}{K}\right)\left[\frac{\pi(K+1, j)}{\displaystyle\sum_{0 \leq i \leq K} \pi(K+1, i)}\right]\right\} \tag{18.25}$$

The first term Eq. (18.25) represents the cost incurred when caching is disabled because the hit ratio threshold exceeds the hit ratio measured in the previous cycle. The second term is the cost when the cache is enabled and consists of two parts, corresponding to calls during which hits occur and calls during which misses occur. The ratio in square brackets is the conditional probability of being in state $\pi(K+1, j)$ during the current cycle.

The numerical computation of $\pi(K+1, j)$ can be done as follows. First, compute $a_{i,j}$ and $b_{i,j}$ where $\pi(i, j) = a_{i,j}\pi^*(1, 1)$ and $\pi^*(i, j) = b_{i,j}\pi^*(1, 1)$. Note that $a_{i,j} = 0(b_{i,j} = 0)$ if $\pi(i, j)[\pi^*(i, j)]$ is not defined in Eqs. (18.18)–(18.24). Since

$$\sum_{i,j}[\pi(i, j) + \pi^*(i, j)] = 1$$

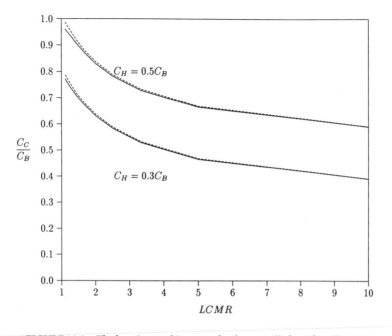

FIGURE 18.8   The location tracking costs for the reset-K algorithm; $K = 20$.

we have

$$\pi^*(1, 1) = \frac{1}{\displaystyle\sum_{i,j}(a_{i,j} + b_{i,j})}$$

and $\pi(K + 1, j)$ can be computed and the location tracking cost for the reset-K algorithm is obtained using Eq. (18.25).

The analysis is validated by a Monte Carlo simulation. In the simulation, the confidence interval for the 98% confidence level of the output measure C_C/C_B is within 3% of the mean value. Figure 18.8 plots curves for Eq. (18.25) (the solid curves) against the simulation experiments (the dashed curves) for $K = 20$ and $C_H = 0.5C_B$ and $0.3C_B$, respectively. The figure indicates that the analysis is consistent with the simulation model.

Comparison of the LCMR Estimation Algorithms

If the distributions for the incoming call process and the user movement process never change, then we would expect the running average algorithm to outperform the reset-K algorithm (especially when K is small) because the measured hit ratio p_M in the running average algorithm approaches the true hit ratio value p in the steady state. Surprisingly, the performance for the reset-K algorithm is roughly the same as the running average algorithm even if K is as small as 10. Figure 18.9 plots the location tracking costs for the running average algorithm and the reset-K algorithm with different K values.

The figure indicates that in steady state, when the distributions for the incoming call process and the user movement process never change, the running average algorithm outperforms reset K, and a large value of K outperforms a small K but the differences are insignificant.

If the distributions for the incoming call process or the user movement process change from time to time, we expect that the reset-K algorithm outperforms the running average algorithm. We have

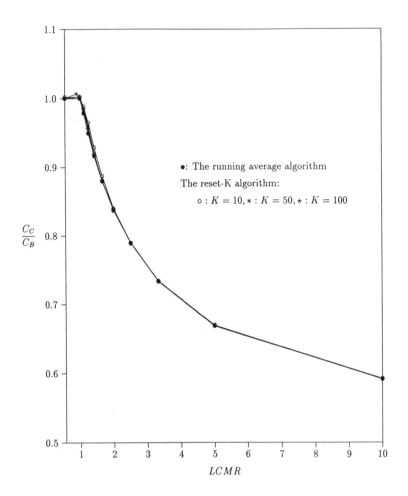

FIGURE 18.9   The location tracking costs for the running average algorithm and the reset-K algorithm; $C_H = 0.5C_B$.

examined this proposition experimentally. In the experiments, 4000 incoming calls are simulated. The call arrival rate changes from 0.1 to 1.0, 0.3, and then 5.0 for every 1000 calls (other sequences have been tested and similar results are observed). For every data point, the simulation is repeated 1000 times to ensure that the confidence interval for the 98% confidence level of the output measure C_C/C_B is within 3% of the mean value. Figure 18.10 plots the location tracking costs for the two algorithms for these experiments. By changing the distributions of the incoming call process, we observe that the reset-K algorithm is better than the running average algorithm for all C_H/C_B values.

18.10 Discussion

In this section we discuss aspects of the caching strategy presented here. Caching in PCS systems raises a number of issues not encountered in traditional computer systems, particularly with respect to architecture and locality in user call and mobility patterns. In addition, several variations in our reference assumptions are possible for investigating the implementation of the caching strategies. Here we sketch some of the issues involved.

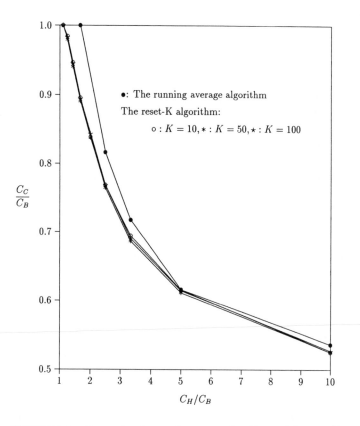

FIGURE 18.10 Comparing the running average algorithm and the reset-K algorithm under unstable call traffic.

Conditions When Caching Is Beneficial

We summarize the conditions for which the auxiliary strategies are worthwhile, under the assumptions of our analysis.

The caching strategy is very promising when the HLR update (or query load) or the remote A link is the performance bottleneck, since a low $LCMR$ ($LCMR > 0$) is required. For caching, the total database load and signalling network traffic is reduced whenever there is a cache hit. In addition, load and traffic is redistributed from the HLR and higher level SS7 network elements (RSTP, D links) to the VLRs and lower levels where excess network capacity may be more likely to exist. If the VLR is the performance bottleneck, the caching strategy is not promising, unless the VLR capacity is upgraded.

The benefits of the caching strategy depend on user call and mobility patterns when the D link, RSTP, and LSTP are the performance bottlenecks. We have used a Poisson call arrival model and exponential intermove time to estimate this dependence. Under very conservative assumptions, for caching to be beneficial requires relatively high $LCMR$ (25–50); we expect that in practice this threshold could be lowered significantly (say, $LCMR > 7$). Further experimental study is required to estimate the amount of locality in user movements for different user populations to investigate this issue further. It is possible that for some classes of users data obtained from active badge location system studies (e.g., Fishman and Mazer, 1992) could be useful. In general, it appears that caching could also potentially provide benefits to some classes of users even when the D link, the RSTP, or the LSTP are the bottlenecks.

We observe that more accurate models of user calling and mobility patterns are required to help resolve the issues raised in this section. We are currently engaged in developing theoretical models for user mobility and estimating their effect on studies of various aspects of PCS performance [Lin and Jain, 1993].

Alternative Network Architectures

The reference architecture we have assumed (Fig. 18.2) is only one of several possible architectures. It is possible to consider variations in the placement of the HLR and VLR functionality, (e.g., placing the VLR at a local SCP associated with the LSTP instead of at the SSP), the number of SSPs served by an LSTP, the number of HLRs deployed, etc. It is quite conceivable that different regional PCS service providers and telecommunications companies will deploy different signalling network architectures, as well as placement of databases for supporting PCS within their serving regions [Russo et al., 1993]. It is also possible that the number and placement of databases in a network will change over time as the number of PCS users increases.

Rather than consider many possible variations of the architecture, we have selected a reference architecture to illustrate the new auxiliary strategies and our method of calculating their costs and benefits. Changes in the architecture may result in minor variations in our analysis but may not significantly affect our qualitative conclusions.

LCMR Estimation and Caching Policy

It is possible that for some user populations estimating the LCMR may not be necessary, since they display a relatively high-average LCMR. For some populations, as we have shown in Sec. 18.9, obtaining accurate estimates of user LCMR in order to decide whether or not to use caching can be important in determining the net benefits of caching.

In general, schemes for estimating the LCMR range from static to dynamic and from distributed to centralized. We have presented two simple distributed algorithms for estimating LCMR, based on a long-range and short-range running calculation; the former is preferable if the call and mobility pattern of users is fairly static, whereas the latter is preferable if it is variable. Tuning the amount of history that is used to determine whether caching should be employed for a particular user is an obvious area for further study but is outside the scope of this chapter.

An alternative approach is to utilize some user-supplied information, by requesting profiles of user movements (e.g., see Tabbane, 1993 and Jain, 1995) and to integrate this with the caching strategy. A variation of this approach is to use some domain knowledge about user populations and their characteristics.

A related issue is that of cache size and management. In practice it is likely that the monetary cost of deploying a cache may limit its size. In that case, cache entries may not be maintained for some users; selecting these users carefully is important to maximize the benefits of caching. Note that the cache hit ratio threshold cannot necessarily be used to determine which users have cache entries, since it may be useful to maintain cache entries for some users even though their hit ratios have temporarily fallen below the threshold. A simple policy that has been found to be effective in computer systems in the least recently used (LRU) policy [Silbershatz and Peterson, 1988] in which cache entries that have been least recently used are discarded; LRU may offer some guidance in this context.

18.11 Conclusions

We began this chapter with an overview of the nuances of PCS, such as personal and terminal mobility, registration, deregistration, call delivery, etc. A tutorial was then provided on the two most common strategies for locating users in PCS, in North American interim standard IS-41 and the Pan-European standard GSM. A simplified analysis of the two standards was then provided to show the reader the extent to which database and signalling traffic is likely to be generated by PCS services. Suggestions were then made that are likely to result in reduced traffic.

Previous studies [Lo, Wolff, and Bernhardt, 1992; Meier-Hellstern and Alonso, 1992; Lo, Mohan, and Wolff, 1993; and Lo and Wolff, 1993] of PCS-related network signalling and data management functionalities suggest a high level of utilization of the signalling network in supporting call and mobility management activities for PCS systems. Motivated by the need to evolve location strategies to reduce signalling and database loads, we then presented an auxiliary strategy, called per-user caching, to augment the basic user location strategy proposed in standards [EIA/TIA, 1991; Mouly and Pautet, 1992].

Using a reference system architecture for PCS, we quantified the criteria under which the caching strategy produces reductions in the network signalling and database loads in terms of users' LCMRs. We have shown that, if the HLR or the remote A link in an SS7 architecture is the performance bottleneck, caching is useful regardless of user call and mobility patterns. If the D link or STPs are the performance bottlenecks, caching is potentially beneficial for large classes of users, particularly if they display a degree of locality in their call reception patterns. Depending on the numbers of PCS users who meet these criteria, the system-wide impacts of these strategies could be significant. For instance, for users with $LCMR \sim 5$ and stable call and move patterns, caching can result in cost reduction of 20–60% over the basic user location strategy *BasicFIND( )* under our analysis. Our results are conservative in that the *BasicFIND( )* procedure we have used for comparison purposes already reduces the network impacts compared to the user location strategy specified in PCS standards such as IS-41.

We have also investigated in detail two simple on-line algorithms for estimating users' LCMRs and examined the call and mobility patterns for which each would be useful. The algorithms allow a system designer to tune the amount of history used to estimate a users' LCMR and, hence, to attempt to optimize the benefits due to caching. The particular values of cache hit ratios and LCMR thresholds will change with variations in the way the PCS architecture and the caching strategy is implemented, but our general approach can still be applied. There are several issues deserving further study with respect to deployment of the caching strategy, such as the effect of alternative PCS architectures, integration with other auxiliary strategies such as the use of user profiles, and effective cache management policies.

Recently, we have augmented the work reported in this paper by a simulation study in which we have compared the caching and basic user location strategies [Harjono, Jain, and Mohan, 1994]. The effect of using a time-based criterion for enabling use of the cache has also been considered [Lin, 1994]. We have proposed elsewhere, for users with low CMRs, an auxiliary strategy involving a system of forwarding pointers to reduce the signalling traffic and database loads [Jain and Lin, 1995], a description of which is beyond the scope of this chapter.

Acknowledgment

We acknowledge a number of our colleagues in Bellcore who have reviewed several previous papers by the authors and contributed to improving the clarity and readability of this work.

References

Awerbuch, B. and Peleg, D. 1991. Concurrent online tracking of mobile users. In *Proc. SIGCOMM Symp. Comm. Arch. Prot.* Oct.

Bellcore. 1991. Advanced intelligent network release 1 network and operations plan, Issue 1. Tech. Rept. SR-NPL-001623. June. Bell Communications Research, Morristown, NJ.

Bellcore. 1993a. Personal communications services (PCS) network access services to PCS providers, Special Report SR-TSV-002459, Oct. Bell Communications Research, Morristown, NJ.

Bellcore. 1992c. Switching system requirements for interexchange carrier interconnection using the integrated services digital network user part (ISDNUP). Tech. Ref. TR-NWT-000394. Dec. Bell Communications Research. Morristown, NJ.

Electronic Industries Association/Telecommunications Industry Association. 1991. Cellular radio telecommunications intersystem operations. Tech. Rept. IS-41. Rev. B. July.

Feller, W. 1966. *An Introduction to Probability Theory and Its Applications.* John Wiley, New York.

Fishman, N. and Mazer, M. 1992. Experience in deploying an active badge system. In *Proc. Globecom Workshop on Networking for Pers. Comm. Appl.* Dec.

Harjono, H., Jain, R. and Mohan, S. 1994. Analysis and simulation of a cache-based auxiliary location strategy for PCS. In *Proc. IEEE Conf. Networks Pers. Comm.*

Jain, R. 1995. A classification scheme for user location strategies in personal communications services systems. Aug. Submitted for publication.

Jain, R. and Lin Y.-B. 1995. An auxiliary user location strategy employing forwarding pointers to reduce network impacts of PCS. *ACM Journal on Wireless Info. Networks (WINET)*, 1(2).

Lin, Y.-B. 1994. Determining the user locations for personal communications networks. *IEEE Trans. Vehic. Tech.* (Aug):466–473.

Lo, C., Mohan, S., and Wolff, R. 1993. A comparison of data management alternatives for personal communications applications. Second Bellcore Symposium on Performance Modeling, SR-TSV-002424, Nov. Bell Communications Research, Morristown, NJ.

Lo, C.N., Wolff, R.S., and Bernhardt, R.C. 1992. An estimate of network database transaction volume to support personal communications services. In *Proc. Intl. Conf. Univ. Pers. Comm.*

Lo, C. and Wolff, R. 1993. Estimated network database transaction volume to support wireless personal data communications applications. In *Proc. Intl. Conf. Comm.* May.

Lycksell, E. 1991. GSM system overview. Tech. Rept. Swedish Telecom. Admin. Jan.

Meier-Hellstern, K. and Alonso, E. 1992. The use of SS7 and GSM to support high density personal communications. In *Proc. Intl. Conf. Comm.*

Mohan, S. and Jain, R. 1994. Two user location strategies for PCS. *IEEE Pers. Comm. Mag.* Premiere issue. (Feb):42–50.

Mouly, M. and Pautet, M.B. 1992. *The GSM System for Mobile Communications.* M. Mouly, 49 rue Louise Bruneau, Palaiseau, France.

Thomas, R., Gilbert, H., and Mazziotto, G. 1988. Influence of the mobile station on the performance of a radio mobile cellular network. In *Proc. 3rd Nordic Seminar.* Sept.

Berman, R.K. and Brewster, J.H. 1992. Perspectives on the AIN architecture. *IEEE Comm. Mag.* 1(2):27–32.

Russo, P., Bechard, K., Brooks, E., Corn, R.L., Honig, W.L., Gove, R., and Young, J. 1993. IN rollout in the United States. *IEEE Comm. Mag.* (March):56–63.

Silberschatz, A. and Peterson, J. 1988. *Operating Systems Concepts.* Addison–Wesley, Reading, MA.

Tabbane, S. 1992. Comparison between the alternative location strategy (AS) and the classical location strategy (CS). Tech. Rept. Rutgers Univ. WINLAB. July. Rutgers, NJ.

Tabbane, S. 1993. Evaluation of an alternative location strategy for future high density wireless communications systems. Tech. Rept. WINALAB-TR-51, Rutgers Univ. WINLAB. Jan. Rutgers, NJ.

19

Cell Design Principles

Michel Daoud Yacoub
University of Campinas

19.1 Introduction

Designing a cellular network is a challenging task that invites engineers to exercise all of their knowledge in telecommunications. Although it may not be necessary to work as an expert in all of the fields, the interrelationship among the areas involved impels the designer to naturally search for a deeper understanding of the main phenomena. In other words, the time for segregation, when radio engineers and traffic engineers would not talk to each other, at least through a common vocabulary, is probably gone.

A great many aspects must be considered in a cellular network planning. The main ones include the following.

Radio Propagation: Here the topography and the morphology of the terrain, the urbanization factor and the clutter factor of the city, and some other aspects of the target geographical region under investigation will constitute the input data for the radio coverage design.

Frequency Regulation and Planning: In most countries there is a centralized organization, usually performed by a government entity, regulating the assignment and use of the radio spectrum. The frequency planning within the assigned spectrum should then be made so that interferences are minimized and the traffic demand is satisfied.

Modulation: As far as analog systems are concerned, the narrowband FM is widely used due to its remarkable performance in the presence of fading. The North American Digital Cellular

Standard IS-54 proposes the $\pi/4$ differential quadrature phase-shift keying ($\pi/4$ DQPSK) modulation, whereas the Global Standard for Mobile Communications (GSM) establishes the use of the Gaussian minimum-shift keying (GMSK).

Antenna Design: To cover large areas and for low-traffic applications omnidirectional antennas are recommended. Some systems at their inception may have these characteristics, and the utilization of omnidirectional antennas certainly keeps the initial investment low. As the traffic demand increases, the use of some sort of capacity enhancement technique to meet the demand, such as replacing the omnidirectional by directional antennas, is mandatory.

Transmission Planning: The structure of the channels, both for signalling and voice, is one of the aspects to be considered in this topic. Other aspects include the performance of the transmission components (power capacity, noise, bandwidth, stability, etc.) and the design or specification of transmitters and receivers.

Switching Exchange: In most cases this consists of adapting the existing switching network for mobile radio communications purposes.

Teletraffic: For a given grade of service and number of channels available, how many subscribers can be accommodated into the system? What is the proportion of voice and signalling channels?

Software Design: With the use of microprocessors throughout the system there are software applications in the mobile unit, in the base station, and in the switching exchange.

Other aspects, such as human factors, economics, etc., will also influence the design.

This chapter outlines the aspects involving the basic design steps in cellular network planning. Topics, such as traffic engineering, cell coverage, and interference, will be covered, and application examples will be given throughout the section so as to illustrate the main ideas. We start by recalling the basic concepts including *cellular principles, performance measures and system requirements*, and *system expansion techniques*.

19.2 Cellular Principles

The basic idea of the cellular concept is *frequency reuse* in which the same set of channels can be reused in different geographical locations sufficiently apart from each other so that *cochannel interference* be within tolerable limits. The set of channels available in the system is assigned to a group of *cells* constituting the *cluster*. Cells are assumed to have a *regular hexagonal* shape and the number of cells per cluster determines the *repeat pattern*. Because of the hexagonal geometry only certain repeat patterns can tessellate. The number N of cells per cluster is given by

$$N = i^2 + ij + j^2 \tag{19.1}$$

where i and j are integers. From Eq. (19.1) we note that the clusters can accommodate only certain numbers of cells such as 1, 3, 4, 7, 9, 12, 13, 16, 19, 21, ..., the most common being 4 and 7. The number of cells per cluster is intuitively related with system capacity as well as with transmission quality. The fewer cells per cluster, the larger the number of channels per cell (higher traffic carrying capacity) and the closer the cocells (potentially more cochannel interference). An important parameter of a cellular layout relating these entities is the D/R ratio, where D is the distance between cocells and R is the cell radius. In a hexagonal geometry it is found that

$$D/R = \sqrt{3N} \tag{19.2}$$

19.3 Performance Measures and System Requirements

Two parameters are intimately related with the grade of service of the cellular systems: carrier-to-cochannel interference ratio and blocking probability.

A high carrier-to-cochannel interference ratio in connection with a low-blocking probability is the desirable situation. This can be accomplished, for instance, in a large cluster with a low-traffic condition. In such a case the required grade of service can be achieved, although the resources may not be efficiently utilized. Therefore, a measure of efficiency is of interest. The **spectrum efficiency** η_s expressed in erlang per square meter per hertz, yields a measure of how efficiently space, frequency, and time are used, and it is given by

$$\eta_s = \frac{\text{number of reuses}}{\text{coverage area}} \times \frac{\text{number of channels}}{\text{bandwidth available}} \times \frac{\text{time the channel is busy}}{\text{total time of the channel}}$$

Another measure of interest is the **trunking efficiency** in which the number of subscribers per channel is obtained as a function of the number of channels per cell for different values of blocking probability. As an example, assume that a cell operates with 40 channels and that the mean blocking probability is required to be 5%. Using the erlang-B formula (refer to the Traffic Engineering section of this chapter), the traffic offered is calculated as 34.6 erlang. If the traffic per subscriber is assumed to be 0.02 erl, a total of $34.6/0.02 = 1730$ subscribers in the cell is found. In other words, the trunking efficiency is $1730/40 = 43.25$ subscribers per channel in a 40-channel cell. Simple calculations show that the trunking efficiency decreases rapidly when the number of channels per cell falls below 20.

The basic specifications require cellular services to be offered with a fixed telephone network quality. Blocking probability should be kept below 2%. As for the transmission aspect, the aim is to provide good quality service for 90% of the time. Transmission quality concerns the following parameters:

- Signal-to-cochannel interference (S/I_c) ratio
- Carrier-to-cochannel interference ratio (C/I_c)
- Signal plus noise plus distortion-to-noise plus distortion $(SINAD)$ ratio
- Signal-to-noise (S/N) ratio
- Adjacent channel interference selectivity (ACS)

The S/I_c is a subjective measure, usually taken to be around 17 dB. The corresponding C/I_c depends on the modulation scheme. For instance, this is around 8 dB for 25-kHz FM, 12 dB for 12.5-kHz FM, and 7 dB for GMSK, but the requirements may vary from system to system. A common figure for $SINAD$ is 12 dB for 25-kHz FM. The minimum S/N requirement is 18 dB, whereas ACS is specified to be no less than 70 dB.

19.4 System Expansion Techniques

The obvious and most common way of permitting more subscribers into the network is by allowing a system performance degradation but within acceptable levels. The question is how to objectively define what is acceptable. In general, the subscribers are more likely to tolerate a poor quality service rather than not having the service at all. Some alternative expansion techniques, however, do exist that can be applied to increase the system capacity. The most widely known are as follows.

Adding New Channels: In general, when the system is set up not all of the channels need be used, and growth and expansion can be planned in an orderly manner by utilizing the channels that are still available.

Frequency Borrowing: If some cells become more overloaded than others, it may be possible to reallocate channels by transferring frequencies so that the traffic demand can be accommodated.

Change of Cell Pattern: Smaller clusters can be used to allow more channels to attend a bigger traffic demand at the expense of a degradation of the transmission quality.

Cell Splitting: By reducing the size of the cells, more cells per area, and consequently more channels per area, are used with a consequent increase in traffic capacity. A radius reduction by a factor of f reduces the coverage area and increases the number of base stations by a factor of f^2. Cell splitting usually takes place at the midpoint of the congested areas and is so planned in order that the old base stations are kept.

Sectorization: A cell is divided into a number of sectors, three and six being the most common arrangements, each of which is served by a different set of channels and illuminated by a directional antenna. The sector, therefore, can be considered as a new cell. The base stations can be located either at the center or at the corner of the cell. The cells in the first case are referred to as center-excited cells and in the second as corner-excited cells. Directional antennas cut down the cochannel interference, allowing the cocells to be more closely spaced. Closer cell spacing implies smaller D/R ratio, corresponding to smaller clusters, i.e., higher capacity.

Channel Allocation Algorithms: The efficient use of channels determines the good performance of the system and can be obtained by different channel assignment techniques. The most widely used algorithm is based on fixed allocation. Dynamic allocation strategies may give better performance but are very dependent on the traffic profile and are usually difficult to implement.

19.5 Basic Design Steps

Engineering a cellular system to meet the required objectives is not a straightforward task. It demands a great deal of information, such as market demographics, area to be served, traffic offered, and other data not usually available in the earlier stages of system design. As the network evolves, additional statistics will help the system performance assessment and replanning. The main steps in a cellular system design are as follows.

Definition of the Service Area: In general, the responsibility for this step of the project lies on the operating companies and constitutes a tricky task, because it depends on the market demographics and, consequently, on how much the company is willing to invest.

Definition of the Traffic Profile: As before, this step depends on the market demographics and is estimated by taking into account the number of potential subscribers within the service area.

Choice of Reuse Pattern: Given the traffic distribution and the interference requirements a choice of the reuse pattern is carried out.

Location of the Base Stations: The location of the first base station constitutes an important step. A significant parameter to be taken into account in this is the relevance of the region to be served. The base station location is chosen so as to be at the center of or as close as possible to the target region. Data, such as available infrastructure and land, as well as local regulations are taken into consideration in this step. The cell radius is defined as a function of the traffic distribution. In urban areas, where the traffic is more heavily concentrated, smaller cells are chosen so as to attend the demand with the available channels. In suburban and in rural areas, the radius is chosen to be large because the traffic demand tends to be small. Once the placement of the first base station has been defined, the others will be accommodated in accordance with the repeat pattern chosen.

Radio Coverage Prediction: Given the topography and the morphology of the terrain, a radio prediction algorithm, implemented in the computer, can be used to predict the signal strength in the geographic region. An alternative to this relies on field measurements with the use of appropriate equipment. The first option is usually less costly and is widely used.

Design Checkup: At this point it is necessary to check whether or not the parameters with which the system has been designed satisfy the requirements. For instance, it may be necessary to re-evaluate the base station location, the antenna height, etc., so that better performance can be attained.

Field Measurements: For a better tuning of the parameters involved, field measurements (radio survey) should be included in the design. This can be carried out with transmitters and towers provisionally set up at the locations initially defined for the base station.

The cost assessment may require that a redesign of the system should be carried out.

19.6 Traffic Engineering

The starting point for engineering the traffic is the knowledge of the required grade of service. This is usually specified to be around 2% during the busy hour. The question lies on defining the busy hour. There are usually three possible definitions: 1) busy hour at the busiest cell, 2) system busy hour, and 3) system average over all hours.

The estimate of the subscriber usage rate is usually made on a demographic basis from which the traffic distribution can be worked out and the cell areas identified. Given the repeat pattern (cluster size), the cluster with the highest traffic is chosen for the initial design. The traffic A in each cell is estimated and, with the desired blocking probability $E(A, M)$, the erlang-B formula as given by Eq. (19.3) is used to determine the number of channels per cell, M

$$E(M, A) = \frac{A^M / M!}{\sum_{i=0}^{M} A^i / i!} \tag{19.3}$$

In case the total number of available channels is not large enough to provide the required grade of service, the area covered by the cluster should be reduced in order to reduce the traffic per cell. In such a case, a new study on the interference problems must be carried out. The other clusters can reuse the same channels according to the reuse pattern. Not all channels need be provided by the base stations of those cells where the traffic is supposedly smaller than that of the heaviest loaded cluster. They will eventually be used as the system grows.

The traffic distribution varies in time and space, but it is commonly bell shaped. High concentrations are found in the city center during the rush hour, decreasing toward the outskirts. After the busy hour and toward the end of the day, this concentration changes as the users move from the town center to their homes. Note that because of the mobility of the users handoffs and roaming are always occurring, reducing the channel holding times in the cell where the calls are generated and increasing the traffic in the cell where the mobiles travel. Accordingly, the erlang-B formula is, in fact, a rough approximation used to model the traffic process in this ever-changing environment. A full investigation of the traffic performance in such a dynamic system requires all of the phenomena to be taken into account, making any traffic model intricate. Software simulation packages can be used so as to facilitate the understanding of the main phenomena as well as to help system planning. This is a useful alternative to the complex modeling, typically present in the analysis of cellular networks, where closed-form solutions are not usually available

On the other hand, conventional traffic theory, in particular, the erlang-B formula, is a handy tool widely used in cellular planning. At the inception of the system the calculations are carried out based on the best available traffic estimates, and the system capacity is obtained by grossly exaggerating the calculated figures. With the system in operation some adjustments must be made so that the requirements are met.

The approach just mentioned assumes the simplest channel assignment algorithm: the fixed allocation. It has the maximum spatial efficiency in channel reuse, since the channels are always assigned at the minimum reuse distance. Moreover, because each cell has a fixed set of channels, the channel assignment control for the calls can be distributed among the base stations.

The main problem of fixed allocation is its inability to deal with the alteration of the traffic pattern. Because of the mobility of the subscribers, some cells may experience a sudden growth in the traffic offered, with a consequent deterioration of the grade of service, whereas other cells may have free channels that cannot be used by the congested cells.

A possible solution for this is the use of dynamic channel allocation algorithms in which the channels are allocated on a demand basis There is an infinitude of strategies using the dynamic assignment principles, but they are usually complex to implement. An interim solution can be exercised if the change of the traffic pattern is predictable. For instance, if a region is likely to have an increase of the traffic on a given day (say, a football stadium on a match day), a mobile base station can be moved toward such a region in order to alleviate the local base.

Another specific solution uses the traffic available at the boundary between cells that may well communicate with more than one base station. In this case, a call that is blocked in its own cell can be directed to the neighboring cell to be served by its base station. This strategy, called *directed retry*, is known to substantially improve the traffic capacity. On the other hand, because channels with marginally acceptable transmission quality may be used, an increase in the interference levels, both for adjacent channel and cochannel, can be expected. Moreover, subscribers with radio access only to their own base will experience an increase in blocking probability.

19.7 Cell Coverage

The propagation of energy in a mobile radio environment is strongly influenced by several factors, including the natural and artificial relief, propagation frequency, antenna heights, and others. A precise characterization of the signal variability in this environment constitutes a hard task. Deterministic methods, such as those described by the *free space, plane earth*, and *knife-edge diffraction* propagation models, are restricted to very simple situations. They are useful, however, in providing the basic mechanisms of propagation. Empirical methods, such as those proposed by many researchers (e.g., Egli, 1957; Okumura et al., 1968; Lee, 1986; Ibrahim and Parsons, 1983; and others), use curves and/or formulas based on field measurements, some of them including deterministic solutions with various correction factors to account for the propagation frequency, antenna height, polarization, type of terrain, etc. Because of the random characteristics of the mobile radio signal, however, a single deterministic treatment of this signal will certainly lead the problem to a simplistic solution. Therefore, we may treat the signal on a statistical basis and interpret the results as random events occurring with a given probability. The cell coverage area is then determined as the proportion of locations where the received signal is greater than a certain threshold considered to be satisfactory.

Suppose that at a specified distance from the base station the *mean signal strength* is considered to be known. Given this we want to determine the cell radius such that the mobiles experience a received signal above a certain threshold with a stipulated probability. The mean signal strength can be determined either by any of the prediction models or by field measurements. As for the statistics of the mobile radio signal, five distributions are widely accepted today: lognormal, Rayleigh, Suzuki [Suzuki, 1977], Rice, and Nakagami. The lognormal distribution describes the variation of the mean signal level (large-scale variations) for points having the same transmitter–receiver antennas separation, whereas the other distributions characterize the instantaneous variations (small-scale variations) of the signal. In the calculations that follow we assume a lognormal environment. The other environments can be analyzed in a like manner; although this may not be of interest if some sort of diversity is implemented, because then the effects of the small-scale variations are minimized.

Propagation Model

Define m_w and k as the mean powers at distances x and x_0, respectively, such that

$$m_w = k\left(\frac{x}{x_0}\right)^{-\alpha} \tag{19.4}$$

where α is the path loss coefficient. Expressed in decibels, $M_w = 10\log m_w$, $K = 10\log k$ and

$$M_w = K - 10\alpha\log\left(\frac{x}{x_0}\right) \tag{19.5}$$

Define the received power as $w = v^2/2$, where v is the received envelope. Let $p(W)$ be the probability density function of the received power W, where $W = 10\log w$. In a lognormal environment, v has a lognormal distribution and

$$p(W) = \frac{1}{\sqrt{2\pi}\sigma_w}\exp\left(-\frac{(W - M_w)^2}{2\sigma_w^2}\right) \tag{19.6}$$

where M_W is the mean and σ_w is the standard deviation, all given in decibels. Define w_T and $W_T = 10\log w_T$ as the threshold above which the received signal is considered to be satisfactory. The probability that the received signal is below this threshold is its *probability distribution function* $P(W_T)$, such that

$$P(W_T) = \int_{-\infty}^{W_T} p(W)\,\mathrm{d}W = \frac{1}{2} + \frac{1}{2}\,\mathrm{erf}\left[\frac{(W_T - M_W)^2}{2\sigma_w^2}\right] \tag{19.7}$$

where erf( ) is the error function defined as

$$\mathrm{erf}(y) = \frac{2}{\sqrt{\pi}}\int_0^y \exp(-t^2)\,\mathrm{d}t \tag{19.8}$$

Base Station Coverage

The problem of estimating the cell area can be approached in two different ways. In the first approach, we may wish to determine the proportion β of locations at x_0 where the received signal power w is above the threshold power w_T. In the second approach, we may estimate the proportion μ of the circular area defined by x_0 where the signal is above this threshold. In the first case, this proportion is averaged over the perimeter of the circumference (cell border); whereas in the second approach, the average is over the circular area (cell area).

The proportion β equals the probability that the signal at x_0 is greater than this threshold. Hence,

$$\beta = \mathrm{prob}(W \geq W_T) = 1 - P(W_T) \tag{19.9}$$

Using Eqs. (19.5) and (19.7) in Eq. (19.9) we obtain

$$\beta = \frac{1}{2} - \frac{1}{2}\mathrm{erf}\left[\frac{W_T - K + 10\alpha\log(x/x_0)}{\sqrt{2}\sigma_w}\right] \tag{19.10}$$

This probability is plotted in Fig. 19.1, for $x = x_0$ (cell border).

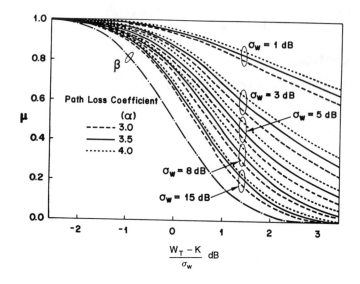

FIGURE 19.1 Proportion of locations where the received signal is above a given threshold; the dashdot line corresponds to the β approach and the other lines to the μ approach.

Let prob$(W \geq W_T)$ be the probability of the received power W being above the threshold W_T within an infinitesimal area dS. Accordingly, the proportion μ of locations within the circular area S experiencing such a condition is

$$\mu = \frac{1}{S} \int_S [1 - P(W_T)] \, dS \qquad (19.11)$$

where $S = \pi r^2$ and $dS = x \, dx d\theta$. Note that $0 \leq x \leq x_0$ and $0 \leq \theta \leq 2\pi$. Therefore, solving for $d\theta$, we obtain

$$\mu = 2 \int_0^1 u\beta \, du \qquad (19.12)$$

where $u = x/x_0$ is the normalized distance.

Inserting Eq. (19.10) in Eq. (19.12) results in

$$\mu = 0.5 \left\{ 1 + \text{erf}\,(a) + \exp\left(\frac{2ab + 1}{b^2}\right) \left[1 - \text{erf}\left(\frac{ab + 1}{b}\right) \right] \right\} \qquad (19.13)$$

where $a = (K - W_T)/\sqrt{2}\sigma_w$ and $b = 10\alpha \log(e)/\sqrt{2}\sigma_w$.

These probabilities are plotted in Fig. 19.1 for different values of standard deviation and path loss coefficients.

Application Examples

From the theory that has been developed it can be seen that the parameters affecting the probabilities β and μ for cell coverage are the path loss coefficient α, the standard deviation σ_w, the required threshold W_T, and a certain power level K, measured or estimated at a given distance from the base station.

The applications that follow are illustrated for two different standard deviations: $\sigma_w = 5$ dB and $\sigma_w = 8$ dB. We assume the path loss coefficient to be $\alpha = 4$ (40 dB/decade), the mobile station receiver sensitivity to be -116 dB (1 mW), and the power level estimated at a given distance from the base station as being that at the cell border, $K = -102$ dB (1 mW). The receiver is considered to operate with a *SINAD* of 12 dB for the specified sensitivity. Assuming that cochannel interference levels are negligible and given that a signal-to-noise ratio S/N of 18 dB is required, the threshold W_T will be -116 dB (1 mW) $+ (18 - 12)$ dB (1 mW) $= -110$ dB (1 mW).

Three cases will be explored as follows.

Case 19.1: We want to estimate the probabilities β and μ that the received signal exceeds the given threshold 1) at the border of the cell, probability β and 2) within the area delimited by the cell radius, probability μ.

Case 19.2: It may be interesting to estimate the cell radius x_0 such that the received signal be above the given threshold with a given probability (say 90%) 1) at the perimeter of the cell and 2) within the cell area. This problem implies the calculation of the mean signal strength K at the distance x_0 (the new cell border) of the base station. Given K and given that at a distance x_0 (the former cell radius) the mean signal strength M_w is known [note that in this case $M_w = -102$ dB (1 mW)], the ratio x_0/x can be estimated.

Case 19.3: To fulfill the coverage requirement, rather than calculating the new cell radius, as in case 19.2, a signal strength at a given distance can be estimated such that a proportion of the locations at this distance, proportion β, or within the area delimited by this distance, proportion μ, will experience a received signal above the required threshold. This corresponds to calculating the value of the parameter K already carried out in case 19.2 for the various situations.

The calculation procedures are now detailed for $\sigma_w = 5$ dB. Results are also shown for $\sigma_w = 8$ dB.

Case 19.1: Using the given parameters we obtain $(W_T - K)/\sigma_w = -1.6$. With this value in Fig. 19.1, we obtain the probability that the received signal exceeds -116 dB (1 mW) for $S/N = 18$ dB given that at the cell border the mean signal power is -102 dB (1 mW) given in Table 19.1.

Note, from Table 19.1, that the signal at the cell border exceeds the receiver sensitivity with 97% probability for $\sigma_w = 5$ dB and with 84% probability for $\sigma_w = 8$ dB. If, on the other hand, we are interested in the area coverage rather than in the border coverage, then these figures change to 100% and 95%, respectively.

Case 19.2: From Fig. 19.1, with $\beta = 90\%$ we find $(W_T - K)/\sigma_w = -1.26$. Therefore, $K = -103.7$ dB (1 mW). Because $M_w - K = -10\alpha \log(x/x_0)$, then $x_0/x = 1.10$. Again, from Fig. 19.1, with $\mu = 90\%$ we find $(W_T - K)/\sigma_w = -0.48$, yielding $K = -107.6$ dB (1 mW). Because $M_w - K = -10\alpha \log(x/x_0)$, then $x_0/x = 1.38$. These results are summarized in Table 19.2, which shows the normalized radius of a cell where the received signal power is above -116 dB (1 mW) with 90% probability for $S/N = 18$ dB, given that at a reference distance from the base station (the cell border) the received mean signal power is -102 dB (1 mW).

Note, from Table 19.2, that in order to satisfy the 90% requirement at the cell border the cell radius can be increased by 10% for $\sigma_w = 5$ dB. If, on the other hand, for the same standard deviation the 90% requirement is to be satisfied within the cell area, rather than at the cell border, a substantial gain in power is achieved. In this case, the cell radius can be increased by a factor of 1.38. For

TABLE 19.1 Case 19.1 Coverage Probability

Standard Deviation, dB	β Approach (Border Coverage), %	μ Approach (Area Coverage), %
5	97	100
8	84	95

TABLE 19.2   Case 19.2 Normalized Radius

Standard Deviation, dB	β Approach (Border Coverage)	μ Approach (Area Coverage)
5	1.10	1.38
8	0.88	1.27

TABLE 19.3   Case 19.3 Signal Power

Standard Deviation dB	β Approach (Border Coverage), dB (1 mW)	μ Approach (Area Coverage), dB (1 mW)
5	−103.7	−107.6
8	−99.8	−106.2

$\sigma_w = 8$ dB and 90% coverage at the cell border, the cell radius should be reduced to 88% of the original radius. For area coverage, an increase of 27% of the cell radius is still possible.

Case 19.3:   The values of the mean signal power K are taken from case 19.2 and shown in Table 19.3, which shows the signal power at the cell border such that 90% of the locations will experience a received signal above −116 dB for $S/N = 18$ dB.

19.8 Interference

Radio-frequency interference is one of the most important issues to be addressed in the design, operation, and maintenance of mobile communication systems. Although both intermodulation and intersymbol interferences also constitute problems to account for in system planning, a mobile radio system designer is mainly concerned about adjacent-channel and cochannel interferences.

Adjacent Channel Interference

Adjacent-channel interference occurs due to equipment limitations, such as frequency instability, receiver bandwidth, filtering, etc. Moreover, because channels are kept very close to each other for maximum spectrum efficiency, the random fluctuation of the signal, due to fading and near–far effect, aggravates this problem.

Some simple, but efficient, strategies are used to alleviate the effects of adjacent channel interference. In narrowband systems, the total frequency spectrum is split into two halves so that the reverse channels, composing the uplink (mobile to base station) and the forward channels, composing the downlink (base station to mobile), can be separated by half of the spectrum. If other services can be inserted between the two halves, then a greater frequency separation, with a consequent improvement in the interference levels, is accomplished. Adjacent channel interference can also be minimized by avoiding the use of adjacent channels within the same cell. In the same way, by preventing the use of adjacent channels in adjacent cells a better performance is achieved. This strategy, however, is dependent on the cellular pattern. For instance, if a seven-cell cluster is chosen, adjacent channels are inevitably assigned to adjacent cells.

Cochannel Interference

Undoubtedly the most critical of all interferences that can be engineered by the designer in cellular planning is cochannel interference. It arises in mobile radio systems using cellular architecture because of the frequency reuse philosophy.

A parameter of interest to assess the system performance in this case is the carrier-to-cochannel interference ratio C/I_c. The ultimate objective of estimating this ratio is to determine the reuse distance and, consequently, the repeat pattern. The C/I_c ratio is a random variable, affected by random phenomena such as 1) location of the mobile, 2) fading, 3) cell site location, 4) traffic distribution, and others. In this subsection we shall investigate the **outage probability**, i.e., the probability of failing to achieve adequate reception of the signal due to cochannel interference. This parameter will be indicated by $p(CI)$. As can be inferred, this is intrinsically related to the repeat pattern.

Cochannel interference will occur whenever the wanted signal does not simultaneously exceed the minimum required signal level s_0 and the n interfering signals, $i_1, i_2, \ldots, i_n$, by some protection ratio r. Consequently, the conditional outage probability, given n interferers, is

$$p(CI \mid n) = 1 - \int_{s_0}^{\infty} p(s) \int_0^{s/r} p(i_1) \int_0^{(s/r)-i_1} p(i_2) \cdots$$

$$\times \int_0^{(s/r)-i_1-\cdots-i_{n-1}} p(i_n)\, di_n \cdots di_2\, di_1\, ds \qquad (19.14)$$

The total outage probability can then be evaluated by

$$p(CI) = \sum_n p(CI \mid n) p(n) \qquad (19.15)$$

where $p(n)$ is the distribution of the number of active interferers.

In the calculations that follow we shall assume an interference-only environment, i.e., $s_0 = 0$, and the signals to be Rayleigh faded. In such a fading environment the probability density function of the signal-to-noise ratio x is given by

$$p(x) = \frac{1}{x_m} \exp\left(-\frac{x}{x_m}\right) \qquad (19.16)$$

where x_m is the mean signal-to-noise ratio. Note that $x = s$ and $x_m = s_m$ for the wanted signal, and $x = i_j$ and $x_m = i_{mj}$ for the interfering signal j, with s_m and i_{mj} being the mean of s and i_j, respectively.

By using the density of Eq. (19.16) in Eq. (19.14) we obtain

$$p(CI \mid n) = \sum_{j=1}^n \prod_{k=1}^j \frac{z_k}{1+z_k} \qquad (19.17)$$

where $z_k = r s_m / i_{mk}$

If the interferers are assumed to be equal, i.e., $z_k = z$ for $k = 1, 2, \ldots, n$, then

$$p(CI \mid n) = 1 - \left(\frac{z}{1+z}\right)^n \qquad (19.18)$$

Define $Z = 10 \log z$, $S_m = 10 \log s_m$, $I_m = 10 \log i_m$, and $R_r = 10 \log r$. Then, $Z = S_m - (I_m + R_r)$. Equation (19.18) is plotted in Fig. 19.2 as a function of Z for $n = 1$ and $n = 6$, for the situation in which the interferers are equal.

If the probability of finding an interferer active is p, the distribution of active interferers is given by the binomial distribution. Considering the closest surrounding cochannels to be the most relevant interferers we then have six interferers. Thus

$$p(n) = \binom{6}{n} p^n (1-p)^{6-n} \qquad (19.19)$$

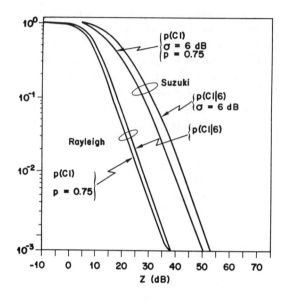

FIGURE 19.2 Conditional and unconditional outage probability for $n = 6$ interferes in a Rayleigh environment and in a Suzuki environment with $\sigma = 6$ dB.

For equal capacity cells and an evenly traffic distribution system, the probability p is approximately given by

$$p = \sqrt[M]{B} \tag{19.20}$$

where B is the blocking probability and M is the number of channels in the cell.

Now Eqs. (19.20), (19.19), and (19.18) can be combined into Eq. (19.15) and the outage probability is estimated as a function of the parameter Z and the channel occupancy p. This is shown in Fig. 19.2 for $p = 75\%$ and $p = 100\%$.

A similar, but much more intricate, analysis can be carried out for the other fading environments. Note that in our calculations we have considered only the situation in which both the wanted signal and the interfering signals experience Rayleigh fading. For a more complete analysis we may assume the wanted signal to fade differently from the interfering signals, leading to a great number of possible combinations. A case of interest is the investigation of the influence of the standard deviation in the outage probability analysis. This is illustrated in Fig. 19.2 for the Suzuki (lognormal plus Rayleigh) environment with $\sigma = 6$ dB.

Note that by definition the parameter z is a function of the carrier-to-cochannel interference ratio, which, in turn, is a function of the reuse distance. Therefore, the outage probability can be obtained as a function of the cluster size, for a given protection ratio.

The ratio between the mean signal power s_m and the mean interfering power i_m equals the ratio between their respective distances d_s and d_i such that

$$\frac{s_m}{i_m} = \left(\frac{d_s}{d_i}\right)^{-\alpha} \tag{19.21}$$

where α is the path loss coefficient. Now, 1) let D be the distance between the wanted and interfering base stations, and 2) let R be the cell radius. The cochannel interference worst case occurs when

TABLE 19.4 Probability of Cochannel Interference in Different Cell Clusters

		Outage Probability, %			
		Rayleigh		Suzuki $\sigma = 6$ dB	
N	$Z + R$, dB	$p = 75\%$	$p = 100\%$	$p = 75\%$	$p = 100\%$
1	−4.74	100	100	100	100
3	10.54	31	40	70	86
4	13.71	19	26	58	74
7	19.40	4.7	7	29	42
12	24.46	1	2.1	11	24
13	25.19	0.9	1.9	9	22

the mobile is positioned at the boundary of the cell, i.e., $d_s = R$ and $d_i = D - R$. Then,

$$\frac{i_m}{s_m} = \left(\frac{D}{R} - 1\right)^{-\alpha} \tag{19.22a}$$

or, equivalently,

$$S_m - I_m = 10\alpha \log\left(\frac{D}{R} - 1\right) \tag{19.22b}$$

In fact, $S_m - I_m = Z + R_r$. Therefore,

$$Z + R_r = 10\alpha\log(\sqrt{3N} - 1) \tag{19.23}$$

With Eq. (19.23) and the curves of Fig. 19.2, we can compare some outage probabilities for different cluster sizes. The results are shown in Table 19.4 where we have assumed a protection ratio $R_r = 0$ dB. The protection ratio depends on the modulation scheme and varies typically from 8 dB (25-kHz FM) to 20 dB [single sideband (SSB) modulation].

Note, from Table 19.4, that the standard deviation has a great influence in the calculations of the outage probability.

19.9 Conclusions

The interrelationship among the areas involved in a cellular network planning is substantial. Vocabularies belonging to topics, such as radio propagation, frequency planning and regulation, modulation schemes, antenna design, transmission, teletraffic, and others, are common to all cellular engineers.

Designing a cellular network to meet system requirements is a challenging task which can only be partially and roughly accomplished at the design desk. Field measurements play an important role in the whole process and constitute an essential step used to tune the parameters involved.

Defining Terms

Outage probability: The probability of failing to achieve adequate reception of the signal due to, for instance, cochannel interference.

Spectrum efficiency: A measure of how efficiently space, frequency, and time are used. It is expressed in erlang per square meter per hertz.

Trunking efficiency: A function relating the number of subscribers per channel and the number of channels per cell for different values of blocking probability.

References

Egli, J. 1957. Radio above 40 Mc over irregular terrain. *Proc. IRE.* 45 (10):1383–1391.

Ibrahim, M.F. and Parsons, J.D. 1983. Signal strength prediction in built-up areas, Part I: median signal strength. *Proc. IEE* Pt. F. (130):377–384.

Hata, M. 1980. Empirical formula for propagation loss in land-mobile radio services. *IEEE Trans. Vehicular Tech.* VT-29:317–325.

Ho, M.J. and Stüber, G.L. 1993. Co-channel interference of microcellular systems on shadowed Nakagami fading channels. *Proc. IEEE Vehicular Tech. Conf.* pp. 568–571.

Lee. W.C.Y., *Mobile Communications Design Fundamentals*, Howard W. Sams, Indianapolis, IN.

Leonardo, E.J. and Yacoub, M.D. 1993a. A statistical approach for cell coverage area in land mobile radio systems. Proceedings of the 7th IEE Conf. on Mobile and Personal Comm., Brighton, UK, Dec. pp. 16–20.

Leonardo, E.J. and Yacoub, M.D. 1993b. (Micro) Cell coverage area using statistical methods. Proceedings of the IEEE Global Telecom. Conf. GLOBECOM'93, Houston, TX, Dec. pp. 1227–1231.

Okumura, Y., Ohmori, E., Kawano, T., and Fukuda, K. 1968. Field strength and its variability in VHF and UHF land mobile service. *Rev. Elec. Comm. Lab.* 16 (Sept.–Oct.):825–873.

Reudink, D.O. 1974. Large-scale variations of the average signal. In *Microwave Mobile Communications*, pp. 79–131, John Wiley, New York.

Sowerby, K.W. and Williamson, A.G. 1988. Outage probability calculations for multiple cochannel interferers in cellular mobile radio systems. *IEE Proc.* Pt. F. 135(3):208–215.

Suzuki, H. 1977. A statistical model for urban radio propagation. *IEEE Trans. Comm.* 25(7):673–680.

Further Information

The fundamentals of mobile radio engineering in connection with many practical examples and applications as well as an overview of the main topics involved can be found in Yacoub, M.D., *Foundations of Mobile Radio Engineering*, CRC Press, Inc. Boca Raton, Fl, 1993.

20

Microcellular Radio Communications

Raymond Steele
*Southampton University
and
Multiple Access
Communications Ltd.*

20.1 Introducing Microcells

In mobile radio communications an operator will be assigned a specific bandwidth W in which to operate a service. The operator will, in general, not design the mobile equipment, but purchase equipment that has been designed and standardized by others. The performance of this equipment will have a profound effect on the number of subscribers the network can support, as we will show later. Suppose the equipment requires a radio channel of bandwidth B. The operator can therefore fit $N_T = W/B$ channels into the allocated spectrum W.

Communications with mobiles are made from fixed sites, known as base stations (BSs). Clearly, if a mobile travels too far from its BS, the quality of the communications link becomes unacceptable. The perimeter around the BS where acceptable communications occur is called a cell and, hence, the term cellular radio. BSs are arranged so that their radio coverage areas, or cells, overlap, and each BS may be given $N = N_T/M$ channels. This implies that there are M BS and each BS uses a different set of channels.

The number N_T is relatively low, perhaps only 1000. As radio channels cannot operate with 100% utilization, the cluster of BSs or cells has fewer than 1000 simultaneous calls. In order to make the business viable, more users must be supported by the network. This is achieved by repeatedly reusing the channels. Clusters of BSs are tessellated with each cluster using the same N_T channels. This means that there are users in each cluster using the same frequency band at the same time, and inevitably there will be interference. This interference is known as cochannel interference. Cochannel cells, i.e., cells using the same channels, must be spaced sufficiently far apart for the interference levels to be acceptable. A mobile will therefore receive the wanted signal of power S and a total interference power of I, and the signal-to-interference ratio (SIR) is a key system design parameter.

0-8493-8573-3/96/$0.00+$.50

Suppose we have large cells, a condition that occurs during the initial stages of deploying a network when coverage is important. For a given geographical area G_A we may have only one cluster of seven cells, and this may support some 800 simultaneous calls in our example. As the subscriber base grows, the number of clusters increases to, say, 100 with the area of each cluster being appropriately decreased. The network can now support some 80,000 simultaneous calls in the area G_A. As the number of subscribers continues to expand, we increase the number of clusters. The geographical area occupied by each cluster is now designed to match the number of potential users residing in that area. Consequently, the smallest clusters and, hence, the highest density of channels per area is found in the center of cities. As each cluster has the same number of channels, the smaller the clusters and, therefore, the smaller the cells, the greater the **spectral efficiency** measured in erlang per hertz per square meter. Achieving this higher spectral efficiency requires a concomitant increase in the infrastructure that connects the small cell BSs to their base station controller (BSC). The BSCs are part of the nonradio part of the mobile network that is interfaced with the public switched telephone network (PSTN) or the integrated service digital network (ISDN).

As we make the cells smaller, we change from locating the BS antennas on top of tall buildings or hills, where they produce large cells or macrocells, to the tops of small buildings or the sides of large buildings, where they form minicells, to lamp post elevations, where they form **street microcells**. Each decrease in cell size is accompanied by a reduction in the radiated power levels from the BSs and from the mobiles. As the BS antenna height is lowered, the neighboring buildings and streets increasingly control the radio propagation. This chapter is concerned with microcells and microcellular networks. We commence with the simplest type of microcells, namely, those used for highways.

20.2 Highway Microcells

Since their conception by Steele and Prabhu, 1985, many papers have been published on **highway microcells**, ranging from propagation measurements to teletraffic issues [Chia et al., 1987; El-Dolil, Wong, and Steele, 1989; Steele and Nofal, 1992; Green, 1990; Keenan and Motley, 1990; Saleh and Valenzula, 1987; Steele, 1992; Merrett, Cooper, and Symington; Steele and Williams, 1993]. Figure 20.1 shows the basic concepts for a highway microcellular system having two cells per cluster. The highway is partitioned into contiguous cigar-shaped segments formed by directional antennas. Omnidirectional antennas can be used at junctions, roundabouts, cloverleaf, and other road intersections. The BS antennas are mounted on poles at elevations of some 6–12 m. Figure 20.2 shows received signal levels as a function of the distance d between BS and MS for different roads

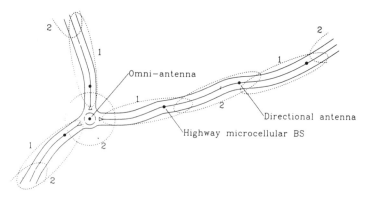

FIGURE 20.1 Highway microcellular clusters. Microcells with the same number use the same frequency set.

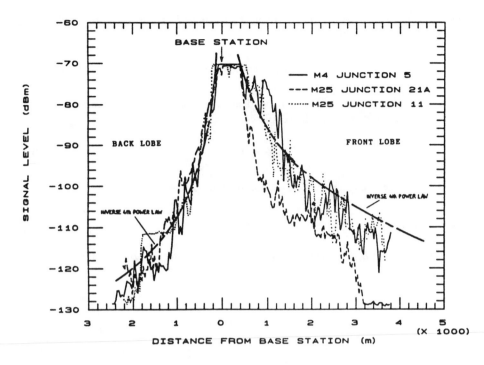

FIGURE 20.2 Overlayed received signal strength profiles of various highway cells including the inverse fourth power law curve for both the front and back antenna lobes. *Source:* Chia et al. 1987. Propagation and bit error ratio measurements for a microcellular system. *JIERE.* 57(6):5255–5266. With permission.

[Chia et al. 1987]. The average loss in received signal level, or path loss, is approximately inversely proportional to d^4. The path loss is associated with a slow fading component that is due to the variations in the terrain, the road curvature and cuttings, and the presence of other vehicles.

The curves in the figure are plotted for an 18-element yagi BS antenna having a gain of 15 dB and a front-to-back ratio of 25 dB. In Fig. 20.2 reference is made to junctions on different motorways, e.g., junction 5 on motorway M4. This is because the BS antennas are mounted at these road junctions with the yagi antenna pointing along the highway in order to create a cigar-shaped cell. The flat part of the curve near the BS is due to the MS receiver being saturated by high-signal levels. Notice that the curve related to M25, junction 11, decreases rapidly with distance when the MS leaves the immediate vicinity of the BS. This is due to the motorway making a sharp turn into a cutting, causing the MS to lose line of sight (LOS) with the BS. Later the path loss exponent is approximately 4. Experiments have shown that using the arrangement just described, with each BS transmitting 16 mW at 900 MHz, 16 kb/s noncoherent frequency shift keying, two-cell clusters could be formed where each cell has a length along the highway ranging from 1 to 2 km. For MSs traveling at 110 km/h the average handover rate is 1.2 per minute [Chia et al. 1987].

Spectral Efficiency of Highway Microcellular Network

Spectral efficiency is a key system parameter. The higher the efficiency, the greater will be the teletraffic carried by the network for the frequency band assigned by the regulating authorities per unit geographical area. We define the spectral efficiency in mobile radio communications in erlang per hertz per square meter as

$$\eta \triangleq A_{CT}/S_T W \tag{20.1}$$

although erlang per megahertz per square kilometer is often used. In this equation, A_{CT} is the total traffic carried by the microcellular network,

$$A_{CT} = C A_C \qquad (20.2)$$

where C is the number of microcells in the network and A_C the carried traffic by each microcellular BS. The total area covered of the tessellated microcells is

$$S_T = C S \qquad (20.3)$$

where S is the average area of a microcell, whereas the total bandwidth available is

$$W = M N B \qquad (20.4)$$

whose terms M, N, and B were defined in Sec. 20.1. Substituting Eqs. 20.2–20.4 into Eq. (20.1), yields

$$\eta = \frac{\rho}{SMB} \qquad (20.5)$$

where

$$\rho = A_C / N \qquad (20.6)$$

is the utilization of each BS channel.

If the length of each microcell is L, there are n up lanes and n down lanes, and each vehicle occupies an effective lane length V, which is speed dependent, then the total number of vehicles in a cell is

$$K = 2nL / V \qquad (20.7)$$

Given that all vehicles have a mobile terminal, the maximum number of mobiles in a cell is K. In a highway microcell we are not interested in the actual area $S = 2nL$ but in how many vehicles can occupy this area, namely, the effective area K. Notice that K is largest in a traffic jam when all vehicles are stationary and V only marginally exceeds the vehicle length. Given that N is sufficiently large, η is increased when the traffic flow is decreased.

Using fixed channel assignment (FCA) with frequency division multiple access (FDMA) or with time division multiple access (TDMA), the cluster size M can be two. Using dynamic channel assignment (DCA) with TDMA, or code division multiple access (CDMA), causes the spectral efficiency η to be very high because for a given traffic utilization ρ and channel bandwidth B, the S is small (as we are considering microcells), and M may be thought of as 1, or less, due to sectorization. The total traffic A_{CT}, given by Eq. (20.2), is also very high because by making L relatively short, C is accordingly high.

The traffic carried by a microcellular BS is

$$A_C = [\lambda_N (1 - P_{bn}) + \lambda_H (1 - P_{fhm})]\overline{T}_H \qquad (20.8)$$

where P_{bn} is the probability of a new call being blocked, P_{fhm} is the probability of handover failure when mobiles enter the microcell while making a call and concurrently no channel is available, λ_N and λ_H are the new call and handover rates, respectively, and $\overline{T}_H$ is the mean channel holding time of all calls. For the simple case where no channels are exclusively reserved for handovers, $P_{bn} = P_{fhm}$, and

$$A_C = \lambda_T \overline{T}_H (1 - P_{bn}) = A(1 - P_{bn}) \tag{20.9}$$

where

$$\lambda_T = \lambda_N + \lambda_H \tag{20.10}$$

and A is the total offered traffic. The mathematical complexity resides in calculating A and P_{bn}, and the reader is advised to consult El-Dolil, Wong, and Steele, 1989, and Steele and Nofal, 1992.

Priority schemes have been proposed whereby N channels are available for handover, but only $N - N_h$ for new calls. Thus N_h channels are exclusively reserved for handover [El-Dolil, Wong, and Steele, 1989]. While P_{bn} marginally increases, P_{fhm} decreases by orders of magnitude for the same average number of new calls per sec per microcell. This is important as people prefer to be blocked while attempting to make a call compared to having a call in progress terminated due to no channel being available on handover. An important enhancement is to use an oversailing macrocellular cluster, where each macrocell supports a microcellular cluster. The role of the macrocell is to provide channels to support microcells that are overloaded and to provide communications to users who are in areas not adequately covered by the microcells [El-Dolil, Wong, and Steele, 1989]. When vehicles are in a solid traffic jam, there are no handovers and so N_h should be zero. When traffic is flowing fast, N_h should be high. Accordingly a useful strategy is to make N_h adaptive to the new call and handover rates [Steele, Nofal, and El-Dolil, 1990].

20.3 City Street Microcells

We will define a city street microcell as one where the BS antenna is located below the lowest building. As a consequence, the diffraction over the buildings can be ignored, and the heights of the buildings are of no consequence. Roads and their attendant buildings form trenches or canyons through which the mobiles travel. If there is a direct line-of-sight path between the BS and a MS and a ground-reflected path, the received signal level vs BS–MS distance is as shown in Fig. 20.3. Should there be two additional paths from rays reflected from the buildings, then the profile for this four-ray situation is also shown in Fig. 20.3. These theoretical curves show that as the MS travels from the

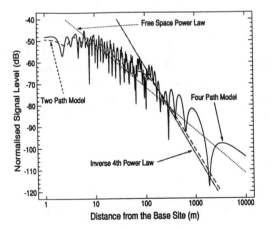

FIGURE 20.3 Signal level profiles for the two- and four-path models. Also shown are the free space and inverse fourth power laws. *Source:* Green. 1990. Radio link design for microcellular system, *British Telecom. Tech. J.* 8(1):85–96. With permission.

BS the average received signal level is relatively constant and then decreases relatively rapidly. This is a good characteristic as it offers a good signal level within the microcell, and the interference into adjacent microcells falls off rapidly with distance.

In practice there are many paths, but there is often a dominant one. As a consequence the fading is Rician [Stelle, 1992]. The Rician distribution approximates to a Gaussian one when the received signal is from a dominant path with the power in the scattered paths being negligible, to a Rayleigh one when there is no dominant path. Macrocells usually have Rayleigh fading, whereas in microcells the fading only occasionally becomes Rayleigh and is more likely to be closer to Gaussian. This means that the depth of the fades in microcells are usually significantly smaller than in macrocells enabling microcellular communications to operate closer to the receiver noise floor without experiencing error bursts and to accommodate higher cochannel interference levels. Because of the small dimensions of the microcells, the delays between the first and last significant paths is relatively small compared to the corresponding delays in macrocells. Consequently, the impulse response is generally shorter in microcells and, therefore, the transmitted bit rate can be significantly higher before intersymbol interference is experienced compared to the situation in macrocells. Microcells are, therefore, more spectrally efficient with an enhanced propagation environment.

There are two types of these city street microcells, one for pedestrians and the other for vehicles. In general, there will be more portables carried by pedestrians than mobile stations in cars. Also, as cars travel more quickly than people, their microcells are accordingly larger than for pedestrians. The handover rates for portables and vehicular MS may be similar, and networks must be capable of handling the many handovers per call that may occur. In addition, the time available to complete a handover may be very short compared to those in macrocells.

City street microcells are irregular when the streets are irregular as demonstrated by the MIDAS[1] plot of a BS in Southampton city area displayed in Fig. 20.4. To achieve a contiguous coverage we site the BSs one at a time. Having sited the first BS and located the microcellular boundary along the streets, we locate adjacent BSs such that their boundaries butt with each other along the main streets. Unless many microcellular BSs are deployed, there will be some secondary streets where there will be insufficient signal levels. Those areas that are not covered by the microcellular BS will be accommodated by an oversailing macrocellular BS that services the complete cluster of microcellular BSs. Figure 20.5 shows a cluster of microcells; the oversailing macrocell could be sited outside the area of this figure. We emphasize that total coverage by microcells in a typical city center is difficult to achieve, and it is vital that oversailing macrocells are used to cover these dead spots. The macrocell also facilitates handovers and efficient microcellular channel utilization.

There are important differences between highway microcells and city microcells, which relate to their one- and two-dimensional characteristics. A similar comment applies to street microcells and hexagonal cells. Basically, the buildings have a profound effect on cochannel interference. The buildings shield much of the cochannel interference, and the double regression path loss law of microcells [Green, 1990] also decreases interference if the break-distance constitutes the notional microcell boundary. City microcellular clusters may have as few as two microcells, but four is more typical, and in some topologies six or more may be required. The irregularity of city streets means that some signals can find paths through building complexes to give cochannel interference where it is least expected.

Teletraffic Issues

Consider the arrangement where each microcellular cluster is overlaid by a macrocell. The macrocells are also clustered. The arrangement is shown in Fig. 20.6. The total traffic carried is

$$A_{CT} = C_m A_{cm} + C_M A_{CM} \qquad (20.11)$$

---

[1]MIDAS is a propriety software outdoor planning tool developed by Multiple Access Communications Ltd.

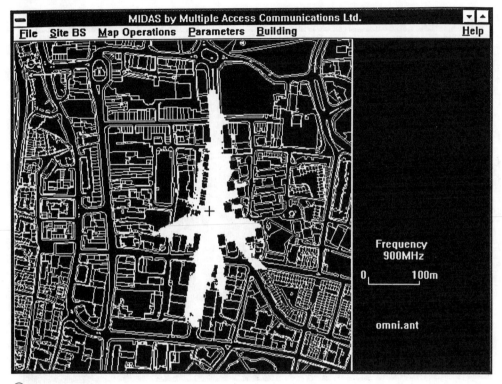

© Crown copyright

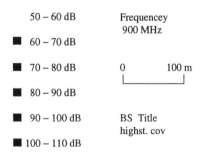

50 – 60 dB	Frequencey 900 MHz
60 – 70 dB	
70 – 80 dB	0 100 m
80 – 90 dB	
90 – 100 dB	BS Title
100 – 110 dB	highst. cov

FIGURE 20.4 MIDAS plot of a city street microcell in Southampton. The plot is normally in colour, with each colour signifying a range of path loss in dB. In this plot the white area represents a path loss of up to 80 dB, and the cross shows the base station location.

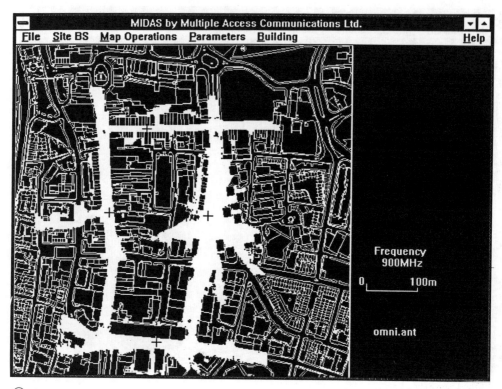

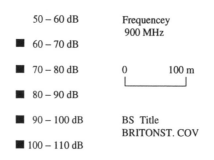

FIGURE 20.5   Cluster of microcells in Southampton city centre. The plot is normally in colour, with each colour signifying a range of path loss in dB. In this plot the white area represents a path loss of up to 80 dB, and the crosses show the microcellular base station locations.

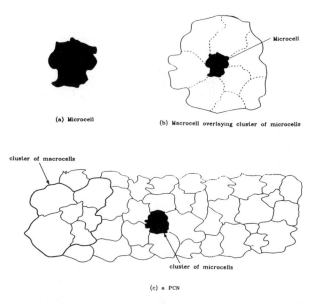

(a) Microcell

(b) Macrocell overlaying cluster of microcells

cluster of macrocells

cluster of microcells

(c) a PCN

FIGURE 20.6 Microcellular clusters with oversailing macrocellular clusters. Each macrocell is associated with a particular microcellular cluster. *Source*: Steele and Williams. 1993. Third generation PCN and the intelligent multimode mobile portable. *IEE Elect. and Comm. Eng. J.* 5(3):147–156. With permission.

where C_m and C_M are number of microcells and macrocells in the network, respectively. Each microcellular BS has N channels and carries A_{cm} erlang. The corresponding values for the macrocellular BSs are N_0 and A_{CM}. The channel utilization for the network is

$$\rho_2 = \frac{C_m A_{cm} + C_M A_{CM}}{C_m N + C_M N_0} = \frac{M A_{cm} + A_{CM}}{MN + N_0} \tag{20.12}$$

where M is the number of microcells per cluster.

The spectral efficiency is found by noting that the total bandwidth is

$$B_T = B_c(MN + M_0 N_0) \tag{20.13}$$

where B_c is the effective channel bandwidth, and M_0 is the number of macrocells per macrocellular cluster. The traffic carried by a macrocellular cluster and its cluster of microcells is

$$A_M = A_{CM} M_0 + M_0(A_{cm} M) \tag{20.14}$$

over an area of

$$S_M = (S_m M) M_0 \tag{20.15}$$

where S_m is the area of each microcell.

The spectral efficiency is, therefore,

$$
\begin{aligned}
\eta &= \frac{A_{CM} + A_{cm}M}{B_c(MN + M_0N_0)S_mM} \\
&= \frac{\rho_2}{B_c S_m M} \left\{ \frac{MN + N_0}{MN + M_0N_0} \right\}
\end{aligned}
\tag{20.16}
$$

We note that by using oversailing macrocells to assist microcells experiencing overloading we are able to operate the microcells at high levels of channel utilization. However, the channel utilization of the macrocells must not be high if the probability of calls forced to terminate due to handover failure is to be minuscule.

20.4 Indoor Microcells

Microcellular BSs may be located within buildings to produce **indoor microcells** whose dimensions may extend from a small office, to part of larger offices, to a complete floor, or to a number of floors. The microcells are box-like for a single office microcell; or may contain many boxes, e.g., when a microcell contains contiguous offices. Furniture, such as bookcases, filing cabinets, and desks may represent large obstacles that may introduce shadowing effects. The signal attenuation through walls, floors, and ceilings may vary dramatically depending on the construction of the building. There is electromagnetic leakage down stairwells and through service ducting, and signals may leave the building and re-enter it after reflection and diffractions from other buildings.

Predicting the path loss in office microcells is, therefore, fraught with difficulties. At the outset it may not be easy to find out the relevant details of the building construction. Even if these are known, an estimation of the attenuation factors for walls, ceilings, and floors from the building construction is far from simple. Then there is a need to predict the effect of the furniture, effect of doors, the presence of people, and so on. Simple equations have been proposed. For example, the one by Keenam and Motley, 1990, who represent the path loss in decibels by

$$
PL = L(V) + 20\log_{10} d + n_f a_f + n_w a_w
\tag{20.17}
$$

where d is the straight line distance between the BS and the MS, a_f and a_w are the attenuation of a floor and a wall, respectively, n_f and n_w are the number of floors and walls along the line d, respectively, and $L(V)$ is a so-called clutter loss, which is frequency dependent. This equation should be used with caution. Researchers have made many measurements and found that even when they use computer ray tracing techniques the results can be considerably disparate. Errors having a standard deviation of 8–12 dB are not unusual at the time of writing. Given the wide range of path loss in mobile communications, however, and the expense of making measurements, particularly when many BS locations are examined, means that there is nevertheless an important role for planning tools to play, albeit their poorer accuracy compared to street microcellular planning tools.

As might be expected, the excess path delays in buildings is relatively small. The maximum delay spread within rooms and corridors may be <200 and 300 ns, respectively [Saleh and Valenzula, 1987]. The digital European cordless telecommunication (DECT) indoor system operates at 1152 kb/s without either channel coding or equalization [Steele, 1992]. This means that the median rms values are relatively low, and 25 ns has been measured [saleh and Valenzula, 1987]. When the delay spread becomes too high resulting in bit errors, the DECT system hops the user to a better channel using DCA.

20.5 Microcellular Infrastructure

An important requirement of first and second generation mobile networks is to restrict the number of base stations to achieve sufficient network capacity with an acceptably low probability of blocking. This approach is wise given the cost of base stations and their associated equipment, plus the cost and difficulties in renting sites. It is somewhat reminiscent of the situation faced by early electronic circuit designers who needed to minimize the number of tubes and later the number of discrete transistors in their equipment. It was the introduction of microelectronics that freed the circuit designer. We are now in an analogous situation where we need to free the network designers of the third generation communication networks, allowing the design to have microcellular BSs in the position where they are required, without being concerned if they are rarely used, and knowing that the cost of the microcellular network is a minor one compared to the overall network cost. This approach is equivalent to installing electric lighting where we are not unduly concerned if not all the lights are switched on at any particular time, preferring to be able to provide illumination where and when it is needed.

To realise high-capacity mobile communications we need to design microcellular BSs of negligible costs, of coffee mug dimensions, and with the ability to connect them at the cost of, say, electrical wiring in streets and buildings. Cordless telecommunication (CT) BSs are already of shoe-box size, and companies are designing coffee mug-size versions. The cost of these BSs will be low in mass production, and many BSs will be equivalent in cost to one first generation analog cellular BS. Microcellular BSs could be miniaturized, fully functional BSs achieved by using microelectronic techniques and by exploiting the low-radiated power levels (<10 mW) required. At the other extreme, the microcells could be formed using distribution points (DPs) that only have optical-to-microwave converters, microwave-to-optical converters, and linear amplifiers, with the remainder of the BS at another location. In between the miniaturized, fully functional BSs and the DPs there is a range of options that depends on how much complexity is built into the microcellular BS and how the intelligence of the network is distributed.

Radio over Fiber

The method of using DPs to form microcells is often referred to as radio over fiber (ROF) [Merrett, Cooper, and Symington]. Figure 20.7(a) shows a microcellular BS transmitting to a MS. When the DP concept is evoked, the microcellular BS contains electrical-to-optical (E/O) and optical-to-electrical (O/E) converters as shown in Fig. 20.7(b). The microwave signal that would have been radiated to the mobile is now applied, after suitable attenuation, to a laser transmitter. Essentially, the microwave signal amplitude modulates the laser, and the modulated signal is conveyed over a single-mode optical fiber to the distribution point. O/E conversion ensues followed by power amplification, and the resulting signal is transmitted to the MS. Signals from the MS are low-noise amplified and applied to the laser transmitter in the DP. Optical signals are sent from the DP to the BS where O/E conversion is performed followed by radio reception.

In general, the BS transceiver will be handling multicarrier signals for many mobiles, and the DP will accordingly be transceiving signals with many mobiles whose power levels may be significantly different, even when power control is used. Care must be exercised to avoid serious intermodulation products arising in the optical components.

Figure 20.7(c) shows the cositing of n microcellular BSs for use with DPs. This cositing may be conveniently done at a mobile switching center (MSC). Shown in the figure are the DPs and their irregular shaped overlapping microcells. The DPs can be attached to lamp posts in city streets, using electrical power from the electric light supply and the same ducting as used by the electrical wiring, or local telephone ducting. The DPs can also be attached to the outside of buildings. DPs within buildings may be conveniently mounted on ceilings.

The DP concept allows small, lightweight equipment in the form of DPs to be geographically distributed to form microcells; however, there are problems. The N radio carriers cause intermod-

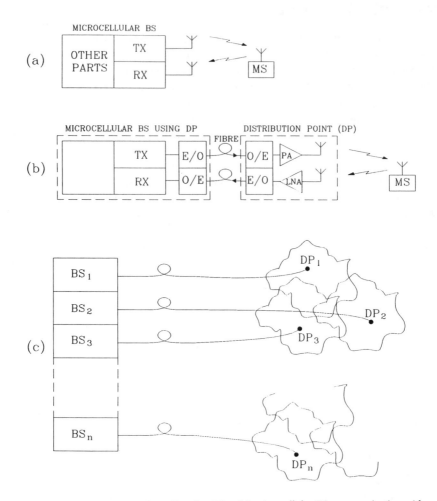

FIGURE 20.7 Creating microcells using DPs: (a) microcellular BS communicating with MS, (b) radio over fiber to distribution point, and (c) microcellular clusters using DPs.

ulation products (IMPs), which may be reduced by decreasing the depth of amplitude modulation for each radio channel. Unfortunately this also decreases the carrier-to-noise ratio (CNR) and the dynamic range of the link. With TDMA having many channels per carrier, we can decrease the number of radio carriers and make the IMPs more controllable. CDMA is particularly adept at coping with IMPs. The signals arriving from the MSs may have different power levels, in spite of power control. Because of the small size cells, the dynamic range of the signals arriving from MSs having power control should be <20 dB. If not, the power levels of the signals arriving at the DP from the MSs may need to be made approximately similar by individual amplification at the DP. We also must be careful to limit the length of the fiber as optical signals propagate along fibers much more slowly than radio signals propagate in free space. This should not be a problem in microcells, unless the fiber makes many detours before arriving at its DP.

The current cost of lasers is not sufficiently low for the ROF DP technique to be deployed. However, there is research into lasers, which are inherently simple, low cost, robust, and provide narrow line widths. There are also the developments in optoelectronic integrated circuits that may ultimately bring costs down. In addition, wavelength division multiplexing will bring benefits.

Miniaturized Microcellular BSs

The low-radiated power levels used by BSs and MSs in microcells have important ramifications on BS equipment design. Small fork combiners can be used along with linear amplifiers. Even FDMA BSs become simple when the high-radiated power levels are abandoned. It is the changes to the size of the RF components that enables the size of the microcellular BSs to be small.

The interesting question that next arises is, how much baseband signal processing complexity should the microcellular BS have? If the microcellular BSs are connected to nodes in an optical LAN, we can convey the final baseband signals to the BS, leaving the BS with the IF and front-end RF components. This means that processing of the baseband signals will be done at the group station (GS), which may be a MSC connected to the LAN. The GS will, therefore, transcode the signals from the ISDN into a suitable format. Using an optical LAN, however, and with powerful microelectronics, the transcoding and full BS operations could be done at each microcellular BS. Indeed, the microcellular BS may eventually execute many of the operations currently handled by the MSC.

20.6 Multiple Access Issues

There are three basic multiple access methods. Time division multiple access, frequency division multiple access, and spread spectrum multiple access (SSMA). SSMA comes in two versions; frequency-hopping SSMA and discrete-sequence SSMA. The latter is usually referred to as CDMA. There are also many hybrids of these systems. The principles of multiple access are described elsewhere in this book and will not be repeated here. Instead, we will comment on key factors that effect the choice of the multiple access method in microcellular environments.

As a preamble, if we observe the equations for spectral efficiency η, we see that η is inversely proportional to the number of microcells per cluster M. The smallest value of M is unity, where every microcell uses the same frequencies. Under these conditions the SIR will be low. Thus to achieve high η, we need a low value of M, and for an acceptable bit error rate (BER), we require the radio link to be able to operate with low values of SIRs. Because cellular radio operates in an intentional jamming environment, whereas CDMA was conceived to operate in a military environment where jamming by the enemy is expected, CDMA is a most appropriate multiple access method for cellular radio. The CDMA system, IS-95, will operate efficiently in single cell ($M = 1$) clusters where each cell is sectorized.

In highway microcells two-cell clusters can be used with TDMA and FDMA. Street microcells have complex shapes, see Figs. 20.4 and 20.5, and when FCA is used with TDMA and FDMA, there is a danger that high levels of interference will be ducted through streets and cause high-interference levels in a small segment of a microcell. To accommodate this phenomenon, the system must have a rapid handover (HO) capability, with HO to either a different channel at the same BS, to the interfering cell, or to an oversailing macrocell. CDMA is much more adept at handling this situation. The irregularity of street microcells, except in regularly shaped cities, such as midtown Manhattan, suggests that FCA should not be used. If it is, it requires $M \geq 4$. Instead, DCA should be employed. For example, when DCA is used with TDMA we abandon the notion of clusters of microcells. We may arrange for all microcells to have the same frequency set and design the system with accurate power control to contain the cochannel interference, and to encourage the MS to switch to another channel at the current BS or switch to a new BS directly when the SIR becomes below a threshold at either end of the duplex link. The application of DCA increases the capacity and can also contend with the situation where a MS suddenly experiences a rapid peak of cochannel interference during its travels.

The interference levels in CDMA in street microcells is mainly from users within its microcell, rather than from users in other microcells due to the shielding of the buildings. For CDMA to operate efficiently in street microcells, it should increase its chip rate to ensure it can exploit path diversity in its RAKE receiver. By increasing the chip rate, higher data rates can be accommodated and, hence, a greater variety of services. CDMA should be used in a similar way in office microcells. If the chip rate cannot be increased, however, the equipment installer can deploy a distributed antenna system where between each antenna a delay element is introduced. By this means path diversity gains are realized.

TDMA/DCA is appropriate for indoor microcells, where the complexity of the DCA is easier to implement compared to street microcells. FDMA should be considered for indoor microcells where it is well suited to provide high-bit-rate services since the transmitted rate is the same as the source rate. It also benefits from the natural shielding that exists within buildings to contain the cochannel interference and the low-power levels that simplify equipment design.

20.7 Discussion

At the time of writing, microcells are used in cordless telecommunications, where indoor microcells and outdoor telepoint microcells are used. There are very few microcells in cellular systems because there are no commercially available microcellular BSs. Nevertheless, operators have formed microcells using existing macrocellular BSs. Microcellular BSs, however, do exist in manufacturer's laboratories, and their entrance into the market is imminent. When microcells are deployed in large numbers, the vast increase in teletraffic will call for new network topologies and protocols.

In our deliberations we focused on highway microcells, city-street microcells, and indoor microcells. Minicells, where the BS antenna is below most of the buildings but above others, are currently being deployed. We may anticipate the fusion of the types of minicells and microcells. We will have microcells of strange shapes, like city street microcells but in three dimensions. Street microcells may serve the lower floors of buildings and vice versa. Microcells, located in minicell environments, may cover the streets as well as floors in neighboring buildings. We may also anticipate very small microcells, the so-called picocells. Indeed, we will have multicellular networks with multimode radio interfaces. This means that an intelligent multimode terminal with its supporting network will be required [Steele and Williams, 1993]. The role of microcells is to carry the high-bit-rate traffic and, hence, support a wide range of services. Our teletraffic equations tell us that microcellular personal communication networks will support orders more teletraffic than current conventional systems. Technology advancements will produce coffee cup size microcellular BSs and facilitate new network architectures that will eventually lead to the widespread concentration of intelligence at the BSs.

Defining Terms

Spectral efficiency: Has a special meaning in cellular radio. It is the traffic carried in erlang per hertz (or kilohertz) per area in square meters (or square kilometeres).

Street microcells: Small cells whose shape are determined by the street topology and their buildings. The base station antennas are below the urban skyline.

Highway microcells: Segments of a highway having a base station and supporting mobile communications.

Indoor microcells: Small volumes of a building, e.g., an office, having a base station and supporting mobile communications.

References

Chia, S.T.S., Steele, R., Green, E., and Baran, A. 1987. Propagation and bit error ratio measurements for a microcellular system. *JIERE*, Supplement 57(6):5255–5266.

El-Dolil, S.A., Wong, W.C., and Steele, R. 1989. Teletraffic performance of highway microcells with overlay macrocell. *IEEE JSAC*, 7(1):71–78.

Green, E. 1990. Radio link design for microcellular systems. *British Telecom Tech. J.* 8(1):85–96.

Keenan, J.M. and Motley, A.J. 1990. Radio coverage in buildings. *British Telecom. Tech. J.* 8(1):19–24.

Merrett, R.P., Cooper, A.J., and Symington, I.C. A cordless access system using radio-over-fiber techniques. IEEE VT-91:921–924.

Saleh, A.A.M. and Valenzula, R.A. 1987. A statistical model for indoor multipath propagation. *IEEE JSAC*, (Feb.):128–137.

Steele, R. 1992. *Mobile Radio Communications*, Pentech Press, London.

Steele, R. and Nofal, M. 1992. Teletraffic performance of microcellular personal communication networks. *IEE Proc-I*, 139(4):448–461.

Steele, R., Nofal, M., and El-Dolil, S. 1990. An adaptive algorithm for variable teletraffic demand in highway microcells. *Electronic Letters*, 26(14):988–990.

Steele R. and Prabhu, V.K. 1985. Mobile radio cellular structures for high user density and large data rates. In *Proc of the IEE*, Pt. F. (5):396–404.

Steele, R. and Williams, J.E.B. 1990. Third generation PCN and the intelligent multimode mobile portable, *IEE Elec. and Comm. Eng. J.* 5(3):147–156.

Further Information

The *IEEE Communications Magazine Special Issue* on an update on personal communications, Vol. 30, No. 12, Dec. 1992 provides a good introduction to microcells, particularly the paper by L.J. Greenstein, et al.

21

Fixed and Dynamic Channel Assignment

Bijan Jabbari
George Mason University

21.1 Introduction

One of the important aspects of frequency reuse-based cellular radio as compared to early land mobile telephone systems is the potential for dynamic allocation of channels to traffic demand. This fact had been recognized from the early days of research (e.g., see chapter 7 in Jakes, 1994; Cox and Reudnik, 1972 and 1973) in this field. With the emergence of wireless personal communications and use of microcell with nonuniform traffic, radio resource assignment becomes essential to network operation and largely determines the available spectrum efficiency. The primary reason for this lies in the use of microcell in dense urban areas where distinct differences exist as compared to large cell systems due to radio propagation and fading effects that affect the interference conditions.

In this chapter, we will first review the channel reuse constraint and then describe methods to accomplish the assignment. Subsequently, we will consider variations of fixed channel assignment and discuss dynamic resource assignment. Finally, we will briefly discuss the traffic modeling aspect.

21.2 The Resource Assignment Problem

The resources in a wireless cellular network are derived either in frequency division multiple access (FDMA), time division multiple access (TDMA), or joint frequency–time (MC-TDMA) [Abramson, 1994]. In these channel derivation techniques, the frequency reuse concept is used throughout the service areas comprised of cells and microcells. The same channel is used by distinct terminals in different cells, with the only constraint of meeting a given interference threshold. In spread spectrum multiple access (SSMA) such as the implemented code division multiple access (CDMA) system (IS-95) [TIA, 1993] each subscriber spreads its transmitted signal over the

0-8493-8573-3/96/$0.00+$.50
© 1996 by CRC Press, Inc.

same frequency band by using a pseudorandom sequence simultaneously. As any active channel is influenced by the others, a new channel can be set up only if the overall interference is below a given threshold. Thus, the problem of resource assignment in CDMA relates to transmission power control in forward (base station to mobile terminal) and reverse (mobile terminal to base station) channels. Of course, the problem of power control applies to TDMA and FDMA as well, but not to the extent that it impacts the capacity of CDMA. Here, we will focus on time- and frequency-based access methods. For preliminary results in CDMA the readers are referred to Everitt, 1994.

Fixed channel assignment (FCA) and dynamic channel assignment (DCA) techniques are the two extremes of allocating radio channels to mobile subscribers. For a specific grade of service and quality of transmission, the assignment scheme provides a tradeoff between spectrum utilization and implementation complexity. The performance parameters from a radio resource assignment point of view are interference constraints (quality of transmission link), probability of call blocking (grade of service), and the system capacity (spectrum utilization) described by busy hour erlang traffic that can be carried by the network. In a cellular system, however, there exist other functions, such as handoff and its execution or radio access control. These functions may be facilitated by the use of specific assignment schemes and, therefore, they should be considered in such a tradeoff [Everitt, 1994].

The problem of channel assignment can be described as the following: Given a set of channels derived from the specified spectrum, assign the channels and their transmission power such that for every set of assigned channels to cell i, $(C/I)_i > (C/I)_0$. Here, $(C/I)_0$ represents the minimum allowed carrier to interference and $(C/I)_i$ represents carrier to interference at cell i.

21.3 Fixed Channel Assignment

In fixed channel assignment the interference constraints are ensured by a frequency plan independent of the number and location of active mobiles. Each cell is then assigned a fixed number of carriers, dependent on the traffic density and cell size. The corresponding frequency plan remains fixed on a long-term basis. In reconfigurable FCA (sometimes referred to as flexible FCA), however, it is possible to reconfigure the frequency plan periodically in response to near/medium term changes in predicted traffic demand.

In FCA, for a given set of communications system parameters, $(C/I)_0$ relates to a specific quality of transmission link (e.g., probability of bit error or voice quality). This parameter in turn relates to the number of channel sets [Jakes, 1994] (or cluster size) given by $K = 1/3 \, (D/R)^2$. Thus, the ratio D/R is determined by $(C/I)_0$. Here D is the cochannel reuse distance and R is the cell radius. For example, in the North American cellular system advanced mobile phone service (AMPS), $(C/I)_0 = 18$ dB, which results in $K = 7$ or $D = 4.6R$. Here, we have used a propagation attenuation proportional to the fourth power of the distance. The radius of the cell is determined mainly by the projected traffic density. In Fig. 21.1 a seven cell cluster with frequency sets F1 through F7 (in cells designated A–G) has been illustrated. It is seen that the same set of frequencies is repeated two cells away.

The number of channels for each cell can be determined through the erlang-B formula (for example, see Cooper, 1990) by knowing the busy hour traffic and the desired probability of blocking (grade of service). Probability of blocking P_B is related to offered traffic A, and the number of channels per cell N by

$$P_B = \frac{A^N/N!}{\displaystyle\sum_{i=0}^{N} A^i/i!}$$

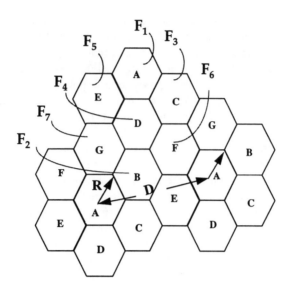

FIGURE 21.1   Fixed channel assignment.

This applies to the case of blocked calls cleared. If calls are delayed, the grade of service becomes the probability of calls being delayed P_Q and is given by the erlang-C formula [Cooper, 1990],

$$P_Q = \frac{\dfrac{A^N}{N!(1 - A/N)}}{\displaystyle\sum_{i=0}^{N-1} A^i/i! + \dfrac{A^N}{N!(1 - A/N)}}$$

FCA is used in almost all existing cellular mobile networks employing FDMA or TDMA. To illustrate resource assignment in FCA, we describe the FCA schemes used in gobal system for mobile communications (GSM or DCS) [Jabbari et al., 1995]. Here, the mobile terminal continuously monitors the received signal strength and quality of a broadcast channel along with the identity of the transmitting base station. This mechanism allows both the mobile and the network to keep track of the subscriber movements throughout the service area. When the mobile terminal tries and eventually succeeds in accessing the network (see Fig. 21.2), a two way control channel is assigned that allows for both authentication and ciphering mode establishment. On completion of these two phases, the setup phase initiates and a traffic channel is eventually assigned to the mobile terminal. Here, the terminal only knows the identity of the serving cell; the (control and traffic) radio channel assignment is a responsibility of the network and fulfills the original carrier to cell association, while simplifying the control functions during normal system operation.

21.4 Enhanced Fixed Channel Assignment

FCA has the advantage of having simple realization. Since the frequency sets are preassigned to cells based on long-term traffic demand, however, it cannot adapt to traffic variation across cells and, therefore, FCA will result in poor bandwidth utilization. To improve the utilization while maintaining the implementation simplicity, various strategies have been proposed as enhancements to FCA and deployed in existing networks. Two often used methods are *channel borrowing* and *directed retry*, which are described here.

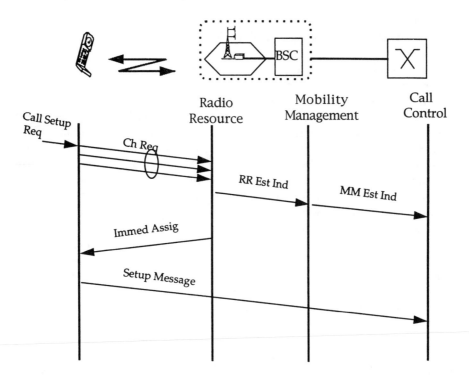

FIGURE 21.2 Radio resource assignment in GSM/DCS. *Source:* Jabbari, B. et al. 1995. Network issues for wireless personal communications. *IEEE Commun. Mag.* 33(1).

In the channel borrowing strategy [Jakes, 1994; Engel and Peritsky, 1973; Elnoubi, Singh, and Gupta, 1982; Zhang and Yum, 1989], channels that are not in use in their cells may be borrowed by adjacent cells with high offered traffic on a call-by-call basis. Borrowing of channels allows the arriving calls to be served in their own cell. This implies that there will be further restrictions in using the borrowed channels in other cells. Various forms of borrowing have been surveyed in Tekinay and Jabbari, 1991.

In directed retry [Eklundh, 1986; Everitt, 1990; 1994], a call to or from a mobile subscriber may try other cells with channels with sufficient signal strength meeting the C/I constraint if there are no channels available in its own cell to be served. In some cases it may be necessary to direct some of the calls in progress in a given congested cell to adjacent lightly loaded cells (those calls that can be served by adjacent cells) in order to accommodate the new calls in that given cell. This is referred to as directed handoff [Karlson and Eklundh, 1989; Everitt, 1990]. The combination of these two capabilities provides a significant increase in bandwidth utilization.

21.5 Dynamic Channel Assignment

In dynamic channel assignment [Cox and Reudnik, 1972; 1973; Engel and Peritsky, 1993; Anderson, 1973; Beck and Panzer, 1989; Panzer and Beck, 1990; Chuang, 1993; Dimitrijevic and Vucetic, 1993], the assignment of channels to cells occurs based on the traffic demand in the cells. In other words, channels are pooled together and assignments are made and modified in real time. Therefore, this assignment scheme has the potential to achieve a significantly improved bandwidth utilization when there are temporal or spatial traffic variations.

In DCA, the interference constraints are ensured by a real-time evaluation of the most suitable (less interfered) channels that can be activated in a given cell (reassignment may be needed). That is, the system behaves as if the frequency plan was dynamically changing to meet the actual radio link quality and traffic loads, realizing an implicit sharing of the frequency band under interference constraints.

The implementation of DCA generally is more complex due to the requirement for system-wide state information where the state refers to which channel in which cell is being used. Obviously, it is impractical to update the system state in a large cellular network at any time, especially those based on microcells, as the controller will be overloaded or call set delay will be unacceptable. Therefore, methods have been devised based on a limited state space centralized control [Everitt and Macfadyen, 1983] or based on a distributed control to perform the necessary updating. A method referred to as maximum packing suggested in Everitt and Macfadyen, 1983, records only the number of channels in use. Distributed control schemes, however, where channel assignments are made at mobile stations or base stations may be attractive. These schemes are particularly suitable for a mobile controlled resource assignment where each mobile measures the actually perceived interference and decides to utilize a radio resource in a completely decentralized way.

Design of DCA algorithms is critical to achieve the potential advantages in efficiency and robustness to traffic heterogeneity throughout cells as compared to FCA and enhanced FCA. Poor DCA algorithms, however, might lead to an uncontrolled global situation, i.e., the locally selected channel might be very good for the specific mobile terminal and at the same time very poor for the interference level induced to other traffic sources.

Two classes of *regulated DCA* and *segregation DCA* are discussed here due to their importance. In a regulated DCA, appropriate thresholds tend to maintain the current channel, avoiding useless handoffs that can tie up channels; in a segregation DCA, channel acquisition obeys priorities assigned to each channel. In the latter, the channels successfully activated in a cell have a higher assignment probability than those found to have a high interference in the cell; of course, priorities change in time depending on changes in the system status. These types of assignment lead to a (frequency) plan still changing dynamically but more slowly and, in general, only because of substantial load imbalance. In steady-state conditions, the plan either tends to be confirmed or fluctuates around a basic configuration, proven to be suboptimal in terms of bandwidth efficiency.

Both digital European cordless telecommunications (DECT) and cordless technology known as CT2 are employing the DCA technique [Tuttlebee, 1992], and the next generation cellular systems are foreseen to deploy it. In CT2 the handset and the base station jointly determine a free suitable channel to serve the call. In DECT, the mobile terminal (portable handset) not only recognizes the visited base station (radio fixed part) through a pilot signal but continuously scans all of the system channels and holds a list of the less interfered ones. The potentially available channels are ordered by the mobile terminal with respect to the measured radio parameters, namely, the radio signal strength indicator, measuring and combining cochannel, adjacent-channel, and intermodulation interference. When a connection has to be established, the best channel is used to communicate on the radio interface.

It is possible to have a hybrid of DCA and FCA in a cellular network in which a fraction of channels are fixed assigned and the remainder are allocated based on FCA. This scheme has less system implementation complexity than the DCA scheme but provides performance improvement (lower probability of call blocking) depending on the DCA–FCA channel partitioning [Kahwa and Georganas, 1978].

In general, DCA schemes cannot be considered independently of the adopted power control mechanism because the transmitted power from both mobile terminal and base station substantially affects the interference within the network. For a detailed discussion of DCA and power control, the readers are referred to [Chuang, Sollenberger, and Cox, 1994]. Despite the availability of several power control algorithms, much work is needed to identify realizable power and channel control algorithms that maximize bandwidth efficiency. Nevertheless, realization of power control mechanisms would require distribution of the system status information and information exchange between mobile terminal and network entities. This in turn will involve overhead and terminal power usage.

The performance of DCA depends on the algorithm implementing this capability [Everitt, 1990; Everitt and Manfield, 1989]. In general, due to interactions between different cells the performance of the system will involve modeling the system as a whole, as opposed to in FCA where cells are treated independently. Therefore, mathematical modeling and performance evaluation of DCA becomes quite complex. Simplifying assumptions may, therefore, be necessary to obtain approximate results [Prabhu and Rappaport, 1974; Everitt and Macfadyen, 1983]. Simulation techniques have been widely used in evaluation of DCA performance. For a representative performance characteristics of DCA and a comparison to enhanced FCA schemes the readers are referred to [Everitt, 1994].

21.6 Conclusion

In this chapter we have classified and reviewed channel assignment techniques. We have emphasized the advantages of DCA schemes over FCA in terms of bandwidth utilization in a heterogeneous traffic environment at the cost of implementation complexity. The DCA schemes are expected to play an essential role in future cellular and microcellular networks.

References

Abramson, N. 1994. Multiple access techniques for wireless networks. *Proc. of IEEE.* 82(9).

Anderson, L.G. 1973. A simulation study of some dynamic channel assignment algorithms in a high capacity mobile telecommunications system. *IEEE Trans. on Comm.* COM-21(11).

Beck, R. and Panzer, H. 1989. Strategies for handover and dynamic channel allocation in microcellular mobile radio systems. *Proceedings of the IEEE Vehicular Technology Conference.*

Chuang, J.C.-I. 1993. Performance issues and algorithms for dynamic channel assignment. *IEEE J. on Selected Areas in Comm.* 11(6).

Chuang, J.C.-I., Sollenberger, N.R., and Cox, D.C. 1994. A pilot-based dynamic channel assignment schemes for wireless access TDMA/FDMA systems. *Internat. J. of Wireless Inform. Networks,* Jan.

Cooper, R.B. 1990. *Introduction to Queueing Theory,* 3rd ed. CEEPress Books.

Cox, D.C. and Reudnik, D.O. 1972. A comparison of some channel assignment strategies in large-scale mobile communications systems. *IEEE Trans. on Comm.* COM-20(2).

Cox, D.C. and Reudnik, D.O. 1973. Increasing channel occupancy in large-scale mobile radio environments: dynamic channel reassignment. *IEEE Trans. on Comm.* COM-21(11).

Dimitrijevic, D. and Vucetic, J.F. 1993. Design and performance analysis of algorithms for channel allocation in cellular networks. *IEEE Trans. on Vehicular Tech.* 42(4).

Eklundh, B. 1986. Channel utilization and blocking probability in a cellular mobile telephone system with directed retry. *IEEE Trans. on Comm.* COM 34(4).

Elnoubi, S.M., Singh, R., and Gupta, S.C. 1982. A new frequency channel assignment algorithm in high capacity mobile communications. *IEEE Trans. on Vehicular Techno.* 31(3).

Engel, J.S. and Peritsky, M.M. 1973. Statistically-optimum dynamic server assignment in systems with interfering servers. *IEEE Trans. on Comm.* COM-21(11).

Everitt, D. 1990. Traffic capacity of cellular mobile communications systems. *Computer Networks and ISDN Systems,* ITC Specialist Seminar, Sept. 25–29, 1989.

Everitt, D. 1994. Traffic engineering of the radio interface for cellular mobile networks. *Proc. of IEEE.* 82(9).

Everitt, D.E. and Macfadyen, N.W. 1983. Analysis of multicellular mobile radiotelephone systems with loss. *BT Tech. J.* 2.

Everitt, D. and Manfield, D. 1989. Performance analysis of cellular mobile communication systems with dynamic channel assignment. *IEEE J. on Selected Areas in Comm.* 7(8).

Jabbari, B., Colombo, G., Nakajima, A., and Kulkarni, J. 1995. Network issues for wireless personal communications. *IEEE Comm. Mag.* 33(1).

Jakes, W.C. ed. 1994. *Microwave Mobile Communications,* Wiley, New York, 1974, reissued by IEEE Press.

Kahwa, T.J. and Georganas, N.D. 1978. A hybrid channel assignment scheme in Large scale, cellular-structured mobile communications systems. *IEEE Trans. on Comm.* COM-26(4).

Karlsson, J. and Eklundh, B. 1989. A cellular mobile telephone system with load sharing—an enhancement of directed retry. *IEEE Trans. on Comm.* COM 37(5).

Panzer, H. and Beck, R. 1990. Adaptive resource allocation in metropolitan area cellular mobile radio systems. *Proceedings of the IEEE Vehicular Technology Conference.*

Prabhu, V. and Rappaport, S.S. 1974. Approximate analysis for dynamic channel assignment in large systems with cellular structure. *IEEE Trans. on Comm.* COM-22(10).

Tekinay, S. and Jabbari, B. 1991. Handover and channel assignment in mobile cellular networks. *IEEE Comm. Mag.* 29(11).

Telecommunications Industry Association. 1993. TIA Interim Standard IS-95, CDMA Specifications.

Tuttlebee, W.H.W. 1992. Cordless personal communications. *IEEE Comm. Mag.* (Dec.).

Zhang, M. and Yum, T.-S.P. 1989. Comparison of channel-assignment strategies in cellular mobile telephone systems. *IEEE Trans. on Vehicular Tech.* 38(4).

22

Propagation Models

Theodore S. Rappaport
*Virginia Polytechnic Institute
and State University*

Rias Muhamed
*Virginia Polytechnic Institute
and State University*

Varun Kapoor
*Virginia Polytechnic Institute
and State University*

22.1 Introduction

The radio channel places fundamental limitations on the performance of mobile communication systems. The transmission path between the transmitter and the receiver can vary from simple direct line of sight to one that is severely obstructed by buildings and foliage. The speed of motion impacts how rapidly the signal level fades as a mobile terminal moves in space. Unlike wired channels that are stationary and predictable, radio channels are extremely random and do not offer easy analysis. In fact, modeling the radio channel has historically been one of the challenging parts of any radio system design and is typically done in a statistical fashion, based on measurements made specifically for an intended communication system.

The mechanisms behind electromagnetic wave propagation are diverse but can generally be attributed to reflection, diffraction, and scattering. Most cellular radio systems operate in urban areas where there is no direct line-of sight path between the transmitter and the receiver, and the presence of high-rise buildings causes severe diffraction loss. Because of the multiple reflections from various objects, the electromagnetic waves travel along different paths of varying lengths. The interaction between these waves causes multipath fading at a specific location, and the strengths of the waves decrease as the distance between the transmitter and receiver increases.

Propagation models have traditionally focused on predicting the average received signal strength at a given distance from the transmitter, as well as the variability of the signal strength in spatial proximity to a particular location. A statistical representation of the mean signal strength for an arbitrary transmitter-receiver (T-R) separation distance is useful in estimating the radio coverage area of a transmitter, whereas radio modem design issues, such as antenna diversity and coding, require models that predict the signal variability over a very small distance. Propagation models that characterize signal strength over large T-R separation distances (several hundreds or thousands of wavelengths) are called **large-scale models**, whereas models that characterize the rapid fluctuations of the received signal strength over short distances (a few wavelengths) or short time durations (on the order of seconds) are called **small-scale models**.

22.2 Free-Space Propagation Model

Free space is an ideal propagation model that can be accurately applied only to satellite communication systems and short line-of-sight radio links. It demonstrates, however, how received power decays as a logarithmic function of the T-R separation, which is a fundamental characteristic of large-scale modeling. The power received by a receiver antenna, which is separated from a radiating transmitter antenna by a distance d, assuming a free space path between the antennas, is given by the Friis free-space equation

$$P_r = \frac{P_t G_t G_r \lambda^2}{(4\pi)^2 d^2 L} \tag{22.1}$$

where P_t the transmitted power, G_t the transmitter antenna gain, G_r the receiver antenna gain, d the T-R separation distance in meters, L the system losses ($L > 1$), and λ the wavelength in meters.

The Friis transmission formula gives an inverse square relationship between the received power and the T-R separation distance. This implies the received power decays at a rate of 20 dB/decade with distance.

The path loss is defined as the difference between the effective transmitted power and the received power and may include the effect of the antenna gains. The path loss PL in decibels for the free space model is given by

$$PL = -10 \log_{10} \frac{P_r}{P_t} = -10 \log_{10} \left[\frac{G_t G_r \lambda^2}{(4\pi)^2 d^2 L} \right] \tag{22.2}$$

22.3 The Three Basic Propagation Mechanisms

Reflection occurs when a propagating electromagnetic wave impinges upon an obstruction whose dimensions are very large when compared to the wavelength of the radio wave. Reflections from the surface of the earth and from buildings or walls produce reflected waves, which may interfere constructively or destructively at the receiver.

Diffraction occurs when the radio path between the transmitter and receiver is obstructed by a surface that has sharp irregularities (edges). The secondary waves resulting from the obstructing surface are present throughout the space and even behind the irregularity, giving rise to bending of waves about the irregularity, even when a line of sight does not exist. At high frequencies, diffraction, like reflection, depends on the geometry of the object, and the amplitude, phase, and polarization of the incident wave at the point of diffraction.

Scattering occurs when the medium through which the wave travels consists of objects with dimensions that are small compared to the wavelength and the number of obstacles per unit volume is quite large.

22.4 Reflection

When a radio wave propagating in one medium impinges upon another medium having a different dielectric constant, permeability, or conductivity, the wave is partially reflected and partially

TABLE 22.1   Ground Parameters at 100 MHz

Material	Permittivity, ε_r	Conductivity σ (s/m)
Poor ground	4	0.001
Typical ground	15	0.005
Good ground	25	0.02
Sea water	81	5.0
Fresh water	81	0.01

transmitted. For the case of a plane wave in air incident normally on a perfect conductor, the wave is completely reflected without loss of energy. If the plane wave is incident on a perfect dielectric, part of the energy is transmitted and part of the energy is reflected, and there is no loss of energy in absorption. The electric field intensity of the reflected wave and the incident wave are related through the Fresnel reflection coefficient Γ. The reflection coefficient is a function of the permittivity of the ground, wave polarization, angle of incidence, and the frequency of the propagating wave. Table 22.1 shows typical values of permittivity and conductivity for a few common types of ground when operating in the VHF frequency range.

Figure 22.1 shows the geometry for calculating the reflection coefficients. The subscripts v and h refer to the vertical and horizontal polarization, respectively, of the E-field, and the subscripts i and r refer to the incident and reflected E-field, respectively. The permittivity, permeability, and conductance of the two media are ε_1, μ_1, σ_1 and ε_2, μ_2, σ_2, respectively. For the case when the first medium is free space, the reflection coefficients for the two cases of vertical (parallel) and horizontal (perpendicular) polarization are given as (for vertical E-field polarization)

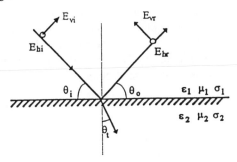

FIGURE 22.1   Geometry for reflection coefficient.

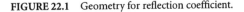

$$\Gamma_v = \frac{E_{vr}}{E_{vi}} = \frac{-\varepsilon_r \sin\theta_i - \sqrt{\varepsilon_r - \cos^2\theta_i}}{\varepsilon_r \sin\theta_i + \sqrt{\varepsilon_r - \cos^2\theta_i}} \tag{22.3}$$

and for horizontal E-field polarization

$$\Gamma_h = \frac{E_{hr}}{E_{hi}} = \frac{\sin\theta_i - \sqrt{\varepsilon_r - \cos^2\theta_i}}{\sin\theta_i + \sqrt{\varepsilon_r - \cos^2\theta_i}} \tag{22.4}$$

where ε_r is the relative permittivity of the second medium. The incident angle θ_i at which the reflection coefficient Γ_v is equal to zero is called the *Brewster angle* and is given by the value of θ_i that satisfies the equation

$$\sin\theta_i = \frac{\sqrt{\varepsilon_r - 1}}{\sqrt{\varepsilon_r^2 - 1}} \tag{22.5}$$

Ground Reflection (Two-Ray) Model

In a mobile radio channel a single direct path between the base station and the mobile seldom exists and, hence, the free-space propagation model of Eq. (22.2) is of limited use. The two-ray ground reflection model shown in Fig. 22.2 is a useful propagation model that is based on geometric optics

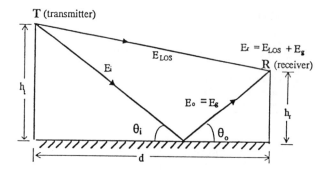

FIGURE 22.2   Two-ray ground reflection model.

and considers both the direct and ground reflected propagation path. This model assumes that the wavelength λ is much smaller than anything else encountered in the channel.

In most mobile communication systems where the separation distance is only a few tens of kilometers, the flat terrain assumption is a valid approximation. The total received E-field E_r is the resultant of the direct line-of-sight component E_{LOS} and a ground reflected component E_g, and is referenced to a close-in E-field measured at a small distance d_0.

Referring to Fig. 22.2, h_t is the height of the transmitter and h_r the height of the receiver. According to laws of reflection,

$$\theta_i = \theta_0 \qquad \text{and} \qquad E_0 = \Gamma E_i \tag{22.6}$$

where Γ is the reflection coefficient for ground. As θ_i approaches $0°$ (i.e., grazing incidence) the reflected wave is equal in magnitude and $180°$ out of phase with the incident wave. The resultant E-field, which is the sum of E_{LOS} and E_g, can be shown to be

$$|E_r(d)| = \frac{2E_{\text{LOS}}d_0}{d} \sin \frac{\theta_\Delta}{2} \tag{22.7}$$

where the phase difference (θ_Δ) is related to the path difference (Δ) between the direct and ground reflected paths by

$$\theta_\Delta = \frac{2\pi \, \Delta}{\lambda} \tag{22.8}$$

At large values of d,

$$\sin \frac{\theta_\Delta}{2} \approx \frac{\theta_\Delta}{2} = \frac{2\pi h_t h_r}{\lambda d} \tag{22.9}$$

and the received E-field in volts per meter is given by

$$E_R(d) \approx 2E_{\text{LOS}} \frac{2\pi h_t h_r d_0}{\lambda d^2} \approx \frac{k}{d^2} \tag{22.10}$$

where k is a constant related to E_{d_0}, the electric field measured at distance d_0 from the transmitter. The power received at d is related to the square of the electric field and can be expressed approximately as

$$P_r = P_t G_t G_r \frac{h_t^2 h_r^2}{d^4} \tag{22.11}$$

At large distances, the received power falls off at a rate of 40 dB/decade, and the received power and path loss become independent of frequency. The path loss in decibels for the two-ray model is approximated as

$$PL = -10\log_{10} G_t - 10\log_{10} G_r - 20\log_{10} h_t - 20\log_{10} h_r + 40\log_{10} d \quad (22.12)$$

22.5 Diffraction

Diffraction around the Earth's curvature makes it possible to transmit radio signals beyond the line of sight and allows an electromagnetic wave to propagate beyond obstructions. Although the field strength decreases rapidly as one moves deeper into the shadowed region, the field is still finite and has sufficient strength to produce a useful signal.

The phenomenon of diffraction can be easily explained using Huygen's principle, which states that all points on a wavefront can be considered as point sources for the production of secondary wavelets, and that these wavelets combine to produce a new wavefront in the direction of propagation. Diffraction is caused by the propagation of secondary wavelets into the shadowed region. The field strength of a diffracted wave in the shadowed region is the vector sum of all of the secondary wavelets.

Knife-Edge Diffraction Model

When shadowing is caused by a single object, such as a mountain or building, the attenuation caused by diffraction over such an object can be estimated by treating the obstruction as a diffracting knife edge. This is the simplest of diffraction models, and the diffraction loss in this case can be readily estimated using the classical Fresnel solution for the field behind a knife edge or half-plane. Figure 22.3 illustrates this approach.

The field strength at point R in the shadowed region (also called the **diffraction zone**) is a vector sum of the fields due to all of the secondary Huygen sources in the plane above the knife edge. The field strength E_d of a knife-edge diffracted wave is given by

$$E_d = E_0 F(v) = E_0 \frac{1+j}{2} \int_v^\infty \exp(-j\pi t^2/2)\, dt \quad (22.13)$$

where E_0 is the free space field strength in the absence of the knife edge and $F(v)$ is the complex Fresnel integral. The Fresnel integral is a function of the Fresnel–Kirchoff diffraction parameter v

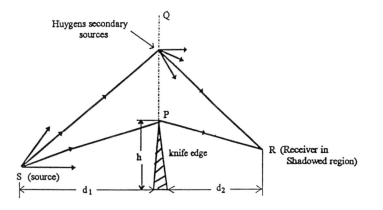

FIGURE 22.3 Knife-edge diffraction.

which is defined as

$$v = h\sqrt{\frac{2(d_1 + d_2)}{\lambda d_1 d_2}} \tag{22.14}$$

where h is the height of the knife edge and d_1 and d_2 are the distances of the knife edge from the transmitter and the receiver, respectively. If the knife edge protrudes above a line drawn from S to R, h and v are positive. If the obstruction lies below the line of sight, then both h and v are negative. The Fresnel integrals are available in tabular form for given values of the diffraction parameter v. The diffraction gain in decibels due to the presence of a knife edge is given by

$$G_d = 20\log_{10}|F(v)| \tag{22.15}$$

Since the expression for diffraction gain involves the Fresnel integral, which is difficult to compute, approximate numerical solutions are often used. An approximate solution, in decibels, provided by Lee, 1985, is given as

$$G_d = 0 \qquad\qquad 1 \le v \tag{22.16a}$$
$$G_d = 20\log_{10}(0.5 + 0.62v) \qquad\qquad 0 \le v \le 1 \tag{22.16b}$$
$$G_d = 20\log_{10}(0.5\exp(0.95v)) \qquad\qquad -1 \le v \le 0 \tag{22.16c}$$
$$G_d = 20\log_{10}(0.4 - \sqrt{0.1184 - (0.1v + 0.38)^2}) \qquad -2.4 \le v \le -1 \tag{22.16d}$$
$$G_d = 20\log_{10}(-0.225/v) \qquad\qquad v \le -2.4 \tag{22.16e}$$

Note that when $v = 0$, the diffraction gain is -6 dB. In many practical situations, especially in hilly terrain, the propagation path may consist of more than one obstruction, and it may be required to calculate the total diffraction losses due to all of the obstacles. Bullington, 1947, suggested that the series of obstacles be replaced by a single equivalent obstacle so that the path loss can be obtained using single knife-edge diffraction models. This method oversimplifies the calculations and often provides very optimistic estimates of the received signal strength. In a more rigorous treatment, Millington, Hewitt, and Immirzi, 1962, gave a wave-theory solution for the field behind two knife edges in series. This solution is very useful and can be applied easily for predicting diffraction losses due to two knife edges. Extending this to more than two knife edges, however, becomes a formidable mathematical problem. Many models that are mathematically less complicated have been developed to estimate the diffraction losses due to multiple obstructions [Epstein and Peterson, 1953; Deygout, 1966].

22.6 Scattering

The measured path loss in a mobile radio environment is often less than what is predicted by reflection and diffraction models alone. This is because when a radio wave impinges on a rough surface, the reflected energy is spread out (diffused) in all directions due to scattering. Objects such as trees, lamp posts, and rough surfaces tend to scatter energy towards a receiver. The roughness of a surface is often tested using the Rayleigh criterion, which defines a critical height h_c of surface protuberances for a given angle of incidence θ_i,

$$h_c = \frac{\lambda}{8\cos\theta_i} \tag{22.17}$$

A surface is considered smooth if its minimum to maximum protuberance h is less than h_c and is considered rough if the protuberance is greater than h_c. For rough surfaces, the reflection coefficient needs to be modified by a scattering loss factor ρ_s to account for the diminished specularly reflected field. Ament, 1953, derived this as

$$\rho_s = \exp\left[-8\left(\frac{\pi \sigma_h \cos \theta_i}{\lambda}\right)^2\right] \qquad (22.18)$$

where σ_h is the standard deviation of the surface height about the mean surface height. The scattering loss factor derived by Ament was modified by Boithias, 1987, to give better agreement with measured results,

$$\rho_s = \exp\left[-8\left(\frac{\pi \sigma_h \cos \theta_i}{\lambda}\right)^2\right] I_0\left[8\left(\frac{\pi \sigma_h \cos \theta_i}{\lambda}\right)^2\right] \qquad (22.19)$$

where I_0 is the Bessel function of the first kind and zeroth order.

Analysis based on the geometric theory of diffraction and physical optics can be used to determine the scattered field strength. For urban mobile radio systems, models based on the bistatic radar equation may be used to compute the scattering losses in the far field. The radar cross section (RCS) of a scattering object is defined as the ratio of the power density of the signal scattered in the direction of the receiver to the power density of the radio wave incident upon the scattering object and has units of square meters. The bistatic radar equation given by Eq. (22.20) describes the propagation of a wave traveling in free space and intercepted by a scattering object, and then reradiated in the direction of the receiver,

$$P_R(\mathrm{dBm}) = P_T(\mathrm{dBm}) + G_T(\mathrm{dBi}) + 20\log_{10}\lambda + \mathrm{RCS}[\mathrm{dBm}^2]$$
$$- 30\log_{10}(4\pi) - 20\log_{10} d_T - 20\log_{10} d_R \qquad (22.20)$$

where d_T and d_R are the distance from the scattering object to the transmitter and receiver, respectively. This equation can only be applied to scatterers in the far field of both the transmitter and receiver.

7 Practical Propagation Models

Most radio propagation models are derived using a combination of analytical and empirical methods. The empirical approach is based on fitting curves or analytical expressions that recreate a set of measured data. This has the advantage of implicitly taking into account all propagation factors, both known and unknown. The validity of empirical models at transmission frequencies and environments other than those used to originally derive the model can only be established by verifying them with data taken from the specific area at the required transmission frequency.

Log-Distance Path Loss Model

Both theoretical and measurement-based propagation models show that the average received signal power decreases with distance raised to some exponent. Here the average path loss for an arbitrary T-R separation is expressed as a function of distance by using a path loss exponent.

$$\overline{PL}(d) \propto \left(\frac{d}{d_0}\right)^n \qquad (22.21)$$

TABLE 22.2 Path Loss Exponents for Different Environments

Environment	Path Loss Exponent, n
Free-space	2
Urban area cellular radio	2.7–4
Shadowed urban cellular radio	5–6
In building line of sight	1.6–1.8
Obstructed in building	4–6
Obstructed in factories	2–3

or in decibels

$$\overline{PL} = \overline{PL}(d_0) + 10n \log_{10}\left(\frac{d}{d_0}\right) \tag{22.22}$$

where n is the path loss exponent that indicates the rate at which the path loss increases with distance, d_0 is the free space close-in reference distance, and d is the T-R separation distance. When plotted on a log-log scale, the path loss is a straight line with a slope equal to n. The value of n depends on the specific propagation environment. For example, in free space, n is equal to 2, and when obstructions are present it will have a larger value. Table 22.2 gives a listing of typical path loss exponents obtained in various mobile radio environments.

It is important to select a free space reference distance that is appropriate for the propagation environment. In large cellular systems, 1 km and 1 mile reference distances are commonly used [Lee, 1985], whereas in microcellular systems, much smaller distances are used. The reference distance should always be in the far field of the antenna ($d_0 > 2D^2/\lambda$, where D is the largest antenna dimension) so that near-field effects do not alter the reference path loss. The reference path loss is calculated using the free space path loss formula given by Eq. (22.2).

Measurements have shown that at any value of d, the path loss $PL(d)$ at a particular location is distributed log-normally (normal in decibels) about the mean distance-dependent value. That is, $PL(d) = \overline{PL}(d) + X_\sigma$, where X_σ is a zero-mean lognormally distributed random variable with standard deviation σ in decibels. The log-normal distribution describes the random shadowing effects, which occur over a large number of measurement locations, which have the same T-R separation but have different levels of clutter on the propagation paths. A path loss exponent n and standard deviation σ will therefore statistically describe the path loss for an arbitrary location having a specific T-R separation. In practice, based on measurement data, the value of n and σ are computed using linear regression such that the difference between the measured and estimated path losses is minimized in a mean square error sense.

Outdoor Propagation Models

Radio transmission in a mobile communications system often takes place over irregular terrain. The terrain profile of a particular area needs to be taken into account while estimating the path loss. The terrain profile may vary from a simple curved earth profile to a highly mountainous profile. The presence of trees, buildings, and other obstacles also need to be taken into account.

A number of propagation models are available to predict path loss over irregular terrain. Although all of these models aim to predict signal strength at a particular receiving point or a specific small area (called a sector), the methods vary widely in their approach, complexity, and accuracy. Most of these models are based on a systematic interpretation of measurement data obtained from the service area. One of the more popular outdoor propagation models is the Longley–Rice [Rice et al., 1967] model. This model is applicable to point-to-point communication systems in the frequency range from 40 MHz to 100 GHz, over different kinds of terrain. The median transmission loss is predicted using the path geometry of the terrain profile and the refractivity of the troposphere.

Geometric-optics techniques (two-ray ground reflection model) are used to predict signal strengths within the radio horizon. Diffraction losses over isolated obstacles are estimated using the Fresnel–Kirchoff knife-edge models. Forward scatter theory is used to make troposcatter predictions over long distances and far-field diffraction losses in double horizon paths are predicted using a modified Van der Pol–Bremmer method. The Longley–Rice propagation prediction model is also referred to as the *ITS irregular terrain model.*

The Longley–Rice model is also available as a computer program [Longley and Rice, 1968] to calculate long-term median transmission loss relative to free space loss over irregular terrain for frequencies between 20 MHz and 10 GHz. For a given transmission path, the program takes as its input the transmission frequency, path length, polarization, antenna heights, surface refractivity, effective radius of Earth, ground conductivity, ground dielectric constant, and climate. The program also operates on path-specific parameters such as horizon distance of the antennas, horizon elevation angle, angular transhorizon distance, terrain irregularity, and other specific inputs.

The Longley–Rice method operates in two modes. When a detailed terrain path profile is available, the path specific parameters mentioned in the preceding paragraph can be easily determined and the prediction is called a *point-to-point mode* prediction. On the other hand, if the terrain path profile is not available, the Longley–Rice method provides techniques to estimate the path specific parameters, and such a prediction is called as an *area mode* prediction.

There have been many modifications and corrections to the Longley–Rice model since its original publication. One important modification [Longley, 1978] deals with radio propagation in urban areas, and this is particularly relevant to mobile radio. This modification introduces an excess term as an allowance for the additional attenuation due to urban clutter near the receiving antenna. This extra term called the urban factor (UF) has been derived by comparing the predictions by the original Longley–Rice model with those obtained by Okumura, 1968.

The major shortcoming of the Longley–Rice model is that it does not detail any method for determining corrections due to environmental factors in the immediate vicinity of the mobile or consider any correction factors to be included to account for the effects of buildings and foliage. Further, multipath is not considered.

The Okumura Model

Okumura's model is one of the most widely used models for signal prediction in urban areas. This model is applicable for frequencies in the range 150–2000 MHz and distances of 1–100 km. It can be used for base station effective antenna heights ranging from 30 to 1000 m.

Okumura developed a set of curves giving the median attenuation relative to free space, A_{mu}, in an urban area over a quasismooth terrain with a base station effective antenna height h_{te} of 200 m and a mobile antenna height h_{re} of 3 m. These curves are plotted as a function of frequency in the range 100–3000 MHz and as a function of distance from the base station in the range 1–100 km. To determine the path loss using Okumura's model, the free space path loss between the points of interest is first determined, and then the value of $A_{mu}(f, d)$ as read from the curves is added to it along with correction factors to account for antennas not at the reference heights. The model can be expressed as

$$L_{50} = L_F + A_{mu}(f, d) + G(h_{te}) + G(h_{re}) \qquad (22.23)$$

where L_{50} is the median propagation loss in decibels, L_F the free space propagation loss, A_{mu} the median attenuation relative to free space, $G(h_{te})$ the base station antenna height gain factor, and $G(h_{re})$ the mobile antenna height gain factor.

Plots of $G(h_{te})$ and $G(h_{re})$ vs the respective effective antenna heights are available. These curves show that $G(h_{te})$ varies at a rate of 20 dB/decade and $G(h_{re})$ for effective antenna heights less than 3 m varies at a rate of 10 dB/decade. Based on the terrain-related parameters, various corrections can be applied to Okumura's model. Some of the important terrain-related parameters are the effective base station antenna height h_{te}, terrain undulation height Δh, isolated ridge height, average slope of the terrain, and the mixed land–sea parameter. Once the terrain-related parameters are calculated, the necessary correction factors can be added or subtracted as required. All of these correction factors are also available as Okumura curves.

Okumura's model is wholly based on measured data and does not provide any analytical explanations. For many situations, extrapolations of the derived curves can be made to obtain values outside the measurement range, and the validity of such extrapolations depends on the circumstances and the smoothness of the curve in question.

Okumura's model is considered to be among the best in terms of accuracy in path loss prediction for mature cellular radio systems, and offers standard deviations within 12 dB of measured data. It is also a very practical model that has become a standard for system planning in today's land mobile radio systems in Japan. The only major disadvantage with the model is its complexity and slow response to rapid changes in radio path profile. Generally, it is found that the model is particularly good in urban and suburban areas but not as good in rural areas over irregular terrain.

The Hata Model

The Hata model [Hata, 1980] is an empirical formulation of the graphical path loss information provided by Okumura. Hata presented the urban area propagation loss as the standard formula and supplied correction equations to the standard formula for application to other situations. The standard formula for median path loss in urban areas is given by

$$L_{50} = 69.55 + 26.16 \log_{10} f_c - 13.82 \log_{10} h_{te} - a(h_{re})$$
$$+ (44.9 - 6.55 \log_{10} h_{te}) \log_{10} d \qquad (22.24)$$

where loss is in decibels, f_c the frequency from 150 to 1500 MHz, h_{te} the effective transmitter (base station) antenna height ranging from 30 to 200 m, h_{re} is the effective receiver (mobile) antenna height ranging from 1 to 10 m, d is the T-R separation distance in kilometers, and $a(h_{re})$ is the correction factor for effective mobile antenna height, which is a function of the size of the service area (city). For a small to medium sized city, the correction factor is given by

$$a(h_{re}) = (1.1 \log_{10} f_c - 0.7)h_{re} - (1.56 \log_{10} f_c - 0.8) \text{ dB} \qquad (22.25)$$

and for a large city, it is given by

$$a(h_{re}) = 8.29(\log_{10} 1.54 h_{re})^2 - 1.1 \text{ dB} \qquad \text{for } f_c \le 200 \text{ MHz} \quad (22.26a)$$
$$= 3.2(\log 11.75 h_{re})^2 - 4.97 \text{ dB} \qquad \text{for } f_c \ge 400 \text{ MHz} \quad (22.26b)$$

To obtain the path loss in decibels in a suburban area the formula is modified as

$$L_{50} = L_{50} \text{ (urban)} - 2[\log_{10}(f_c/28)]^2 - 5.4 \qquad (22.27)$$

and for the path loss in decibels in open areas the formula is modified as

$$L_{50} = L_{50} \text{ (urban)} - 4.78(\log_{10} f_c)^2 - 18.33 \log_{10} f_c - 40.98 \qquad (22.28)$$

Although Hata's formulation does not have any of the path-specific corrections available in the Okumura's model, the preceding expressions have significant practical value. The predictions of the Hata model compare very closely with that of the original Okumura's model as the difference is generally within 1 dB, so long as d exceeds 1 km.

Walfish and Bertoni Model

A recent model developed by Walfish and Bertoni, 1988, considers the impact of rooftops and building height by using diffraction to predict average signal strength at street level. The model considers the path loss S to be a product of three factors.

$$S = P_0 Q^2 P_1 \qquad (22.29)$$

where P_0 represents free space path loss, which is the ratio of received to radiated power for isotropic antennas in free space, and is given by

$$P_0 = \left(\frac{\lambda}{4\pi d}\right)^2 \qquad (22.30)$$

The factor Q^2 gives the reduction in the roof top signal at the row of buildings, which immediately shadows the receiver at the street level. The P_1 term is based on wedge diffraction and determines the signal loss from the rooftop to the street.

Indoor Propagation Models

With the advent of Personal Communication Systems (PCS), there is a great deal of interest in characterizing the radio communication channel inside a building. The indoor channel differs from the traditional mobile radio channel in two aspects, the interference effects and the fading rate. The interference caused by the presence of electronic equipment can be unpredictable. The dynamic range of fading experienced by a hand-held unit inside the building is often smaller than that experienced by a moving vehicle in an urban setting. Because of secondary effects, such as motion of people and doors being opened and closed, the channel characteristics change with time, although slowly. It also has been observed that propagation within buildings is strongly influenced by local features, such as the layout of the building, the construction materials, and the building type. This section outlines models and measurement results for propagation within buildings.

Saleh Model

Saleh and Valenzula, 1987, reported the results of indoor propagation measurements between two vertically polarized omnidirectional antennas located on the same floor of a medium sized office building. Measurements were made using 10-ns, 1.5-GHz, radar-like pulses. The method involved averaging the square law detected received pulse response while sweeping the frequency of the transmitted pulse. Using this method, multipath components within 5 ns were resolvable.

The results obtained by Saleh and Valenzula show that 1) the indoor channel is quasistatic or very slowly time varying and 2) the statistics of the channel's impulse response are independent of transmitting and receiving antenna polarization, if there is no line-of-sight path between them. They reported a maximum multipath delay spread of 100–200 ns within the rooms of a building and 300 ns in hallways. The measured rms delay spread within rooms had a median of 25 ns and a maximum of 50 ns. The signal attenuation with no line-of-sight path was found to vary over 60-dB range and obey a log-distance power law [Eqs. (22.21) and (22.22)] with an exponent between 3 and 4.

Saleh and Valenzuela developed a simple multipath model for indoor channels based on measurement results. The model assumed that the rays arrive in clusters. The amplitudes of the received ray are independent Rayleigh random variables with variances that decay exponentially with cluster delay as well as ray delay within a cluster. The corresponding phase angles are independent uniform random variables over $[0, 2\pi]$. The clusters and rays within a cluster form a Poisson arrival process with different rates. The clusters and rays have exponentially distributed interarrival times. The formation of the clusters is related to the building structure, whereas the rays within the cluster are formed by multiple reflections from objects in the vicinity of the transmitter and the receiver.

SIRCIM Model

Rappaport et al., 1989, reported results of measurements at 1300 MHz in five factory buildings and carried out subsequent measurements in other types of buildings. Multipath delays ranged from 40 to 800 ns. Mean multipath delay and rms delay spread values ranged from 30 to 300 ns, with median values of 96 ns in LOS paths and 105 ns in obstructed paths. Delay spreads were found to be uncorrelated with T-R separation but were affected by factory inventory, building construction materials, building age, wall locations, and ceiling heights. Measurements in a food processing factory that manufactures dry-goods and has considerably less metal inventory than other factories had an rms delay spread that was half of those observed in factories producing metal products. Newer factories, which incorporate steel beams and steel reinforced concrete in the building structure, have stronger multipath signals than older factories, which used wood and brick for perimeter walls. The data suggested that radio propagation in buildings may be described by a hybrid geometric/statistical model that accounts for both specular reflections from walls and ceilings and random scattering from inventory and equipment. The authors developed an elaborate empirically derived statistical model and computer code called Simulation of Indoor Radio Channel Impulse-response Models (SIRCIM) that recreates realistic samples of indoor channel measurements [Rappaport et al., 1991], and subsequent work developed models for predicting coverage between floors using building blue prints [Seidel and Rappaport, 1992].

22.8 Small-Scale Fading

In a mobile radio environment, the short-term fluctuations caused by multipath propagation is called small-scale fading, to distinguish it from the large-scale variation in mean signal level, which is dependent on T-R separation. Small-scale fading is caused by wave interference between two or more multipath components that arrive at the receiver while the mobile travels a short distance (a few wavelengths) or over a short period of time. These waves combine vectorally at the receiver antenna to give the resultant signal, which can vary widely in amplitude, depending on the distribution of phases of the waves and the bandwidth of the transmitted signal.

Different channel conditions produce different types of small-scale fading. The type of fading experienced by the mobile depends on the following factors.

- *Speed of the mobile unit:* The relative motion between the base transmitter and the mobile results in random frequency modulation due to different Doppler shifts on different multipath components.
- *The transmission bandwidth of the signal:* If the transmitted signal bandwidth is greater than the flat-fading bandwidth of the multipath channel, the received signal is distorted.
- *The time delay spread of the received signal:* The presence of reflecting objects and scatterers in the channel constitutes a constantly changing environment that dissipates the signal energy. These effects result in multiple signals that arrive at the receiving antenna displaced with respect to one another in time and space, resulting in different times of arrival and time delays.
- *Random phase and amplitude:* The random phase and amplitudes of the different multipath components arriving at the receiving antenna cause rapid fluctuations in signal strength.
- *Rate of change of the channel:* The temporal variations of the channel impulse response caused by the changing multipath geometry results in signal fading. In a mobile environment, channel variations may arise due to movements of the transmitter, receiver or objects in their vicinity.

Flat Fading

Small-scale fading is generally classified as being either *flat* or *frequency selective*. If the mobile radio channel has a constant gain and a linear phase response over a bandwidth that is greater than the bandwidth of the transmitted signal, then the received signal will undergo *flat fading*. In this type of fading, the spectral characteristics of the transmitted signal are preserved. The strength of the received signal, however, will change with time, due to fluctuations in the gain of the channel caused by multipath. Flat fading channels are also known as **amplitude varying channels** and are sometimes referred to as *narrowband channels* since the bandwidth of the applied signal is narrow when compared to the channel bandwidth.

Rayleigh fading: In a flat fading mobile radio channel, where either the transmitter or the receiver is immersed in cluttered surroundings, the envelope of the received signal will typically have a Rayleigh distribution. The Rayleigh distribution has a probability density function given as

$$p(r) = \frac{r e^{-(\frac{r^2}{2\sigma^2})}}{\sigma^2} \qquad 0 \leq r \leq \infty \tag{22.31}$$

where σ^2 is the variance of the of the received signal r.

The *level crossing rate* and the *average fade duration* of a Rayleigh fading signal are two important statistics for determining error control codes and diversity schemes to be used in a communication system. The *level crossing rate* (LCR) is defined as the expected rate at which the received signal envelope, normalized to the local mean signal, crosses a specified level R, in a positive going direction. The number of level crossings per second is given by

$$N_R = \sqrt{2\pi} f_m \rho e^{-\rho^2} \tag{22.32}$$

where $f_m = (v/\lambda) \cos\theta$ is the maximum Doppler frequency and $\rho = R/R_{rms}$ is the value of the specified signal level R normalized to the local rms amplitude of the fading envelope. The *average fade duration* is defined as the average period of time for which the received signal is below a specified level R. It depends primarily on the speed of the mobile and is given by

$$\bar{\tau} = \frac{e^{\rho^2} - 1}{\rho f_m \sqrt{2\pi}} \tag{22.33}$$

The duration of a signal fade below a specified value of ρ determines the most likely number of signalling bits that will be lost during a fadeoutage event.

Rician fading: When there is a dominant signal component, such as a line-of-sight propagation path, the small-scale fading distribution is Rician. In such a situation, random components arriving at different angles are superimposed on a stationary signal. At the output of an envelope detector, this has the effect of adding a dc component to the random multipath. The Rician distribution is given by

$$p(r) = \frac{r}{\sigma^2} \exp - \left(\frac{r^2 + A^2}{2\sigma^2} \right) I_0 \left(\frac{A_r}{\sigma^2} \right), \qquad \text{for } \{A \geq 0, \quad r \geq 0\} \tag{22.34a}$$

$$= 0 \qquad\qquad\qquad \text{for } r < 0 \tag{22.34b}$$

The Rician distribution is often described in terms of a parameter K, which is defined as the ratio between the deterministic signal power and the variance of the multipath,

$$K\,(\text{dB}) = 10 \log_{10} \frac{A^2}{2\sigma^2} \tag{22.35}$$

The parameter A denotes the peak to peak amplitude of the dominant sine wave about the carrier in decibels. The parameter K completely specifies the Rician distribution. As $A \to 0$ and $K \to 0$, the dominant path diminishes and the Rician distribution degenerates to a Rayleigh distribution.

Frequency Selective Fading

If the channel has a constant gain and linear phase over a bandwidth that is much **smaller** than the bandwidth of the transmitted signal, then the channel induces **frequency selective fading**. Under such conditions, the channel impulse response has a multipath delay spread that is greater than the time duration of the transmitted message waveform. When this occurs, the received signal includes multiple versions of the transmitted waveform, each of which are delayed in time, and hence the received signal is distorted. Viewed in the frequency domain, this amounts to certain frequency components in the received signal spectrum having greater gains than others. Frequency selective fading channels are more difficult to model than flat fading channels since each multipath signal must be modeled, and the channel must be considered as a linear filter. When analyzing mobile communication systems, statistical impulse response models such as SIRCIM (for indoor channels) and SURP [Turin et al., 1972] or SMRCIM [Rappaport et al., 1993] (for outdoor channels) are generally used to study the effects of frequency selective small-scale fading.

Defining Terms

Large-scale models: Radio propagation models that characterize signal strength variations over large transmitter–receiver separation distances.

Small-scale models: Radio propagation models that characterize fluctuations of the received signal strength over small distances or short time durations.

Reflection: Phenomenon by which part of an electromagnetic wave traveling from one medium to another is turned back at the surface of the boundary.

Diffraction: Bending of electromagnetic waves around obstructions with sharp irregularities.

Scattering: Diffusion of reflected electromagnetic energy in all directions.

Rayleigh fading: Random variations in received signal envelope which follow a Rayleigh probability distribution.

Rician fading: Random variations in received signal envelope which follow a Rician probability distribution.

References

Ament, W.S. 1953. Toward a theory of reflection by a rough surface. *Proceedings of the IRE.* 41(1):142–146.

Boithias, L. 1987. *Radio Wave Propagation*, McGraw-Hill, New York.

Bullington, K. 1947. Radio propagation at frequencies above 30 megacycles. *Proceedings of the IEEE.* 35:1122–1136.

Deygout J. 1966. Multiple knife-edge diffraction of microwaves. *IEEE Trans. on Antennas and Propagation.* AP-14(4):480–489.

Epstein J. and Peterson, D.W. 1953. An exerimental study of wave propagation at 840 MC. *Proceedings of the IRE.* 41(5):595–611.

Hashemi, H. 1993. Indoor radio propagation channel. *Proceedings of the IEEE.* 81(7):941–968.

Hata, M. 1980. Empirical formula for propagation loss in land mobile radio services. *IEEE Trans. on Vehicular Tech.* VT-29(3):317–325.

Lee, W.C.Y. 1985. *Mobile Communications Engineering.* McGraw–Hill New York.

Longley, A.G. and Rice, P.L. 1968. Prediction of tropospheric radio transmission loss over irregular terrain; A computer method. ESSA Tech. Rept. ERL 79-ITS 67.

Longley, A.G. 1978. Radio propagation in urban areas. OT Rept. 78–144, April.

Millington, G., Hewitt, R., and Immirzi, F.S. 1962. Double knife edge diffraction in field strength predictions. *Proceedings of the IEE.* 109C:419–429.

Molkdar, D. 1991. Review on radio propagation into and within buildings. *IEE Proceedings,* 138(1).

Okumura, Y., Ohmori, E., Kawano, T., and Fakuda, K. 1968. Field strength and its variability in VHF and UHF land mobile radio service. *Rev. Elec. Communication Lab.* 16:825–873.

Rappaport, T.S., et al. 1991. Statistical channel impulse response models for factory and open plan building radio communication system design. *IEEE Trans. on Comm.* COM-39(5):794–806.

Rappaport, T.S., et al. 1993. Performance of decision feedback equalizers is simulated urban and indoor channels. Special issue on land/mobile/portable propagation, *IEICE Transactions on Communications,* Vol. E76-B, No. 2, Feb. 1993, Japan.

Rice, P.L., Longley, A.G., Norton, K.A., and Barsis, A.P. 1967. Transmission loss predictions for tropospheric communication circuits, NBS Tech Note 101, 2 vol. iss. May 7, 1965, rev. May 1, 1966, rev. Jan.

Saleh, A.M. and Valenzula, R.A. 1987. A statistical model for indoor multipath propagation. *IEEE J. Selected Areas in Comm.* SAC-5(2):128–137.

Seidel, S.Y. and Rappaport, T.S. 1992. 914 Mhz path loss prediction models for indoor wireless communications in multifloored buildings. *IEEE Trans. on Antenna and Propagation.* AP-40(2):1–11.

Turin, L.G. et al. 1972. A statistical model of urban multipath propagation. *IEEE Trans. on Vehicular Tech.* VT-21(1).

Walfish, J. and Bertoni, H.L. 1988. A theoretical model of UHF propagation in urban environments. *IEEE Trans. on Antennas and Propagation.* AP-36(Oct.):1788–1796.

Further Information

Recently, several books and survey papers on mobile radio propagation have been published. Two particularly well-written texts that cover the topic are *The Mobile Radio Propagation Channel*, by Parsons (Wiley 1992) and *Radio Wave Propagation*, by Griffiths (Wiley 1989). Two recent survey papers that cover the subject of indoor radio propagation are those by Molkdar, 1991 and Hashemi, 1993. The textbook *Wireless Communications* by Rappaport (Prentice-Hall) treats many propagation issues for wireless communications.

23

Power Control

Roman Pichna
University of Victoria

Qiang Wang
University of Victoria

23.1 Introduction

The growing demand for mobile communications is pushing the technological barriers of wireless communications. The available spectrum is becoming crowded and the old analog frequency division multiple access (FDMA) cellular systems no longer meet the growing demand for new services, higher quality, and spectral efficiency. A second generation of digital cellular mobile communication systems are on the horizon. The second generation systems are represented by two standards, the IS-54 and **global system for mobility (GSM)**. Both are time division multiple access (TDMA) based digital cellular systems and offer a significant increase in spectral efficiency and quality of service. Concurrently, another digital cellular standard has been proposed and accepted, known as IS-95, which is based on direct sequence code division multiple access (DS/CDMA) technology and promises further increase in spectral efficiency.

The channel capacity of a cellular system is significantly influenced by the cochannel interference. To minimize the cochannel interference, several techniques are proposed: frequency reuse patterns, which ensure that the same frequencies are not used in adjacent cells; efficient power control, which minimizes the transmitted power; cochannel interference cancellation techniques; and orthogonal signalling (time, frequency, or code). All of these are being intensively researched, and some have already been implemented.

This chapter provides a short overview of power control. Since power control is a very broad topic, it is not possible to exhaustively cover all facets associated with power control. The interested reader can find additional information in the recommended reading found at the end of this chapter.

The following section (Sec. 23.2) provides a brief introduction into cellular networks and demonstrates the necessity of power control. The various types of power control are presented. The next section (Sec. 23.3) illustrates some applications of power control employed in various systems such as analog **advanced mobile phone service (AMPS)**, GSM, DS/CDMA cellular standard IS-95, and digital cordless telephone standard **cordless telephone second generation (CT2)**. A list of defining terms is provided at the end of the chapter.

0-8493-8573-3/96/$0.00+$.50
© 1996 by CRC Press, Inc.

23.2 Cellular Systems and Power Control

In cellular communication systems, the service area is divided into cells, each covered by a single base station. If, in the forward link (base station to mobile), all users served by all base stations share the same frequency, each communication between a base station and a particular user would also reach all other users in the form of cochannel interference. The greater the distance between the mobile and the interfering transmitter, however, the weaker the interference becomes due to the propagation loss. To ensure a good quality of service throughout the cell, the received signal in the fringe area of the cell must be strong. Once the signal has crossed the boundary of a cell, however, it becomes interference and is required to be as weak as possible. Since this is difficult, the channel frequency is usually not reused in adjacent cells in most of the cellular systems. If the frequency is reused, the cochannel interference damages the signal reception in the adjacent cell, and the quality of service severely degrades unless other measures are taken to mitigate the interference. Therefore, a typical reuse pattern reuses the frequency in every seventh cell (frequency reuse factor = 1/7). The only exception is for CDMA-based systems where the users are separated by codes, and the allocated frequency may be shared by all users in all cells.

Even if the frequency is reused in every seventh cell, there is still some cochannel interference arriving at the receiver. It is, therefore, very important to maintain a minimal transmitted level at the base station to keep the cochannel interference low, frequency reuse factor high, and therefore the capacity of the system and quality of service high.

The same principle applies in the reverse link (mobile to base station); the power control maintains the minimum necessary transmitted power for reliable communication. Several additional benefits can be gained from this strategy. The lower transmitted power conserves the battery energy allowing the mobile terminal (the portable) to be lighter and stay on the air longer. Furthermore, recent concerns about health hazards caused by the portable's electromagnetic emissions are also alleviated.

In the reverse link, the power control also serves to alleviate the near–far effect. If all mobiles transmitted at the same power level, the signal from a near mobile would be received as the strongest. The difference between the received signal strength from the nearest and the farthest mobile can be in the range of 100 dB, which would cause saturation of the weaker signals' receivers or an excessive amount of adjacent channel interference. To avoid this, the transmitted power at the mobile must be adjusted inversely proportional to the effective distance from the base station. The term effective distance is used since a closely located user in a propagation shadow or in a deep fade may have a weaker signal than a more distant user having excellent propagation conditions.

In a DS/CDMA system, power control is a vital necessity for system operation. The capacity of a DS/CDMA cellular system is interference limited since the channels are neither separated in frequency nor separated in time, and the cochannel interference is inherently strong. A single user exceeding the limit on transmitted power could inhibit the communication of all other users.

The power control systems have to compensate not only for signal strength variations due to the varying distance between base station and mobile but must also attempt to compensate for signal strength fluctuations typical of a wireless channel. These fluctuations are due to the changing propagation environment between the base station and the user as the user moves across the cell or as some elements in the cell move. There are two main groups of channel fluctuations: slow (i.e., **shadowing**) and fast **fading**.

As the user moves away from the base station, the received signal becomes weaker because of the growing propagation attenuation with the distance. As the mobile moves in uneven terrain, it often travels into a propagation shadow behind a building or a hill or other obstacle much larger than the wavelength of the frequency of the wireless channel. This phenomenon is called shadowing. Shadowing in a land-mobile channel is usually described as a stochastic process having log-normal distributed amplitude. For other types of channels other distributions are used, e.g., Nakagami.

Electromagnetic waves transmitted from the transmitter may follow multiple paths on the way from the transmitter to the receiver. The different paths have different delays and interfere at the antenna of the receiver. If two paths have the same propagation attenuation and their delay differs in an odd number of half-wavelengths (half-periods), the two waves may cancel each other at the antenna completely. If the delay is an even multiple of the half-wavelengths (half-periods), the two waves may constructively add, resulting in a signal of double amplitude. In all other cases (nonequal gains, delays not a multiple of half-wavelength), the resultant signal at the antenna of the receiver is between the two mentioned limiting cases. This fluctuation of the channel gain is called fading. Since the scattering and reflecting surfaces in the service area are randomly distributed (houses, trees, furniture, walls, etc.), the amplitude of the resulting signal is also a random variable. The amplitude of fading is usually described by a Rayleigh, Rice, or Nakagami distributed random variable.

Since the mobile terminal may move at the velocity of a moving car or even of a fast train, the rate of channel fluctuations may be quite high and the power control has to react very quickly in order to compensate for it. The rate of fading is usually expressed in terms of Doppler frequency.

The performance of the **reverse link** of DS/CDMA systems is most affected by the near–far effect and, therefore, very sophisticated power control systems in the reverse link that attempt to alleviate the effects of channel fluctuations must be used. Together with other techniques, such as micro- and macrodiversity, interleaving, and coding, the DS/CDMA cellular system is able to cope with the wireless channel extremely well.

The effective use of the **power control** in DS/CDMA cellular system enables the frequency to be reused in every cell, which in turn enables features such as the soft hand-off and base station diversity. All together, these help enhance the capacity of the system.

In the **forward link** of a DS/CDMA system, power control may also be used. It may vary the transmitted power to the mobile, but the dynamic range is smaller due to the shared spectrum and, thus, shared interference.

We can distinguish between two kinds of power control, the open-loop power control and the closed-loop power control. The open-loop power control estimates the channel and adjusts the transmitted power accordingly but does not attempt to obtain feedback information on its effectiveness. Obviously, the open-loop power control is not very accurate, but since it does not have to wait for the feedback information it may be relatively fast. This can be advantageous, in the case of a sudden channel fluctuation, such as a mobile driving from behind a big building. This fast action is required, for instance, in the reverse link of a DS/CDMA system where the sudden increase of received strength at the base station may suppress all other signals.

The principle operation of open-loop power control is shown in Fig. 23.1. The open-loop power control must base its action on the estimation of the channel state. In the reverse link it estimates the channel by measuring the received power level of the pilot from the base station in the forward link and sets the transmitted power level inversely proportional to it. Ideally, this ensures that the average power level received from the mobile at the base station remains constant irrespective of the channel variations. This approach, however, assumes that the forward and the reverse link signal strengths are closely correlated. Although forward and reverse link may not share the same frequency and, therefore, the fading is significantly different, the long-term channel fluctuations due to shadowing and propagation loss are basically the same.

The closed-loop power control system [Fig. 23.2(a)] may base its decision on an actual communication link performance metric, e.g., received signal power level, received signal-to-noise ratio, received bit-error rate, or received frame-error rate. In the case of the reverse link power control, this metric may be forwarded to the mobile as a base for an autonomous power control decision, or the metric may be evaluated at the base station and only a power control adjustment command is transmitted to the mobile. If the reverse link power control decision is made at the base station, it

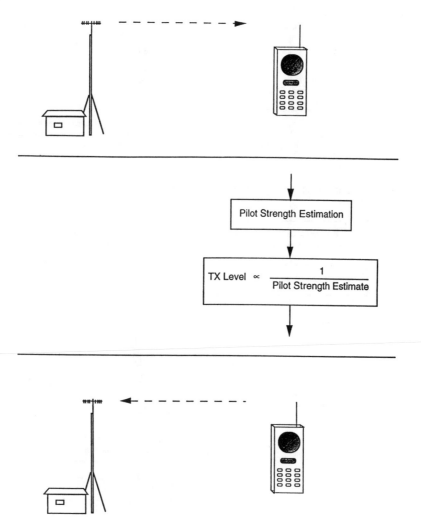

FIGURE 23.1   Reverse link open-loop power control.

may be based on the additional knowledge of the particular mobile's performance and/or a group of mobiles' performance (such as mobiles in a sector, cell, or even in a cluster of cells). If the power control decision for a particular mobile is made at the base station or at the switching office for all mobiles and is based on the knowledge of all other mobile's performance, it is called a centralized power control system. A centralized power control system may be more accurate than a distributed power control system, but it is much more complex in design, more costly, and technologically challenging.

In principle, the same categorization may be used for the power control in the forward link [Fig. 23.2(b)] except that in the forward link pilots from the mobiles are usually unavailable and only closed-loop power control is applied.

In the ideal case, power control compensates for the propagation loss, shadowing, and fast fading. There are many effects, however, that prevent the power control from becoming ideal. Fast fading rate, finite delays of the power control system, nonideal channel estimation, error in the power control command transmission, limited dynamic range, etc., all contribute to degrading the performance of the power control system. It is very important to examine the performance of power

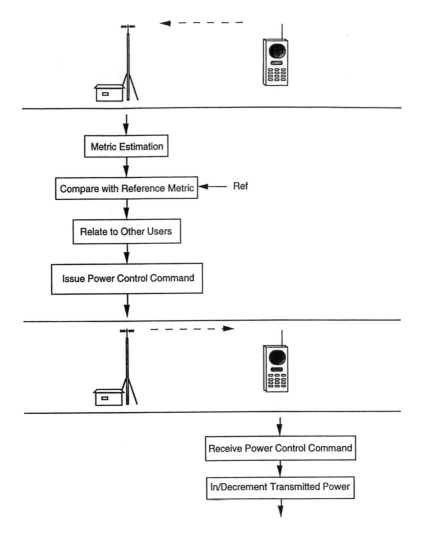

FIGURE 23.2(a)   Reverse link closed-loop power control.

control under nonideal conditions since the research done has shown that the power control system is quite sensitive to some of these conditions [Viterbi and Zehavi, 1993]. Kudoh, 1993, simulated a nonideal closed-loop power control system. Errors in the system were represented by a log-normal distributed control error with standard deviation σ_E in decibels. Some results on capacity reduction are presented in Table 23.1.

The effects of Doppler and delay and feedback errors in power control loop on power control have also been studied. [Pichna et al., 1993].

TABLE 23.1   Capacity Reduction Versus Power Control Error

	$\sigma_E = 0.5$ dB, %	$\sigma_E = 1$ dB, %	$\sigma_E = 2$ dB, %	$\sigma_E = 3$ dB, %
Forward link	10	29	64	83
Reverse link	10	31	61	81

Source: Kudoh, E. 1993. On the capacity of DS/CDMA cellular mobile radios under imperfect transmitter power control. *IEICE Trans. Commun*, E76-B(April):886–893.

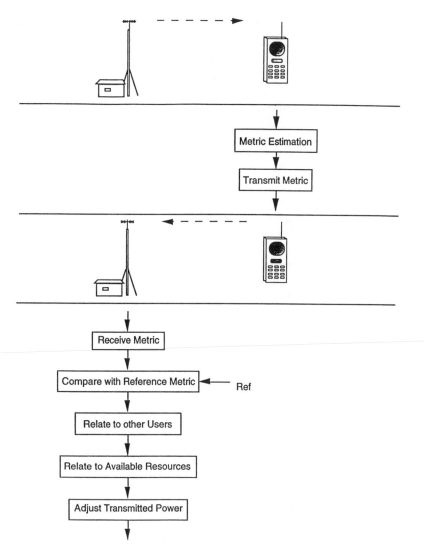

FIGURE 23.2(b)   Forward link closed-loop power control.

23.3 Power Control Examples

In the following section, several applications of power control of analog and digital cellular systems are presented.

In the analog networks we may see power control implemented in both the reverse link and forward link [Lee, 1989]. Power control in the reverse link 1) reduces the chance of receiver saturation by a closely located mobile, 2) reduces the co-channel interference and thus increases the frequency-reuse factor and capacity, and 3) reduces the average transmitted power at the mobile thus conserving battery energy at the mobile.

The power control in the forward link 1) reduces co-channel interference and thus increases the frequency reuse factor and capacity and 2) reduces adjacent-channel interference and improves the quality of service.

One example of a power control system shown by Lee, 1993, was of an air-to-ground communication system. The relevant airspace is divided into six zones based on the aircraft altitude. The transmitted power at the aircraft is then varied in six steps based on the zone in which the aircraft is located. The power control system exhibits a total of approximately 28 dB of dynamic range. This reduces the co-channel interference and, due to the excellent propagation conditions in the free air, has a significant effect on the capacity of the system.

Another example of a power control system in an analog wireless network is in the analog part of the TIA standard IS-95 [TIA, 1993]. IS-95 standardizes a dual-mode FDMA/CDMA cellular system compatible with the present day AMPS analog FDMA cellular system.

The analog part of IS-95 divides the mobiles into three classes according to nominal effective radiated power (ERP) with respect to half-wave dipole at the mobile. For each class, the standard specifies eight power levels. Based on the propagation conditions, the mobile station may receive a power control command that specifies at what power level the mobile should transmit. The maximum change is 4 dB per step. See Table 23.2.

IS-95 supports further discontinuous transmission. This feature allows the mobile to vary its transmitted power between two states: low and high. These two states must be at least 8 dB apart.

As for the power control in a digital wireless system, three examples will be shown: GSM [Balston and Macario, 1993], CT2/CT2PLUS standard [DOC, 1993] for digital cordless telephones of second generation, and the IS-95 standard for digital cellular DS/CDMA system [TIA, 1993].

TABLE 23.2 Nominal ERP of the Mobile

Power level	Nominal ERP (dBW) of mobile		
	I	II	III
0	6	2	−2
1	2	2	−2
2	−2	−2	−2
3	−6	−6	−6
4	−10	−10	−10
5	−14	−14	−14
6	−18	−18	−18
7	−22	−22	−22

Source: Telecommunications Industry Association/Electronic Industries Association. 1993. Mobile Station-Base Station Compatibility Standard for Dual-Mode Wideband Spread Spectrum Cellular System, *TIA/EIA/IS-95 Interim Standard.* Arlington, VA.

TABLE 23.3 GSM Transmitter Classes

Power class	Base station power, W	Mobile station power, W
1	320	20
2	160	8
3	80	5
4	40	2
5	20	0.8
6	10	
7	5	
8	2.5	

Source: Balston, D.M. and Macario, R.C.V. 1993. *Cellular Radio Systems,* Artech House, Norwood, MA.

GSM is a Pan-European digital cellular system that was introduced in many countries during the 1992–1993 period. GSM is a digital TDMA system with a frequency hopping feature. The power control in GSM ensures that the mobile station uses only the minimum power level necessary for reliable communication with the base station. GSM defines eight classes of base stations and five classes of mobiles according to their power output, as shown on Table 23.3.

The transmitted power at the base station is controlled, nominally in 2-dB steps. The adjustment of the transmitted power reduces the intercell interference and, thus, increases the frequency reuse factor and capacity. The transmitted power at the base station may be decremented to a minimum of 13 dBm.

The power control of the mobile station is a closed-loop system controlled from the base station. The power control at the mobile sets the transmitted power to one of 15 transmission power levels spaced by 2 dB. Any change can be made only in steps of 2-dB during each time slot. Another task for the power control in GSM is to control graceful rampon and rampoff of the TDMA bursts since too steep slopes would cause spurious frequency emissions.

The dynamic range of the received signal at the base station may be up to 116 dB [Balston and Macario, 1993] and, thus, the near–far problem may also by experienced, especially if the problem occurs in adjacent time slots. In addition to power control, a careful assignment of adjacent slots can also alleviate the near–far effect.

The CT2PLUS standard [DOC, 1993] is a Canadian enhancement of the ETI CT2 standard. Both these standards allow power control in the forward and in the reverse link. Because of the expected small cell radius and relatively slow signal level fluctuation rate, given by the fact that the user of the portable is a pedestrian, the power control specifications are relatively simple. The transmission at the portable can have two levels: normal (full) and low. The low–normal difference is up to 20 dB.

The IS-95 standard represents a second generation digital wireless cellular using system using DS/CDMA. Since in a DS/CDMA system all users have the same frequency allocation, the cochannel interference is crucial for the performance of the system [Gilhousen et al., 1991]. The near–far effect may cause the received signal level to change up to 100 dB [Viterbi, 1994]. This considerable dynamic range is disastrous for a DS/CDMA where the channels are separated by a finite correlation between spreading sequences. This is further aggravated by the shadowing and the fading. The fading may have a relatively high rate since the mobile terminal is expected to move at the speed of a car. Therefore, the power control system must be very sophisticated. Power control is employed in both the reverse link and in the forward link. The reverse link power control serves the following two functions.

1. It equalizes the received power level from all mobiles at the base station. This function is vital for system operation. The better the power control performs, the more it reduces the cochannel interference and, thus, increases the capacity. The power control compensates for the near–far effect, shadowing, and partially for slow fading.
2. It minimizes the necessary transmission power level to achieve good quality of service. This reduces the cochannel interference, which increases the system capacity and alleviates health concerns. In addition, it saves the battery power. Viterbi, 1994, has shown up to 20–30-dB average power reduction compared to the AMPS mobile user as measured in field trials.

The forward link power control serves the following three functions.

1. It equalizes the system performance over the service area (good quality signal coverage of the worst-case areas).
2. It provides load shedding between unequally loaded cells in the service areas (e.g., along a busy highway) by controlling the intercell interference to the heavy loaded cells.
3. It minimizes the necessary transmission power level to achieve good quality of service. This reduces the cochannel interference in other cells, which increases the system capacity and alleviates health concerns in the area around the base station.

The reverse link power control system is composed of two subsystems: the closed-loop and the open-loop. The system operates as follows. Prior to the application to access, closed-loop power control is inactive. The mobile estimates the mean received power of the received pilot from the base station and the open-loop power control estimates the mean output power at the access channel [TIA, 1993]. The system then sets the closed-loop probing and estimates the mean output power,

$$
\begin{aligned}
\text{mean output power (dBm)} = \ & - \text{mean input power (dBm)} \\
& - 73 \\
& + \text{NOM\_PWR (dB)} \\
& + \text{INIT\_PWR (dB)}
\end{aligned}
\tag{23.1}
$$

where NOM_PWR and INIT_PWR are parameters obtained by the mobile prior to transmission. Subsequent probes are sent at increased power levels in steps until a response is obtained. The initial transmission on the reverse traffic channel is estimated as

$$
\begin{aligned}
\text{mean output power (dBm)} = \ & - \text{mean input power (dBm)} \\
& - 73 \\
& + \text{NOM\_PWR (dB)} \\
& + \text{INIT\_PWR (dB)} \\
& + \text{the sum of all access probe} \\
& \quad \text{corrections (dB)}
\end{aligned}
\tag{23.2}
$$

Once the first closed-loop power control bit is received the mean output power is estimated as

$$
\begin{aligned}
\text{mean output power (dBm)} = \ & - \text{ mean input power (dBm)} \\
& - 73 \\
& + \text{NOM\_PWR (dB)} \\
& + \text{INIT\_PWR (dB)} \\
& + \text{the sum of all access probe corrections (dB)} \\
& + \text{the sum of all closed-loop power control} \\
& \qquad \text{corrections (dB)} \hspace{3cm} (23.3)
\end{aligned}
$$

The ranges of the parameters NOM_PWR and INIT_PWR are shown in Table 23.4.

TABLE 23.4 NOM_PWR and INIT_PWR Parameters

	Nominal value, dB	Range, dB
NOM_PWR	0	−8–7
INIT_PWR	0	−16–15

Source: Telecommunications Industry Association/Electronic Industries Association. 1993. Mobile station-base station compatibility standard for dual-mode wideband spread spectrum cellular system. *TIA/EIA/ IS-95 Interim Standard.* Arlington, VA.

The closed-loop power control command arrives at the mobile every 1.25 ms (i.e., 800 b/s). Therefore, the base station estimates the received power level for approximately 1.25 ms. A closed-loop power control command can have only two values: 0 to increase the power level and 1 to decrease the power level. The mobile must respond to the power control command by setting the required transmitted power level within 500 μs. The total range of the closed-loop power control system is ± 24 dB. The total supported range of power control (closed loop and open loop) must be at least ± 32 dB.

The behavior of the closed-loop power control system while the mobile receives base station diversity transmissions is straightforward. If all diversity transmitting base stations request the mobile to increase the transmitted power (all power control commands are 0), the mobile increases the power level. If at least one base station requests the mobile to decrease its power, the mobile decreases its power level.

The system also offers a feature of gated transmitted power for variable rate transmission mode. The gate-off state reduces the output power by at least 20 dB within 6 μs. This reduces the interference to the other users at the expense of transmitted bit rate. This feature may be used together with variable rate voice encoder or voice activated keying of the transmission.

The forward link power control works as follows. The mobile monitors the errors in the frames arriving from the base station. It reports the frame-error rate to the base station periodically. (Another mode of operation may report the error rate only if the error rate exceeds a preset threshold.) The base station evaluates the received frame-error rate reports and slightly adjusts its transmitting power. In this way, the base station may equalize the performance of the forward links in the cell or sector.

A system conforming with the standard has been field tested, and the results show that the power control is able to combat the channel fluctuation (together with other techniques such as RAKE reception) and achieve the bit energy to interference power density (E_b/I_0) necessary for a reliable service [Viterbi, 1994]. The histogram of the achieved E_b/I_0 is shown in Fig. 23.3.

Power control together with soft handoff determines the feasibility of the DS/CDMA cellular system and is crucial to its performance. QUALCOMM, Inc. has shown on field trials that their system conforms with the theoretical predictions and surpasses the capacity of other currently proposed cellular systems [Viterbi, 1994].

23.4 Summary

We have shown the basic principles of power control in wireless cellular networks as well as some examples of power control systems employed in some networks.

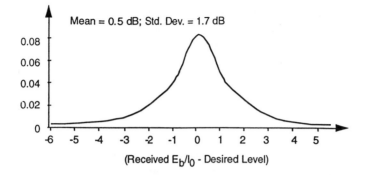

FIGURE 23.3 Differential E_b/I_0 Histogram. *Source:* Viterbi, A.J. 1994. The orthogonal-random waveform dichotomy for digital mobile personal communication. *IEEE Personal Comm.* 1(1st Quarter): 18–24.

In a wireless channel the channel transmission or channel gain is a random variable. If all transmitters in the system transmitted at equal and constant power levels, the received powers would be random.

In the reverse link (mobile to base station) each user has its own wireless channel, generally uncorrelated with all other users. The received signals at the base station are independent and random. Furthermore, since the users are randomly distributed over the cell, the distance between the mobiles and the base station may vary and so does the propagation loss. The differences between the strongest and the weakest received signal level may approach the order of 100 dB. This power level difference may cause saturation of the receivers at the base station even if they are allocated a different frequency or time slot. This phenomenon is called the near–far effect.

The near–far effect is especially detrimental for a DS/CDMA system where the frequency band is shared by all users and, for any given user, all other users' transmissions form the cochannel interference. Therefore, for the DS/CDMA system it is vitally important to efficiently mitigate the near–far effect.

The most natural way to mitigate the near–far effect is to power control the transmission in such a way that the transmitted power counterfollows the channel fluctuations and compensates for them. Then the received signal at the base station arrives at a constant amplitude.

The use of power control is not limited to the reverse link but is also employed in the forward link. The controlled transmission maintaining the transmitted level at the minimum acceptable level reduces the cochannel interference, which translates into an increased capacity of the system.

Since the DS/CDMA systems are most vulnerable to the near–far effect they have a very sophisticated power control system. In giving examples, we have concentrated on the DS/CDMA cellular system. We have also shown the power control used in other systems such as GSM, AMPS, and CT2.

Although there are more techniques available for mitigation of the near–far effect, power control is the most efficacious. As such, power control forms the core in the effort in combatting the near–far effect and channel fluctuations in general [Viterbi, 1994].

Defining Terms

Advanced mobile phone service (AMPS):   Analog cellular system in North America.

Cordless telephone, second generation (CT2):   A digital FDMA/TDD system.

Fading: Fast varying fluctuations of the wireless channel mainly due to the interference of time-delayed multipaths.

Forward link: Link from the base (fixed) station to the mobile (user, portable).

Groupe Spéciale Mobile (GSM):   Recently referred to as the **global system for mobility**. An ETSI standard for digital cellular and microcellular systems.

Power control: Control system for controlling the transmission power. Used to reduce the co-channel interference and mitigate the near–far effect in the reverse link.

Reverse link: Link from the mobile (user, portable) to the base (fixed) station.

Shadowing: Slowly varying fluctuations of the wireless channel due mainly to the shades in propagation of electromagnetic waves. Often described by log-normal probability density function.

References

Balston, D.M. and Macario, R.C.V. 1993. *Cellular Radio Systems*, Artech House, Norwood, MA.

Department of Communications. 1993. ETI Interim Standard # I-ETS 300 131, Annex 1, Issue 2, Attachment 1. In *CT2PLUS Class 2: Specification for the Canadian Common Air Interface for Digital Cordless Telephony, Including Public Access Services, RS-130*, Communications Canada, Ottawa, ON.

Gilhousen, K.S., Jacobs, I.S., Padovani, R, Viterbi, A.J., Weaver, L.A., and Wheatley C.E., III. 1991. On the capacity of cellular CDMA system. *IEEE Trans. Veh. Tech.* 40(May):303–312.

Kudoh, E. 1993. On the capacity of DS/CDMA cellular mobile radios under imperfect transmitter power control. *IEICE Trans. Commun.* E76-B(April):886–893.

Lee, W.C.Y. 1993. *Mobile Communications Design Fundamentals*, 2nd ed., John Willey & Sons, New York.

Lee, W.C.Y. 1989. *Mobile Cellular Telecommunications Systems*, McGraw-Hill, New York.

Pichna, R., Kerr, R., Wang, Q., Bhargava, V.K., and Blake, I.F. 1993. CDMA cellular network analysis software. Final Rep. Ref. No. 36-001-2-3560/01-ST, prepared for Department of Communications, Communications Research Centre, Ottawa, ON. March.

Simon, M.K., Omura, J.K., Scholtz, R.A, and Levitt, B.K., 1994. *Spread Spectrum Communication Handbook*, McGraw-Hill, New York.

Telecommunications Industry Association/Electronic Industries Association. 1993. Mobile station-base station compatibility standard for dual-mode wideband spread spectrum cellular system. *TIA/EIA/IS-95 Interim Standard*, Washington, D.C. July.

Viterbi, A.J. and Zehavi, E. 1993. Performance of power-controlled wideband terrestrial digital communication. *IEEE Trans. Comm.* 41(April):559–569.

Viterbi, A.J. 1994. The orthogonal-random waveform dichotomy for digital mobile personal communication. *IEEE Personal Comm.* 1(1st Quarter):18–24.

Further Information

For general information see the following overview books.

Balston, D.M., and Macario, R.C.V. 1993. *Cellular Radio Systems*, Artech House, Norwood, MA.
Simon, M.K., Omura, J.K. Scholtz, R.A, Levitt, B.K., 1994. *Spread Spectrum Communication Handbook*, McGraw-Hill, New York.

For more details on power control in DS/CDMA systems consult the following.

Gilhousen, K.S., Jacobs, I.S., Padovani, R., Viterbi, A.J., Weaver, L.A., and Wheatley C.E., III. 1991. On the capacity of cellular CDMA system. *IEEE Trans. Veh. Tech.* 40(May):303–312.
Viterbi, A.J. and Zehavi, E. 1993. Performance of power-controlled wideband terrestrial digital communication. *IEEE Trans. Comm.* 41(April):559–569.

Readers deeply interested in power control are recommended to see *IEEE Transactions on Communications*, *IEEE Transactions on Vehicular Technology*, and relevant issues of *IEEE Journal on Selected Areas in Communications*.

24

Second Generation Systems

Marc Delprat
Alcatel Mobile Communication

Vinod Kumar
Alcatel Mobile Communication

24.1 Introduction

Designed during the 1980s, all of the so-called second generation mobile communication systems are digital. For voice calls, digitally encoded speech is transmitted on the radio interface using one of the many available digital modulation schemes. In view of the processing complexity required for these digital systems, two offered advantages are 1) the possibility of using spectrally efficient radio transmission schemes (e.g., time division multiple access [TDMA] or code division multiple access [CDMA]) in comparison to the analog frequency division multiple access (FDMA) schemes previously employed and 2) the facility of implementation of a wide variety of (integrated) speech and data services and security features (e.g., encryption).

Standardization has played an essential role in the development of second generation systems. In Europe, the need of a pan-European system to replace a large variety of disparate analog **cellular** systems was the major motivating factor behind the creation of the Global System for Mobile communications (GSM). In North America and Japan, where unique analog systems existed, the need to standardize, respectively IS-54, IS-95, and personal digital cellular (PDC) for digital cellular applications arose from the lack of spectrum to serve the high traffic density areas [Cox,

1992]. Additionally, some of the second generation systems, such as Digital European **Cordless** Telecommunications (DECT) and Personal Handy Phone Systems (PHPS) are the result of a need to offer wireless services in residential and office environments with low-cost subscriber equipment [Tuttlebee, 1992]. Both Trans European Trunked Radio systems (TETRA) and Associated Public Safety Communications Officers (APCO) Project 25 aim at providing some sort of unified systems for professional applications, especially security networks. Although second generation systems have also been designed for satellite and paging applications, this chapter concentrates on bidirectional land mobile radio systems.

The physical layer characteristics of all of these systems offer robust radio links paired with good spectral efficiency. The network related functionalities have been designed to offer secure communication to authenticated users even when roaming between various networks based on the same system.

After a brief description of the basic characteristics and generic system architecture of second generation wireless systems, this chapter presents the salient features of cellular, cordless, and **professional mobile radio** (PMR) systems from Europe, North America, and Japan. Interworking scenarios between some systems, problems related to coexistence of multiple radio interfaces, and future possible enhancements of wireless systems are finally presented.

24.2 Basic Features and System Architecture in Cellular, Cordless, and PMR

Essential Features

Services: Cellular and cordless systems are telephony-oriented and provide integrated services digital network (ISDN) type services, whereas the typical PMR operation mode is half-duplex group calls within fleets of users. A large variety of data services and supplementary services, tailored to system type, are offered by second generation systems.

Security: Enhanced security features (authentication, encryption) are currently offered by the cellular and cordless systems. PMR systems, for public-safety applications, are required to exhibit a high level of security.

Capacity: Cellular systems are designed for medium to high traffic density and cordless systems must provide very high capacity, up to 10,000 Erlang per square kilometer (E/km^2). PMR systems generally encounter low traffic density (<1 E/km^2). For optimized spectral efficiency, most cellular and cordless systems use medium- to wide-band TDMA or CDMA access, whereas PMR systems use narrow-band TDMA or FDMA access.

Range: In cellular systems the cell dimensions may vary from 0.3 km (in microcellular urban systems) to 30 km and above in low-density rural areas. In cordless systems it is typically limited to a few tens of meters (<200 m). In PMR systems it may vary from 2 km (urban portable coverage) to 30 km and above (rural mobile coverage).

Radio Resource Management: In cellular and in PMR, predefined channel allocation and frequency reuse patterns are employed. In high-capacity, interference limited cordless systems dynamic channel selection, dynamic channel selection (DCS) is preferred. This is based on radio measurements performed regularly by the mobile station (MS) and the base station (BS). Moreover, DCS helps to alleviate some frequency planning issues in very small cell environment.

Mobility: Sophisticated mobility management (location updating, roaming, handover) is essential for cellular, highly desirable for some cordless applications (large business or Telepoint), and less important in most PMR applications (large cells and limited mobility users). In addition, the already standardized cordless systems have been optimized for slowly moving mobiles (<10 km/h).

Type of Traffic: Cellular and cordless systems are characterized mainly by individual calls between mobile and fixed users. The average call duration can be several minutes. PMR systems have mainly local group calls with short duration (<1 min) between mobile users and involving line stations of the PMR network (e.g., dispatcher). Here short call setup time is a major requirement (<0.5 s). Specific operational modes of PMR are open channel (permanent allocation of a channel with late entry facility) and direct mode (mobile-to-mobile direct communication).

Speech Quality and Delay: Since cordless systems are considered an extension of the public switched telephone network (PSTN), they require an equivalent speech quality and, therefore, use toll quality codecs such as the 32-kb/s adaptive differential pulse code modulation (ADPCM) scheme of the International Telecommunication Union (ITU-T) G.721 recommendation. Such low transmission delay codecs are helpful to avoid the need for additional echo control devices in the network. A relatively lower quality could be acceptable for cellular and PMR systems, which, therefore, use medium to low bit rate vocoders (4–16 kb/s).

Terminal Type: Cordless systems aims at providing small, simple, and low-cost portable stations by limiting the technical characteristics (low power, no equalization, simple speech codec). Cellular and PMR systems support both vehicular and portable stations. A clear tendency towards size reduction can be observed. Because of their professional use, PMR terminals are generally designed with tighter mechanical and environment constraints.

Level of Standardization: In addition to the air interface, cordless standards must specify at least PSTN access, and cellular standards generally specify the interfaces between network elements. PMR networks are often delivered as a complete system so that it is more important to standardize the intersystem interfaces than the internal network architecture.

Architecture

Typical cellular, cordless, and PMR network architectures are represented in Fig. 24.1. These are reference models which may not always fully apply to all systems described hereafter.

In cellular systems, the base station subsystem (BSS) comprises a controller (BSC) and radio transceivers (BS or BTS) which provide radio communication with MS in the covered area. The network subsystem (NSS) includes dedicated mobile switching equipment (MSC) linking all system elements through leased lines to PSTN, ISDN, and Packet Switched Public Data Network (PSPDN). The home and visitor location registers (HLR/VLR) are databases containing mobile subscriber data and used for subscriber registration and mobility management. Copies of the subscribers' secret keys are stored in the authentication center (AuC) and the mobile equipment serial numbers are stored in the equipment identity register (EIR). ITU-T signaling system no. 7 (SS#7) and related application protocols are often used in the mobile network. All system elements are operated, controlled, and maintained by the operation and maintenance center (OMC).

The cordless network architecture depends on application. For residential use, the portable station (PS) behaves like a regular telephone and has direct access to the PSTN through the private BS. In a public-access system, BS are connected to a local exchange (LE) containing a local database (DB) used for subscriber registration and mobility management in the covered area. The LEs are connected to the PSTN/ISDN (for the purpose of traffic routing) and to centralized elements of the cordless system through the PSPDN (for signaling exchange). These centralized elements perform control functions (user identification, charging, network management) and may contain a centralized database that stores location updates of the cordless subscribers and, therefore, enables routing of incoming calls. For business applications, the same private automatic branch exchange (PABX)

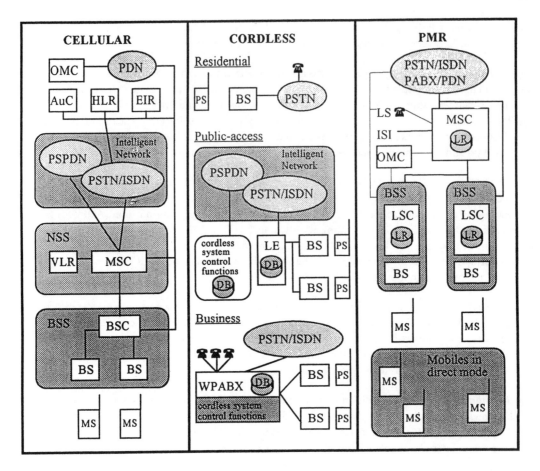

FIGURE 24.1 Network reference models.

may be used for both wire and wireless access. The wireless PABX (WPABX) interconnects the BSs of the private network. Cordless subscribers can therefore access other private wired subscribers or the PSTN/ISDN. The WPABX generally incorporates the subscriber DB and control functions of the cordless system.

PMR networks' architecture is somewhat similar to that of cellular networks, but the BSS is generally constituted of a single piece of equipment incorporating the BS and the local station controller (LSC). Since the LSC contains a copy of the subscribers location register (LR), local calls (which represent a significant part of the traffic) can be established and routed locally in the BSS, thus achieving a short call setup time and maintaining local operation even in fallback mode if the BSS-MSC link is interrupted. Intersite calls are routed through the MSC and access to other networks (PSTN/ISDN/PDN), or devices (PABX) may be provided either at MSC or BSS level. Line stations ([LS], e.g., dispatcher) may be connected directly to BSS or MSC, or through an intervening network (e.g., ISDN). Interworking of different PMR networks of the same standard is provided through the intersystem interface (ISSI). Mobiles may also use direct simplex communications (direct mode), either autonomously or while keeping contact with the network (dual watch).

24.3 Systems Description

In this section, the major second generation systems are briefly described, namely, cellular: GSM/DCS1800, IS-54, IS-95, and PDC; cordless: CT2, DECT, and PHPS; and PMR: TETRA and APCO25. Table 24.1 provides a comparative summary of system characteristics.

TABLE 24.1   Air Interface Characteristics of Second Generation Systems

	Cellular				Cordless			PMR	
Standard	GSM1800 (DCS)	IS-54	IS-95	PDC	CT2	DECT	PHPS	TETRA	APCO Project 25
Frequency band (MHz)	Europe,	USA	USA	Japan	Europe & Asia	Europe	Japan	Europe	USA
Uplink	890–915 (1710–1785)	824–849	824–849	940–956 (1429–1441, 1453–1465)	864–868	1880–1900	1895–1907	380–400?	Various bands, e.g. 150–170
Downlink	935–960 (1805–1880)	869–894	869–894	810–826 (1477–1489, 1501–1513)					~800
Duplex spacing (MHz)	45 (95)	45	45	130 (48)	—	—	—	10?	?
Carrier spacing (kHz)	200	30	1250	25	100	1728	300	25	12.5 (6.25)
No. of radio channels in the frequency band	124 (DCS: 374)	832	20	640	40	10	77	?	several hundreds of channel pairs
Multiple access	TDMA	TDMA	CDMA	TDMA	FDMA	TDMA	TDMA	TDMA	FDMA
Duplex mode	FDD	FDD	FDD	FDD	TDD	TDD	TDD	FDD	FDD
Number of channels per carrier	8 (half rate: 16)	3 (half rate: 6)	MABC	3 (half rate: 6)	1	12	4	4	1
Modulation	GMSK	Π/4 DQPSK	QPSK BPSK	Π/4 DQPSK	GFSK	GFSK	Π/4 DQPSK	Π/4 DQPSK	C4FM or CQPSK
Carrier bit rate (kb/s)	270.8	48.6	1288	42	72	1152	384	36	9.6
Speech coder Net bit rate kb/s	RPE-LTP 13	VSELP 7.95	QCELP (var. rate: 8, 4, 2, 1)	VSELP 6.7	ADPCM 32	ADPCM 32	ADPCM 32	ACELP 4.5	IMBE 4.4
Channel coder for speech channels	1/2 rate convol. + CRC	1/2 rate convol. + CRC	1/2 (down) 1/3 (up) convol. + CRC	1/2 rate convol. + CRC	no	no	no	2/3 & 4/9 rates convol. + CRC	Golay & Hamming codes
Gross bit rate speech+channel coding (kb/s)	22.8	13	var. rate 19.2, 9.6, 4.8, 2.4	11.2	—	—	—	7.2	7.2
Frame size (ms)	4.6	40	20	20	2	10	5	57	20
MS transmission power(W)	Peak Aver. 20 2.5 8 1 5 0.625 2 0.25 DCS1800 1 0.125 0.25 0.031	Peak Aver. 9 3 4.8 1.6 1.8 1.6	0.6	Peak Aver. 2 0.66	Peak Aver. 0.01 0.005	Peak Aver. 0.25 0.01	Peak Aver. 0.08 0.01	Peak Aver. 10 2.5 3 0.75 1 0.25	?
Power control MS control BS	Y Y	Y Y	Y Y	Y Y	low power mode in MS	N N	Y Y	Y N	N N
Operational C/I (dB)	9	16	6	17	20	12	26	19	?
Equalizer	needed	needed	Rake receiver	option	no	option	no	option	no
Handover	Y	Y	Soft handoff	Y	N	Y	Y	option	option

Global System for Mobile Communications/Digital Cellular System1800 (GSM/DCS1800)

The GSM standard was specified by the European Telecommunications Standards Institute (ETSI) for pan-European digital cellular mobile radio services [European Telecommunications Standards Institute, 1990]. It was designed to answer the need for a common mobile communications standard throughout Europe, where a variety of incompatible analog cellular systems like the Nordic Mobile Telephone (NMT) and the Total Access Communications Systems (TACS) existed. After some years of intensive research efforts in the early to mid-1980s, GSM standard specification started. This was preceded with a comprehensive evaluation of several system implementations. The allocation of a dedicated pan-European frequency band around 900 MHz and the signature of a memorandum of understanding (MoU) between countries, which committed to launch nation-wide GSM networks with internetworking capabilities, were the other important steps leading to an European network.

Phase 1 of the GSM standard was completed by ETSI in 1990 and is the basis of currently implemented networks. It provides a variety of speech and data services. These services being offered progressively to the users include telephony, emergency calls, conference calls, fax transmission, short messages, and data transmission at various rates up to 9600 bit/s. Supplementary services like call forwarding, call barring, and connected line identification are also defined.

The GSM network architecture closely follows the general principles introduced in the preceding section. All interfaces between network elements have been standardized, including MSC–BSC (A) and BSC–BTS (Abis) interfaces. Interfaces to AuC/HLR/EIR and to PSTN/ISDN use SS#7 with the mobile application part (MAP) protocol for noncircuit related signaling. This architecture, with a clear breakdown between the different machines and different functional parts (e.g., radio resource management in the BSC), facilitates evolutions.

The subscriber identity module (SIM) is the key to personal mobility, whereby a user can use any GSM terminal equipment just by inserting his SIM card. This smart card contains all subscriber data and is also used for basic security functions, such as subscriber identity authentication and key generation for traffic encryption on the air interface. This prevents fraudulous use of the system and ensures call privacy.

The GSM air interface is characterized by an eight-order TDMA scheme with frequency division duplex (FDD). The available frequency band in Europe is 2×25 MHz, with a radio channel spacing of 200 kHz. Data are modulated at 270 kb/s using Gaussian minimum shift keying (GMSK) modulation and transmitted in bursts of 577 μs. Each TDMA frame is constituted of eight time slots corresponding to eight separate physical channels. Each of these physical channels supports a combination of logical channels that are used in turn to carry signaling or traffic data. Slow frequency hopping is used to combat adverse propagation conditions, and most infrastructure implementations also include BTS receiver antenna diversity. The relatively low carrier to interference ratio (C/I) operational value (9 dB) is achieved through powerful channel coding, interleaving, and equalization techniques. Speech transmission is based on a linear prediction coder called regular pulse excited–long term prediction (RPE–LTP), which yields a net bit rate of 13 kb/s and a gross bit rate of 22.8 kb/s after channel coding. The air interface protocol follows a classical layered structure and includes a number of advanced features that are specific to mobile radio applications, such as mobile assisted handover (MAHO), power control (in both up and down links), and discontinuous transmission (DTX) based on voice activity detection (VAD).

GSM standardization in ETSI is still an ongoing process, and a number of additional services and features (multiparty calls, half-rate coder, general packet data service, etc.) will be available in phase 2. Standardization of a GSM adaptation tailored to railways application has also recently started. This will include specific features such as group calls or support of high-speed mobiles.

One important GSM extension is the digital cellular system-1800 (DCS1800) standard designed for personal communication networks ([PCN], optimized for urban and suburban use) and for which various licences have already been granted in Europe. The main differences with GSM are the frequency band (around 1800 MHz), national roaming capabilities, and a reduced transmission power (hence a reduced cell size). It is also a candidate standard for the U.S. Personal Communication Services (PCS) in the 1900-MHz band.

Originally focused on the European market, the GSM/DCS1800 standard has now achieved world-wide credibility. More than 65 countries have already adopted the GSM standard, and at least 40 GSM/DCS1800 networks are currently in service around the world. These figures are constantly growing.

Interim Standard (IS-54)

The main driving force for the definition of a second generation digital standard in North America was the rapidly growing demand for cellular services during the 1980s. This would have easily exceeded the capacity of the analog networks based on the advanced mobile phone system ([AMPS] developed in the 1970s by Bell Laboratories) networks. Accordingly, the new digital standard was specified by the Telecommunications Industry Association (TIA) upon request of the Cellular Telecommunications Industry Association (CTIA), with the major constraints to provide a significant increase in system capacity while maintaining upward compatibility with widespread AMPS (e.g., through dual mode BS and MS). The Federal Communications Commission (FCC), however, decided to open the existing cellular band (2×25 MHz in the 800-MHz range) to any suitable technology.

The IS-54 standard was finally selected from several proposals and published in January 1991 [Electronic Industries Association/Telecommunications Industry Association, 1991]. Both dual mode (AMPS/IS-54) mobile station and base station are specified in this standard, thus enabling the design of equipment capable of analog or digital operation. Other related standards have also been defined: IS-55 and IS-56 were defined for the performance specifications and measurement methods for mobile station and base station, respectively. Concerning network aspects, several standards have been developed by TIA since 1988 independently from the air interface design and are, hence, applicable to analog AMPS as well as IS-54 and other systems. The network reference model is as represented in Fig. 24.1 and the interfaces between network elements are specified in IS-41, except for the MSC-BSS interface, which is under-going evolution. IS-41 deals with automatic roaming (including validation and authentication of roamers), intersystem handover, and operation, administration, and maintenance, and others. IS-41 procedures are mapped on the transaction capabilities application part (TCAP) protocol (for transaction handling and message packaging) and at lower layers the transmission of network messages may use either X.25 or SS#7 formats. Other network related standards are IS-52, numbering plan; IS-53, supplementary services; IS-93, interfaces to other systems, e.g., PSTN; and IS-124, on-line exchange of call records.

Though less ISDN oriented than GSM, the IS-54 standard also supports several services, e.g., telephone service, short message service, and data services with a maximum transmission rate of 9.6 kb/s. Supplementary services include call forwarding, three-party call, and call barring. Security features include a personal identification number (PIN), subscriber authentication upon connection to the system, and voice as well as subscriber data encryption.

The IS-54 air interface uses TDMA/FDD technology with three channels per 30-kHz AMPS carrier. The modulation bit rate is 48.6 kb/s. $\Pi/4$ shifted differential quadrature phase shift keying (DQPSK) modulation is employed. For each (full rate) channel, the gross bit rate is 13 kb/s, and speech is encoded at 7.95 kb/s using a vector sum excited linear prediction (VSELP) algorithm. Advanced radio-link control based on power control and DTX improves spectrum efficiency. The air interface protocol, compatible with the AMPS protocol, includes an optional extended mode to allow for the addition of new system features and operational capabilities.

The traffic capacity achievable with IS-54 is expected to be three to four times that of existing AMPS systems. This capacity will be doubled with the introduction of a half-rate codec, currently under standardization. Using the original frequency band of the AMPS system, digital channels compliant to the IS-54 standard are now progressively replacing analog channels, thereby alleviating the shortage of spectrum while enabling a smooth transition from analog to digital. Network and terminal equipment are already available from several manufacturers and commercial service is being offered in the largest U.S. cities.

Based on the IS-54 standard, Hughes has developed a new technology called E-TDMA, with operational networks in the U.S. and recently adopted for several regional networks in Russia and China. It is claimed to achieve a significantly higher capacity, taking advantage of advanced features such as half-rate coding, digital speech interpolation (DSI) and channel pools.

Integrated Services-95

In 1991, Qualcomm demonstrated a CDMA digital cellular validation system, compliant with CTIA requirements for second generation cellular technology. The results of the field trials, conducted publicly with the support of a number of manufacturers and carriers, incited CTIA to request that TIA start the development of a wide-band (i.e., spread spectrum technology) digital cellular standard. IS-95 was then specified by TIA based on the Qualcomm proposal. Companion standards IS-96 and IS-97 have also been designed for the performance specifications and measurement methods for mobile station and base station, respectively.

IS-95 [Electronic Industries Association/Telecommunications Industry Association; 1993] is an air interface specification, meeting requirements similar to IS-54 (e.g., significant increase over analog system capacity, ease of transition, and compatibility with existing analog system) but with completely different technical choices, based on direct sequence (DS) CDMA. In DS-CDMA, a wide-band frequency channel is shared between several overlapping signals, each characterized by a specific pseudorandom binary sequence that spreads the initial spectrum of the data to be transmitted.

The waveform design uses a pseudorandom noise (PN) spreading sequence with a chip rate of 1.23 MHz. The transmitted signal bandwidth is about 1.25 MHz, i.e., one-tenth of the total bandwidth allocated to one cellular service carrier. Smooth transition from analog to digital can be achieved by removing initially only one or a small number of 1.25-MHz channels from the present analog service to provide digital service.

A combination of open- and closed-loop power control enables the mobile to operate at minimum required transmission (Tx) power level. A less sophisticated power control is also available in downlink. Path diversity is achieved with "soft handoff," a technique used during transitions between cells. This permits the instantaneous selection (on a frame-by-frame basis) of the best paths between a mobile and two or more cell sites. It requires precise time synchronization ($\sim 1~\mu s$), however, among all cell sites and extra network resources.

Wide-band transmission allows the use of powerful forward error correction (convolutional encoding with a constraint length $K = 9$) and modulation (PN modulation, quadriphase in downlink and biphase in uplink, based on orthogonal Walsh sequences in 64 dimensions). Rake receivers are implemented to process and combine signal (multipath) components.

The vocoder is an 8-kb/s specific code excited linear prediction (CELP) scheme called QCELP that incorporates variable bit rate capabilities. Selection of one of four rates (8-, 4-, 2-, or 1-kb/s) is based on adaptive energy thresholds on the input signal. For speech calls (voice activity factor <50%), a significant decrease in radio interference and in MS power consumption is thus achievable.

IS-95 employs a specific layered protocol with extension capabilities. The security functions are similar to those of IS-54 (e.g., same authentication mechanisms). Concerning the network aspects, IS-41 Revision C shall support intersystem procedures for IS-95 terminals. Several field trials using the Qualcomm system (origin of IS-95) have been successfully performed. Moreover, Korea has selected this technology for its second generation cellular system, and a full-scale commercial service is expected to start by end 1995.

Personal Digital Cellular (PDC)

As in North America and in Europe, the design of the PDC standard in Japan was motivated in the late 1980s by the saturation of analog cellular networks and by the need for new and enhanced services. After a study phase initiated by the Japanese Ministry of Posts and Telecommunications in April 1989, the PDC air interface standard was issued in April 1991 by the Research and Development Center for Radio Systems (RCR) under the name STD27 [RCR-STD 27, 1991]. It was complemented by network interface specifications providing the basis for a unified digital cellular system in Japan and enabling connectivity with fixed ISDN.

RCR STD27 is a common air interface specification. Though there are some similarities with the American IS-54 standard in terms of technical features, no compatibility with existing analog systems was required here. As a matter of fact, the new digital systems in Japan will benefit from a specific spectrum allocation, initially in the 800-MHz band and later in the 1.5-GHz band.

The carrier spacing is 25 kHz and the multiple access scheme is a TDMA/FDD of order 3 with the current full-rate codec and of order 6 with the future half-rate codec. The carrier bit rate is 42 kb/s. $\Pi/4$ DQPSK modulation is used. The full-rate speech codec employs a VSELP algorithm with a gross bit rate of 11.2 kb/s and a net bit rate of 6.7 kb/s, with forward error correction based on a convolutional code (rate $\frac{1}{2}$) and a cyclic redundancy check (CRC). A specific channel assignment procedure (flexible channel reuse from one BS to another) enables an increase of the system capacity. The standard also specifies power control and MAHO type handover procedures.

PDC systems will provide numerous services, including speech transmission, data transmission (G3-facsimile, modem, videotex), and short message service. Supplementary services such as calling line identification, call forwarding, or three-party call are also foreseen. The air interface protocol is ISDN oriented with a layered structure following the open systems interconnection (OSI) principles, including a link access protocol called LAPDM at layer 2 and a layer 3 divided into radio transmission management (RT), mobility management (MM), and call control (CC, based on ITU-T I.451). Security features include authentication and encryption.

Interfaces between network elements of a PDC system have been defined by cellular operators in Japan, except the A interface (BSS–MSC) which is left open for implementation. The network architecture follows the reference model of Fig. 24.1, ITU-T SS#7, if used between network elements and on the interface to other networks. The application protocols are an enhanced version of ISDN user part (ISUP) for circuit related signaling and MAP, developed as an application service element on TCAP, for noncircuit related signaling.

Commercial service with a PDC network was initiated by NTT in 1993 for the 800-MHz band and in 1994 for the 1.5-GHz band. Two other operators have launched digital cellular services in the 800-MHz band in 1994, and the government has recently decided to allow two new operators to offer digital service in the 1.5-GHz band. Enhancements of the PDC standard are also foreseen (half-rate codec, packet data, etc.).

Cordless Telecommunications 2 (CT2)

European research in digital cordless telephony started in the U.K. and Sweden in the early 1980s, mainly for WPABX applications. The first digital cordless standards were published in the U.K in 1987. Standards designed as a limited set of coexistence specifications (frequency band, transmitter power, interface to PSTN) resulted in several proprietary products with different and incompatible air interface characteristics. In 1989, the U.K. government issued four operator licences for public-access cordless systems (i.e., Telepoint), requesting the design of a common air interface (CAI) to allow interworking between systems. The CT2 CAI standard, developed in cooperation between U.K. manufacturers and various operators, was published in May 1989. Despite the initial ETSI choice for a different technology (DECT, described below), some manufacturers considered CT2 suitable for short term products. Thus, the CT2 CAI standard was finally endorsed by ETSI in November 1991 as an Interim European Telecommunication Standard (I-ETS) [European Telecommunications Standard Institute, 1991].

The CT2 CAI specifies incoming and outgoing calls for business and residential applications, whereas the Telepoint application was initially limited to outgoing calls only. Speech services and, recently, data services have also been specified, including asynchronous and synchronous data protocols with user rates up to 19.2 and 32 kb/s, respectively.

The air interface is designed for operation in the 800-MHz band, with a carrier spacing of 100 kHz. It employs a FDMA/time division duplex (TDD) access scheme where each carrier supports a single bidirectional communication using data blocks of 2-ms duration. Each block contains reserved fields for signaling information, with different multiplex formats for in-call signaling, normal call setup, or Telepoint access. The modulation is Gaussian frequency shift keying (GFSK) with a carrier bit rate of 72 kb/s. The speech codec is a 32-kb/s ADPCM scheme following ITU-T G.721. A 16-bit CRC enables detection of errors on signaling messages. Traffic channels are allocated using a dynamic channel assignment (DCA) technique. Simplified power control (normal/low power modes) can be implemented.

The protocol has an OSI-layered structure and includes a simplified and recoded version of Q.931 at layer 3, useful for expansion to new services and facilities. Security features include mobile authentication (mandatory for a public-access system) and network authentication (optional) using the UKF1 algorithm. The standard was subsequently enhanced (Revision 1) with additional features such as link re-establishment and location tracking (thus enabling the routing of incoming calls in a public-access system).

After several years of operation, the situation of CT2 is moderate, with some commercial success for Telepoint applications in Asia and more recently in Europe, contrasted to the failure of initial U.K. networks. An enhanced version, called CT2+, has also been introduced in Canada. Residential CT2 cordless products are essentially found as part of Telepoint packages. WPABX applications are emerging and several products have already been launched.

Digital European Cordless Telecommunications (DECT)

The standardization of a second generation cordless system covering a wide range of existing or emerging cordless applications and providing enhanced services with increased spectrum efficiency was initiated in Europe in 1985. A decision in 1988 favored the TDMA/TDD access scheme (based on CT3 designed in Sweden). The DECT standard, finally published in March 1992, benefits from a pan-European frequency allocation just below the 2-GHz frequency [European Telecommunications Standard Institute, 1992].

Basic capabilities (standard telephony features) and enhancements (ISDN interface, data transmission, privacy) common to all applications are specified by the DECT standard. Additional features are available depending on the cordless application. Residential applications can be enhanced with an intercom function. For public use, outgoing calls and authentication are basic whereas incoming call (log-on) and handover are enhancement features. The standard defines data transmission bearer services at basic bit rate multiples of 32 kb/s and up to 320 kb/s. This makes it particularly suitable for the implementation of ISDN connection-based services as well as X.25 or IEEE.802 services. The DECT protocol allows activation of ISDN supplementary services (through stimulus procedures), plus a few specific supplementary services (queue management, cost information). As security features, it includes user and network authentication, possibly using a smart card and encryption of user and/or signaling information.

System operation is based on ten carriers, spaced 1.728 MHz apart, in the 1880–1900-MHz range. GFSK modulation with a bit rate of 1152 kb/s is employed. The frame structure defines 24 time slots per 10-ms frame, therefore allowing 12 bidirectional speech calls per carrier with the 32-kb/s ADPCM codec (ITU-T G.721). Higher data rate links can be obtained in a multiple time slots configuration and/or using both uplink and downlink time slots (for asymmetric data transmission). Various logical channels are defined for in-call signaling, the control information being embedded with traffic data in each time slot. Error detection is provided for signaling information using a 16-bit CRC. DECT uses dynamic channel selection, which is equivalent to DCA in CT2, but here the TDMA mode offers more monitoring capabilities and, therefore, facilitates channel reallocation for intra- or intercell handover. The standard specifies a "make before break" seamless handover when BSs are synchronized. The protocol is OSI layered and draws extensively from ISDN and GSM protocols for layers 2 and 3.

Various DECT products for WPABX exist, and successful field trials for wireless access applications have been reported. ETSI is still in the process of enhancing the DECT standard with additional network access protocols, namely, the generic access profile ([GAP], aiming to provide a unique air interface protocol for all telephony applications), the data interoperability profile (for radio LAN applications), the ISDN/DECT interworking profile, and the GSM interoperability profile (for DECT extension to GSM networks).

Personal Handy Phone Systems (PHPS)

In 1989, the Japanese Ministry of Posts and Telecommunications (MPT) set up a group to define the requirements for the introduction of PHPS for digital cordless telephonic applications. The definition of a common air interface (CAI) started in 1990, and after some field tests to validate the transmission method, the Telecommunication Technology Committee (TTC) of MPT reported on technical specifications in June 1992. The PHPS standard was finally published in March 1993 by the RCR under the name STD28 [Research and Development Center for Radio Systems, 1993]. Voice and data services have been defined, including circuit and packet data transmission with a maximum bit rate of 128 kb/s. Around 20 supplementary services are also specified (e.g., automatic call back, call forwarding, calling number identification) and the protocol incorporates expansion capabilities for additional features. For residential use, both incoming and outgoing calls are possible and, in addition, two terminals may communicate directly (without going through the base station) if they are close enough.

For public access, outgoing calls are supported and incoming calls may be offered through location registration. The capabilities of a public-access system also depend on the network architecture, an efficient configuration should benefit from intelligent network (IN) facilities of the public ISDN. BSs may also be connected to a local switch, performing call control and mobility management (both handover and internetwork roaming). Control functions of the cordless network (user identification, charging, network management) are performed in centralized equipment connected to the public ISDN, possibly through a PSPDN. SS#7 protocol is used between elements of the cordless network. A user PIN, validation and authentication procedures, as well as speech and data encryption ensure security in network operation.

The PHPS standard benefits from a frequency allocation of 77 carriers in the 1900-MHz band. The air interface is based on a TDMA/TDD access method with eight time slots per frame and allows four bidirectional communications per 300 kHz wide carrier. The modulation is $\Pi/4$ DQPSK with a carrier bit rate of 384 kb/s. Speech coding uses the G.721 ADPCM technique at 32 kb/s. Traffic data are transmitted in bursts containing 5 ms of coded speech or data embedded with associated signaling. The frame structure allows future introduction of half-rate (16-kb/s) and fourth-rate (8-kb/s) codecs. Error detection is performed on signaling using a 16-bit CRC.

Improved spectral efficiency is achievable through antenna diversity in the BS with postdetection selection for uplink and transmitter antenna selection for downlink. Power control is also available in both uplink and downlink. The BS starts its operation by quasistatic autonomous frequency assignment for the carrier supporting the main control channel. Dynamic channel selection (DCS) is then performed per call for traffic channels. Synchronization of the BSs is desirable for fast handover and interference reduction. The air interface protocol architecture relies on a three-phase link setup process: radio channel access (with a simplified signaling structure), link connection, and communication (using a fully OSI-layered architecture).

In June 1993, MPT launched a vast system validation campaign involving three coordinating agencies, 40 manufacturers, and six operator groups. PHPS terminals are now available for residential use and several operators are expected to open commercial service for public-access systems in spring 1995.

Trans European Trunked Radio (TETRA)

Considering the growing demand for PMR systems and the need to cover the wide range of user requirements, the European Commission and ETSI decided, in 1988, to launch the standardization of a new digital trunked PMR system called TETRA. In fact, TETRA will be a set of standards including three air interface specifications for voice plus data (V+D), packet data optimized (PDO), and direct mode operation (DMO), all based on similar technical characteristics [European Telecommunications Standard Institute, 1994].

TETRA offers a wide range of services and facilities, allowing both typical PMR operation (i.e., group calls within fleets of users) and more advanced applications (secure speech and data, facsimile, file transfer, or fleet management). Circuit mode and packet mode data (connection oriented and connectionless) services have been defined with a maximum user rate of 19.2 kb/s, as well as a short message service. Around 30 supplementary services are being specified, including typical fixed network services (line identification, call forwarding) and more PMR-oriented services (various types of priority, late entry, discreet listening). The internal network architecture is not for standardization, but a number of interfaces will be specified (air interface, intersystem interface, terminal equipment interface, and PSTN/ISDN/PABX/PDN interfaces).

TETRA is capable of operating in frequency bands between 150 and 900 MHz, and its design is tailored to the 400-MHz band. A pan-European harmonized frequency allocation in the 380–400 MHz band will progressively become available for public safety applications. The V+D air interface uses a TDMA/FDD access scheme with four channels per carrier and a carrier spacing of 25 kHz. The modulation is Π/4 DQPSK and the carrier bit rate is 36 kb/s. The speech codec uses the algebraic CELP (ACELP) technique at 4.5 kb/s, and forward error correction is achieved through convolutional coding at various rates and a CRC, yielding a gross bit rate of 7.2-kb/s per channel. The protocol is OSI layered and exhibits some similarities with GSM and DECT protocols but is optimized to support the specific TETRA features. The V+D air interface will also include a number of advanced features, such as call re-establishment (handover), transmission trunking (whereby radio resource is allocated only for the duration of a transaction and then released), and uplink power control.

The PDO air interface is very similar to the V+D one, except that traffic data transmission is not based on a fixed time slot allocation, but on a random access technique coupled with a contention protocol (either Reservation-ALOHA or data-sense multiple access, depending on system load). The DMO air interface, currently under specification, will enable simplex communication between mobiles (without any relaying in the infrastructure). It will be derived from the V+D air interface, though allowing only one channel per 25-kHz carrier.

The TETRA standard will be suitable for small and large systems and for open trunks (with a commercial service) as well as for closed networks (dedicated to particular private users). It should also contribute to the development of the emerging mobile data market through its PDO version. Since European police and safety organizations are expected to be early TETRA customers, the standard fully takes into account the specific requirements of security networks (e.g., open channel, end-to-end encryption of speech and data, signaling encryption, user and network authentication).

The core of an approved TETRA standard (V+D and PDO air interfaces) should be published in 1995. Other parts of the standard are still under specification (DMO air interface, intersystem interface, supplementary services protocol). First pilot systems are expected in 1996, and commercial products should be available in 1998.

Associated Public Safety Communications Officers Project 25 (APCO25)

The U.S. Project 25 is a joint project between APCO, the National Association of State Telecommunications Directors (NASTD), and a number of federal agencies. It is the first ever standard setting effort involving public safety agencies at the local, state, and federal level. APCO Canada and the British Home Office also contribute to this effort.

The purpose of Project 25 is to develop standards for digital land mobile radio, tailored to public safety requirements. Unlike APCO16, the current requirement specification that covers existing analog systems, Project 25 will establish open system standards enabling interoperability in a multivendor environment. The objectives are to provide enhanced functionalities, to maximize spectrum efficiency, and to allow efficient and reliable intraagency and interagency communications. The work began in 1989, following a FCC notice on advanced technology for the public safety radio services. Project 25 received support from many telecommunications companies and a MoU was developed with the TIA, allowing an industry advisory group (TIA–25) to carry out the technical specification work [Electronic Industries Association/Telecommunications Industry Association/Associated Public Safety Communications Officers, 1994].

The services defined are digital voice (individual, group, and broadcast calls), circuit data (protected or unprotected), packet data (acknowledged or unacknowledged) and a set of nine supplementary services including encryption. The standard allows traffic data encryption (four different types), signaling encryption and electronic serial number check (but no authentication). The air interface has been defined for two different operational modes: the conventional mode and the trunked mode. A talk-around functionality, allowing direct communication between mobiles has also been specified as part of the conventional mode.

The air interface relies on an FDMA/FDD access scheme with initially 12.5-kHz channel spacing and future migration to 6.25-kHz spacing. The standard will be suitable for operation in various frequency bands from below 100 to 1000 MHz. The modulation is a four-state continuous frequency (C4FM) scheme with a carrier bit rate of 9.6 kb/s; continuous quadrature PSK modulation is considered for the second phase with 6.25-kHz spacing. The frame structure is based on a 20-ms speech frame, associated signaling information being embedded with traffic data in each frame. The speech codec uses the improved multiband excitation technique at 4.4 kb/s, and forward error correction is provided through Hamming and Golay codes, yielding a gross bit rate of 7.2 kb/s. Handover mechanisms have also been defined.

The 12.5-kHz air interface standard is now published, and other interfaces have been specified (data port, data host, network management). The 6.25-kHz air interface and other specifications (telephone interconnect and intersystem interfaces) are still under consideration. Project 25 has defined a migration strategy that not only has backward compatibility to today's analog systems (APCO16) but also forward compatibility between receivers in the first and second generation of future radios.

The potential market includes U.S. public safety agencies and similar agencies in other countries. In the U.S., bids for public safety systems in the 800-MHz band asking for compliance with Project 025 have already been issued, and several manufacturers are developing Project 25 products and have recently demonstrated equipment for the conventional mode.

24.4 Interworking and Compatibility

A large variety of standardized second generation systems exists today. The spectrum in the 800–900-MHz and 1.5–1.9-GHz ranges has started to be pretty congested. Diversified service applications can be implemented at an optimized equipment/radio spectrum cost using products designed to one or the other standard.

Network level interworking between different systems coupled with the use of multistandard subscriber equipment can offer two major advantages, namely, suitable service quality in every environment at reasonable cost and completely seamless radio coverage and subscriber mobility from one application environment to the other.

In North America, IS-54 and IS-95 have been designed as compatibility standards with analog AMPS; at network level, interworking between the various cellular systems will be possible with future versions of IS-41. At another level, in Europe, applications of interconnection between DECT-based WPABX and MSC of a GSM/DCS1800 network have been already standardized. Figure 24.2 is a schematic representation of one such setup, wherein WPABXs

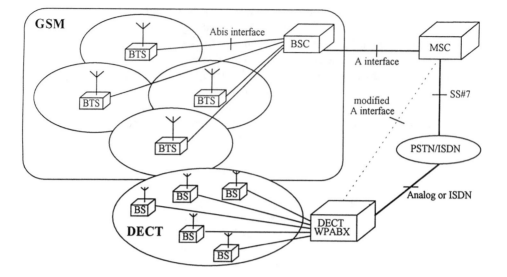

FIGURE 24.2   DECT/GSM interworking at network level.

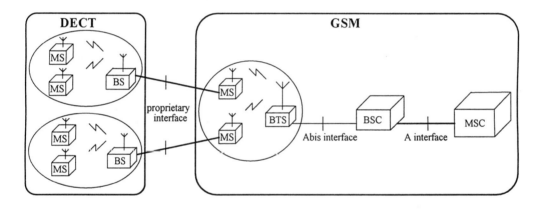

FIGURE 24.3   DECT/GSM interworking through interconnection of a GSM MS with a DECT BS.

controlling an island of DECT base stations for indoors or dense office coverage are shown to be interworking with MSC of a large cell GSM network. Figure 24.3 shows another such application where a GSM MS is connected, through a proprietary interface, to a DECT base station, which can handle multiple DECT subscribers simultaneously. This type of "moving cell" is of interest for handling multiple subscribers using low-cost handsets in fast moving vehicles like trains.

However this interworking may eventually lead to the coexistence of networks based on different standards in overlapping or geographically adjacent areas. Such radio coverage can create some intersystem, near–far interference problems either if two networks operate on the extremities of adjacent bands (e.g., DCS1800 and DECT) or if harmonic or other spurious emissions from a system are uncontrolled (e.g., disturbance of 1.8–1.9-GHz systems by 800–900-MHz ones).

Various scenarios of the type where the MS of a high-power cellular system can saturate the BS receiver of a lower power small-cell system have been analyzed. European standards have designed specifications on spurious emissions, Tx power classes, and minimum transmission/reception coupling loss in order to avoid any degradation in service quality. Similarly, in the U.S., all of the spectrum allocations for new services/systems by the FCC are accompanied with requirements on limited spurious emissions.

24.5 Further Improvements of Radio Interface Performances

Many ideas for the evolution of second generation systems have been put forward, and some of those are already under standardization and/or implementation. Enhanced service quality and further improvements in spectrum efficiency are the two most sought after goals. Most systems, especially cellular ones, have already embraced one or more of the following ideas

Variety of Speech Codecs

With the recent advances in speech coding techniques, low-delay toll quality at 8 kb/s and near-toll quality below 6 kb/s are now available. This enables the introduction of half-rate codecs in cellular systems, though with an increased complexity. In GSM/DCS1800, IS-54, and PDC this evolution was anticipated at an early stage, so the need for a half-rate speech channel was taken into account in the design of the TDMA frame structure. Standardization of half-rate codecs in the case of these three systems is now well advanced or completed. This will double the systems capacity (in terms of number of voice circuits per cell site) while maintaining a speech quality comparable to that available from related full-rate codecs. Alternatively, requirements for an enhanced speech quality have also been identified, especially for some emerging cellular applications. Standardization of an enhanced full-rate GSM codec has recently started, with the aim of achieving toll quality in good propagation environments.

Packet Mode Services

To benefit from better spectral efficiency properties of packet mode transmission, certain new services are being implemented. Cellular digital packet data (CDPD) and the implementation of relatively high-speed (>9.6 kb/s) data transmission in IS-95 are already well known. For GSM/DCS1800, the standardization of general packet radio service (GPRS) for packet mode data transmission is progressing well. New information multiplexing schemes (based on GSM/DCS1800 air interface), useful for high-reliability packet data transmission in transport applications are under development.

Air Interface Adaptivity for Better Spectrum Efficiency

TDMA burst and frame structure, channel coding, and interleaving are designed for worst-case channel distortions, i.e., large and very large cells, low-C/I ratio, and fast moving mobile. Handheld terminals are common in areas of high traffic density and usually operate in a small cell coverage. Good receiver performance can be ensured with shorter interburst guard time and shortened receiver training sequences usually included in the middle of each burst, and so the user information bit rate per hertz can be increased. A similar achievement is possible by decreasing the channel coding overhead wherever permissible. As an extreme case, various consecutive bursts of optimized size could be concatenated with a well-designed receiver training sequence, and the increase in MS to BS information rate can be exploited to offer high bit rate data services. "Allocation of bandwidth on demand" based on constant observation of quality of each communication is an adaptivity measure helpful to increase the spectral efficiency of the system. Also, variable rate speech coding techniques, either source or channel dependent, seem promising to improve quality and/or spectral efficiency. In particular, flexible allocation of a given gross bit rate between speech and channel coding on the basis of channel quality has been proposed. A set of parameters from the above list and some others like Tx power can be adapted dynamically. Both fast adaptation during a call and slow adaptation during call establishment on the basis of required and available quality of service are being considered for future enhancements of second generation systems.

Enhanced Quality of Service Through Improved Radio Coverage

The quality of service in digital wireless systems is based on the end-to-end bit error rate and on the continuity of radio links between the two ends. Interference free radio coverage with sufficient desired signal strength needs to be provided to achieve the above. Moreover, communication continuity has to be ensured between coverage areas with high traffic density (microcells) and low/medium traffic density (macrocells). Various second generation systems provide the possibility of implementing mechanisms such as slow frequency hopping, antenna diversity (also called microdiversity), and macrodiversity or multisite transmission. In addition, slow and fast and sometimes seamless handover methods are available through soft handover based on antenna diversity or frequency diversity, dynamic channel selection (as in DECT), fast mobile decided handovers between synchronized base stations using multisite operation, and overlaid cell structures whereby macrocells are superimposed on islands of microcells mainly to handle fast moving mobile subscribers. Combinations of these techniques can be implemented to further optimize overall radio coverage quality provided that the complexity of radio resource management is not unduly increased.

24.6 Conclusion

The availability of cost competitive digital and radio technology and constantly increasing demand for wireless services formed the two major factors for the success of second generation mobile communication systems described in this chapter. The fast world-wide acceptance and deployment of GSM is illustrative of this situation. Today all of the indicators point towards a constant growth of this demand in terms of quantity, quality, and variety of wireless services. Various international organizations are involved in the preparation of telecommunications for the 21st century.

Standardization of a universal mobile telecommunication system (UMTS) is one of the major working items in ETSI. Moreover, various research projects coordinated by the European Union are busy evaluating various technological possibilities for the implementation of UMTS. Support of voice, data, and multimedia services in the mobile environments is one of the major objectives of UMTS. Several industry partners believe that network interworking based on the enhanced second generation cellular and cordless systems such as GSM and DECT will offer a valid and economically viable platform for achieving the above stated objective. Integration of satellite systems in UMTS is being considered as well.

International Telecommunication Union (ITU) and, particularly, the Task Group 8/1 (TG 8/1) of ITU-R (radio) is working towards the standardization of a world-wide third generation mobile communication system called the Future Public Land Mobile System (FPLMTS). The more recently adopted acronym IMT-2000 for International Mobile Telecommunications-2000 better illustrates the purpose of this world-wide standardization effort. Like UMTS, the FPLMTS could be largely based on the interworking between various existing and newly created terrestrial or satellite mobile communication systems. The presently existing regular two-way flow of information between ETSI and ITU-R should be helpful in achieving technically coordinated standards for UMTS and FPLMTS.

The World Administration Radio Conference (WARC 92) earmarked some 220 MHz of spectrum for FPLMTS after the year 2000. Some portions of this spectrum are already being allocated on regional basis for immediate use (e.g., PCS in the U.S.). The second generation systems starting to be popular today may thus have many long years of prosperous life, especially if a transition to so-called feature rich third generation systems is possible through their evolutions.

Defining Terms

Cellular: Refers to public land mobile radio networks for generally wide area, e.g., national, coverage, to be used with medium- or high-power vehicular mobiles or portable stations and for providing mobile access to the Public Switched Telephone Network (PSTN). The network implementation exhibits a cellular architecture which enables frequency reuse in nonadjacent cells.

Cordless: These are systems to be used with simple low-power portable stations operating within a short range of a base station and providing access to fixed public or private networks. There are three main applications, namely, residential (at home, for plain old telephone service [POTS]), public access (in public places and crowded areas, also called telepoint), and wireless private automatic branch exchange ([WPABX], providing cordless access in the office environment), plus emerging applications like radio access for local loop.

Professional (or private) mobile radio (PMR): Covers a large variety of land mobile radio systems designed for professional users. This includes small local systems as well as regional or national networks, and open systems (with a commercial service) or closed systems (dedicated to particular private users). In "conventional systems," each radio channel is permanently allocated to a given fleet of users (generally for low-density systems). In "trunked systems," radio resources are shared among users and allocated on a per-call or per-transaction basis.

References

Cox, D.C. 1992. Wireless Network Access for Personal Communications. *IEEE Communications Magazine*, (12):96–115.

Electronic Industries Association/Telecommunications Industry Association. 1991. Cellular System, Dual–Mode Mobile Station-Base Station Compatibility Standard, EIA/TIA Interim Standard-54, Electronic Industries Association, Washington, D.C.

Electronic Industries Association/Telecommunications Industry Association. 1993. Mobile Station-Base Station Compatibility Standard for Dual-Mode Wideband Spread Spectrum Cellular System, EIA/TIA Interim Standard-95, Electronic Industries Association, Washington, D.C.

Electronic Industries Association/Telecommunications Industry Associations and Associated Public Safety Communications Officers. 1994. APCO Project 25, IS-102 and related Telecommunications Systems Bulletin TSB-102. Electronic Industries Association, Washington, D.C.

European Telecommunications Standards Institute. 1990. GSM Recommendations Series 01–12. Secretariat, Sophia Antipolis Cedex, France.

European Telecommunications Standards Institute. 1991. Common Air Interface Specification to be Used for the Interworking Between Cordless Telephone Apparatus in the Frequency Band 864.1 MHz to 868.1 MHz, Including Public Access Services, I-ETS 300 131, ETSI, Secretariat, Sophia Antipolis Cedex, France.

European Telecommunications Standards Institute. 1992. Digital European Cordless Telecommunications Common Interface, ETS 300 175, ETSI, Secretariat, Sophia Antipolis Cedex, France.

European Telecommunications Standards Institute. 1994. Trans European Trunked Radio System, draft prETS 300 392, draft prETS 300 393, draft prETS 300 394. ETSI, Secretariat, Sophia Antipolis Cedex, France.

Research and Development Center for Radio Systems. 1991. Personal Digital Cellular System Common Air Interface, RCR-STD27B. Research and Development Center for Radio Systems, Tokyo, Japan.

Research and Development Center for Radio Systems. 1993. Personal Handy Phone System: Second Generaration Cordless Telephone System Standard, RCR-STD28. Research and Development Center for Radio Systems, Tokyo, Japan.

Tuttlebee, W.H.W. 1992. Cordless Personal Communications. *IEEE Communications Magazine*, (12):42–53.

Further Information

European standards (GSM, CT2, DECT, TETRA) are published by ETSI Secretariat, 06921 Sophia Antipolis Cedex, France.

US standards (IS-54, IS-95, APCO) are published by Electronic Industries Association, Engineering Department, 2001 Eye Street, N.W. Washington D.C. 20006, USA.

Japanese standards (PDC, PHPS) are published by RCR (Research and Development Center for Radio Systems) 1-5-16, Toranomon, Minato-ku, Tokyo 105, Japan.

25

The Pan-European Cellular System

Lajos Hanzo
University of Southampton

25.1 Introduction

Following the standardization and launch of the Pan-European digital mobile cellular radio system known as GSM, it is of practical merit to provide a rudimentary introduction to the system's main features for the communications practitioner. Since GSM operating licences have been allocated to 126 service providers in 75 countries, it is justifiable that the GSM system is often referred to as the Global System of Mobile communications.

The GSM specifications were released as 13 sets of recommendations [ETSI, 1988], which are summarized in Table 25.1, covering various aspects of the system [Hanzo and Stefanov, 1992].

After a brief system overview in Sec. 25.2 and the introduction of physical and logical channels in Sec. 25.3, we embark upon describing aspects of mapping logical channels onto physical resources for speech and control channels in Secs. 25.4 and 25.5, respectively. These details can be found in recommendations R.05.02 and R.05.03. These recommendations and all subsequently enumerated ones are to be found in ETSI, 1988. Synchronization issues are considered in Sec. 25.6. Modulation (R.05.04), transmission via the standardized wideband GSM channel models (R.05.05), as well as adaptive radio link control (R.05.06 and R.05.08), discontinuous transmission (DTX) (R.06.31), and voice activity detection (VAD) (R.06.32) are highlighted in Secs. 25.7–25.10, whereas a summary of the fundamental GSM features is offered in Sec. 25.11.

0-8493-8573-3/96/$0.00+$.50
© 1996 by CRC Press, Inc.

399

TABLE 25.1   GSM recomendations [R.01.01]

R.00	*Preamble* to the GSM recommendations
R.01	*General structure* of the recommendations, description of a GSM network, associated recommendations, vocabulary, etc.
R.02	*Service aspects*: bearer-, tele- and supplementary services, use of services, types and features of mobile stations (MS), licensing and subscription, as well as transferred and international accounting, etc.
R.03	*Network aspects*, including network functions and architecture, call routing to the MS, technical performance, availability and reliability objectives, handover and location registration procedures, as well as discontinuous reception and cryptological algorithms, etc.
R.04	*Mobile/base station (BS) interface and protocols*, including specifications for layer 1 and 3 aspects of the open systems interconnection (OSI) seven-layer structure.
R.05	*Physical layer on the radio path*, incorporating issues of multiplexing and multiple access, channel coding and modulation, transmission and reception, power control, frequency allocation and synchronization aspects, etc.
R.06	*Speech coding specifications*, such as functional, computational and verification procedures for the speech codec and its associated voice activity detector (VAD) and other optional features.
R.07	*Terminal adaptors for MSs*, including circuit and packet mode as well as voiceband data services.
R.08	*Base station and mobile switching center* (MSC) *interface*, and transcoder functions.
R.09	*Network interworking* with the public switched telephone network (PSTN), integrated services digital network (ISDN) and, packet data networks.
R.10	*Service interworking, short message service.*
R.11	*Equipment specification and type approval specification* as regards to MSs, BSs, MSCs, home (HLR) and visited location register (VLR), as well as system simulator.
R.12	*Operation and maintenance*, including subscriber, routing tariff and traffic administration, as well as BS, MSC, HLR and VLR maintenance issues.

25.2 Overview

The system elements of a GSM public land mobile network (PLMN) are portrayed in Fig. 25.1, where their interconnections via the standardized interfaces A and Um are indicated as well. The mobile station (MS) communicates with the serving and adjacent base stations (BS) via the radio interface Um, whereas the BSs are connected to the mobile switching center (MSC) through the network interface A. As seen in Fig. 25.1, the MS includes a mobile termination (MT) and a terminal equipment (TE). The TE may be constituted, for example, by a telephone set and fax machine. The MT performs functions needed to support the physical channel between the MS and the base station, such as radio transmissions, radio channel management, channel coding/decoding, speech encoding/decoding, and so forth.

The BS is divided functionally into a number of base transceiver stations (BTS) and a base station controller (BSC). The BS is responsible for channel allocation (R.05.09), link quality and power budget control (R.05.06 and R.05.08), signalling and broadcast traffic control, frequency hopping (FH) (R.05.02), handover (HO) initiation (R.03.09 and R.05.08), etc. The MSC represents the gateway to other networks, such as the public switched telephone network (PSTN), integrated services digital network (ISDN) and packet data networks using the interworking functions standardized in recommendation R.09. The MSC's further functions include paging, MS location updating (R.03.12), HO control (R.03.09), etc. The MS's mobility management is assisted by the home location register (HLR) (R.03.12), storing part of the MS's location information and routing incoming calls to the visitor location register (VLR) (R.03.12) in charge of the area, where the paged MS roams. Location update is asked for by the MS, whenever it detects from the received and decoded broadcast control channel (BCCH) messages that it entered a new location area. The HLR contains, amongst a number of other parameters, the international mobile subscriber identity (IMSI), which is used for the authentication (R.03.20) of the subscriber by his authentication center (AUC). This enables the system to confirm that the subscriber is allowed to access it. Every subscriber belongs to a home network and the specific services that the subscriber is allowed to use are entered into his HLR. The equipment identity register (EIR) allows for stolen, fraudulent, or faulty mobile stations

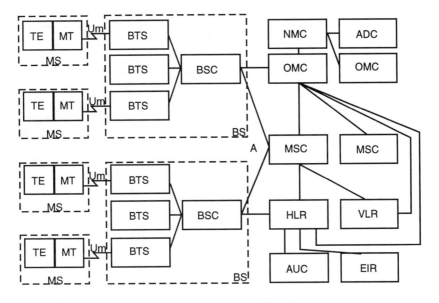

FIGURE 25.1 Simplified structure of GSM PLMN © ETT [Hanzo and Steele, 1994].

to be identified by the network operators. The VLR is the functional unit that attends to a MS operating outside the area of its HLR. The visiting MS is automatically registered at the nearest MSC, and the VLR is informed of the MSs arrival. A roaming number is then assigned to the MS, and this enables calls to be routed to it. The operations and maintenance center (OMC), network management center (NMC) and administration center (ADC) are the functional entities through which the system is monitored, controlled, maintained and managed (R.12).

The MS initiates a call by searching for a BS with a sufficiently high received signal level on the BCCH carrier; it will await and recognize a frequency correction burst and synchronize to it (R.05.08). Now the BS allocates a bidirectional signalling channel and also sets up a link with the MSC via the network. How the control frame structure assists in this process will be highlighted in Sec. 25.5. The MSC uses the IMSI received from the MS to interrogate its HLR and sends the data obtained to the serving VLR. After authentication (R.03.20) the MS provides the destination number, the BS allocates a traffic channel, and the MSC routes the call to its destination. If the MS moves to another cell, it is reassigned to another BS, and a handover occurs. If both BSs in the handover process are controlled by the same BSC, the handover takes place under the control of the BSC, otherwise it is performed by the MSC. In case of incoming calls the MS must be paged by the BSC. A paging signal is transmitted on a paging channel (PCH) monitored continuously by all MSs, and which covers the location area in which the MS roams. In response to the paging signal, the MS performs an access procedure identical to that employed when the MS initiates a call.

25.3 Logical and Physical Channels

The GSM logical traffic and control channels are standardized in recommendation R.05.02, whereas their mapping onto physical channels is the subject of recommendations R.05.02 and R.05.03. The GSM system's prime objective is to transmit the logical traffic channel's (TCH) speech or data information. Their transmission via the network requires a variety of logical control channels. The set of logical traffic and control channels defined in the GSM system is summarized in Table 25.2. There are two general forms of speech and data traffic channels: the full-rate traffic channels (TCH/F), which carry information at a gross rate of 22.8 kb/s, and the half-rate traffic channels (TCH/H), which communicate at a gross rate of 11.4 kb/s. A physical channel carries either a

TABLE 25.2   GSM Logical Channels ©ETT [Hanzo and Steele, 1994]

Logical Channels					
Duplex BS ↔ MS Traffic Channels: TCH		Control Channels: CCH			
FEC-coded Speech	FEC-coded Data	Broadcast CCH BCCH BS → MS	Common CCH CCCH	Stand-alone Dedicated CCH SDCCH BS ↔ MS	Associated CCH ACCH BS ↔ MS
TCH/F 22.8 kb/s	TCH/F9.6 TCH/F4.8 TCH/F2.4 22.8 kb/s	Freq. Corr. Ch: FCCH	Paging Ch: PCH BS → MS	SDCCH/4	Fast ACCH: FACCH/F FACCH/H
TCH/H 11.4 kb/s	TCH/H4.8 TCH/H2.4 11.4 kb/s	Synchron. Ch: SCH	Random Access Ch: RACH MS → BS	SDCCH/8	Slow ACCH: SACCH/TF SACCH/TH SACCH/C4 SACCH/C8
		General Inf.	Access Grant Ch: AGCH BS → MS		

full-rate traffic channel, or two half-rate traffic channels. In the former, the traffic channel occupies one timeslot, whereas in the latter the two half-rate traffic channels are mapped onto the same timeslot, but in alternate frames.

For a summary of the logical control channels carrying signalling or synchronisation data, see Table 25.2. There are four categories of logical control channels, known as the BCCH, the common control channel (CCCH), the stand-alone dedicated control channel (SDCCH), and the associated control channel (ACCH). The purpose and way of deployment of the logical traffic and control channels will be explained by highlighting how they are mapped onto physical channels in assisting high-integrity communications.

A physical channel in a time division multiple access (TDMA) system is defined as a timeslot with a timeslot number (TN) in a sequence of TDMA frames. The GSM system, however, deploys TDMA combined with frequency hopping (FH) and, hence, the physical channel is partitioned in both time and frequency. Frequency hopping (R.05.02) combined with interleaving is known to be very efficient in combatting channel fading, and it results in near-Gaussian performance even over hostile Rayleigh-fading channels. The principle of FH is that each TDMA burst is transmitted via a different RF channel (RFCH). If the present TDMA burst happened to be in a deep fade, then the next burst most probably will not be. Consequently, the physical channel is defined as a sequence of radio frequency channels and timeslots. Each carrier frequency supports eight physical channels mapped onto eight timeslots within a TDMA frame. A given physical channel always uses the same TN in every TDMA frame. Therefore, a timeslot sequence is defined by a TN and a TDMA frame number FN sequence.

25.4 Speech and Data Transmission

The speech coding standard is recommendation R.06.10, whereas issues of mapping the logical speech traffic channel's information onto the physical channel constituted by a timeslot of a certain carrier are specified in recommendation R.05.02. Since the error correction coding represents part of this mapping process, recommendation R.05.03 is also relevant to these discussions. The example of the full-rate speech traffic channel (TCH/FS) is used here to highlight how this logical channel is mapped onto the physical channel constituted by a so-called normal burst (NB) of the TDMA frame structure. This mapping is explained by referring to Figs. 25.2 and 25.3. Then this example will

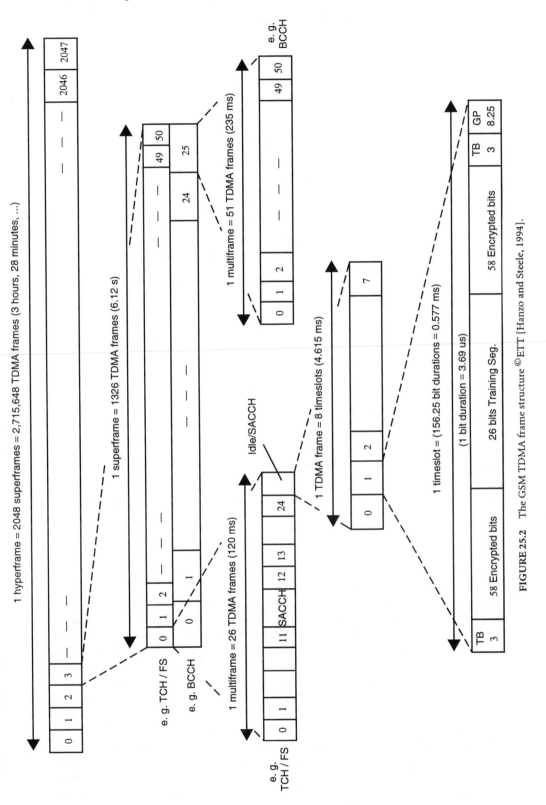

FIGURE 25.2 The GSM TDMA frame structure ©ETT [Hanzo and Steele, 1994].

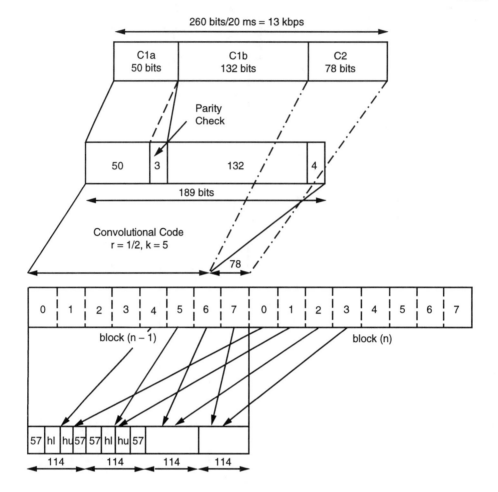

FIGURE 25.3 Mapping the TCH/FS logical channel onto a physical channel, ©ETT [Hanzo and Steele, 1994].

be extended to other physical bursts such as the frequency correction (FCB), synchronization (SB), access (AB), and dummy burst (DB) carrying logical control channels, as well as to their TDMA frame structures, as seen in Figs. 25.2 and 25.6.

The regular pulse excited (RPE) speech encoder is fully characterized in the following references: Vary and Sluyter, 1986; Salami et al., 1992; and Hanzo and Stefanov, 1992. Because of its complexity, its description is beyond the scope of this chapter. Suffice to say that, as it can be seen in Fig. 25.3, it delivers 260 b/20 ms at a bit rate of 13 kb/s, which are divided into three significance classes: class 1a (50 b), class 1b (132 b) and class 2 (78 b). The class-1a bits are encoded by a systematic (53, 50) cyclic error detection code by adding three parity bits. Then the bits are reordered and four zero tailing bits are added to perodically reset the memory of the subsequent half-rate, constraint length five convolutional codec (CC) CC(2, 1, 5), as portrayed in Fig. 25.3. Now the unprotected 78 class-2 bits are concatenated to yield a block of 456 b/20 ms, which implies an encoded bit rate of 22.8 kb/s. This frame is partitioned into eight 57-b subblocks that are block diagonally interleaved before undergoing intraburst interleaving. At this stage each 57-b subblock is combined with a similar subblock of the previous 456-b frame to construct a 116-b burst, where the flag bits *hl* and *hu* are included to classify whether the current burst is really a TCH/FS burst or it has been stolen by an urgent fast associated (FACCH) control channel message. Now the bits are encrypted and positioned in a NB, as depicted at the bottom of Fig. 25.2, where three tailing bits (TB) are added at both ends of the burst to reset the memory of the Viterbi channel equalizer (VE), which is responsible for removing both the channel-induced and the intentional controlled intersymbol interference [Steele ed., 1992].

The 8.25-b interval duration guard period (GP) at the bottom of Fig. 25.2 is provided to prevent burst overlapping due to propagation delay fluctuations. Finally, a 26-b equalizer training segment is included in the center of the normal traffic burst. This segment is constructed by a 16-b Viterbi channel equalizer training pattern surrounded by five quasiperiodically repeated bits on both sides. Since the MS has to be informed about which BS it communicates with, for neighboring BSs one of eight different training patterns is used, associated with the so-called BS color codes, which assist in identifying the BSs.

This 156.25-b duration TCH/FS NB constitutes the basic timeslot of the TDMA frame structure, which is input to the Gaussian minimum shift keying (GMSK) modulator to be highlighted in Sec. 25.7, at a bit rate of approximately 271 kb/s. Since the bit interval is $1/(271 \text{ kb/s}) = 3.69 \ \mu s$, the timeslot duration is $156.25 \cdot 3.69 \approx 0.577$ ms. Eight such normal bursts of eight appropriately staggered TDMA users are multiplexed onto one (RF) carrier giving, a TDMA frame of $8 \cdot 0.577 \approx 4.615$-ms duration, as shown in Fig. 25.2. The physical channel as characterized earlier provides a physical timeslot with a throughput of 114 b/4.615 ms = 24.7 kb/s, which is sufficiently high to transmit the 22.8 kb/s TCH/FS information. It even has a reserved capacity of $24.7 - 22.8 = 1.9$ kb/s, which can be exploited to transmit slow control information associated with this specific traffic channel, i.e., to construct a so-called slow associated control channel (SACCH), constituted by the SACCH TDMA frames, interspersed with traffic frames at multiframe level of the hierarchy, as seen in Fig. 25.2.

Mapping logical data traffic channels onto a physical channel is essentially carried out by the channel codecs [Wong and Hanzo, 1992], as specified in recommendation R.05.03. The full- and half-rate data traffic channels standardized in the GSM system are: TCH/F9.6, TCH/F4.8, TCH/F2.4, as well as TCH/H4.8, TCH/H2.4, as was shown earlier in Table 25.2. Note that the numbers in these acronyms represent the data transmission rate in kilobits per second. Without considering the details of these mapping processes we now focus our attention on control signal transmission issues.

25.5 Transmission of Control Signals

The exact derivation, forward error correcting (FEC) coding and mapping of logical control channel information is beyond the scope of this chapter, and the interested reader is referred to ETSI, 1988 (R.05.02 and R.05.03) and Hanzo and Stefanov, 1992, for a detailed discussion. As an example, the mapping of the 184-b SACCH, FACCH, BCCH, SDCCH, PCH, and access grant control channel (AGCH) messages onto a 456-b block, i.e., onto four 114-b bursts is demonstrated in Fig. 25.4. A double-layer concatenated FIRE-code/convolutional code scheme generates 456 bits, using an overall coding rate of $R = 184/456$, which gives a stronger protection for control channels than the error protection of traffic channels.

Returning to Fig. 25.2 we will now show how the SACCH is accommodated by the TDMA frame structure. The TCH/FS TDMA frames of the eight users are multiplexed into multiframes of 24 TDMA frames, but the 13th frame will carry a SACCH message, rather than the 13th TCH/FS frame, whereas the 26th frame will be an idle or dummy frame, as seen at the left-hand side of Fig. 25.2 at the multiframe level of the traffic channel hierarchy. The general control channel frame structure shown at the right of Fig. 25.2 is discussed later. This way 24-TCH/FS frames are sent in a 26-frame multiframe during $26 \cdot 4.615 = 120$ ms. This reduces the traffic throughput to $(24/26) \cdot 24.7 = 22.8$ kb/s required by TCH/FS, allocates $(1/26) \cdot 24.7 = 950$ b/s to the SACCH and wastes 950 b/s in the idle frame. Observe that the SACCH frame has eight timeslots to transmit the eight 950-b/s SACCHs of the eight users on the same carrier. The 950-b/s idle capacity will be used in case of half-rate channels, where 16 users will be multiplexed onto alternate frames of the

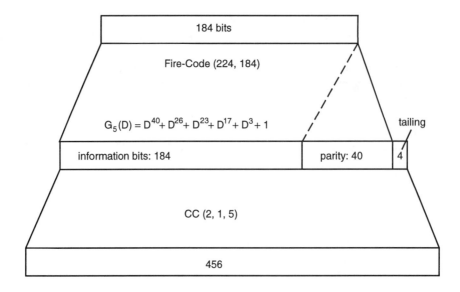

FIGURE 25.4 FEC in SACCH, FACCH, BCCH, SDCCH, PCH and AGCH, ©ETT [Hanzo and Steele, 1994].

TDMA structure to increase system capacity. Then 16, 11.4-kb/s encoded half-rate speech TCHs will be transmitted in a 120-ms multiframe, where also 16 SACCHs are available.

The FACCH messages are transmitted via the physical channels provided by bits stolen from their own host traffic channels. The construction of the FACCH bursts from 184 control bits is identical to that of the SACCH, as also shown in Fig. 25.4, but its 456-b frame is mapped onto eight consecutive 114-b TDMA traffic bursts, exactly as specified for TCH/FS. This is carried out by stealing the even bits of the first four and the odd bits of the last four bursts, which is signalled by setting $hu = 1$, $hl = 0$ and $hu = 0$, $hl = 1$ in the first and last bursts, respectively. The unprotected FACCH information rate is 184 b/20 ms = 9.2 kb/s, which is transmitted after concatenated error protection at a rate of 22.8 kb/s. The repetition delay is 20 ms, and the interleaving delay is $8 \cdot 4.615 = 37$ ms, resulting in a total of 57-ms delay.

In Fig. 25.2 at the next hierarchical level, 51-TCH/FS multiframes are multiplexed into one superframe lasting $51 \cdot 120$ ms = 6.12 s, which contains $26 \cdot 51 = 1326$-TDMA frames. In the case of 1326-TDMA frames, however, the frame number would be limited to $0 \le FN \le 1326$ and the encryption rule relying on such a limited range of FN values would not be sufficiently secure. Then 2048 superframes were amalgamated to form a hyperframe of $1326 \cdot 2048 = 2,715,648$-TDMA frames lasting $2048 \cdot 6.12$ s ≈ 3 h 28 min, allowing a sufficiently high FN value to be used in the encryption algorithm. The uplink and downlink traffic-frame structures are identical with a shift of three timeslots between them, which relieves the MS from having to transmit and receive simultaneously, preventing high-level transmitted power leakage back to the sensitive receiver. The received power of adjacent BSs can be monitored during unallocated timeslots.

In contrast to duplex traffic and associated control channels, the simplex BCCH and CCCH logical channels of all MSs roaming in a specific cell share the physical channel provided by timeslot zero of the so-called BCCH carriers available in the cell. Furthermore, as demonstrated by the right-hand side section of Fig. 25.2, 51 BCCH and CCCH TDMA frames are mapped onto a $51 \cdot 4.615 = 235$-ms duration multiframe, rather than on a 26-frame, 120-ms duration multiframe. In order to compensate for the extended multiframe length of 235 ms, 26 multiframes constitute a 1326-frame superframe of 6.12-s duration. Note in Fig. 25.5 that the allocation of the uplink and downlink frames is different, since these control channels exist only in one direction.

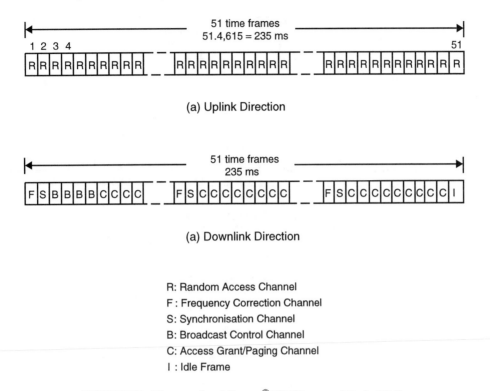

FIGURE 25.5 The control multiframe, ©ETT [Hanzo and Steele, 1994].

Specifically, the random access channel (RACH) is only used by the MSs in the uplink direction if they request, for example, a bidirectional SDCCH to be mapped onto an RF channel to register with the network and set up a call. The uplink RACH has a low capacity, carrying messages of 8-b/235-ms multiframe, which is equivalent to an unprotected control information rate of 34 b/s. These messages are concatenated FEC coded to a rate of 36 b/235 ms = 153 b/s. They are not transmitted by the NB derived for TCH/FS, SACCH, or FACCH logical channels, but by the AB, depicted in Fig. 25.6 in comparison to a NB and other types of bursts to be described later. The FEC coded, encrypted 36-b AB messages of Fig. 25.6, contain among other parameters, the encoded 6-b BS identifier code (BSIC) constituted by the 3-b PLMN color code and 3-b BS color code for unique BS identification. These 36 b are positioned after the 41-b synchronization sequence, which has a high wordlength in order to ensure reliable access burst recognition and a low probability of being emulated by interfering stray data. These messages have no interleaving delay, while they are transmitted with a repetition delay of one control multiframe length, i.e., 235 ms.

Adaptive time frame alignment is a technique designed to equalize propagation delay differences between MSs at different distances. The GSM system is designed to allow for cell sizes up to 35 km radius. The time a radio signal takes to travel the 70 km from the base station to the mobile station and back again is 233.3 μs. As signals from all the mobiles in the cell must reach the base station without overlapping each other, a long guard period of 68.25 b (252 μs) is provided in the access burst, which exceeds the maximum possible propagation delay of 233.3 μs. This long guard period in the access burst is needed when the mobile station attempts its first access to the base station or after a handover has occurred. When the base station detects a 41-b random access synchronization sequence with a long guard period, it measures the received signal delay relative to the expected signal from a mobile station of zero range. This delay, called the timing advance, is signalled using a 6-b number to the mobile station, which advances its timebase over the range of 0–63 b, i.e., in units

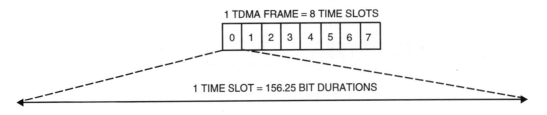

NORMAL BURST

TAIL BITS	ENCRYPTED BITS	TRAINING SEQUENCE	ENCRYPTED BITS	TAIL BITS	GUARD PERIOD
3	58	26	58	3	8.25

FREQUENCY CORRECTION BURST

TAIL BITS	FIXED BITS	TAIL BITS	GUARD PERIOD
3	142	3	8.25

SYNCHRONISATION BURST

TAIL BITS	ENCRYPTED SYNC BITS	EXTENDED TRAINING SEQUENCE	ENCRYPTED SYNC BITS	TAIL BITS	GUARD PERIOD
3	39	64	39	3	8.25

ACCESS BURST

TAIL BITS	SYNCHRO SEQUENCE	ENCRYPTED BITS	TAIL BITS	GUARD PERIOD
8	41	36	3	68.25

FIGURE 25.6   GSM burst structures, ©ETT, [Hanzo and Steele, 1994].

of 3.69 μs. By this process the TDMA bursts arrive at the BS in their correct timeslots and do not overlap with adjacent ones. This process allows the guard period in all other bursts to be reduced to 8.25 · 3.69 μs ≈ 30.46 μs (8.25 b) only. During normal operation, the BS continously monitors the signal delay from the MS and, if necessary, it will instruct the MS to update its time advance parameter. In very large traffic cells there is an option to actively utilize every second timeslot only to cope with higher propagation delays, which is spectrally inefficient, but in these large, low-traffic rural cells it is admissible.

As demonstrated by Fig. 25.2, the downlink multiframe transmitted by the BS is shared amongst a number of BCCH and CCCH logical channels. In particular, the last frame is an idle frame (I), whereas the remaining 50 frames are divided in five blocks of ten frames, where each block starts with a frequency correction channel (FCCH) followed by a synchronization channel (SCH). In the first block of ten frames the FCCH and SCH frames are followed by four BCCH frames and by either four AGCH or four PCH. In the remaining four blocks of ten frames, the last eight frames are devoted to either PCHs or AGCHs, which are mutually exclusive for a specific MS being either paged or granted a control channel.

The FCCH, SCH, and RACH require special transmission bursts, tailored to their missions, as depicted in Fig. 25.6. The FCCH uses frequency correction bursts (FCB) hosting a specific 142-b pattern. In partial response GMSK it is possible to design a modulating data sequence, which results in a near-sinusoidal modulated signal imitating an unmodulated carrier exhibiting a fixed frequency offset from the RF carrier utilized. The synchronization channel transmits SB hosting a 16 · 4 = 64-b extended sequence exhibiting a high-correlation peak in order to allow frame alignment with a quarter-bit accuracy. Furthermore, the SB contains 2 · 39 = 78 encrypted FEC-coded synchronization bits, hosting the BS and PLMN color codes, each representing one of eight legitimate identifiers. Lastly, the AB contain an extended 41-b synchronization sequence, and they are invoked to facilitate initial access to the system. Their long guard space of 68.25-b duration prevents frame overlap, before the MS's distance, i.e., the propagation delay becomes known to the BS and could be compensated for by adjusting the MS's timing advance.

25.6 Synchronization Issues

Although some synchronization issues are standardized in recommendations R.05.02 and R.05.03, the GSM recommendations do not specify the exact BS-MS synchronization algorithms to be used, these are left to the equipment manufacturers. A unique set of timebase counters, however, is defined in order to ensure perfect BS-MS synchronism. The BS sends FCB and SB on specific timeslots of the BCCH carrier to the MS to ensure that the MS's frequency standard is perfectly aligned with that of the BS, as well as to inform the MS about the required initial state of its internal counters. The MS transmits its uniquely numbered traffic and control bursts staggered by three timeslots with respect to those of the BS to prevent simultaneous MS transmission and reception, and also takes into account the required timing advance (TA) to cater for different BS-MS-BS round-trip delays.

The timebase counters used to uniquely describe the internal timing states of BSs and MSs are the quarter-bit number ($QN = 0$–624) counting the quarter-bit intervals in bursts, bit number ($BN = 0$–156), timeslot number ($TN = 0$–7) and TDMA Frame Number ($FN = 0$–26·51·2048), given in the order of increasing interval duration. The MS sets up its timebase counters after receiving a SB by determining QN from the 64-b extended training sequence in the center of the SB, setting $TN = 0$ and decoding the 78-encrypted, protected bits carrying the 25-SCH control bits.

The SCH carries frame synchronization information as well as BS identification information to the MS, as seen in Fig. 25.7, and it is provided solely to support the operation of the radio subsystem. The first 6 b of the 25-b segment consist of three PLMN color code bits and three BS color code bits supplying a unique BS identifier code (BSIC) to inform the MS which BS it is communicating with. The second 19-bit segment is the so-called reduced TDMA frame number RFN derived from the full TDMA frame number FN, constrained to the range of $[0–(26 \cdot 51 \cdot 2048) - 1] = (0–2,715,647)$ in terms of three subsegments $T1$, $T2$, and $T3$. These subsegments are computed as follows: $T1(11\,b) = [FN \text{ div } (26 \cdot 51)]$, $T2(5\,b) = (FN \text{ mod } 26)$ and $T3'(3b) = [(T3 - 1) \text{ div } 10]$, where $T3 = (FN \text{ mod } 5)$, whereas div and mod represent the integer division and modulo operations, respectively. Explicitly, in Fig. 25.7 $T1$ determines the superframe index in a hyperframe, $T2$ the multiframe index in a superframe, $T3$ the frame index in a multiframe, whereas $T3'$ is the so-called signalling block index [1–5] of a frame in a specific 51-frame control multiframe, and their roles are best understood by referring to Fig 25.2. Once the MS has received the SB, it readily computes the FN required in various control algorithms, such as encryption, handover, etc., as

$$FN = 51[(T3 - T2) \text{ mod } 26] + T3 + 51 \cdot 26 \cdot T1, \qquad \text{where } T3 = 10 \cdot T3' + 1$$

PLMN colour 3 bits	BS colour 3 bits	T1 : superframe index 11 bits	T2 : multiframe index 5 bits	T1 : block frame index 3 bits

BSIC 6 bits ◄──────► RFN 19 bits ◄──────────────────────────►

FIGURE 25.7 Synchronization channel (SCH) message format, ©ETT [Hanzo and Steele, 1994].

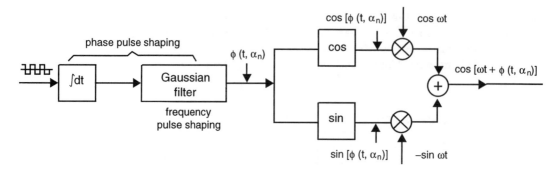

FIGURE 25.8   GMSK modulator schematic diagram, ©ETT [Hanzo and Steele, 1994].

25.7 Gaussian Minimum Shift Keying Modulation

The GSM system uses constant envelope partial response GMSK modulation [Steele ed., 1992] specified in recommendation R.05.04. Constant envelope, continuous-phase modulation schemes are robust against signal fading as well as interference and have good spectral efficiency. The slower and smoother are the phase changes, the better is the spectral efficiency, since the signal is allowed to change less abruptly, requiring lower frequency components. The effect of an input bit, however, is spread over several bit periods, leading to a so-called partial response system, which requires a channel equalizer in order to remove this controlled, intentional intersymbol interference (ISI) even in the absence of uncontrolled channel dispersion.

The widely employed partial response GMSK scheme is derived from the full response minimum shift keying (MSK) scheme. In MSK the phase changes between adjacent bit periods are piecewise linear, which results in discontinuous-phase derivative, i.e., instantaneous frequency at the signalling instants, and hence widens the spectrum. Smoothing these phase changes, however, by a filter having a Gaussian impulse response [Steele ed., 1992], which is known to have the lowest possible bandwidth, this problem is circumvented using the schematic of Fig. 25.8, where the GMSK signal is generated by modulating and adding two quadrature carriers. The key parameter of GMSK in controlling both bandwidth and interference resistance is the 3-dB down filter-bandwidth × bit interval product $(B \cdot T)$, referred to as normalized bandwidth. It was found that as the $B \cdot T$ product is increased from 0.2 to 0.5, the interference resistance is improved by approximately 2 dB at the cost of increased bandwidth occupancy, and best compromise was achieved for $B \cdot T = 0.3$. This corresponds to spreading the effect of 1 b over approximately 3-b intervals. The spectral efficiency gain due to higher interference tolerance and, hence, more dense frequency reuse was found to be more significant than the spectral loss caused by wider GMSK spectral lobes.

The channel separation at the TDMA burst rate of 271 kb/s is 200 kHz, and the modulated spectrum must be 40 dB down at both adjacent carrier frequencies. When TDMA bursts are transmitted in an on-off keyed mode, further spectral spillage arises, which is mitigated by a smooth power ramp up and down envelope at the leading and trailing edges of the transmission bursts, attenuating the signal by 70 dB during a 28- and 18-μs interval, respectively.

25.8 Wideband Channel Models

The set of 6-tap GSM impulse responses [Greenwood and Hanzo, 1992] specified in recommendation R.05.05 is depicted in Fig. 25.9, where the individual propagation paths are independent Rayleigh fading paths, weighted by the appropriate coefficients h_i corresponding to their relative powers portrayed in the figure. In simple terms the wideband channel's impulse response is measured by transmitting an impulse and detecting the received echoes at the channel's output in every D-spaced so-called delay bin. In some bins no delayed and attenuated multipath component

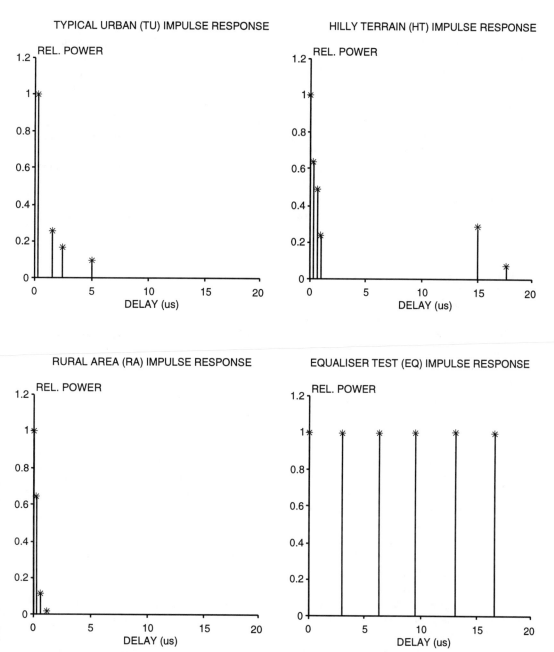

FIGURE 25.9 Typical GSM channel impulse responses, ©ETT [Hanzo and Steele, 1994].

is received, whereas in others significant energy is detected, depending on the typical reflecting objects and their distance from the receiver. The path delay can be easily related to the distance of the reflecting objects, since radio waves are travelling at the speed of light. For example, at a speed of 300, 000 km/s, a reflecting object situated at a distance of 0.15 km yields a multipath component at a round-trip delay of 1 μs.

The typical urban (TU) impulse response spreads over a delay interval of 5 μs, which is almost two 3.69-μs bit-intervals duration and, therefore, results in serious ISI. In simple terms, it can be treated as a two-path model, where the reflected path has a length of 0.75 km, corresponding to a reflector located at a distance of about 375 m. The hilly terrain (HT) model has a sharply decaying short-delay section due to local reflections and a long-delay path around 15 μs due to distant reflections. Therefore, in practical terms it can be considered a two- or three-path model having reflections from a distance of about 2 km. The rural area (RA) response seems the least hostile amongst all standardized responses, decaying rapidly inside 1-b interval and, therefore, is expected to be easily combated by the channel equalizer. Although the type of the equalizer is not standardized, partial response systems typically use VEs. Since the RA channel effectively behaves as a single-path nondispersive channel, it would not require an equalizer. The fourth standardized impulse response is artificially contrived in order to test the equalizer's performance and is constituted by six equidistant unit-amplitude impulses representing six equal-powered independent Rayleigh-fading paths with a delay spread over 16 μs. With these impulse responses in mind, the required channel is simulated by summing the appropriately delayed and weighted received signal components. In all but one case the individual components are assumed to have Rayleigh amplitude distribution, whereas in the RA model the main tap at zero delay is supposed to have a Rician distribution with the presence of a dominant line-of-sight path.

25.9 Adaptive Link Control

The adaptive link control algorithm portrayed in Fig. 25.10 and specified in recommendation R.05.08 allows for the MS to favor that specific traffic cell which provides the highest probability of reliable communications associated with the lowest possible path loss. It also decreases interference with other cochannel users and, through dense frequency reuse, improves spectral efficiency, whilst maintaining an adequate communications quality, and facilitates a reduction in power consumption, which is particularly important in hand-held MSs. The handover process maintains a call in progress as the MS moves between cells, or when there is an unacceptable transmission quality degradation caused by interference, in which case an intracell handover to another carrier in the same cell is performed. A radio-link failure occurs when a call with an unacceptable voice or data quality cannot be improved either by RF power control or by handover. The reasons for the link failure may be loss of radio coverage or very high-interference levels. The link control procedures rely on measurements of the received RF signal strength (RXLEV), the received signal quality (RXQUAL), and the absolute distance between base and mobile stations (DISTANCE).

RXLEV is evaluated by measuring the received level of the BCCH carrier which is continuously transmitted by the BS on all time slots of the B frames in Fig. 25.5 and without variations of the RF level. A MS measures the received signal level from the serving cell and from the BSs in all adjacent cells by tuning and listening to their BCCH carriers. The root mean squared level of the received signal is measured over a dynamic range from -103 to -41 dBm for intervals of one SACCH multiframe (480 ms). The received signal level is averaged over at least 32 SACCH frames ($\approx$15 s) and mapped to give RXLEV values between 0 and 63 to cover the range from -103 to -41 dBm in steps of 1 dB. The RXLEV parameters are then coded into 6-b words for transmission to the serving BS via the SACCH.

RXQUAL is estimated by measuring the bit error ratio (BER) before channel decoding, using the Viterbi channel equalizer's metrics [Steele ed., 1992] and/or those of the Viterbi convolutional decoder [Wong and Hanzo, 1992]. Eight values of RXQUAL span the logarithmically scaled BER range of 0.2–12.8% before channel decoding.

The absolute DISTANCE between base and mobile stations is measured using the timing advance parameter. The timing advance is coded as a 6-b number corresponding to a propagation delay from 0 to $63 \cdot 3.69 \ \mu s = 232.6 \ \mu s$, characteristic of a cell radius of 35 km.

While roaming, the MS needs to identify which potential target BS it is measuring, and the BCCH carrier frequency may not be sufficient for this purpose, since in small cluster sizes the same BCCH frequency may be used in more than one surrounding cell. To avoid ambiguity a 6-b BSIC is transmitted on each BCCH carrier in the SB of Fig. 25.6. Two other parameters transmitted in the BCCH data provide additional information about the BS. The binary flag called

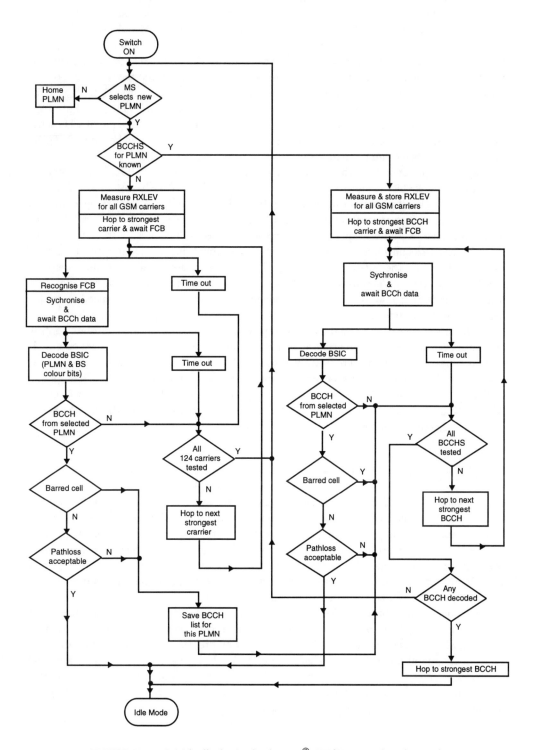

FIGURE 25.10   Initial cell selection by the MS, ©ETT [Hanzo and Steele, 1994].

PLMN_PERMITTED indicates whether the measured BCCH carrier belongs to a PLMN that the MS is permitted to access. The second Boolean flag, CELL_BAR_ACCESS, indicates whether the cell is barred for access by the MS, although it belongs to a permitted PLMN. A MS in idle mode, i.e., after it has just been switched on or after it has lost contact with the network, searches all 125 RF channels and takes readings of RXLEV on each of them. Then it tunes to the carrier with the highest RXLEV and searches for FCB in order to determine whether or not the carrier is a BCCH carrier. If it is not, then the MS tunes to the next highest carrier, and so on, until it finds a BCCH carrier, synchronizes to it and decodes the parameters BSIC, PLMN_PERMITTED and CELL_BAR_ACCESS in order to decide whether to continue the search. The MS may store the BCCH carrier frequencies used in the network accessed, in which case the search time would be reduced. Again, the process described is summarized in the flowchart of Fig. 25.10.

The adaptive power control is based on RXLEV measurements. In every SACCH multiframe the BS compares the RXLEV readings reported by the MS or obtained by the base station with a set of thresholds. The exact strategy for RF power control is determined by the network operator with the aim of providing an adequate quality of service for speech and data transmissions while keeping interferences low. Clearly, adequate quality must be achieved at the lowest possible transmitted power to keep cochannel interferences low, which implies contradictory requirements in terms of transmitted power. The criteria for reporting radio link failure are based on the measurements of RXLEV and RXQUAL performed by both the mobile and base stations, and the procedures for handling link failures result in the re-establishment or the release of the call, depending on the network operator's strategy.

The handover process involves the most complex set of procedures in the radio-link control. Handover decisions are based on results of measurements performed both by the base and mobile stations. The base station measures RXLEV, RXQUAL, DISTANCE, and also the interference level in unallocated time slots, whereas the MS measures and reports to the BS the values of RXLEV and RXQUAL for the serving cell and RXLEV for the adjacent cells. When the MS moves away from the BS, the RXLEV and RXQUAL parameters for the serving station become lower, whereas RXLEV for one of the adjacent cells increases.

25.10 Discontinuous Transmission

Discontinuous transmission (DTX) issues are standardized in recommendation R.06.31, whereas the associated problems of voice activity detection VAD are specified by R.06.32. Assuming an average speech activity of 50% and a high number of interferers combined with frequency hopping to randomize the interference load, significant spectral efficiency gains can be achieved when deploying discontinuous transmissions due to decreasing interferences, while reducing power dissipation as well. Because of the reduction in power consumption, full DTX operation is mandatory for MSs, but in BSs, only receiver DTX functions are compulsory.

The fundamental problem in voice activity detection is how to differentiate between speech and noise, while keeping false noise triggering and speech spurt clipping as low as possible. In vehicle-mounted MSs the severity of the speech/noise recognition problem is aggravated by the excessive vehicle background noise. This problem is resolved by deploying a combination of threshold comparisons and spectral domain techniques [ETSI, 1988; Hanzo and Stefanov, 1992]. Another important associated problem is the introduction of noiseless inactive segments, which is mitigated by comfort noise insertion (CNI) in these segments at the receiver.

25.11 Summary

Following the standardization and launch of the GSM system its salient features were summarized in this brief review. Time division multiple access (TDMA) with eight users per carrier is used at a multiuser rate of 271 kb/s, demanding a channel equalizer to combat dispersion in large cell environments. The error protected chip rate of the full-rate traffic channels is 22.8 kb/s, whereas in half-rate channels it is 11.4 kb/s. Apart from the full- and half-rate speech traffic channels, there are 5 different rate data traffic channels and 14 various control and signalling channels to support the system's operation. A moderately complex, 13 kb/s regular pulse excited speech codec with long term predictor (LTP) is used, combined with an embedded three-class error correction codec and multilayer interleaving to provide sensitivity-matched unequal error protection for the speech bits. An overall speech delay of 57.5 ms is maintained. Slow frequency hopping at 217 hops/s yields substantial performance gains for slowly moving pedestrians.

Constant envelope partial response GMSK with a channel spacing of 200 kHz is deployed to support 125 duplex channels in the 890–915-MHz up-link and 935–960-MHz down-link bands, respectively. At a transmission rate of 271 kb/s a spectral efficiency of 1.35-bit/s/Hz is achieved. The controlled GMSK-induced and uncontrolled channel-induced intersymbol interferences are removed by the channel equalizer. The set of standardized wideband GSM channels was introduced in order to provide bench markers for performance comparisons. Efficient power budgeting and minimum cochannel interferences are ensured by the combination of adaptive power and handover control based on weighted averaging of up to eight up-link and down-link system parameters. Discontinuous transmissions assisted by reliable spectral-domain voice activity detection and comfort-noise insertion further reduce interferences and power consumption. Because of ciphering, no unprotected information is sent via the radio link. As a result, spectrally efficient, high-quality mobile communications with a variety of services and international roaming is possible in cells of up to 35 km radius for signal-to-noise and interference ratios in excess of 10–12 dBs. The key system features are summarized in Table 25.3.

TABLE 25.3 Summary of GSM features

System feature	Specification
Up-link bandwidth, MHz	890–915 = 25
Down-link bandwidth, MHz	935–960 = 25
Total GSM bandwidth, MHz	50
Carrier spacing, KHz	200
No. of RF carriers	125
Multiple access	TDMA
No. of users/carrier	8
Total No. of channels	1000
TDMA burst rate, kb/s	271
Modulation	GMSK with BT = 0.3
Bandwidth efficiency, b/s/Hz	1.35
Channel equalizer	yes
Speech coding rate, kb/s	13
FEC coded speech rate, kb/s	22.8
FEC coding	Embedded block/ convolutional
Frequency hopping, hop/s	217
DTX and VAD	yes
Maximum cell radius, km	35

Defining Terms

A3: Authentication algorithm
A5: Cyphering algorithm
A8: Confidential algorithm to compute the cyphering key
AB: Access burst
ACCH: Associated control channel
ADC: Administration center
AGCH: Access grant control channel
AUC: Authentication center
AWGN: Additive Gaussian noise
BCCH: Broadcast control channel
BER: Bit error ratio
BFI: Bad frame indicator flag
BN: Bit number
BS: Base station

BS-PBGT: BS powerbudget: to be evaluated for power budget motivated handovers
BSIC: Base station identifier code
CC: Convolutional codec
CCCH: Common control channel
CELL_BAR_ACCESS: Boolean flag to indicate, whether the MS is permitted to access the specific traffic cell
CNC: Comfort noise computation
CNI: Comfort noise insertion
CNU: Comfort noise update state in the DTX handler
DB: Dummy burst
DL: Down link
DSI: Digital speech interpolation to improve link efficiency
DTX: Discontinuous transmission for power consumption and interference reduction
EIR: Equipment identity register
EOS: End of speech flag in the DTX handler
FACCH: Fast associated control channel
FCB: Frequency correction burst
FCCH: Frequency correction channel
FEC: Forward error correction
FH: Frequency hopping
FN: TDMA frame number
GMSK: Gaussian minimum shift keying
GP: Guard space
HGO: Handover in the VAD
HLR: Home location register
HO: Handover
HOCT: Handover counter in the VAD
HO_MARGIN: Handover margin to facilitate hysteresis
HSN: Hopping sequence number: frequency hopping algorithm's input variable
IMSI: International mobile subscriber identity
ISDN: Integrated services digital network
LAI: Location area identifier
LAR: Logarithmic area ratio
LTP: Long term predictor
MA: Mobile allocation: set of legitimate RF channels, input variable in the frequency hopping algorithm
MAI: Mobile allocation index: output variable of the FH algorithm
MAIO: Mobile allocation index offset: intial RF channel offset, input variable of the FH algorithm
MS: Mobile station
MSC: Mobile switching center
MSRN: Mobile station roaming number
MS_TXPWR_MAX: Maximum permitted MS transmitted power on a specific traffic channel in a specific traffic cell
MS_TXPWR_MAX(n): Maximum permitted MS transmitted power on a specific traffic channel in the nth adjacent traffic cell
NB: Normal burst
NMC: Network management center
NUFR: Receiver noise update flag
NUFT: Noise update flag to ask for SID frame transmission
OMC: Operation and maintenance center
PARCOR: Partial correlation
PCH: Paging channel

PCM: Pulse code modulation

PIN: Personal identity number for MSs

PLMN: Public land mobile network

PLMN_ PERMITTED: Boolean flag to indicate whether the MS is permitted to access the specific PLMN

PSTN: Public switched telephone network

QN: Quarter bit number

R: Random number in the authentication process

RA: Rural area channel inpulse response

RACH: Random access channel

RF: Radio frequency

RFCH: Radio frequency channel

RFN: Reduced TDMA frame number: equivalent representation of the TDMA frame number that is used in the synchronization channel

RNTABLE: Random number table utilized in the frequency hopping algorithm

RPE: Regular pulse excited

RPE-LTP: Regular pulse excited codec with long term predictor

RS-232:   Serial data transmission standard equivalent to CCITT V24. interface

RXLEV: Received signal level: parameter used in handovers

RXQUAL: Received signal quality: parameter used in handovers

S: Signed response in the authentication process

SACCH: Slow associated control channel

SB: Synchronization burst

SCH: Synchronization channel

SCPC: Single channel per carrier

SDCCH: Stand-alone dedicated control channel

SE: Speech extrapolation

SID: Silence identifier

SIM: Subscriber identity module in MSs

SPRX: Speech received flag

SPTX: Speech transmit flag in the DTX handler

STP: Short term predictor

TA: Timing advance

TB: Tailing bits

TCH: Traffic channel

TCH/F: Full-rate traffic channel

TCH/F2.4:   Full-rate 2.4-kb/s data traffic channel

TCH/F4.8:   Full-rate 4.8-kb/s data traffic channel

TCH/F9.6:   Full-rate 9.6-kb/s data traffic channel

TCH/FS: Full-rate speech traffic channel

TCH/H: Half-rate traffic channel

TCH/H2.4:   Half-rate 2.4-kb/s data traffic channel

TCH/H4.8:   Half-rate 4.8-kb/s data traffic channel

TDMA: Time division multiple access

TMSI: Temporary mobile subscriber identifier

TN: Time slot number

TU: Typical urban channel inpulse response

TXFL: Transmit flag in the DTX handler

UL: Up link

VAD: Voice activity detection

VE: Viterbi equalizer

VLR: Visiting location register

References

European Telecommunications Standardization Institute. 1988. Group Speciale Mobile or Global System of Mobile Communication (GSM) Recommendation, ETSI Secretariat, Sophia Antipolis Cedex, France.

Greenwood, D. and Hanzo, L. 1992. Characterisation of mobile radio channels, In *Mobile Radio Communications*. ed. R. Steele, Chap. 2, pp. 92–185. IEEE Press–Pentech Press, London.

Hanzo, L. and Stefanov, J. 1992. The Pan-European digital cellular mobile radio system—known as GSM. In *Mobile Radio Communications*, ed. R. Steele, Chap. 8, pp. 677–773, IEEE Press–Pentech Press, London.

Hanzo, L. and Steele, R. 1994. The Pan-European mobile radio system, Pts. 1 and 2, *European Trans. on Telecomm.*, 5(2):245–276.

Salami, R.A., Hanzo, L. et al. 1992. Speech coding. In *Mobile Radio Communications*, ed. R. Steele, Chap. 3, pp. 186–346. IEEE Press–Pentech Press, London.

Steele, R. ed., 1992. *Mobile Radio Communications*, IEEE Press–Pentech Press, London.

Vary, P. and Sluyter, R.J. 1986. MATS-D speech codec: Regular-pulse excitation LPC, *Proceedings of Nordic Conference on Mobile Radio Communications*. pp. 257–261.

Wong, K.H.H and Hanzo, L. 1992. Channel coding. In *Mobile Radio Communications*, ed. R. Steele, Chap. 4, pp. 347–488. IEEE Press–Pentech Press, London.

26

The IS-54 Digital Cellular Standard

Paul Mermelstein
INRS-Télécommunications
University of Québec

26.1 Introduction

The goals of this chapter are to give the reader a tutorial introduction and high-level understanding of the techniques employed for speech transmission by the IS-54 digital cellular standard. It builds on the information provided in the standards document but is not meant to be a replacement for it. Separate standards cover the control channel used for the setup of calls and their handoff to neighboring cells, as well as the encoding of data signals for transmission. For detailed implementation information the reader should consult the most recent standards document [TIA, 1992].

IS-54 provides for encoding bidirectional speech signals digitally and transmitting them over cellular and microcellular mobile radio systems. It retains the 30-kHz channel spacing of the earlier advanced mobile telephone service (AMPS), which uses analog frequency modulation for speech transmission and frequency shift keying for signalling. The two directions of transmission use frequencies some 45 MHz apart in the band between 824 and 894 MHz. AMPS employs one channel per conversation in each direction, a technique known as frequency division multiple access (FDMA). IS-54 employs time division multiple access (TDMA) by allowing three, and in the future six, simultaneous transmissions to share each frequency band. Because the overall 30-kHz channelization of the allocated 25 MHz of spectrum in each direction is retained, it is also known as a FDMA-TDMA system. In contrast, the later IS-95 standard employs code division multiple access (CDMA) over bands of 1.23 MHz by combining several 30-kHz frequency channels.

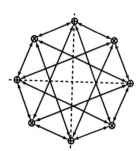

FIGURE 26.1 Constellation for $\pi/4$ shifted QPSK modulation. *Source*: TIA, 1992. Cellular System Dual-mode Mobile Station–Base Station Compatibility Standard TIA/EIA IS-54. With permission.

Each frequency channel provides for transmission at a digital bit rate of 48.6 kb/s through use of differential quadrature-phase shift key (DQPSK) modulation at a 24.3-kBd channel rate. The channel is divided into six time slots every 40 ms. The full-rate voice coder employs every third time slot and utilizes 13 kb/s for combined speech and channel coding. The six slots provide for an eventual half-rate channel occupying one slot per 40 ms frame and utilizing only about 6.5 kb/s for each call. Thus, the simultaneous call carrying capacity with IS-54 is increased by a factor 3(factor 6 in the future) above that of AMPS. All digital transmission is expected to result in a reduction in transmitted power. The resulting reduction in intercell interference may allow more frequent reuse of the same frequency channels than the reuse pattern of seven cells for AMPS. Additional increases in erlang capacity (the total call-carrying capacity at a given blocking rate) may be available from the increased trunking efficiency achieved by the larger number of simultaneously available channels. The first systems employing dual-mode AMPS and TDMA service were put into operation in 1993.

26.2 Modulation of Digital Voice and Data Signals

The modulation method used in IS-54 is $\pi/4$ shifted differentially encoded quadrature phase-shift keying (DPSK). Symbols are transmitted as changes in phase rather than their absolute values. The binary data stream is converted to two binary streams X_k and Y_k formed from the odd- and even-numbered bits, respectively. The quadrature streams I_k and Q_k are formed according to

$$I_k = I_{k-1} \cos[\Delta\phi(X_k, Y_k)] - Q_{k-1} \sin[\Delta\phi(X_k, Y_k)]$$

$$Q_k = I_{k-1} \sin[\Delta\phi(X_k, Y_k)] + Q_{k-1} \cos[\Delta\phi(X_k, Y_k)]$$

where I_{k-1} and Q_{k-1} are the amplitudes at the previous pulse time. The phase change $\Delta\phi$ takes the values $\pi/4$, $3\pi/4$, $-\pi/4$, and $-3\pi/4$ for the dibit (X_k, Y_k) symbols (0,0), (0,1), (1,0) and (1,1), respectively. This results in a rotation by $\pi/4$ between the constellations for odd and even symbols. The differential encoding avoids the problem of 180° phase ambiguity that may otherwise result in estimation of the carrier phase.

The signals I_k and Q_k at the output of the differential phase encoder can take one of five values, $0, \pm 1, \pm 1/\sqrt{2}$ as indicated in the constellation of Fig. 26.1. The corresponding impulses are applied to the inputs of the I and Q baseband filters, which have linear phase and square root raised cosine frequency responses. The generic modulator circuit is shown in Fig. 26.2. The rolloff factor α determines the width of the transition band and its value is 0.35,

$$|H(f)| = \begin{cases} 1, & 0 \le f \le (1-\alpha)/2T \\ \sqrt{1/2\{1 - \sin[\pi(2fT - 1)/2\alpha]\}}, & (1-\alpha)/2T \le f \le (1+\alpha)/2T \\ 0, & f > (1+\alpha)/2T \end{cases}$$

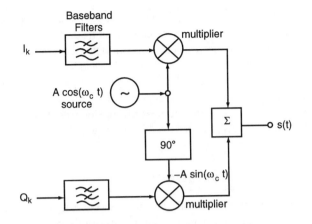

FIGURE 26.2 Generic modulation circuit for digital voice and data signals. *Source*: TIA, 1992. Cellular System Dual-mode Mobile Station–Base Station Compatibility Standard TIA/EIA IS-54.

26.3 Speech Coding Fundamentals

The IS-54 standard employs a vector-sum excited linear prediction (VSELP) coding technique. It represents a specific formulation of the much larger class of code-excited linear preduction (CELP) coders [Atal and Schroeder, 1984] that have proved effective in recent years for the coding of speech at moderate rates in the range 4–16 kb/s. VSELP provides reconstructed speech with a quality that is comparable to that available with frequency modulation and analog transmission over the AMPS system. The coding rate employed is 7.95 kb/s. Each of the six slots per frame carry 260 b of speech and channel coding information for a gross information rate of 13 kb/s. The 260 b correspond to 20 ms of real time speech, transmitted as a single burst.

For an excellent recent review of speech coding techniques for transmission, the reader is referred to Gersho, 1994. Most modern speech coders use a form of analysis by synthesis coding where the encoder determines the coded signal one segment at a time by feeding candidate excitation segments into a replica of a synthesis filter and selecting the segment that minimizes the distortion between the original and reproduced signals. Linear prediction coding (LPC) techniques [Atal and Hanauer, 1971] encode the speech signal by first finding an optimum linear filter to remove the short-time correlation, passing the signal through that LPC filter to obtain a residual signal, and encoding this residual using much fewer bits than would have been required to code the original signal with the same fidelity. In most cases the coding of the residual is divided into two steps. First, the long-time correlation due to the periodic pitch excitation is removed by means of an optimum one-tap filter with adjustable gain and lag. Next, the remaining residual signal, which now closely resembles a white-noise signal, is encoded. Code-excited linear predictors use one or more **codebooks** from which they select replicas of the residual of the input signal by means of a closed-loop error-minimization technique. The index of the codebook entry as well as the parameters of all the filters are transmitted to allow the speech signal to be reconstructed at the receiver. Most code-excited coders use trained codebooks. Starting with a codebook containing Guassian signal segments, entries that are found to be used rarely in coding a large body of speech data are iteratively eliminated to result in a smaller codebook that is considered more effective.

The speech signal can be considered quasistationary or stationary for the duration of the speech frame, of the order of 20 ms. The parameters of the short-term filter, the LPC coefficients, are determined by analysis of the autocorrelation function of a suitably windowed segment of the input signal. To allow accurate determination of the time-varying pitch lag as well as simplify the computations, each speech frame is divided into four 5-ms subframes. Independent pitch filter computations and residual coding operations are carried out for each subframe.

The speech decoder attempts to reconstruct the speech signal from the received information as best possible. It employs a codebook identical to that of the encoder for excitation generation and, in the absence of transmission errors, would produce an exact replica of the signal that produced the minimized error at the encoder. Transmission errors do occur, however, due, to signal fading and excessive interference. Since any attempt at retransmission would incur unacceptable signal delays, sufficient error protection is provided to allow correction of most transmission errors.

26.4 Channel Coding Considerations

The sharp limitations on available bandwidth for error protection argue for careful consideration of the sensitivity of the speech coding parameters to transmission errors. Pairwise interleaving of coded blocks and convolutional coding of a subset of the parameters permit correction of a limited number of transmission errors. In addition, a cyclic redundancy check (CRC) is used to determine whether the error correction was successful. The coded information is divided into three blocks of varying sensitivity to errors. Group 1 contains the most sensitive bits, mainly the parameters of the LPC filter and frame energy, and is protected by both error detection and correction bits. Group 2 is provided with error correction only. The third group, comprising mostly the fixed codebook indices, is not protected at all.

The speech signal contains significant temporal redundancy. Thus, speech frames within which errors have been detected may be reconstructed with the aid of previously correctly received information. A bad-frame masking procedure attempts to hide the effects of short fades by extrapolating the previously received parameters. Of course, if the errors persist, the decoded signal must be muted while an attempt is made to hand off the connection to a base station to/from which the mobile may experience better reception.

26.5 VSELP Encoder

A block diagram of the VSELP speech encoder [Gerson and Jasiuk, 1990] is shown in Fig. 26.3. The excitation signal is generated from three components, the output of a long term or pitch filter, as well as entries from two codebooks. A weighted synthesis filter generates a synthesized approximation to the frequency-weighted input signal. The weighted mean square error between these two signals is used to drive the error minimization process. This weighted error is considered to be a better approximation to the percpetually important noise components than the unweighted mean square error. The total weighted square error is minimized by adjusting the pitch lag and the codebook indices as well as their gains. The decoder follows the encoder closely and generates the excitation signal identically to the encoder but uses an unweighted linear-prediction synthesis filter to generate the decoded signal. A spectral postfilter is added after the synthesis filter to enhance the quality of the reconstructed speech.

The precise data rate of the speech coder is 7950 b/s or 159 b per time slot, each corresponding to 20 ms of signal in real time. These 159 b are allocated as follows: 1) short-term filter coefficients, 38 bits; 2) frame energy, 5 bits; 3) pitch lag, 28 bits; 4) codewords, 56 bits; and 5) gain values, 32 bits.

26.6 Linear Prediction Analysis and Quantization

The purpose of the LPC analysis filter is to whiten the spectrum of the input signal so that it can be better matched by the codebook outputs. The corresponding LPC synthesis filter $A(z)$ restores the short-time speech spectrum characteristics to the output signal. The transfer function of the

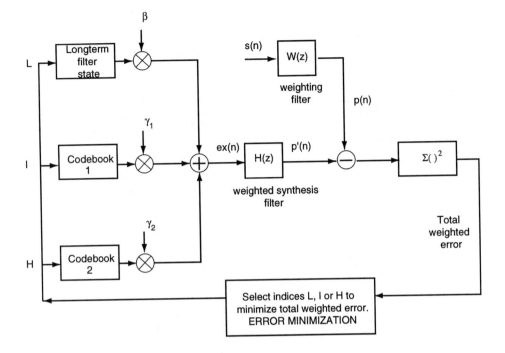

FIGURE 26.3 Black diagram of the speech encoder in VSELP. TIA. 1992. Cellular system Dual-mode Mobile Station–Base Station Compatibility Standard. TIA/EIA IS-54.

tenth-order synthesis filter is given by

$$A(z) = \frac{1}{1 - \sum_{i=1}^{N_p} \alpha_i z^{-i}}$$

The filter predictor parameters $\alpha_1, \ldots, \alpha_{N_p}$ are not transmitted directly. Instead, a set of reflection coefficients $r_1, \ldots, r_{N_p}$ are computed and quantized. The predictor parameters are determined from the reflection coefficients using a well-known backward recursion algorithm [Makhoul, 1975].

A variety of algorithms are known that determine a set of reflection coefficients from a windowed input signal. One such algorithm is the fixed point **covariance lattice**, FLAT, which builds an optimum inverse lattice stage by stage. At each stage j, the sum of the mean-squared forward and backward residuals is minimized by selection of the best reflection coefficient r_j. The analysis window used is 170 samples long, centered with respect to the middle of the fourth 5-ms subframe of the 20-ms frame. Since this centerpoint is 20 samples from the end of the frame, 65 samples from the next frame to be coded are used in computing the reflection coefficient of the current frame. This introduces a lookahead delay of 8.125 ms.

The FLAT algorithm first computes the covariance matrix of the input speech for $N_A = 170$ and $N_p = 10$,

$$\phi(i, k) = \sum_{n=N_p}^{N_A-1} s(n-i)s(n-k), \qquad 0 \le i, \quad k \le N_p,$$

Define the forward residual out of stage j as $f_j(n)$ and the backward residual as $b_j(n)$. Then the autocorrelation of the initial forward residual $F_0(i, k)$ is given by $\phi(i, k)$. The autocorrelation of the initial backward residual $B_0(i, k)$ is given by $\phi(i + 1, k + 1)$ and the initial cross correlation of

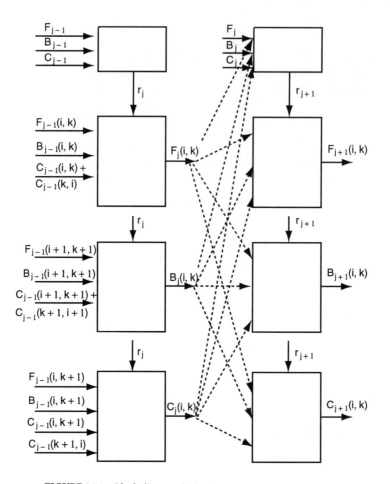

FIGURE 26.4   Block diagram for lattice covariance computations.

the two residuals is given by $C_0(i, k) = \phi(i, k + 1)$ for $0 \leq i, k \leq N_{p-1}$. Initially j is set to 1. The **reflection coefficient** at each stage is determined as the ratio of the cross correlation to the mean of the autocorrelations. A block diagram of the computations is shown in Fig. 26.4. By quantizing the reflection coefficients within the computation loops, reflection coefficients at subsequent stages are computed taking into account the quantization errors of the previous stages. Specifically,

$$C'_{j-1} = C_{j-1}(0, 0) + C_{j-1}(N_p - j, N_p - j)$$
$$F'_{j-1} = F_{j-1}(0, 0) + F_{j-1}(N_p - j, N_p - j)$$
$$B'_{j-1} = B_{j-1}(0, 0) + B_{j-1}(N_p - j, N_p - j)$$

and

$$r_j = \frac{-2C'_{j-1}}{F'_{j-1} + B'_{j-1}}$$

Use of two sets of correlation values separated by $N_p - j$ samples provides additional stability to the computed reflection coefficients in case the input signal changes form rapidly.

Once a quantized reflection coefficient r_j has been determined, the resulting auto- and cross correlations can be determined iteratively as

$$F_j(i, k) = F_{j-1}(i, k) + r_j[C_{j-1}(i, k) + C_{j-1}(k, i)] + r_j^2 B_{j-1}(i, k)$$
$$B_j(i, k) = B_{j-1}(i + 1, k + 1) + r_j[C_{j-1}(i + 1, k + 1) + C_{j-1}(k + 1, i + 1)]$$
$$+ r_j^2 F_{j-1}(i + 1, k + 1)$$

and

$$C_j(i, k) = C_{j-1}(i, k + 1) + r_j[B_{j-1}(i, k + 1) + F_{j-1}(i, k + 1)]$$
$$+ r_j^2 C_{j-1}(k + 1, i)$$

These computations are carried out iteratively for r_j, $j = 1, \ldots, N_p$.

26.7 Bandwidth Expansion

Poles with very narrow bandwidths may introduce undesirable distortions into the synthesized signal. Use of a binomial window with effective bandwidth of 80 Hz suffices to limit the ringing of the LPC filter and reduce the effect of the LPC filter selected for one frame on the signal reconstructed for subsequent frames. To achieve this, prior to searching for the reflection coefficients, the $\phi(i, k)$ is modified by use of a window function $w(j)$, $j = 1, \ldots, 10$, as follows:

$$\phi'(i, k) = \phi(i, k)w(|i - k|)$$

26.8 Quantizing and Encoding the Reflection Coefficients

The distortion introduced into the overall spectrum by quantizing the reflection coefficients diminishes as we move to higher orders in the reflection coefficients. Accordingly, more bits are assigned to the lower order coefficients. Specifically, 6, 5, 5, 4, 4, 3, 3, 3, 3, and 2 b are assigned to $r_1, \ldots, r_{10}$, respectively. Scalar quantization of the reflection coefficients is used in IS-54 because it is particularly simple. **Vector quantization** achieves additional quantizing efficiencies at the cost of significant added complexity.

It is important to preserve the smooth time evolution of the linear prediction filter. Both the encoder and decoder linearly interpolate the coefficients α_i for the first, second and third subframes of each frame using the coefficients determined for the previous and current frames. The fourth subframe uses the values computed for that frame.

26.9 VSELP Codebook Search

The codebook search operation selects indices for the long-term filter (pitch lag L) and the two codebooks I and H so as to minimize the total weighted error. This closed-loop search is the most computationally complex part of the encoding operation, and significant effort has been invested to minimize the complexity of these operations without degrading performance. To reduce complexity, simultaneous optimization of the codebook selections is replaced by a sequential optimization procedure, which considers the long-term filter search as the most significant and therefore executes it first. The two vector-sum codebooks are considered to contribute less and less to the minimization of the error, and their search follows in sequence. Subdivision of the total codebook into two vector sums simplifies the processing and makes the result less sensitive to errors in decoding the individual bits arising from transmission errors.

Entries from each of the two vector-sum codebooks can be expressed as the sum of basis vectors. By orthogonalizing these basis vectors to the previously selected codebook component(s), one ensures that the newly introduced components reduce the remaining errors. The subframes over which the codebook search is carried out are 5 ms or 40 samples long. An optimal search would need exploration of a 40-dimensional space. The vector-sum approximation limits the search to 14 dimensions after the optimal pitch lag has been selected. The search is further divided into two stages of 7 dimensions each. The two codebooks are specified in terms of the fourteen, 40-dimensional basis vectors stored at the encoder and decoder. The two 7-b indices indicate the required weights on the basic vectors to arrive at the two optimum codewords.

The codebook search can be viewed as selecting the three best directions in 40-dimensional space, which when summed result in the best approximation to the weighted input signal. The gains of the three components are determined through a separate error minimization process.

26.10 Long-Term Filter Search

The long-term filter is optimized by selection of a lag value that minimizes the error between the weighted input signal $p(n)$ and the past excitation signal filtered by the current weighted synthesis filter $H(z)$. There are 127 possible coded lag values provided corresponding to lags of 20–146 samples. One value is reserved for the case when all correlations between the input and the lagged residuals are negative and use of no long term filter output would be best. To simplify the convolution operation between the impulse response of the weighted synthesis filter and the past excitation, the impulse response is truncated to 21 samples or 2.5 ms. Once the lag is determined, the untruncated impulse response is used to compute the weighted long-term lag vector.

26.11 Orthogonalization of the Codebooks

Prior to the search of the first codebook, each filtered basis vector may be made orthogonal to the long-term filter output, the zero-state response of the weighted synthesis filter $H(z)$ to the long-term prediction vector. Each orthogonalized filtered basis vector is computed by subtracting its projection onto the long-term filter output from itself.

Similarly, the basis vectors of the second codebook can be orthogonalized with respect to both the long-term filter output and the first codebook output, the zero-state response of $H(z)$ to the previously selected summation of first-codebook basis vectors. In each case the codebook excitation can be reconstituted as

$$u_{k,i}(n) = \sum_{m=1}^{M} \theta_{im} v_{k,m}(n)$$

where $k = 1, 2$ for the two codebooks, $i = I$ or H the 7-b code vector received, $v_{k,m}$ are the two sets of basis vectors, and $\theta_{im} = +1$ if bit m of codeword $i = 1$ and -1 if bit m of codeword $i = 0$. Orthogonalization is not required at the decoder since the gains of the codebooks outputs are determined with respect to the weighted nonorthogonalized code vectors.

26.12 Quantizing the Excitation and Signal Gains

The three codebook gain values β, γ_1, and γ_2 are transformed to three new parameters GS, $P0$ and $P1$ for quantization purposes. GS is an energy offset parameter that equalizes the input and output signal energies. It adjusts the energy of the output of the LPC synthesis filter to equal the energy computed for the same subframe at the encoder input. $P0$ is the energy contribution of the long-term prediction vector as a fraction of the total excitation energy within the subframe.

Similarly, $P1$ is the energy contribution of the code vector selected from the first codebook as a fraction of the total excitation energy of the subframe. The transformation reduces the dynamic range of the parameters to be encoded. An 8-b vector quantizer efficiently encodes the appropriate $(GS, P0, P1)$ vectors by selecting the vector which minimizes the weighted error. The received and decoded values β, γ_1, and γ_2 are computed from the received $(GS, P0, P1)$ vector and applied to reconstitute the decoded signal.

26.13 Channel Coding and Interleaving

The goals of channel coding are to reduce the impairments in the reconstructed speech due to transmission errors. The 159 b characterizing each 20-ms block of speech are divided into two classes, 77 in class 1 and 82 in class 2. Class 1 includes the bits in which errors result in a more significant impairment, whereas the speech quality is considered less sensitive to the class- 2 bits. Class 1 generally includes the gain, pitch lag, and more significant reflection coefficient bits. In addition, a 7-b cyclic redundancy check is applied to the 12 most perceptually significant bits of class 1 to indicate whether the error correction was successful. Failure of the CRC check at the receiver suggests that the received information is so erroneous that it would be better to discard it than use it. The error correction coding is illustrated in Fig. 26.5.

The error correction technique used is rate 1/2 convolutional coding with a constraint length of 5 [Lin and Costello, 1983]. A tail of 5 b is appended to the 84 b to be convolutionally encoded to result in a 178-b output. Inclusion of the tail bits ensures independent decoding of successive time slots and no propagation of errors between slots.

Interleaving the bits to be transmitted over two time slots is introduced to diminish the effects of short deep fades and to improve the error-correction capabilities of the channel coding technique. Two speech frames, the previous and the present, are interleaved so that the bits from each speech block span two transmission time slots separated by 20 ms. The interleaving attempts to separate

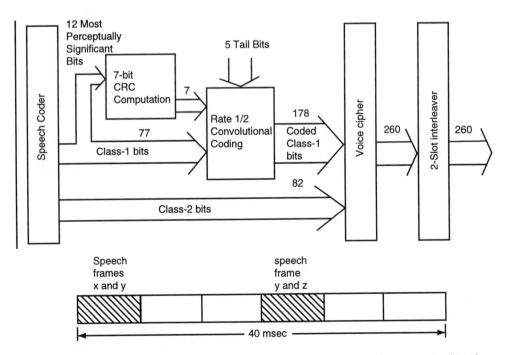

FIGURE 26.5 Error correction insertion for speech coder. Source TIA, 1992. Cellular Systems Dual-Mode Mobile Station–Base Station Compatibility Standards. TIA/EIA IS-54. With permission.

the convolutionally coded class-1 bits from one frame as much as possible in time by inserting noncoded class-2 bits between them.

26.14 Bad Frame Masking

A CRC failure indicates that the received data is unusable, either due to transmission errors resulting from a fade, or from pre-emption of the time slot by a control message (fast associated control channel, FACCH). To mask the effects that may result from leaving a gap in the speech signal, a masking operation based on the temporal redundancy between adjacent speech blocks has been proposed. Such masking can at best bridge over short gaps but cannot recover loss of signal of longer duration. The bad frame masking operation may follow a finite state machine where each state indicates an operation appropriate to the elapsed duration of the fade to which it corresponds. The masking operation consists of copying the previous LPC information and attenuating the gain of the signal. State 6 corresponds to error sequences exceeding 100 ms, for which the output signal is muted. The result of such a masking operation is generation of an extrapolation in the gap to the previously received signal, significantly reducing the perceptual effects of short fades. No additional delay is introduced in the reconstructed signal. At the same time, the receiver will report a high frequency of bad frames leading the system to explore handoff possibilities immediately. A quick successful handoff will result in rapid signal recovery.

26.15 Conclusions

The IS-54 digital cellular standard specifies modulation and speech coding techniques for mobile cellular systems that allow the interoperation of terminals built by a variety of manufacturers and systems operated across the country by a number of different service providers. It permits speech communication with good quality in a transmission environment characterized by frequent multipath fading and significant intercell interference. Generally, the quality of the IS-54 decoded speech is better at the edges of a cell than the corresponding AMPS transmission due to the error mitigation resulting from channel coding. Near a base station or in the absence of significant fading and interference, the IS-54 speech quality is reported to be somewhat worse than AMPS due to the inherent limitations of the analysis–synthesis model in reconstructing arbitrary speech signals with limited bits. At this time the TIA is considering introducing an alternative higher quality 8-kb/s speech coding technique for applications where higher speech quality is desirable. The International Telecommunications Union (ITU) is in the process of selecting such a coding algorithm, which promises speech quality equivalent to wireline telephony links in the absence of transmission errors [Salami et al., 1994]. The selection of the standard placed significant emphasis on the ability to implement the transceiver and encoder/decoder in a variety of mobile and portable terminals at low cost and with low-power dissipation. The standard selected reflects a practically reasonable compromise between high performance and low complexity. The low-weight terminals permitting hours of talk time in mobile environments attest to the benefits of the compromises arrived at in the definition of the standard.

Defining Terms

Codebook: A set of signal vectors available to both the encoder and decoder.
Covariance lattice algorithm: An algorithm for reduction of the covariance matrix of the signal consisting of several lattice stages, each stage implementing an optimal first-order filter with a single coefficient.
Reflection coefficient: A parameter of each stage of the lattice linear prediction filter that determines 1) a forward residual signal at the output of the filter-stage by subtracting from the forward residual at the input a linear function of the backward residual, also 2) a backward

residual at the output of the filter stage by subtracting a linear function of the forward residual from the backward residual at the input.

Vector quantizer: A quantizer that assigns quantized vectors to a vector of parameters based on their current values by minimizing some error criterion.

References

Atal, B.S. and Hanauer, S.L. 1971. Speech analysis and synthesis by linear prediction of the speech wave. *J. Acoust. Soc. Am.* 50:637–655.

Atal, B.S. and Schroeder, M. 1984. Stochastic coding of speech signals at very low bit rates. *Proc. Int. Conf. Comm.* pp. 1610–1613.

Gersho, A. 1994. Advances in speech and audio compression. *Proc. IEEE.* 82:900–918.

Gerson, I.A. and Jasiuk, M.A. 1990. Vector sum excited linear prediction (VSELP) speech coding at 8 kbps. *Int. Conf. Acoust. Speech and Sig. Proc.* ICASSP90. pp. 461–464.

Lin S. and Costello, D. 1983. *Error Control Coding: Fundamentals and Application*, Prentice Hall, Englewood Cliffs NJ.

Makhoul, J. 1975. Linear prediction, a tutorial review. *Proc. IEEE.* 63:561–580.

Salami, R., Laflamme, C., Adoul, J.P., and Massaloux, D. 1994. A toll quality 8 kb/s speech codec for the personal communication system (PCS). *IEEE Trans. Vehic. Tech.* 43:808–816.

Telecommunications Industry Association, 1992. EIA/TIA Interim Standard, Cellular System Dual-mode Mobile Station–Base Station Compatibility Standard IS-54B, TIA/EIA, Washington, D.C.

Further Information

For a general treatment of speech coding for telecommunications, see N.S. Jayant and P. Noll, *Digital Coding of Waveforms*, Prentice Hall, Englewood, NJ, 1984. For a more detailed treatment of linear prediction techniques, see J. Markel and A. Gray, *Linear Prediction of Speech*, Springer–Verlag, NY, 1976.

27

CDMA Technology and the IS-95 North American Standard

Arthur H.M. Ross
QUALCOMM Incorporated

Klein S. Gilhousen
QUALCOMM Incorporated

27.1 General Overview

Code division multiple access (CDMA) provides a superior, spectrally efficient, digital solution for the second generation of cellular, wireless telephony, and personal communications systems (PCS) services. The CDMA air interface is near optimum in its use of the subscriber station transmitter power, enabling the widespread commercial use of low-cost, lightweight, hand-held portable units that have vastly superior battery life. The technology is also near optimum in its link budgets, minimizing the number of base stations required for an excellent grade of service coverage. As customer penetration in a given service area increases, the system is called upon to deliver more capacity. CDMA has demonstrated a capacity increase over advanced mobile phone service (AMPS) of at least a factor of ten, which means that up to ten times fewer base stations will be required when

0-8493-8573-3/96/$0.00+$.50

customer demand for service increases. The use of soft handoff nearly eliminates the annoyance of dropped calls, fading, and poor voice quality. In short, no other air interface technology comes close to its performance.

CDMA, in this context, means not just generic code division multiple access, but the specific implementation of it described in the air interface standard: *TIA/EIA/IS-95-A: Mobile Station–Base Station Compatibility Standard for Dual-Mode Wideband Spread Spectrum Cellular System.* Compliance with the requirements in this document assures subscribers and service providers that equipment of all types will interoperate satisfactorily.

IS-95-A originated with a system design pioneered by QUALCOMM, beginning in April 1989. The early requirements were defined by the cellular carriers with input from manufacturers and carriers and by public testing. After the initial proposal and presentation of a draft air interface by QUALCOMM, a standards committee composed of system operators, subscriber equipment vendors, infrastructure equipment vendors, and test equipment vendors reviewed, revised, and formalized the air interface. The formal adoption of what is now the IS-95-A air interface took place in December of 1993.

This overview provides an overall understanding of the basic principles of CDMA system design and facilitates comprehension of the IS-95-A standard.

27.2 A Short History

In September 1988 the Cellular Telecommunications Industry Association (CTIA), out of concern that the burgeoning subscriber population would soon overwhelm the capacity of the already stressed AMPS facilities, even with cell subdivision, adopted a resolution they called the user performance requirements (UPR). The UPR expressed a need for a new all-digital air interface design that would achieve the following key goals.

- Tenfold increase over analog capacity
- Long life and adequate growth of second-generation technology
- Ability to introduce new features
- Quality improvements
- Privacy
- Ease of transition and compatibility with existing analog system
- Early availability and reasonable costs for dual mode radios and cells
- Cellular open network architecture (CONA)

An engineering subcommittee of the Telecommunications Industry Association was tasked with the development of such a standard. The result, after about two years of development, was the IS-54 air interface, sometimes known as digital AMPS (D-AMPS). D-AMPS uses time division multiple access techniques, in conjunction with low-bit-rate speech coding, to support not one but three conversations on each 30-kHz radio channel. The three-way time division of each channel, although an improvement over analog AMPS, is not responsive to the ten times capacity goal.

Meanwhile, QUALCOMM engineers, skeptical of even the modest capacity claims made on behalf of D-AMPS, began investigating an alternative technology, based on spread-spectrum techniques. The properties of spread spectrum are well understood, but it had at the time seen little, if any, commercial application. The primary applications had been military, usually motivated either by antijam or low-probability-of-intercept requirements.

For D-AMPS to achieve even the modest capacity gain claimed for it, the carrier-to-interference (C/I) ratio needed must be no worse than that required for AMPS. QUALCOMM believed that the C/I was no better than AMPS, and might actually be worse, leading to capacity gain less than three times. As little public information is available about the few D-AMPS systems in operation, the real performance achieved by this technology cannot be easily determined.

Recognizing that management of interference is the key to achieving the dramatic capacity improvement goal of the UPR, QUALCOMM took the radical step of dropping the 30-kHz channelization. As discussed later in this chapter, if a large number of conversations can be somehow overlapped in a common spectral band, then averaging of the interference takes place and the frequency reuse rules are radically altered. In addition to the use of common spectrum, it would be necessary to be near optimum in all other aspects of signal design, modulation, and coding, and to use a high-quality variable rate speech coder with full error protection of the coded speech.

Initial estimates of the potential CDMA capacity gain were very encouraging. They ranged up to, perhaps, 20 times AMPS, well in excess of the UPR goal, and almost seven times better than the purported three-times capacity increase of D-AMPS.

With extensive cellular industry support, QUALCOMM developed a demonstration CDMA system compliant with the CTIA requirements as defined by the UPR. This system's field trials were conducted publicly with the support and participation of infrastructure equipment manufacturers including Northern Telecom, AT&T, and Motorola; subscriber equipment manufacturers including Motorola, OKI Telecom, Clarion, Sony, Alps Electric, Nokia and Matsushita-Panasonic; and major cellular carries including PacTel Cellular (now AirTouch), Ameritech Mobile, NYNEX Mobile, GTE Mobile Communications, Bell Atlantic Mobile Systems, US West New Vector, Group, and Bell Cellular of Canada.

In July, 1990 a draft of the proposed new standard was reviewed by many major carriers and equipment manufactures. The document was revised as a result of their comments, and on October 1, 1990, Revision 1.0 of the standard was released.

The field trial results were formally presented to the CTIA on Dec. 5, 1991. Subsequently, on Jan. 6, 1992 the CTIA Board of Directors unanimously adopted a resolution that stated "CTIA further requests that the Telecommunications Industry Association (TIA) prepare 'structurally' to accept contributions regarding wideband (cellular) systems. This should be a separate effort not diluting the IS-54 revision process."

On Feb. 11, 1992 the TIA's cellular and common carrier radio section recommended, by unanimous vote, that the TR45 Committee address standardization activities regarding wideband spread-spectrum digital technologies. Accordingly, and responding to the desires of the user and service provider communities, the TIA TR45 Committee in March, 1992 created a new engineering subcommittee, TR45.5, to develop spread spectrum digital cellular standards.

The original QUALCOMM common air interface document was formally submitted to TR45.5 in April of 1992. After extensive discussion, detailed critical review, and revision, the CDMA common air interface, was formally adopted as North American digital cellular standard IS-95 on July 16, 1993. Subsequent minor improvements resulted in publication of Revision A in May of 1995.

27.3 Overview of CDMA

CDMA is a modulation and multiple access scheme based on spread-spectrum communication, a well-established technology that QUALCOMM has been applying to digital cellular radio communications and advanced wireless technologies. The approach will solve the near-term capacity concerns of major markets and the industry's long-term need for an economic, efficient, and truly portable communications.

Ever since the second pair of wireless telegraphs came into existence, we have been confronted with the problem of multiple access to the frequency spectrum without mutual interference. In the early days of wireless telegraphy, both frequency division in the form of resonant antennas and time division in the form of schedules, as well as netted operations were employed. As the number of wireless radios in operation increased and as the technology allowed, it became necessary to impose some discipline on the process in the form of frequency allocations. This has grown over the years to the complex process we have today for world-wide frequency allocations and licensing by service type.

The multiple access problem can be thought of as a filtering problem. There are many simultaneous users that want to use the same electromagnetic spectrum, and there is a choice of an array of filtering and processing techniques that allow the different signals to be separately received and demodulated without excessive mutual interference. The techniques that have long been used include: propagation mode selection, spatial filtering with directive antennas, frequency filtering, and time sharing. Over the last 40 years, techniques involving spread spectrum modulation have evolved in which more complex waveforms and filtering processes are employed.

Propagation mode selection involves a proper choice of operating frequency and antenna so that signals propagate between the intended communicators but not between (very many) other communicators. Frequency reuse in cellular mobile telephone systems is an example of this technique carried to a great degree of sophistication.

Spatial filtering uses the properties of directive antenna arrays to maximize response in the direction of desired signals and to minimize response in the direction of interfering signals. The current analog cellular system uses sectorization to a good advantage to reduce interference from cochannel users in nearby cells.

With frequency division multiple access (FDMA), a traffic channel is a relatively narrow band in the frequency domain into which a signal's transmission power is concentrated. Different signals are assigned different frequency channels. Interference to and from adjacent channels is limited by the use of bandpass filters that pass signal energy within the specified narrow frequency band while rejecting signals at other frequencies. The analog FM cellular system uses FDMA.

FDMA spectral efficiency in a cellular system is determined by the modulation spectral efficiency (the information bit rate per hertz of bandwidth) and the frequency reuse factor. The U.S. analog cellular system, divides the allocated spectrum into 30-kHz bandwidth channels; narrowband FM modulation is employed, resulting in a modulation efficiency of 1 call per 30 kHz of spectrum. Because of interference, the same frequency cannot be used in every cell. The frequency reuse factor is a number representing how often the same frequency can be reused. To provide acceptable call quality, a carrier-to-interference ratio (C/I) of 18 dB or greater is needed. Empirical results have shown that in most cases this level of C/I requires a reuse factor of seven. The resulting capacity is one call per 210 kHz of spectrum in each cell. Note that by increasing the number of cells, an arbitrarily high capacity can be obtained but with increased equipment costs. In addition, there is also a cost of increasing handoff rates as mobile stations move through smaller coverage areas.

With time division multiple access (TDMA), a traffic channel consists of a time slot in a periodic train of time intervals making up a frame. A given signal's energy is confined to one of these time slots. Adjacent channel interference is limited by the use of a time gate that only passes signal energy that is received at the proper time. Some systems use a combination of FDMA and TDMA. The TIA Digital Cellular Standard, IS-54-B, now IS-136, uses 30-kHz FDMA channels that are subdivided into three time slots for TDMA transmissions. One time slot is required for each call when employing 8 kb/s vocoders.

TDMA spectral efficiency is determined in a manner similar to that used for FDMA. The IS-54-B TDMA standard provides a basic modulation efficiency of three voice calls per 30 kHz of bandwidth. The currently accepted frequency reuse criteria is similar to the analog design. The resulting capacity is one call per 70 kHz of spectrum or three times that of the analog FM system.

With CDMA, (see Fig. 27.1) each signal consists of a different pseudorandom binary sequence that modulates a carrier, spreading the spectrum of the waveform. A large number of CDMA signals share the same frequency spectrum. If CDMA is viewed in either the frequency or time domain, the multiple access signals appear to be on top of each other. The signals are separated in the receivers by using a correlator that accepts only signal energy from the selected binary sequence and despreads its spectrum. The other users' signals, whose codes do not match, are not despread in bandwidth and, as a result, contribute only to the noise and represent a self-interference generated by the

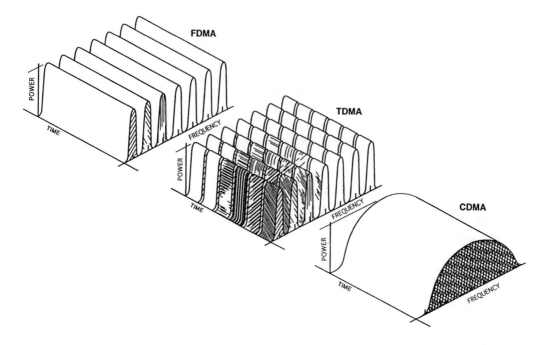

FIGURE 27.1 *Frequency and time domain representations of FDMA, TDMA, and CDMA.* Unlike FDMA or TDMA, CDMA has multiple users simultaneously sharing the same wideband channel. Individual users are selected by correlation processing of the pseudonoise waveform.

system. All of the desired signal's energy will pass through a narrow-bandwidth filter following the correlator, while the interfering signals energy is reduced by the ratio of the bandwidth before the correlator to the bandwidth after the correlator, greatly improving the signal-to-noise ratio for the desired signal. This improvement ratio is known as the processing gain.

The increased signal-to-noise ratio for the desired signal is shown in Fig. 27.2. The signal-to-interference ratio is determined by the ratio of desired signal power to the sum of the power of all of the other signals and is enhanced by the system processing gain or the ratio of spread bandwidth to baseband data rate. The major parameters that determine the CDMA digital cellular system capacity are processing gain, required E_b/N_0, voice duty cycle, frequency reuse efficiency, and the number of sectors in the cell.[1] The CDMA cellular telephone system achieves a spectral efficiency of up to 20 times the analog FM system efficiency when serving the same area with the same antenna system when the antenna system has three sectors per cell. This is a capacity of up to one call per 10 kHz of spectrum.

In the cellular radio frequency reuse concept, interference is accepted but controlled with the goal of increasing system capacity. CDMA does this effectively because it is inherently an excellent anti-interference waveform. Since all calls use the same frequencies, CDMA frequency reuse efficiency is determined by a small reduction in the signal-to-noise ratio caused by system users in neighboring cells. CDMA frequency reuse efficiency is approximately 2/3 compared to 1/7 for narrowband FDMA systems. The CDMA system can also be a hybrid of FDMA and CDMA techniques where the total system bandwidth is divided into a set of wideband channels, each of which contains a large number of CDMA signals.

[1] E_b/N_0 is defined as the bit energy to noise power spectral density: comparable to C/I.

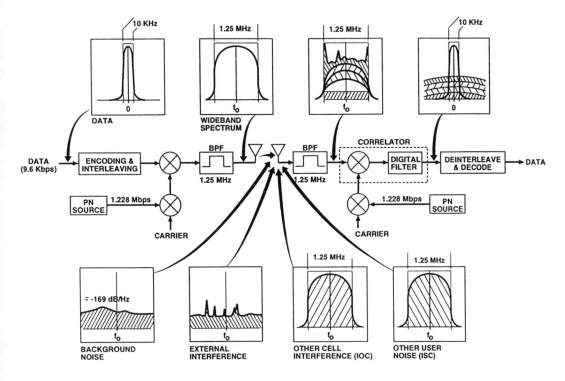

FIGURE 27.2 View of the CDMA concept. The desired signal is selected from four different sources of interference. The dominant source is system self-interference produced by other users of the same cell. This source is controlled by closed-loop power control.

27.4 The CDMA System

The multiple access scheme exploits isolation provided by the antenna system, geometric spacing, power gating of transmissions by voice activity, power control, a very efficient modem, and a signal design that uses very powerful error correction coding.

A combination of open-loop and closed-loop power control (through measurements of the received power at the mobile station and the base station) commands the mobile station to make power adjustments in order to maintain only the power level required for adequate performance. This minimizes interference to other users, helps to overcome fading, and conserves battery power in the mobile station.

The CDMA digital cellular waveform design uses a pseudorandom noise (PN) spread spectrum carrier. The chip rate of the PN spreading sequence was chosen so that the resulting bandwidth is about 1.25 MHz after filtering or approximately one-tenth of the total bandwidth allocated to one cellular service carrier.

The Federal Communications Commission (FCC) has allocated a total of 25 MHz for mobile station to cell site and 25 MHz for cell site to mobile station for the provision of cellular services. The FCC has divided this allocation equally between two service providers, the A and the B carriers, in each service area. Because of the time sequence of the FCC's actions in allocating the cellular spectrum, the 12.5 MHz allocated to each carrier for each direction of the link is further subdivided into two subbands. For the B carriers, the subbands are 10 MHz and 2.5 MHz each. For the A carriers, the subbands are 11 MHz and 1.5 MHz each. A signal bandwidth of less than 1.5 MHz fits into any of the subbands, whereas a bandwidth of less than 2.5 MHz fits into all but one subband.

A set of ten 1.25-MHz-bandwidth CDMA channels can be used by each operator if the entire allocation is converted to CDMA. Initially, only one or a small number of 1.25-MHz channels needs to be removed from the present FM analog service to provide digital service. This facilitates the deployment by introducing a gradual reduction in analog capacity. Each 1.25-MHz CDMA segment can provide about twice the capacity of the entire 12.5-MHz allocation using the present FM system. Some frequency guard band is necessary if there are adjacent high-power cellular (or other) frequencies in use, and the maximum capacity of the CDMA cell is required. Capacity can be sacrificed for decreased guard band if desired. Adjacent CDMA channels need not employ a guard band.

27.5 Multiple Forms of Diversity

In relatively narrowband modulation systems, such as analog FM modulation employed by the first-generation cellular phone system, the existence of multiple paths causes severe fading. With wideband CDMA modulations, however, the different paths may be independently received greatly reducing the severity of the multipath fading. Multipath fading is not completely eliminated because multipaths that cannot be independently processed by the demodulator occasionally occur. This will result in some fading behavior.

Diversity is the favored approach to mitigate fading. There are three major types of diversity: time, frequency, and space. Time diversity can best be obtained by the use of interleaving and error correction coding. Wideband CDMA offers a form of frequency diversity by spreading the signal energy over a wide bandwidth; frequency selective fading usually affects only a 200–300 kHz portion of the signal bandwidth. Space or path diversity is obtained three different ways by providing the following.

- Multiple signal paths through simultaneous links from the mobile station to two or more cell sites (soft handoff).
- Exploitation of the multipath environment through spread-spectrum processing (rake receiver), allowing signals arriving with different propagation delays to be received separately and combined.
- Multiple antennas at the cell site.

The following are different types of diversity employed in the CDMA system to greatly improve performance.

- Time diversity: symbol interleaving, error detection, and correction coding
- Frequency diversity: 1.25-MHz wideband signal
- Space (path) diversity: dual cell site receive antennas, multipath rake receivers, and multiple cell sites (soft handoff)

Antenna diversity can easily be provided in FDMA and TDMA systems. Time diversity can be provided in all digital systems that can tolerate the required higher transmitted symbol rate needed to make the required error correction process effective. The remaining methods, however, can only be provided easily with CDMA. A unique feature of direct sequence CDMA is the ability to provide extensive path diversity; the greater the order of diversity in a system, the better the performance in this difficult propagation environment. Additional diversity of a different mode is much more powerful than additional numbers of the same type of diversity because the fading processes are more likely to be independent with different diversity modes.

Multipath processing takes the form of parallel correlators for the PN waveform. The mobile and cell receivers employ three and four parallel correlators, respectively. Receivers using parallel correlators (sometimes called rake receivers) allow individual path arrivals to be tracked independently, and the sum of their received signal strengths is then used to demodulate the signal. Although there is fading on each arrival, the fades are independent. Demodulation based on the sum of the signals is then much more reliable. The multiplicity of correlators is also the basis for the simultaneous tracking of signals from two different cells and allows the subscriber unit to control the soft handoff.

27.6 Mobile Power Control for CDMA Digital Cellular

Spread-spectrum techniques, long established for antijam and multipath rejection applications, have also been proposed for CDMA to support simultaneous digital communication among a large community of relatively uncoordinated users. It has frequently been pointed out that the mobile to base link in such a system is subject to a near far problem in which a mobile station close to the base has a much lower path loss to the station than far away mobile stations. If all of the mobiles were to use the same transmitter power, then the close-by mobile would apparently jam the far-away mobile stations. Thus, the need for a mobile power control system is postulated in order mitigate this problem.

Before proceeding further, it is interesting to determine the performance of a CDMA system with no power control at all in order to have a standard with which to compare various power control techniques. We will use as an example system, the system used by IS-95. This system uses a 1.25-MHz-bandwidth (W) direct sequence spread-spectrum modulation, a maximum data rate of 9600 b/s (R_b) and requires an bit energy to noise density ratio, E_b/N_0, of 6 dB in nominal conditions, depending on the fading environment, for good quality reception.

The near–far problem can be reduced to a classical jamming problem for which the J/S equation[2] applies,

$$\frac{J}{S} = \frac{W/R_b}{E_b/N_0} \qquad (27.1)$$

With the given values, the J/S is 15 dB. This means that a jamming or interfering signal can be 15 dB stronger than a desired signal before the desired signal quality is affected (15 dB is a factor of 30 in power). This is a result of CDMA's unique ability to discriminate against undesired signals whose spectrum spreading codes do not match the desired signal's code.

Consider a case with only two mobiles in one sector of a cell, an adequate link will be obtained for both mobiles as long as one mobile is not more than 15 dB closer (in path loss) to the base station than the other mobile. In an environment in which the path loss is proportional to the fourth power of distance (the usual case in cellular), this corresponds to a factor of 2.4 in distance. For mobiles randomly and uniformly distributed in the sector area of a cell, the probability that both will achieve an acceptable quality link is 83%. Note that a narrowband cellular system, such as the analog FM cellular system, will achieve an acceptable quality with about the same probability when operating under the same propagation assumptions (fourth power, 8-dB σ, frequency reuse = 1/7, three sectored cell). Since both provide two active calls per 1.25 MHz of spectrum within the sectored cell, one could conclude that CDMA with no power control at all has about the same capacity as the existing analog FM system.

The heart of the CDMA capacity advantage is that the jamming can be the aggregate effect of the other users of the system. If each station optimally controls its transmitter power so that all signals arrive at the cell with the equal power, then Eq. (27.1) defines a limit on system capacity.

The Mobile Fading Environment

First, consider the dynamic range requirement. A cellular mobile may be anywhere in the cell from right under the base station antenna tower to perhaps 5–10 miles distant. In an environment where

---

[2] J/S is the jamming margin or the factor by which a jammer can exceed the signal strength of a desired signal before interference results.

propagation goes as the fourth power of distance, as in most cellular service areas, it has been found that the total dynamic range of path loss is on the order of 80 dB. This means that the mobile transmitter must vary its power from a few tens of nanowatts up to the order of 1 W.

Another problem with power control is that the path loss can vary rapidly because of multipath induced Rayleigh fading. In multipath fading, the signal arrives at the receiver after traveling directly between antennas and also after being reflected from building, hills, etc., in the physical environment. When such multiple signals arrive at the receive antenna, the RF phase difference between the signals can be such that the signals may cancel at one moment and be enhanced the next. Such fading commonly causes fluctuations of 20–30 dB while the mobile travels a distance of only 1 ft. If the mobile is traveling on the freeway at speeds of 100 ft/s, clearly the fade rate can exceed 100 Hz. Actually, such deep and fast fades, although relatively common with narrowband waveforms is relatively uncommon with CDMA because of the mitigating effects of the various diversity modes in the system.

An additional complicating factor is that the multipath fading of signals to the mobile is not necessarily the same as fading from the mobile. This is caused primarily by the fact that the signals to and from the mobile are separated by 45 MHz in the frequency domain. This is usually great enough to decouple any dependency between fading in the two directions.

Open-Loop Control

The wide dynamic range portion of the problem is best dealt with using an open-loop power control technique in which the mobile determines an estimate of the path loss between the cell and the mobile. This is done by measuring the received signal level at the mobile. This is accomplished by utilizing the automatic gain control (AGC) circuitry of the receiver. The AGC circuits operate on the receiver's IF frequency amplifiers so that the input to the receiver's A/D converters is held constant. The mobile transmitter circuits use IF amplifier circuits identical to those used in the receiver. The AGC control voltage, offset by another control signal, is used to control the gain of the transmitter IF amplifiers exactly in step with the receiver's IF gain. Thus, if the mobile moves closer to the cell increasing the received signal level, the receiver AGC will reduce the receiver IF amplifier gain, and the transmitter IF gain. This will result in a proportionally lower mobile transmitter power. This is as it should be since the mobile is now closer to the cell and must reduce transmitter power in order to maintain a constant receive power at the cell base station.

The equation which governs the operation of the open-loop control (with power in decibels referred to 1 mW) is

$$\text{transmit power in dBm} = -73 - \text{receive power in dBm} + \text{parameters}$$

Thus, when the mobile received power is −90 dBm, the mobile transmitter power will be +17 dB (1 mW) in a nominal sized cell. The parameters are used to adjust the open-loop power control for different sized cells and different cell ERP (Effective radiated power) and receiver sensitivities.

The circuitry for mechanizing the open-loop power control has proven to be quite simple and reliable. The desired dynamic range has been achieved with an accuracy of on the order of ±6 dB. As stated earlier however, even with perfect open-loop control, because of the lack of reciprocity of path loss due to multipath fading, the signals arriving at the cell may still be significantly different from the desired power level.

Closed-Loop Control

The closed-loop control has the function of controlling the mobile transmit power so that the desired SNR is received at the cell. Note that this is SNR control not just power control. The difference will become clear further on.

Each cell receiver (there is one for every call in progress being received by this cell) forms an estimate of the received SNR of its mobile's signal. This estimate is made every 1.25 ms. The SNR measurement is compared with the set point SNR. If the received SNR is too high, a decrease power command is sent. If the SNR is too low, an increase power command is sent. A power command is sent every 1.25 ms providing an 800 b/s control stream rate.

The power commands are inserted into the data stream being transmitted to the mobile being controlled. The command bits are simply written over data channel symbols. (The data channel error detection and correction can easily fill in the missing symbols.) The power control commands themselves are not protected by error detection and correction, primarily because the necessary delay involved in error correction is intolerable to the closed-loop control.

The mobile receiver picks off the power control commands as they emerge from the demodulator. The bits are accumulated in a digital word that represents the total accumulated correction to the mobile's power. The value represented by the accumulated correction is converted to an analog voltage and then added to the open-loop control voltage derived from the receiver AGC signal and applied to the transmitter gain control circuits. A total dynamic range of closed-loop control is limited to ± 24 dB. This is adequate to correct the combined errors already described. The loop is also fast enough with the 800 b/s control stream rate to keep up with most of the fast changing multipath induced Rayleigh fading.

The power control system has been provided with a number of setable parameters to adapt its use to different conditions. The open-loop control parameter is used primarily to adjust for different cell sizes and ERP. Primary closed-loop parameters include the power control step size and the dynamic range. Nominal step size is 1.0 dB. Note that every cell can have a different parameter set. The parameter values are communicated to each mobile by a broadcast message transmitted on the sync channel.

Other Benefits and Features

Besides controlling the mobile power received by each cell and, therefore, the self-interference environment, the power control system also provides an automatically adaptive anti-interference function. For example, suppose that a jammer transmitter suddenly starts transmission near a cell site. The immediate effect is that all of the mobiles' SNRs are degraded. The closed-loop power control system will respond by transmitting a series of power-up commands until the desired SNR is restored. Possibly, not all mobiles will be able to comply with the power increase commands because of already being at extreme range. These calls will be lost unless they can be handed off to another cell. This is facilitated by reducing the cell's transmit power when interference is present so as to shrink the cell and move the handoff boundary closer, thus accomplishing the desired objective. Usually, however, all mobiles will be able to continue without interruption of their calls. When the interference ceases, the mobiles will automatically be commanded to reduce power to the normal level.

A similar automatic adaption occurs when a cell has an unusually heavy demand compared to its neighbors. The power control system in the heavily loaded cell will command its mobiles to increase power. This will result in de-emphasizing the interference received from neighbor cells, allowing temporarily higher capacity in this cell. This will, of course, increase interference into the neighbor cells, but the premise was that a heavily loaded cell was surrounded by more lightly loaded cells and, thus, the added interference could be well tolerated. This flexible borrowing of neighboring cells' capacity has no counterpart in analog FM or in digital TDMA systems.

Another advantage of the power control system for handheld portable units is that battery drain is minimized by the power control system. Clearly, if high power need not be transmitted most of the time, then the corresponding battery drain need not be incurred. Note that CDMA phones average only 2 mW transmit power compared to analog FM of 600 mW.

Yet another capability of the power control system is in the use of adaptive SNR thresholds. Rather than fixing the SNR threshold at a constant value, the system allows the SNR to vary as a

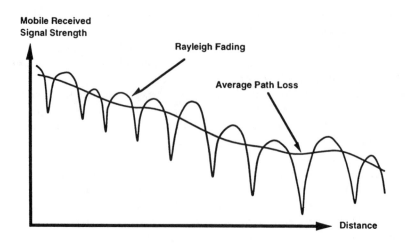

FIGURE 27.3   Mobile received signal strength in log-normal shadowing and Rayleigh fading.

function of system loading and of link quality. Consider that if the system load drops to 50% of maximum then up to 3 dB additional SNR could be allowed to every mobile. This much of an increase in a coded digital system such as CDMA will produce virtually perfect links.

It has also been noted that less SNR is required by mobiles that are motionless, or nearly so. This can be used to advantage by reducing SNR to those mobiles that are achieving a low-channel frame error rate, while increasing the SNR to those mobiles that are moving and, thus, achieving a higher frame error rate. This better balances the grade of service seen by the subscribers, while maximizing system capacity. Figure 27.3 shows a typical situation as a mobile drives away from the cell-site. Depicted are both log-normal shadowing and Rayleigh fading.

27.7 Low Transmit Power

Besides directly improving capacity, one of the more important results of reducing the required E_b/N_0 (signal-to-interference level) is the reduction of transmitter power required to overcome noise and interference. This reduction means that mobile stations also have reduced transmitter

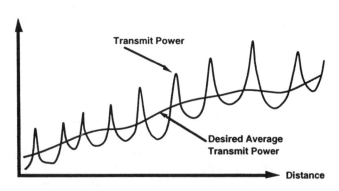

FIGURE 27.4   Transmit power, open-loop control only.

output requirements that reduce cost and allow lower power units to operate at larger ranges than the similarly powered analog or TDMA units. Furthermore, a reduced transmitter output requirement increases coverage and penetration and may also allow a reduction in the number of cells required for coverage. (See Figs. 27.4 and 27.5.)

An even greater gain is the reduction of average (rather than peak) transmitted power that is realized because of the power control used in CDMA. Most of the time, propagation conditions are benign. But because of occasional severe fading, narrowband systems must always transmit with enough power to override the occasional fades. CDMA uses power control to provide only the power required at the time it is actually needed, and thus reduces the average power by transmitting at high levels only during fades. In effect, the link margin in CDMA is kept in reserve ready for use when needed but when not needed does not contribute interference to other users or increase battery drain.

27.8 Privacy

The scrambled form of CDMA signals provides for a very high degree of privacy and makes this digital cellular system inherently more immune to cross talk, inexpensive scanning receivers, and air-time fraud. The standard includes the authentication and voice privacy features specified in EIA/TIA/IS-54-B even though the CDMA architecture inherently provides voice privacy and provisions for extended protection. The digital voice channel is, of course, amenable to direct encryption using Data encryption standards (DES) or other standard encryption techniques.

27.9 Mobile Station Assisted Soft Handoff

Soft handoff allows both the original cell and a new cell to temporarily serve the call during the handoff transition. The transition is from the original cell to both cells and then to the new cell. Not only does this greatly minimize the probability of a dropped call, but it also makes the handoff virtually undetectable by the user. In this regard, the analog system (and the digital TDMA-based systems) provides a break-before-make switching function, whereas the CDMA-based soft handoff system provides a make-before-break switching function.

After a call is initiated, the mobile station continues to scan the neighboring cells to determine if the signal from another cell becomes comparable to that of the original cell. When this happens, it indicates to the mobile station that the call has entered a new cell's coverage area and that a handoff can be initiated. The mobile station transmits a control message to the mobile switching center (MSC), which states that the new cell site is now strong and identifies the new cell site. The MSC initiates the handoff by establishing a link to the mobile station through the new cell while maintaining the old link. While the mobile station is located in the transition region between the two cell sites, the call is supported by communication through both cells; thereby eliminating the ping-ponging effect, or repeated requests to hand the call back and forth between two cell sites. The original cell site will only discontinue handling the call when the mobile station is firmly established in the new cell.

A soft handoff may persist for the entire duration of a call if the mobile remains in the overlapping coverage area of two cells. In this sense, it should be thought of more as a mode of operation than as a discrete event. Besides providing undetectable handoff, it also greatly improves the effectiveness of coverage in the difficult areas between base stations. In coverage design, sufficient coverage margin must be provided to cope with shadow fading. This is slow fading that occurs primarily because of signal path blockage by physical objects. It is usually modeled as a log-normal process with sigma of 8 dB. Coverage is designed to provide a certain probability of coverage, say 90%. To obtain 90% probability with a single log-normal, 8-dB sigma process, 10.25 dB of margin is required. But with soft handoff, there are two paths available, which means that if either path is adequate then the coverage will be satisfactory. This means that the margin can be reduced to about 4 dB, a very

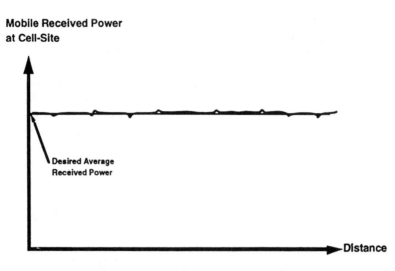

FIGURE 27.5   Mobile power received at cell site.

significant gain in coverage. This allows cells to be much larger in radius and greatly reduces the number of cells necessary to provide a certain quality of coverage.

27.10 Capacity

The primary parameters that determine CDMA digital cellular system capacity are processing gain, E_b/N_0 (with the required margin for fading), voice duty cycle, frequency reuse efficiency, and the number of sectors in the cell site antenna. Additionally, for a given blocking probability, the larger number of voice circuits provided by CDMA results in a significant increase in trunking efficiency, which serves a larger number of subscribers per voice circuit.

For example, if a spread spectrum bandwidth of 1.25 MHz is utilized by mobile stations transmitting continuously at 9600 b/s and if the modulation and coding technique utilized requires an E_b/N_0 of 6 dB, then up to 32 mobile stations could transmit simultaneously, as long as they are each power controlled to provide equal received power at the receiving location. In a CDMA cellular system, this capacity is reduced by interference received from neighboring cells and increased by other factors.

In CDMA, frequency reuse efficiency is determined by the signal-to-interference ratio that results from all of the system users within range, instead of the users in any given cell. Since the total capacity becomes quite large, the statistics of all of the users are more important than those of a single user. The law of large numbers can be said to apply. This means that the net interference to any given signal is the average of all of the users' received power times the number of users. As long as the ratio of received signal power to the average noise power density is greater than a threshold value, the channel will provide an acceptable signal quality. With TDMA and FDMA, interference is governed by a law of small numbers in which worst-case situations determine the percentage of time in which the desired signal quality will not be achieved.

Voice Activity Detection

In a typical full duplex two-way voice conversation, the duty cycle of each voice is less than 35%. It is difficult to exploit the voice activity factor in either FDMA or TDMA systems because of the time delay associated with reassigning the channel resource during the speech pauses. With CDMA, it is possible to reduce the transmission rate when there is no speech, and thereby substantially reduce interference to other users. Since the level of other user interference directly determines capacity, the capacity is increased by approximately a factor of two. This also reduces average mobile station transmit power requirements by approximately a factor of two.

Frequency Reuse

In CDMA, the wideband channel is reused in every cell. The total interference at the cell site to a given inbound mobile station signal is comprised of interference from other mobile stations in the same cell plus interference from mobile stations in neighboring cells. The frequency reuse efficiency of omnidirectional cells, defined as the ratio of interference from mobile stations within a cell to the total interference from all cells, is about 65%. Figure 27.6 shows, in an idealized geometry, the distribution of this interference from neighboring cells. Each cell in the first tier contributes about 6% of the total interference and so the entire first tier contributes an average of six times 6% or 36%; cells in the second and greater tiers contribute less than 4%.

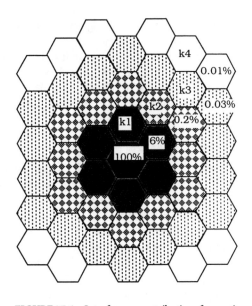

FIGURE 27.6 Interference contributions from neighboring cells. Significant contribution to interference seen by a cell as a result of activity in other cells is limited to the first tier of cells surrounding the cell.

Sectorization

Sectorization of CDMA cells is far more effective in increasing capacity than it is for narrowband FDMA and TDMA systems. Cells are typically sectored three ways, with 120°-beamwidth antennas. Because of the broad antenna patterns and frequent propagation anomalies, the coverage areas of these antennas overlap considerably. The isolation from one sector to another cannot be counted on to ensure good separation of narrowband signals. Frequency planning must take into account worst-case interference, which is only slightly reduced by the sectorization. The cellwise frequency reuse obtained in narrowband systems, thus, does not change when the cells are sectorized. A typical seven way reuse pattern remains seven way over all three sectors. That is, viewed sector-wise, it becomes 21-way reuse: each AMPS channel is reused in every 21st sector. The six channels that an AMPS omnidirectional cell has available in 1.25 MHz become two channels per sector in a three-way sectored cell.

Sector overlap in CDMA also has a deleterious effect due to interference. But the interference affects planning through its average value, not its worst-case, single station value. Therein lies a substantial advantage for CDMA. The increase in average interference due to sector overlap is small. Moreover it is partially mitigated by gain due to soft handoff between sectors. The result is that N-way sectorization of CDMA cells increases capacity nearly N-fold. Each sector has approximately the same capacity regardless of the degree of sectorization. Sectorization in AMPS, by contrast, entails little or no increase in capacity, only an increase in range.

Low E_b/N_0 (or C/I) and Error Protection

E_b/N_0 is the ratio of energy per bit to the noise power spectral density and is the standard figure of merit by which digital modulation and coding schemes are compared. It is directly analogous to the carrier-to-noise ratio C/N for analog FM modulation. Because of the wide channel bandwidth employed in the CDMA system, it is possible to use extremely powerful, high-redundancy error correction coding techniques. With narrowband digital modulation techniques, a much higher E_b/N_0 is required compared to CDMA because less powerful, low-redundancy error correction codes must be used to conserve channel bandwidth. The CDMA system employs the most powerful combination of forward error correction coding ever used in a mobile radio system together with an extremely efficient digital demodulator in its implementation of the CDMA digital cellular system. The lower E_b/N_0 increases capacity, increases cell radius of coverage, and decreases transmitter output power requirements.

Coding is provided for all transmitted bits not just for the most important bits as in other systems. Additionally, each frame is protected by a cyclic redundancy check code (CRCC) that allows data frames in error to be identified so that the vocoder can mitigate the effects of errored frames. The result is that isolated errored frames are essentially not noticeable by the user.

Effect of Thermal Noise

At each base station receiver, noise comes from two primary sources, thermal noise and mutual interference. The base station receiver power control system will adjust each mobile's transmitted power so that the desire SNR is obtained. As users are added, the interference to all users increases, and to maintain the desired SNR, the mobiles' power must be increased slightly. In the limit, one more mobile cannot be added without causing all mobiles' power to increase without bound.

In practice, the system is operated at a maximum capacity loading such that about half of the total receiver noise is from mutual interference and half is from thermal noise. Another way of looking at it is that we dedicate half of the theoretic capacity to the maintenance of stability in the power control system and to prevent excess mobile power from being used.

Net Capacity Calculation

The net capacity of the system can be calculated from all of the discussed considerations and Eq. (27.1). Taking F to be the fraction of same-sector interference and v to be the voice activity factor, the total interference disregarding thermal noise is

$$J = (N - 1)SvF^{-1} \tag{27.2}$$

From Eq. (27.1) the system capacity limit per sector is approximately

$$N_p \approx \frac{W/R_b}{E_b/N_0} \frac{F}{v} \tag{27.3}$$

For each 1.25-MHz bandwidth CDMA channel, and 9600 b/s voice coding, the processing gain W/R_b is a factor of 128. This is reduced by the required E_b/N_0, a factor of 4 (6 dB), and by the neighbor cell interference, an additional factor of $F^{-1} \approx 3/2$. The voice activity gain contributes a factor of $v^{-1} \approx 2$, yielding $N_p \approx 128/3$ calls per sector. Unlike narrowband systems, antenna sectorization increases cell capacity in approximate proportion to the number of sectors, usually taken to be 3. The theoretical maximum capacity is, thus, about 128 calls per 1.25-MHz channel per cell.

For a practical system the power rise over thermal must be limited. From Eq. (27.1) it is readily shown that the relationship between rise over thermal is related to loading by

$$\frac{S_{\text{total}} + N_0 W}{N_0 W} \approx \frac{1}{1 - N/N_p} \tag{27.4}$$

A reasonable practical limit is a twofold increase, which by Eq. (27.4) reduces the capacity to $0.5N_p \approx 64/3$ calls per 1.25 MHz channel per sector, or 64 calls per cell. This is greater than ten times the capacity of the analog AMPS system, which provides only 6 calls per 1.25 MHz per cell. When blocking effects are considered, the capacity advantage widens significantly.

The calculated capacity has been verified in many, many field trials conducted all over the world. Never has the capacity been less than the desired minimum factor of ten improvement over AMPS.

27.11 Soft Capacity

In the present U.S. cellular environment, the FCC has allocated 25 MHz of spectrum that is equally split between two system operators in each service area. The spectrum is further divided between the cells, with a maximum of 57 analog FM channels in a three sector cell site. When demand for service is at a peak, the 58th caller in a given cell must be given a system busy signal. There is no way to add even one more signal to a fully occupied system. This call blocking behavior results in about a 35% loss of capacity. With the CDMA system, however, there is a much softer relationship between the number of users and the grade of service. For example, the system operator could decide to allow a small degradation in the bit error rate and increase the number of available channels during peak hours.

This capability is especially important for avoiding dropped calls at handoff because of a lack of channels. In the analog system and in digital TDMA, if a channel is not available, the call must be reassigned to a second candidate or it will be dropped at the handoff. With CDMA, however, the call can be accommodated if it is acceptable to slightly raise the users bit error rates until another call is completed.

It is also possible to offer a higher grade of service (at a higher cost to the user) where the high-grade user would obtain a larger fraction of the available power (capacity) than the low-grade user. Handoffs for high-grade users can be given priority over those for other users.

27.12 Transition to CDMA

In the initial introduction of CDMA service, a band segment of approximately 1.25 MHz is occupied by the CDMA operation because the spread-spectrum modulation requires this minimum bandwidth. An additional guard band of about 600 kHz must be provided for the first CDMA channel. The total represents only 15% of the present FDMA/FM system capacity, or about two analog FM channels per sector in a three sector cell. In return, however, the introductory single channel CDMA system allows up to 20 calls per sector.

Initially, a set of cells capable of covering the entire geographic area will be identified and equipped with omnidirectional or multisectored CDMA cell site equipment. This should be far fewer cells than required by the existing FM system. Although only the selected cells are equipped with CDMA cell site equipment, the 1.25-MHz segment is cleared out in all cells in the local area (i.e., the area of coverage), to prevent mutual interference between the FM and CDMA segments of the system.

As demand for CDMA service grows, additional omnidirectional cell sites are added and existing omnidirectional cells are converted to multisectored cells to increase capacity or improve coverage in the more difficult areas. Frequency planning is not necessary to support the change as additional cells are added or converted by sectorization; as demand for CDMA service grows beyond the capacity provided by the initial service, an additional 10% of the band segments can be removed from analog service and dedicated to the CDMA service. Each 1.25-MHz band segment requires an additional RF chain and power amplifier per sector. Additional modems are required to support the new channels.

27.13 Overview of the IS-95 CDMA Standard

The common air interface standard, IS-95-A, prescribes, in considerable detail, the behavior of dual-mode CDMA/AMPS subscriber stations, and, to a lesser extent, the base stations. Use of compliant subscriber stations assures base station manufacturers of predictable behavior on which

TABLE 27.1   IS-95 CDMA Standard Documentation

Section	Title	Topic
1	General	Defines the terms and numeric indications used in the document, including the time reference used in CDMA systems and the tolerances used throughout.
2	Requirements for mobile station analog operation	Describes the operation of the CDMA-analog dual-mode mobile stations operating in analog mode.
3	Requirements for base station analog operation	Describes the requirements for analog base stations.
4	Requirements for mobile station analog options	Describes the requirements for CDMA-analog dual-mode mobile stations which use the 32-digit dialing option on the reverse analog control channel and also describes mobile station requirements for use of the optional extended protocol.
5	Requirements for base station analog options	Describes the base station requirements for CDMA-analog dual-mode mobile stations operating the CDMA mode.
6	Requirements for mobile station CDMA operation	Describes the requirements for CDMA-analog dual-mode mobile stations operating in the CDMA mode.
7	Requirements for base station CDMA operation	Describes the requirements for CDMA base stations.

TABLE 27.2   Appendices to IS-95 CDMA Standard

Appendix	Title	Topic
A	Message encryption and voice privacy	Describes the requirements for message encryption and voice privacy. Note: This appendix is governed by U.S. International Traffic and Arms Regulation (TIARA) and the Export Administration Regulations.
B	CDMA call flow examples	Provides examples of simple call flow in the CDMA system.
C	CDMA system layering	Describes the layers of the CDMA system: physical layer, link layer, multiplex sublayer and the control process layer.
D	CDMA constants	Values for the constant identifiers found in Secs. 6 and 7.
E	CDMA retrievable and setable parameters	Describes the mobile station parameters that the base station can set and retrieve.
F	Mobile station database	Database model that can be used for dual-mode mobile stations complying with this document.
G	Bibliography	This appendix is not considered part of the standard but lists documents that may be useful in implementing the standard.

to base system designs. An IS-95-A compliant subscriber station can obtain service by communicating with either an AMPS (analog FM) base station or with a CDMA base station. System selection depends on the availability of either system in the geographic area of the station, as well as its programmed preference. Minimum performance requirements for dual-mode base stations, IS-97, and subscriber stations, IS-98, supplement the basic air interface.

IS-95-A emphasizes subscriber station requirements because the subscriber side reflects all call processing features, and because the specification is simpler in the context of a single user. In contrast to the very detailed subscriber station requirements, base station requirements are sketchy and incomplete. Generally the standard prescribes only those base station requirements that are important for the design of subscriber stations, leaving unspecified behavior to the discretion of the vendors. Base stations are fielded in much smaller quantities, but at much greater cost per unit. Marketplace considerations therefore incentivize good designs. Minimal requirements facilitate innovation in those designs.

The standard specifies that mobile stations operating with analog base stations meet the analog compatibility provisions for mobile stations as specified in EIA/TIA/IS-54-B, *Dual-Mode Mobile Station–Base Station Compatibility Specification,* January 1992. The incorporation of the analog portions of EIA/TIA/IS-54-B instead of EIA/TIA-553 (*Mobile Station–Land Station Compatibility Specification,* September 1989) accommodates all the changes to analog operation imposed by the EIA/TIA/IS-54-B dual-mode standard. (See Tables 27.1 and 27.2.)

Defining Terms

Access channel: A reverse CDMA channel used by subscriber stations for communicating to the base station. The access channel is used for short signalling message exchanges, such as call originations, responses to pages, and registrations.

CDMA channel: The set of channels transmitted between the base station and the subscriber stations within a given CDMA frequency assignment. See also forward CDMA channel and reverse CDMA channel.

Code channel: A subchannel of a forward CDMA channel. A forward CDMA channel contains 64 code channels. Code channel zero is assigned to the pilot channel. Code channels 1–7 may be assigned to the either paging channels or the traffic channels. Code channel 32 may be assigned to either a sync channel or a traffic channel. The remaining code channels may be assigned to traffic channels.

Code division multiple access (CDMA):   A technique for spread-spectrum multiple-access digital communications that creates channels through the use of unique code sequences.

Forward CDMA channel: A CDMA channel from a base station to subscriber stations. The forward CDMA channel contains one or more code channels that are transmitted on a CDMA frequency assignment using a particular pilot time offset. The code channels are associated with the pilot channel, sync channel, paging channels, and traffic channels.

Forward traffic channel: A code channel used to transport user and signalling traffic from the base station to the subscriber station.

Handoff: The act of transferring communication with a subscriber station from one base station to another. Hard handoff is characterized by temporary disconnection of the traffic channel. Soft handoff is characterized by simultaneous communication with a subscriber by more than one base station.

Paging channel: A code channel in a forward CDMA channel used for transmission of control information and pages from a base station to a subscriber station.

Paging: The act of seeking a subscriber station when a call has been placed to that subscriber station.

Pilot channel: An unmodulated, direct-sequence spread-spectrum signal transmitted continuously by each CDMA base station. The pilot channel allows a subscriber station to acquire the timing of the forward CDMA channel, provides a phase reference for coherent demodulation, and provides a means for signal strength comparisons between base stations for determining when to handoff.

Reverse CDMA channel: The CDMA channel from the subscriber station to the base station. From the base stations perspective, the reverse CDMA channel is the sum of all subscriber station transmissions on a CDMA frequency assignment.

Reverse link power control: Process ensuring that all subscriber signals arrive at a base station at their setpoint powers.

Reverse traffic channel: A reverse CDMA channel used to transport user and signalling traffic from a single subscriber station to one or more base stations.

Sync channel: Code channel 32 in the forward CDMA channel that transports the synchronization message to the subscriber station.

Traffic channel: A communication path between a subscriber station and a base station used for user and signalling traffic. The term traffic channel implies a forward traffic channel and reverse traffic channel pair. See also forward traffic channel and reverse traffic channel.

References

Bello, P.L. 1963. Characterization of randomly time-variant linear channels. *IEEE Trans. Comm. Syst.* CS-11:360–393.

Gilhousen, K.S., Jacobs, I.M., Padovani, R., Viterbi, A.J., Weaver, L.A., and Wheatley, C. A. 1991. On the capacity of a cellular CDMA system. *IEEE Trans. Veh. Tech.* VT-40(2):303–312.

Jakes, W.C., Jr. ed. 1974. *Microwave Mobile Communications,* John Wiley & Sons, New York.

Lee, W.C.Y. 1989. *Mobile Cellular Telecommunications Systems,* McGraw–Hill, New York.

Parsons, D. 1992. *The Mobile Radio Propagation Channel,* Wiley, New York.

Peterson, R.L., Ziemer, R.E., and Borth, D.E., *Introduction to Spread Spectrum Communications,* Prentice Hall, Englewood Cliffs, NJ.

Shannon, C.E. 1949. Communication in the presence of noise. *Proc. IEEE.* 37:10–21.

Simon, M.K., Omura, J.K., Scholtz, R.A., and Levitt, B.A. 1985. *Spread Spectrum Communications,* Vol. I, II, III. Computer Science Press, Rockville, MD.

Telecommunications Industry Association. 1995. Mobile Station-Base Station Compatibility Standard for Dual-Mode Wideband Spread Spectrum Cellular System. TIA/EIA/IS-95-A.

Turin, G.L. 1980. Introduction to spread spectrum antimultipath techniques and their application to urban digital radio. *Proc. IEEE.* 68:328–354.

Viterbi, A.J. 1995. *CDMA Principles of Spread Spectrum Communication,* Addison–Wesley, Reading, MA.

Viterbi, A.M. and Viterbi, A.J. 1993. Erlang capacity of a power controlled CDMA system. *IEEE J. on Selected Areas in Comm.* 11(6):892–900.

Viterbi, A.J., Viterbi, A.M., Gilhousen, K.S., and Zehavi, E. 1994. Soft handoff extends CDMA cell coverage and increases reverse link capacity. *IEEE J. on Selected Areas in Comm.* 12(8): 1281–1288.

Viterbi, A.J., Viterbi, A.M., and Zehavi, E. 1993. Performance of power-controlled wideband terrestrial digital communication. *IEEE Trans. on Comm.* COM-41(4):559–569.

Further Information

Gilhousen, K.S., Jacobs, I.M., Padovani, R., Viterbi, A.J., Weaver, L.A., and Wheatley, C. A. 1991. On the capacity of a cellular CDMA system. *IEEE Trans. Veh. Tech.* VT-40(2):303–312.

Telecommunications Industry Association. 1995. Mobile Station-Base Station Compatibility Standard for Dual-Mode Wideband Spread Spectrum Cellular System. TIA/EIA/IS-95-A.

Turin, G.L. 1980. Introduction to spread spectrum antimultipath techniques and their application to urban digital radio. *Proc. IEEE.* 68:328–354.

Viterbi, A.J. 1995. *CDMA Principles of Spread Spectrum Communication,* Addison–Wesley, Reading, MA.

28

Japanese Cellular Standard

28.1 Cellular Market in Japan

Recently amazing growth has been experienced in the analog market in Japan. Since 1988 this growth has been spurred by the competition provided by the **new common carrier (NCC)** cellular networks. As a result, the total number of cellular subscribers reached 1.3 million in 1992. It is presumed that growth will continue to accelerate, and the total is expected to reach 10 million by the year 2000. Therefore, the current analog system will soon meet saturation with the given frequency bands.

On the other hand, Japanese analog cellular telephone service is provided by two different systems, NTT systems and Japanese TACS systems. It is very important, however, that the terminals work anywhere and in any cellular network in Japan in order to make cellular telephones more popular. To realize roaming ability nationwide between several different cellular networks, standardizing air-interfaces between base stations and terminals is necessary.

28.2 Standardization Process

As just mentioned it is very important to standardize the air interface in developing digital systems. Therefore, in April 1989 the Ministry of Posts and Telecommunications (MPT) organized a research and study committee on digital cellular systems in Japan, and the committee started to study the technical requirements of the system. The Japanese digital cellular radio system committee, which was formed in July 1989 under the Telecommunications Technology Council organized by the consultative body of the minister of MPT, further studied the requirements after receiving the report from the research and study committee. The Japanese digital cellular radio system committee submitted the technical requirements for the Japanese digital cellular radio system to the MPT in June 1990. Under these requirements, radio equipment regulations provided by the MPT should be changed.

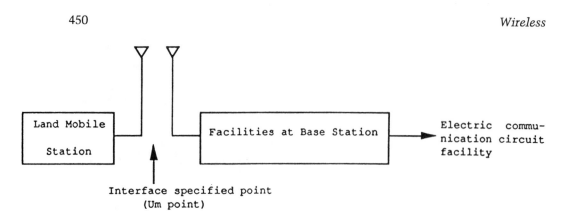

FIGURE 28.1   Structure of the digital cellular telecommunications system.

On the other hand, the Japanese Research and Development Center for Radio Systems (RCR) Standards Committee organized subcommittees in May 1990 for the standardization of Japanese digital cellular systems and for the treatment of intellectual proprietary rights (IPR) issues under the standards committee. The RCR standards committee is a private organization for standardization of Japanese radio systems comprising people from universities, users, telecommunication and consumer electronics manufacturers (including overseas companies), network operators, etc. This RCR standards committee draws up the detailed standard specifications for the connections between base stations and terminals in realizing the standard **Japanese digital cellular (PDC) system** based on the technical requirements.

28.3 General

Scope of Application

The digital cellular telecommunication system consists of land mobile stations (MS) and base station facilities as shown in Fig. 28.1. The Japanese standard regulates the radio interfaces and services for the digital cellular telecommunication system.

Principle of Standardization

In terms of mutual connectivity and compatibility, the Japanese standard defines the minimum level of specifications required for basic connections and services as the essential requirement and the minimum level of specifications required for whatever free choice is permitted, such as protocols, as optional standard to provide options and future expansion. Further, in order to provide options and future expansion capabilities as much as possible, care has been taken not to place restrictions on nonstandardized specifications.

Figure 28.2 outlines the relationship between standardized services and optional protocols used.

28.4 System Overview

Definition of the Interface

The points where the interface occurs in digital cellular telecommunication systems exist at the four locations, Um, R, S, and C, shown in Fig. 28.3. The definition of each interface point in Fig. 28.3 is as follows.

1. Interface Um: Interface between the MS and base station system. This interface shall conform to this standard.

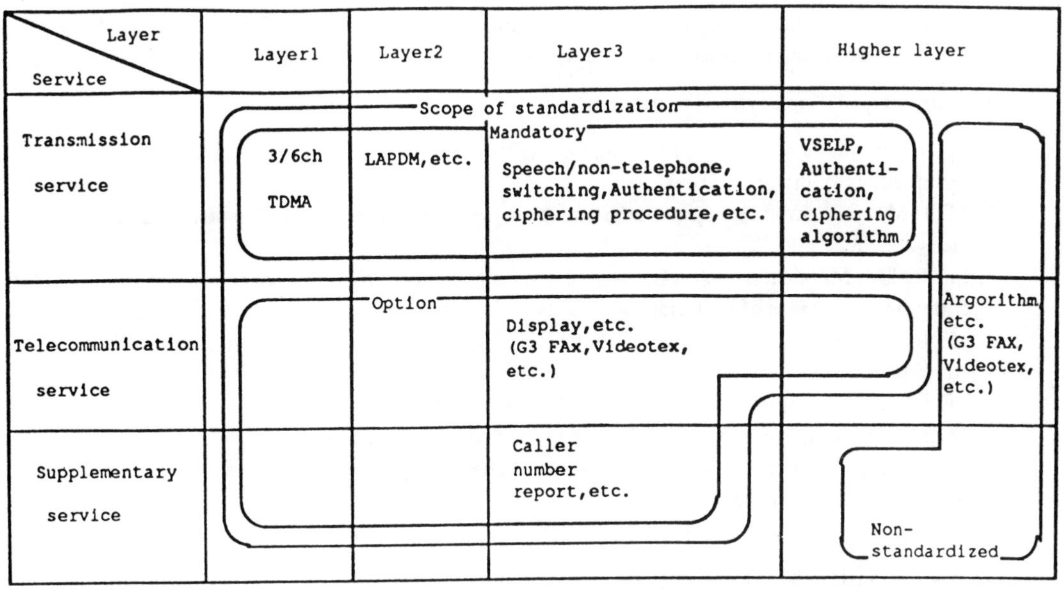

Layer
Service
Layer1
Layer2
Layer3
Higher layer
Transmission service
Scope of standardization
Mandatory
3/6ch TDMA
LAPDM,etc.
Speech/non-telephone, switching,Authentication, ciphering procedure,etc.
VSELP, Authenti-cation, ciphering algorithm
Telecommunication service
Option
Display,etc. (G3 FAx,Videotex, etc.)
Argorithm, etc. (G3 FAX, Videotex, etc.)
Supplementary service
Caller number report,etc.
Non-standardized

FIGURE 28.2 The relationship between services and protocols.

C
S
R
Um
Handset
TE1
TE2
TA
TE2
MT0
MT1
MT2
MT3
MT4
Base station system
Base station system
Base station system
Base station system
Base station system
General switched network
General switched network
General switched network
General switched network
General switched network

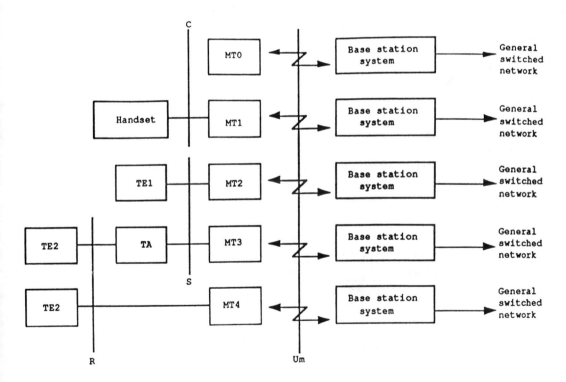

FIGURE 28.3 Interface points.

2. Interface R: Interface between non-I interface (I interface is conformed to integrated services digital network, ISDN) terminal equipment and either mobile terminal equipment or terminal adapters. This interface is not specified in this standard.
3. Interface S: Interface between either I interface terminal equipment or terminal adapters and the mobile terminal equipment. Conditions of this interface shall conform to the I interface standards.

4. Interface C: Interface between the mobile terminal equipment and handset within the MS. This interface is not specified in this standard.

Elements in Fig. 28.3 are defined as follows.

1. MT0: Mobile terminal equipment that contain man–machine interface facilities, such as terminals, as an integral part of it.
2. MT1: Mobile terminal equipment that can be connected with an external handset via interface C.
3. MT2, MT3: Mobile terminal equipment that offer a connection point with external terminal equipment by means of I interface standards.
4. MT4: Mobile terminal equipment that offer a connection point with external terminal equipment by means of non-I interface standards.
5. TE1: Terminal equipment conforming to I interface standards.
6. TE2: Terminal equipment not conforming to I Interface standards.
7. TA: Terminal adapters that perform conversation between the non-I interface condition and the I interface condition.

Services Provided by the System

Service Features

The services provided by Japanese digital cellular telecommunication system consist of several attributes listed in Table 28.1. For the service attribute consisting of more than one service item, one of those items can be selected.

Service Types

Bearer Services. The bearer services provided through the information transfer channel are listed in Table 28.2.

Teleservices. The teleservices provided via the information transfer channel are listed in Table 28.3.

Supplementary Services (Circuit Switched). Supplementary services that are provided for circuit switched service are listed in Tables 28.4–28.12. The numbers in the item column of Table 28.4–28.12 are the referenced CCITT recommendation numbers.

TABLE 28.1   Service Attributes

Service Attribute	Service Item
Information transfer capability	Speech Alternative speech or nontelephone Unrestricted digital information 3.1-KHz audio
Information transfer mode	Circuit Packet
Information transfer rate, kb/s	11.2 5.6
Frame structure, Hz	50 25
Communication configuration	Point to point Point to multipoint
Establishment of communication	Demand Reserved

TABLE 28.2   Bearer services

Item	Service
11.2-kb/s speech*	Transmission function suitable for speech communications. 11.2 kb/s VSELP codec is used. Bit transparency is not guaranteed.
5.6-kb/s speech*	Transmission function suitable for speech communications. 5.6 kb/s PSI-CELP codec is used. Bit transparency is not guaranteed.
8-kb/s unrestricted digital	Enables communications between a user terminal at MS with transfer rate of 8 kb/s and ISDN terminal using the 8 kb/s subrate.
64-kb/s unrestricted digital	Enables communications between a user terminal at MS with transfer rate of 64 kb/s and ISDN terminal.
Alternate 11.2-kb/s* speech/data	Supports voice and data. After a call is established, the transmission function of the network becomes switchable between 11.2-kb/s speech and 11.2-kb/s data according to requests by the user.
Alternate 5.6-kb/s* speech/data	Supports voice and data. After a call is established, the transmission function of the network becomes switchable between 5.6-kb/s speech and 11.2-kb/s data according to requests by the user.
Alternate 11.2-kb/s speech/8-kb/s unrestricted digital	Supports voice and 8-kb/s unrestricted digital information. After a call is established, the network's transmission function can be switched between 11.2-kb/s speech and 8-kb/s unrestricted digital information by the user.
Alternate 5.6-kb/s speech/8-kb/s unrestricted digital	Supports voice and 8-kb/s unrestricted digital information. After a call is established, the network's transmission function can be switched between 5.6-kb/s speech and 8-kb/s unrestricted digital information by the user.
Packet	ISDN packet service(X.31) and X.25 packet services are available.

*Items originally stipulated by Japanese standard (PDC system).

TABLE 28.3   Teleservices

Item	Service
G3 facsimile*	Enables communications between G3 fax terminals according to CCITT.T. 30 procedure.
G4 facsimile	Enables communications between G4 fax terminals.
Videotex	Video information transmission service using the Captain.
JUST-PC	Enables data communications between personal computers using the MPT recommended method.
JUST-MHS	Message handling service using a higher layer of MPT recommended method for PC data communications.
Modem (V.42 ANNEX)*	Enables data communications between personal computers using a modem, conforms to V.42 ANNEX.
Short message	Sends a message to a single user or broadcasts a message to multiple users and subsequently reports a reception acknowledgment to the party who transmitted the message.
Automobile location information service	Reports location of the specified MS to the user requesting such information.

*Items originally stipulated by Japanese standard (PDC system).

TABLE 28.4   Supplementary Services: Number Identification

Item	Service
Direct dial in (DDI) (I.251.1)	Enables a user to dial another user directly on an ISPBX or a private network.
Multiple subscriber number (MSN) (I.251.2)	Enables assigning multiple ISDN numbers to one interface.
Calling line identification presentation* (CLIP) (I.251.3)	Reports the calling user's number (including the subaddress if one exists) to the caller user.
Calling line identification restriction* (CLIR) (I.251.4)	Inhibits reporting of the calling user's number (including the subaddress if one exists) to the called user.

(*continues*)

TABLE 28.4   Supplementary Services: Number Identification (*continued*)

Item	Service
Connected line identification presentation (COLP) (I.251.5)	Reports the called user's number (including the subaddress if one exists) to the calling user.
Connected line identification restriction (COLR) (I.251.6)	Inhibits reporting of the called user's number (including the subaddress if one exists) to the calling user.
Malicious call identification (MCI) (I.251.7)	Allows the user to request the network to identify and memorize the information of the originator of the calls that are terminated by the user.
Subaddressing (SUB) (I.251.8)	Enables network to transmit the subaddress between users transparently.

*Items originally stipulated by Japanese standard (PDC system).

TABLE 28.5   Supplementary Services: Call Offering

Item	Services
Call transfer* (CT) (I.252.1)	Allows the user to transfer an active call to a third party. This service applies to both originating calls and terminating calls. It also differs from the call forwarding service that transfers a call from the called party before call establishment.
Call forwarding busy (CFB) (I.252.2)	Whereby a call is forwarded to another user when the called user is busy. The served user's originating service is unaffected.
Call forwarding no reply (CFNR) (I.252.3)	Whereby an unanswered call to a user is forwarded to another user. The served user's originating service is unaffected.
Call forwarding unconditional (CFU) (I.252.4)	Whereby the network forwards the call of a registered user to another user, regardless of the condition of the termination.
Call deflection* (CD) (I.252.5)	Upon receiving a call allows the user to choose if the call should be forwarded to another user or not.
Call forwarding no page response	Mobile communication specific service that forwards all incoming calls or incoming calls of specified basic service to another user when a paging response is not received.
Call forwarding not registered	Mobile communication specific service that forwards all incoming calls or incoming calls or specified basic service to another user when the location registration of the MS is not registered.
Call forwarding no radio resource	Mobile communication specific service that forwards all incoming calls or incoming calls of specified basic service to another user when the radio channel is congested.
Voice messaging function*	Function that on alerting, transfers the call to a voice messaging equipment to record a message instead of answering the call.
Line hunting (LH) (I.252.6)	Enables reception of a call by using a specific number of an interface featuring multiple channels and numbers.

*Items originally stipulated by this standard.

TABLE 28.6   Supplementary Services: Call Completion

Item	Service
Call waiting* (CW) (I.253.1)	Notifies the user of an incoming call on a call basic when no traffic channel is available.
Call hold (HOLD) (I.253.2)	Interrupts the existing call by setting it to the hold state. The held call may be reactivated if desired. After the call is interrupted, the traffic channel used for the call may be set on hold for use by newly incoming call.
Completion of calls to busy subscribers (CCBS) (I.253.3)	When the called user is busy, the network realerts the called user after it becomes idle. When the called user is available for the call, the network reports this to the originating user and may subsequently set up a call if necessary.

*Items originally stipulated by this standard (PDC system).

TABLE 28.7   Supplementary Services: Multiparty

Item	Service
Conference calling (CONF) (I.254.1)	Enables the user to communicate with several other users simultaneously.
Three-party service* (3PTY) (I.245.2)	Allows the user to hold the active call and make an additional call to the third user. It subsequently allows switching between the two calls, and/or the release of one call while maintaining the other. Optionally, this service enables the conference calling so that the three parties can talk simultaneously.

*Items originally stipulated by this standard (PDC system).

TABLE 28.8   Supplementary Services: Community of Interest

Item	Service
Closed user group (CUG) (I.255.1)	Enables users to form a group to/from which user access is restricted. One user can be a member of one or more CUGs. Generally, a member user of a CUG can only communicate with other users in the same CUG and cannot communicate with users outside the group. Specific CUG members are additionally allowed to originate calls outside the group or terminate calls from outside the group.
Private numbering plan (PNP) (I.255	Allows users to originate or terminate calls using user defined private numbers.

TABLE 28.9   Supplementary Services: Charging

Item	Service
Credit card calling (CRED) (I.256.1)	Puts call charges on a credit card account.
Advice of charge* (AOC) (I.256.2)	Advises the user of charging information on a call-by-call basis.
Reverse charging (REV) (I.256.3)	Puts call charges on the called party upon request by the originating party and at the consent of the called party.
Free phone	Puts call charge on the called party throughout the nation or a specified region when free phone number is dialed.

*Items originally stipulated by this standard (PDC system).

TABLE 28.10   Supplementary Services: Additional Information Transfer

Item	Service
User-to-user signalling (UUS) (I.257.1)	Allows the user to transfer the user-to-user information through the signalling channel in association with the call.

TABLE 28.11   Supplementary Services: Origination and Termination Restriction

Item	Service
Outgoing call barring*	Restricts outgoing calls based on the called party number or the location of the called terminal. This service can be set for all or for specified basic services; however, it does not restrict termination of a call and origination of an emergency call.
Incoming call barring	Restricts incoming calls. This service is set for all or for specified basic services. Outgoing calls from a terminal are not restricted.

*Items originally stipulated by this standard (PDC system).

Access method

Core parameters

See Table 28.13.

Time Division Multiple Access (TDMA) system

The three-channel multiplex TDMA system is used as the access method for the radio channel in the digital cellular telecommunication system. The six-channel multiplex TDMA system is an optional function for mobile stations whereas the three-channel multiplex TDMA system is a mandatory function (see Table 28.14).

Functional Structure of Radio Channel

The functional structure of radio channel is shown in Fig. 28.4.

Broadcast Channel (BCCH). BCCH is a unidirectional channel used by the base station system to broadcast the system control information related to location registration, channel structure, system state, etc., to land mobile stations.

TABLE 28.12   Supplementary Services: Other

Item	Service
Priority connection and channel hold (CH)	Allows the following operation by setting priority classes to an MS. 1. If an MS or a terminal of higher priority class originates a call when all radio CHs are busy, a radio channel used for an MS or a terminal of lower priority class is disconnected for subsequent use for higher priority class. 2. The radio channel used for the higher priority or important communication is held after such communications have been completed.

TABLE 28.13   Core Parameters

Frequency bands, MHz	Base station transmit frequency	810–826
		1477–1501
	Mobile station transmit frequency	940–956
		1429–1453
Send/receive distance, MHz	130/48	
Error correction mode	CCH	BCH
	TCH	Convolutional code
Zone composition	3 sector 4 iteration	
Others	Equalizer (option)*	
	Diversity (option)	

*According to propagation experiments, the average delay spread in Tokyo area is 1 μs. Even in the Kofu area, which is supposed to get more delay spread, only about 10% of the area has a result of over 5-μs delay spread. It has been concluded that it is not always necessary to use an equalizer with diversity or adequate assignment of base stations.

TABLE 28.14   TDMA system parameters

Item	At full-rate codec	At half-rate codec
Multiplexed No. of channels	3	6
Carrier frequency separation, KHz	50 (25 interleave)	
Modulation system	$\pi/4$ shift QPSK (Rolloff factor = 0.5)	
Transmission rate, kb/s	42	
Info. bit rate, kb/s	11.2	5.6

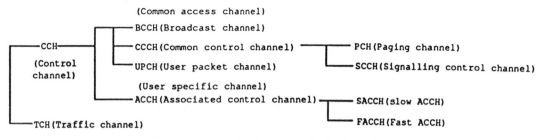

FIGURE 28.4   Functional structure of radio channels.

Common Control Channel (CCCH).   CCCH is a bidirectional channel for transmitting signalling information. The following two types of CCCH exist.

1. Paging channel (PCH) is a point-to-multipoint unidirectional channel used for transmitting the common information from the base station system to mobile stations within a wide area, i.e., paging area, which is composed of multiple calls. It is used for paging and grouping control for intermittent reception by the MS.
2. Signalling control channel (SCCH) SCCH is a point-to-multipoint bidirectional channel used for transmitting information from/to the base station system to/from mobile stations when a cell area within which the mobile station is located is known to the base station system. A SCCH is prepared for transfer of the cell specific information by using different frequencies on a cell-by-cell basis. The uplink channel (from MS to base station system) is operating in the random access mode.

User Packet Channel (UPCH).   UPCH is a point-to-multipoint bidirectional channel that transfers the control signal and user packet data. The uplink channel is operating in the random access mode.

Associated Control Channel (ACCH).   ACCH is a point-to-point bidirectional channel associated with the TCH and is used for transferring signalling information and user packet data. The normal ACCH is called a slow ACCH (SACCH). In addition to the SACCH, there is a fast ACCH (FACCH), which is established by temporarily stealing the TCH to perform high-speed data transfer.

Traffic Channel (TCH).   TCH is a point-to-point bidirectional channel that transfers the user information and its control signal. The TCH carries voice and facsimile signals.

Radio Circuit Control

Control Procedure.   The control procedure is specified to enable the MS originating and terminating call connections, the location registration by mobile stations, the channel handover during a call, the service identification, etc. These controls shall be exactly performed by using commonly and independently assigned slots.

Slot Configuration.   The slot configuration shall be in accordance with Fig. 28.5. The configuration is designed to meet the following requirements.

1. Uplink reception processing and downlink transmission processing can be carried out in a time sequential manner at a base station.
2. The base station, which enables collision control of the random access channel, can transmit a downlink signal after having confirmed the reception of uplink signal.
3. Duplexer at the MS can be simplified.
4. Antenna switching diversity can easily be implemented at the MS.

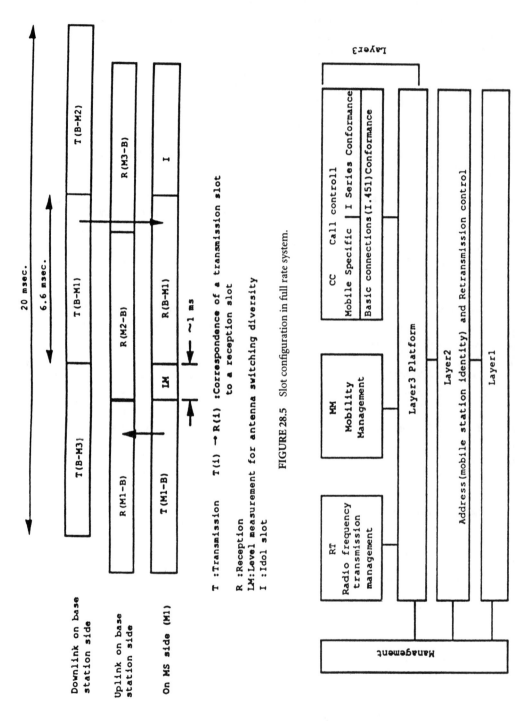

FIGURE 28.5   Slot configuration in full rate system.

FIGURE 28.6   Signalling system structure.

Signalling System

Signal System Structure

The signalling system structure of the digital cellular telecommunication system is shown in Fig. 28.6. The signalling system will have a layered structure made up of layers 1–3, which conforms to the OSI reference model.

Layered Structure

The definitions of layers 1–3 of the signalling system are shown in Table 28.15. The layer-3 will be classified into call control (CC), mobility management (MM), and radio frequency transmission (RT) functions according to CCIR and CCITT recommendations.

Characteristics of the Signalling System

The structure of layers 1–3 will feature an expandable design, while ensuring high serviceability, such as connection quality, etc., and system economy, i.e., signalling efficiency. The signalling format for each layer is depicted in Fig. 28.7.

1. Layer 1 assembles and disassembles layers 2 and 3 using the error correction and bit interleaving. Signals of layers 2–3, along with the preamble, synchronization (sync.) word, and supplementary information, form a slot.
2. Layer 2 will consist of address part and control part.
3. Layer 3 will feature a common platform used by CC, MM, and RT functions. This platform makes efficiency of signal transmission high, as well as shortening the time required for the service. For example, call origination by an MS, RT and CC of layer 3 will report radio frequency condition information and setup information, respectively, to the base station. The layer-3 common platform allows layers 1 and 2 to deal with information such as a signal so as to increase the efficiency of signal transmission and to shorten the service time.
4. Layer-3 messages will be configured as: (message) + (supplementary information)

TABLE 28.15 Definitions of Layers 1–3

Layer	Functional definitions		Examples
Layer 1	Ensures transmission of bit sequence using a communication circuit consisting of physical entities.		Radio signal transmission, frequency assignment, radio channel packet random access control, etc.
Layer 2	Located on layer 1 and offers highly reliable and transparent data transfer using the bit transmission function provided by layer 1.		Frame structure, procedural elements, data field, procedural specifications, etc.
Layer 3	Performs end-to-end data transfers between end system entities using the data transfer function provided by layer 2	RT	Specifies items related to radio frequency transmission control and performs establishment, maintenance, handover, etc., of the radio channel
		MM	Specifies items related to mobility management control, and performs location registration and authentication
		CC	Specifies items related to call connection control. Basic call connections will be in accordance with the CCITT I-recommendations. For supplementary services which require a large number of signals, mobile specific sequences of I-series recommendations, leading to reduction of time required for services.

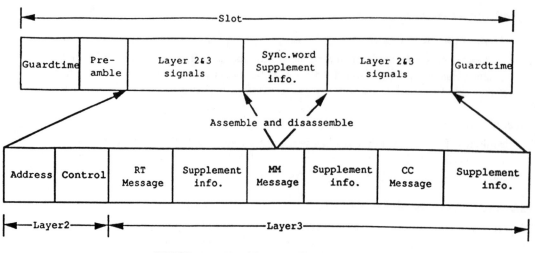

FIGURE 28.7   Signal format outline for layers 1–3.

5. MM and RT messages will have a fixed length, taking into account that these functions are rather limited.
6. CC will conform to the ISDN user-network interface layer 3 specifications (I.451), with emphasis on the harmonization with ISDN. The CC messages will be based on the I.451 format. The mobile specific format, however, will be used for coding of bearer capabilities, because the information transfer capability of a digital cellular telecommunication system differs from that of fixed networks. In addition, the caller number reporting is shortened from a full octet to 4 b.

Defining Terms

Common carrier: A company who possesses a big public telecommunication network, such as NTT.
New common carrier (NCC): A company of common carriers founded after the privatization in 1980s such as IDO.
Japanese digital cellular system (JDC): Operations began in 1993 under the standard explained in this chapter.
Personal digital cellular system (PDC): Formerly called JDC, named changed in 1994.
Personal handy phone system (PHS): Personal mobile telecommunications system using microcells starts its operation in 1995. This type of microcellular system will be the first one in the world.
Future public land mobile telecommunication system (FPLMTS): Japan will propose some systems including time division multiple access (TDMA) and code division multiple access (CDMA) for the future world standard microcellular system.

References

Kuramoto, M., Kinosita, K., Nakajima, A., Utano, T., and Murase, A. 1993. Overall system performance in a digital cellular system based on Japanese standard. *43rd IEEE Vehicular Technology Conference*, pp. 168–171.
Research & Development Center for Radio Systems (RCR). Personal Digital Cellular Telecommunication System RCR Standard-27C.

Further Information

See "Overview of Wireless Personal Communications," by J.E. Padgett, C.G. Gunther, and T. Hattori, *IEEE Communications Magazine*, Jan. 1995, or K. Kinoshita et al., "Development of a TDMA Digital Cellular System Based on Japanese Standard," *Proc. 41st. IEEE Vehicular Technology Conf.* pp. 642–645, 1991.

29

The British Cordless Telephone Standard: CT-2

Lajos Hanzo
Univ. of Southampton

29.1 History and Background

Following a decade of world-wide research and development (R&D), cordless telephones (CT) are now becoming widespread consumer products, and they are paving the way towards ubiquitous, low-cost personal communications networks (PCN) [Tuttlebee ed., 1990; Steele ed., 1992]. The two most well-known European representatives of CTs are the digital European cordless telecommunications (DECT) system [Ochsner, 1990; Asghar, 1995] and the CT-2 system [Steedman, 1990; Gardiner, 1990]. Three potential application areas have been identified, namely, domestic, business, and public access, which is also often referred to as telepoint (TP).

In addition to conventional voice communications, CTs have been conceived with additional data services and local area network (LAN) applications in mind. The fundamental difference between conventional mobile radio systems and CT systems is that CTs have been designed for small to very small cells, where typically benign low-dispersion, dominant line-of-sight (LOS) propagation conditions prevail. Therefore, CTs can usually dispense with channel equalizers and complex low-rate speech codecs, since the low-signal dispersion allows for the employment of higher bit rates before the effect of channel dispersion becomes a limiting factor. On the same note, the LOS propagation scenario is associated with mild fading or near-constant received signal level, and when combined with appropriate small-cell power-budget design, it ensures a high average signal-to-noise ratio (SNR). These prerequisites facilitate the employment of high-rate, low-complexity speech codecs, which maintain a low battery drain. Furthermore, the deployment of forward error correction codecs can often also be avoided, which reduces both the bandwidth requirement and the power consumption of the portable station (PS).

A further difference between public land mobile radio (PLMR) systems [Hanzo, 1995] and CTs is that whereas the former endeavor to standardize virtually all system features, the latter seek to offer a so-called access technology, specifying the common air interface (CAI), access and signalling protocols, and some network architecture features, but leaving many other characteristics unspecified. By the same token, whereas PLMR systems typically have a rigid frequency allocation scheme and fixed cell structure, CTs use dynamic channel allocation (DCA) [Jabbari, 1995]. The DCA principle allows for a more intelligent and judicious channel assignment, where the base station (BS) and PS select an appropriate traffic channel on the basis of the prevailing traffic and channel quality conditions, thus minimizing, for example, the effect of cochannel interference or channel blocking probability.

In contrast to PLMR schemes, such as the Pan-European global system of mobile communications (GSM) system [Hanzo, 1995], CT systems typically dispense with sophisticated mobility management, which accounts for the bulk of the cost of PLMR call charges, although they may facilitate limited hand-over capabilities. Whereas in residential applications CTs are the extension of the public switched telephone network (PSTN), the concept of omitting mobility management functions, such as location update, etc., leads to telepoint CT applications where users are able to initiate but not to receive calls. This fact drastically reduces the network operating costs and, ultimately, the call charge at a concomittant reduction of the services rendered.

Having considered some of the fundamental differences between PLMR and CT systems let us now review the basic features of the the CT-2 system.

29.2 The CT-2 Standard

The European CT-2 recommendation has evolved from the British standard MPT-1375 with the aim of ensuring the compatibility of various manufacturers' systems as well as setting performance requirements, which would encourage the development of cost-efficient implementations. Further standardization objectives were to enable future evolution of the system, for example, by reserving signalling messages for future applications and to maintain a low PS complexity even at the expense of higher BS costs. The CT-2 or MPT 1375 CAI recommendation is constituted by the four following parts.

1. *Radio interface:* Standardizes the radio frequency (RF) parameters, such as legitimate channel frequencies, the modulation method, the transmitter power control, and the required receiver sensitivity as well as the carrier-to-interference ratio (CIR) and the time division duplex (TDD) multiple access scheme. Furthermore, the transmission burst and master/slave timing structures to be used are also laid down, along with the scrambling procedures to be applied.
2. *Signalling layers one and two:* Defines how the bandwidth is divided among signalling, traffic data, and synchronization information. The description of the first signalling layer includes the dynamic channel allocation strategy, calling channel detection, as well as link setup and establishment algorithms. The second layer is concerned with issues of various signalling message formats, as well as link establishment and re-establishment procedures.
3. *Signalling layer three:* The third signalling layer description includes a range of message sequence diagrams as regards to call setup to telepoint BSs, private BSs, as well as the call clear down procedures.

4. *Speech coding and transmission:* The last part of the standard is concerned with the algorithmic and performance features of the audio path, including frequency responses, clipping, distortion, noise, and delay characteristics.

Having briefly reviewed the structure of the CT-2 recommendations let us now turn our attention to its main constituent parts and consider specific issues of the system's operation.

29.3 The Radio Interface

Transmission Issues

In our description of the system we will adopt the terminology used in the recommendation, where the PS is called cordless portable part (CPP), whereas the BS is referred to as cordless fixed part (CFP). The channel bandwidth and the channel spacing are 100 kHz, and the allocated system bandwidth is 40 MHz, which is hosted in the range of 864.15–868.15 MHz. Accordingly, a total of 40 RF channels can be utilized by the system.

The accuracy of the radio frequency must be maintained within ±10 kHz of its nominal value for both the CFP and CPP over the entire specified supply voltage and ambient temperature range. To counteract the maximum possible frequency drift of 20 kHz, automatic frequency correction (AFC) may be used in both the CFP and CPP receivers. The AFC may be allowed to control the transmission frequency of only the CPP, however, in order to prevent the misalignment of both transmission frequencies.

Binary frequency shift keying (FSK) is proposed, and the signal must be shaped by an approximately Gaussian filter in order to maintain the lowest possible frequency occupancy. The resulting scheme is referred to as Gaussian frequency shift keying (GFSK), which is closely related to Gaussian minimum shift keying (GMSK) [Steele ed., 1992] used in the DECT [Asghar, 1995] and GSM [Hanzo, 1995] systems.

Suffice to say that in M-arry FSK modems the carrier's frequency is modulated in accordance with the information to be transmitted, where the modulated signal is given by

$$S_i(t) = \sqrt{\frac{2E}{T}} \cos[\omega_i t + \Phi] \qquad i = 1, \ldots, M$$

and E represents the bit energy, T the signalling interval length, ω_i has M discrete values, whereas the phase Φ is constant.

Multiple Access and Burst Structure

The so-called TDD multiple access scheme is used, which is demonstrated in Fig. 29.1. The simple principle is to use the same radio frequency for both uplink and downlink transmissions between

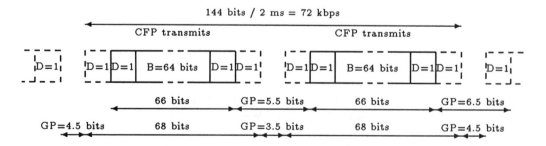

FIGURE 29.1   M1 burst and TDD frame structure.

the CPP and the CFP, respectively, but with a certain staggering in time. This figure reveals further details of the burst structure, indicating that 66 or 68 b per TDD frame are transmitted in both directions.

There is a 3.5- or 5.5-b duration guard period (GP) between the uplink and downlink transmissions, and half of the time the CPP (the other half of the time the CFP) is transmitting with the other part listening, accordingly. Although the guard period wastes some channel capacity, it allows a finite time for both the CPP and CFP for switching from tranmission to reception and vice versa. The burst structure of Fig. 29.1 is used during normal operation across an established link for the transmission of adaptive differential pulse code modulated (ADPCM) speech at 32 kb/s according to the CCITT G721 standard in a so-called B channel or bearer channel. The D channel, or signalling channel, is used for the transmission of link control signals. This specific burst structure is referred to as a multiplex one (M1) frame.

Since the speech signal is encoded according to the CCITT G721 recommendation at 32 kb/s the TDD bit rate must be in excess of 64 kb/s in order to be able to provide the idle guard space of 3.5- or 5.5-b interval duration plus some signalling capacity. This is how channel capacity is sacrificed to provide the GP. Therefore, the transmission bit rate is stipulated to be 72 kb/s and the transmission burst length is 2 ms, during which 144-b intervals can be accommodated. As it was demonstrated in Fig. 29.1, 66 or 68 b are transmitted in both the uplink and downlink burst, and taking into account the guard spaces, the total transmission frame is constituted by $(2 \cdot 68) + 3.5 + 4.5 = 144$ b or equivalently, by $(2 \cdot 66) + 5.5 + 4.5 = 144$ b. The 66-b transmission format is compulsory, whereas the 68-b format is optional. In the 66-b burst there is one D bit dedicated to signalling at both ends of the burst, whereas in the 68-b burst the two additional bits are also assigned to signalling. Accordingly, the signalling rate becomes 2 b/2 ms or 4 b/2 ms, corresponding to 1 kb/s or 2 kb/s signalling rates.

Power Ramping, Guard Period, and Propagation Delay

As mentioned before and suggested by Fig. 29.1, there is a 3.5- or 5.5-b interval duration GP between transmitted and received bursts. Since the signalling rate is 72 kb/s, the bit interval becomes about $1/(72 \text{ kb/s}) \approx 13.9 \ \mu s$ and, hence, the GP duration is about 49 μs or 76 μs. This GP serves a number of purposes. Primarily, the GP allows the transmitter to ramp up and ramp down the transmitted signal level smoothly over a finite time interval at the beginning and end of the transmitted burst. This is necessary, because if the transmitted signal is toggled instantaneously, that is equivalent to multiplying the transmitted signal by a rectangular time-domain window function, which corresponds in the frequency domain to convolving the transmitted spectrum with a sinc function. This convolution would result in spectral side-lobes over a very wide frequency range, which would interfere with adjacent channels. Furthermore, due to the introduction of the guard period, both the CFP and CPP can tolerate a limited propagation delay, but the entire transmitted burst must arrive within the receivers' window, otherwise the last transmitted bits cannot be decoded.

Power Control

In order to minimize the battery drain and the cochannel interference load imposed upon cochannel users, the CT-2 system provides a power control option. The CPPs must be able to transmit at two different power levels, namely, either between 1 and 10 mW or at a level between 12 and 20 dB lower. The mechanism for invoking the lower CPP transmission level is based on the received signal level at the CFP. If the CFP detects a received signal strength more than 90 dB relative to 1 μV/m, it may instruct the CPP to drop its transmitted level by the specified 12–20 dB. Since the 90-dB gain factor corresponds to about a ratio of 31,623, this received signal strength would be equivalent for a 10-cm antenna length to an antenna output voltage of about 3.16 mV. A further beneficial ramification of using power control is that by powering down CPPs that are in the vicinity of a telepoint-type multiple-transceiver CFP, the CFP's receiver will not be so prone to being desensitised by the high-powered close-in CPPs, which would severely degrade the reception quality of more distant CPPs.

29.4 Burst Formats

As already mentioned in the previous section on the radio interface, there are three different subchannels assisting the operation of the CT-2 system, namely, the *voice/data channel* or *B channel*, the *signalling channel* or *D channel*, and the *burst synchronization channel* or *SYN channel*. According to the momentary system requirements, a variable fraction of the overall channel capacity or, equivalently, a variable fraction of the bandwidth can be allocated to any of these channels. Each different channel capacity or bandwidth allocation mode is associated with a different burst structure and accordingly bears a different name. The corresponding burst structures are termed as multiplex one (M1), multiplex two (M2), and multiplex three (M3), of which multiplex one used during the normal operation of established links has already been described in the previous section. Multiplex two and three will be extensively used during link setup and establishment in subsequent sections, as further details of the system's operation are unravelled.

Signalling layer one (L1) defines the burst formats multiplex one–three just mentioned, outlines the calling channel detection procedures, as well as link setup and establishment techniques. *Layer two (L2)* deals with issues of acknowledged and unacknowledged information transfer over the radio link, error detection and correction by retransmission, correct ordering of messages, and link maintenance aspects.

The burst structure multiplex two is shown in Fig. 29.2. It is constituted by two 16-b D-channel segments at both sides of the 10-b *preamble (P)* and the 24-b frame synchronization pattern (SYN), and its signalling capacity is 32 b/2 ms = 16 kb/s. Note that the M2 burst does not carry any B-channel information, it is dedicated to synchronization purposes. The 32-b D-channel message is split in two 16-b segments in order to prevent that any 24-b fraction of the 32-b word emulates the 24-b SYN segment, which would result in synchronization misalignment.

Since the CFP plays the role of the master in a telepoint scenario communicating with many CPPs, all of the CPP's actions must be synchronized to those of the CFP. Therefore, if the CPP attempts to initiate a call, the CFP will reinitiate it using the M2 burst, while imposing its own timing structure. The 10-b preamble consists of an alternate zero/one sequence and assists in the operation of the clock recovery circuitry, which has to be able to recover the clock frequency before the arrival of the SYN sequence, in order to be able to detect it. The SYN sequence is a unique word determined by computer search, which has a sharp autocorrelation peak, and its function is discussed later. The way the M2 and M3 burst formats are used for signalling purposes will be made explicit in our further discussions when considering the link setup procedures.

The specific SYN sequences used by the CFP and the CPP are shown in Table 29.1 along with the so-called *channel marker (CHM)* sequences used for synchronization purposes by the M3 burst format. Their differences will be made explicit during our further discourse. Observe from the table that the sequences used by the CFP and CPP, namely, SYNF, CHMF and SYNP, CHMP, respectively,

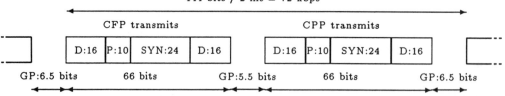

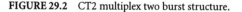

FIGURE 29.2   CT2 multiplex two burst structure.

TABLE 29.1 CT-2 Synchronization Patterns

	MSB (sent last)					LSB (sent first)
CHMF	1011	1110	0100	1110	0101	0000
CHMP	0100	0001	1011	0001	1010	1111
SYNCF	1110	1011	0001	1011	0000	0101
SYNCP	0001	0100	1110	0100	1111	1010

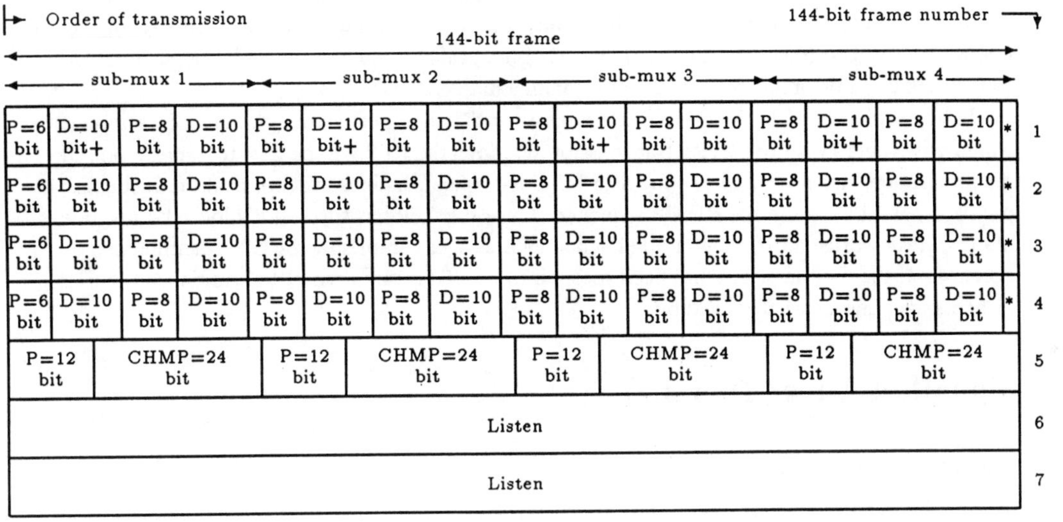

Order of transmission →   144-bit frame number →

144-bit frame

sub-mux 1 ——→ ◄— sub-mux 2 ——→ ◄— sub-mux 3 ——→ ◄— sub-mux 4 ——→

Frame	Structure
1	P=6 bit / D=10 bit+ / P=8 bit / D=10 bit / P=8 bit / D=10 bit+ / P=8 bit / D=10 bit / P=8 bit / D=10 bit+ / P=8 bit / D=10 bit / P=8 bit / D=10 bit+ / P=8 bit / D=10 bit *
2	P=6 bit / D=10 bit / P=8 bit / D=10 bit / P=8 bit / D=10 bit / P=8 bit / D=10 bit / P=8 bit / D=10 bit / P=8 bit / D=10 bit / P=8 bit / D=10 bit / P=8 bit / D=10 bit *
3	P=6 bit / D=10 bit / P=8 bit / D=10 bit / P=8 bit / D=10 bit / P=8 bit / D=10 bit / P=8 bit / D=10 bit / P=8 bit / D=10 bit / P=8 bit / D=10 bit / P=8 bit / D=10 bit *
4	P=6 bit / D=10 bit / P=8 bit / D=10 bit / P=8 bit / D=10 bit / P=8 bit / D=10 bit / P=8 bit / D=10 bit / P=8 bit / D=10 bit / P=8 bit / D=10 bit / P=8 bit / D=10 bit *
5	P=12 bit / CHMP=24 bit / P=12 bit / CHMP=24 bit / P=12 bit / CHMP=24 bit / P=12 bit / CHMP=24 bit
6	Listen
7	Listen

* 2 bit P

Notes:

1/ Transmission is continuous for five bursts or 10 ms, then off for two burst periods or 4 ms

2/ The 20 bits of D chan are repeated in each of the 4 sub-mux's before D changes

3/ The D chan sync. word (SYNCD) always begins at the start of the slots marked +

FIGURE 29.3 CT2 multiplex three burst structure.

are each other's bit-wise inverses. This was introduced in order to prevent CPPs and CFPs from calling each other directly. The CHM sequences are used, for instance, in residential applications, where the CFP can issue an M2 burst containig a 24-b CHMF sequence and a so-called poll message mapped on to the D-channel bits in order to wake up the specific CPP called. When the called CPP responds, the CFP changes the CHMF to SYNF in order to prevent waking up further CPPs unnecessarily.

Since the CT-2 system does not entail mobility functions, such as registration of visiting CPPs in other than their own home cells, in telepoint applications all calls must be initiated by the CPPs. Hence, in this scenario when the CPP attempts to set up a link, it uses the so-called multiplex three burst format displayed in Fig. 29.3. The design of the M3 burst reflects that the CPP initiating the call is oblivious of the timing structure of the potentially suitable target CFP, which can detect access attempts only during its receive window, but not while the CFP is transmitting. Therefore, the M3 format is rather complex at first sight, but it is well structured, as we will show in our further discussions. Observe in the figure that in the M3 format there are five consecutive 2-ms long 144-b transmitted bursts, followed by two idle frames, during which the CPP listens in order to determine whether its 24-b CHMP sequence has been detected and acknowledged by the CFP. This process can be followed by consulting Fig. 29.6, which will be described in depth after considering the detailed construction of the M3 burst.

The first four of the five 2-ms bursts are identical D-channel bursts, whereas the fifth one serves as a synchronization message and has a different construction. Observe, furthermore, that both the first four 144-b bursts as well as the fifth one contain four so-called submultiplex segments, each of which hosts a total of $(6 + 10 + 8 + 10 + 2) = 36$ b. In the first four 144-b bursts there are $(6 + 8 + 2) = 16$ one/zero clock-synchronizing P bits and $(10 + 10) = 20$ D bits or signalling bits. Since the D-channel message is constituted by two 10-b half-messages, the first half of the D-message is marked by the $+$ sign in the figure. As mentioned in the context of M2, the D-channel bits are split in two halves and interspersed with the preamble segments in order to ensure that these bits do not emulate valid CHM sequences. Without splitting the D bits this could happen upon concatenating the one/zero P bits with the D bits, since the tail of the SYNF and SYNP sequences is also a one/zero segment. In the fifth 144-b M3 burst, each of the four submultiplex segments is constituted by 12 preamble bits and 24 CPP channel marker (CHMP) bits.

The four-fold submultiplex M3 structure ensures that irrespective of how the CFP's receive window is aligned with the CPP's transmission window, the CFP will be able to capture one of the four submultiplex segments of the fifth M3 burst, establish clock synchronization during the preamble, and lock on to the CHMP sequence. Once the CFP has successfully locked on to one of the CHMP words, the corresponding D-channel messages comprising the CPP identifier can be decoded. If the CPP identifier has been recognized, the CFP can attempt to reinitialize the link using its own master synchronization.

29.5 Signalling Layer Two (L2)

General Message Format

The signalling L2 is responsible for acknowledged and un-acknowledged information transfer over the air interface, error detection and correction by retransmission, as well as for the correct ordering of messages in the acknowledged mode. Its further functions are the link end-point identification and link maintenance for both CPP and CFP, as well as the definition of the L2 and L3 interface.

Compliance with the L2 specifications will ensure the adequate transport of messages between the terminals of an established link. The L2 recommendations, however, do not define the meaning of messages, this is specified by L3 messages, albeit some of the messages are undefined in order to accommodate future system improvements.

The L3 messages are broken down to a number of standard packets, each constituted by one or more codewords (CW), as shown in Fig. 29.4. The codewords have a standard length of eight octets, and each packet contains up to six codewords. The first codeword in a packet is the so-called address codeword (ACW) and the subsequent ones, if present, are data codewords (DCW). The first octet of the ACW of each packet contains a variety of parameters, of which the binary flag **L3_END** is indicated in Fig. 29.4, and it is set to zero in the last packet. If the L3 message transmitted is mapped onto more than one packet, the packets must be numbered up to N. The address codeword is always preceded by a 16-b D-channel frame synchronization word **SYNCD**. Furthermore, each eight-octet CW is protected by a 16-b parity-check word occupying its last two octets. The binary Bose–Chaudhuri–Hocquenghem BCH(63,48) code is used to encode the first six octets or 48 b by adding 15 parity b to yield 63 b. Then bit 7 of octet 8 is inverted and bit 8 of octet 8 added such that the 64-b codeword has an even parity. If there are no D-channel packets to send, a 3-octet idle message **IDLE_D** constituted by zero/one reversals is transmitted. The 8-octet format of the ACWs and DCWs is made explicit in Fig. 29.5, where the two parity check octets occupy octets 7 and 8. The first octet hosts a number of control bits. Specifically, bit 1 is set to logical one for an ACW and to zero for a DCW, whereas bit 2 represents the so-called format type FT bit. $FT = 1$ indicates that variable length packet format is used for the transfer of L3 messages, whereas $FT = 0$ implies that a fixed length link setup is used for link end point addressing end service requests. FT is only relevant to ACWs, and in DCWs it has to be set to one.

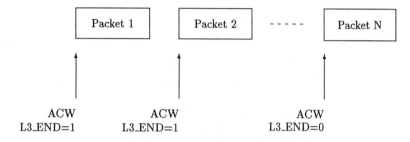

a/ L2 packet: Contains at least an ACW and 0...5 DCWs

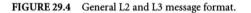

Packet 1	Packet 2	- - - - -	Packet N

ACW
L3_END=1

ACW
L3_END=1

ACW
L3_END=0

b/ L3 message: N packets

FIGURE 29.4 General L2 and L3 message format.

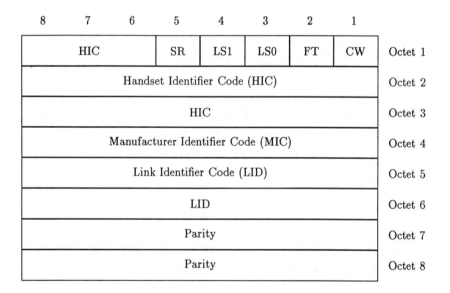

8	7	6	5	4	3	2	1	
HIC			SR	LS1	LS0	FT	CW	Octet 1
Handset Identifier Code (HIC)								Octet 2
HIC								Octet 3
Manufacturer Identifier Code (MIC)								Octet 4
Link Identifier Code (LID)								Octet 5
LID								Octet 6
Parity								Octet 7
Parity								Octet 8

FIGURE 29.5 Fixed format packets mapped on M1, M2, and M3 during link initialization and on M1 and M2 during handshake.

Fixed Format Packet

As an example, let us focus our attention on the fixed format scenario associated with $FT = 0$. The corresponding codeword format defined for use in M1, M2, and M3 for link initiation and in M1 and M2 for handshaking is displayed in Fig. 29.5. Bits 1 and 2 have already been discussed, whereas the 2-bit link status (LS) field is used during call setup and handshaking. The encoding of the four possible LS messages is given in Table 29.2. The aim of these LS messages will

TABLE 29.2 Encoding of Link Status Messages

LS1	LS0	Message
0	0	Link_request
0	1	Link_grant
1	0	ID_OK
1	1	ID_lost

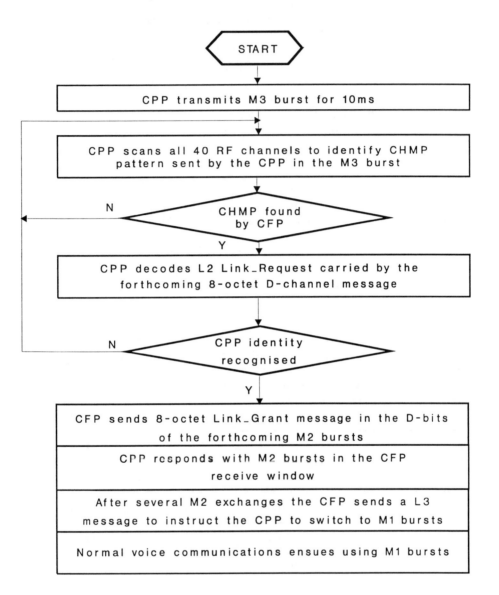

FIGURE 29.6   Flowchart of the CT-2 link initialization by the CPP.

become more explicit during our further discussions with reference to Fig. 29.6 and Fig. 29.7. Specifically, **link_request** is transmitted from the CPP to the CFP either in an M3 burst as the first packet during CPP-initiated call setup and link re-establishment, or returned as a poll response in an M2 burst from the CPP to the CFP, when the CPP is responding to a call. **Link_grant** is sent by the CFP in response to a link_request originating from the CPP. In octets 5 and 6 it hosts the so-called link identification (LID) code, which is used by the CPP, for example, to address a specific CFP or a requested service. The LID is also used to maintain link reference during handshake exchanges and link re-establishment. The two remaining link status handshake messages, namely, ID_OK and ID_lost, are used to report to the far end whether a positive confirmation of adequate link quality has been received within the required time-out period. These issues will be revisited during our further

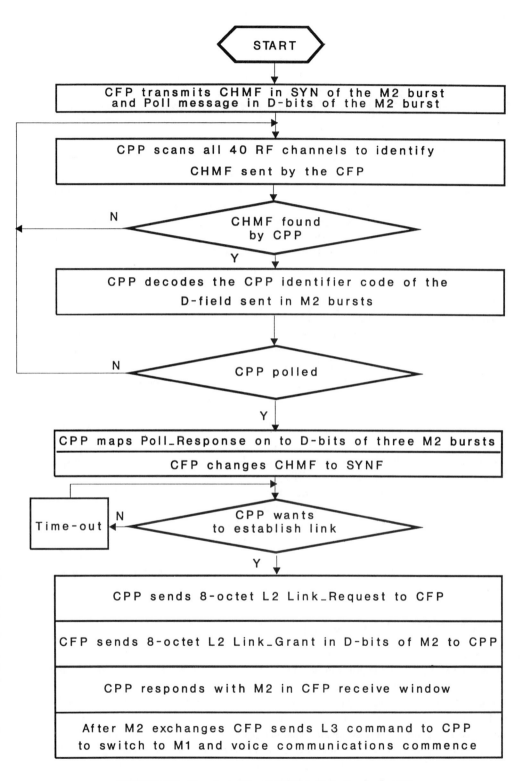

FIGURE 29.7   Flowchart of the CT-2 link initialization by the CFP.

elaborations. Returning to Fig. 29.5, we note that the fixed packet format ($FT = 0$) also contains a 19-b handset identification code (HIC) and an 8-b manufacturer identification code (MIC). The concatenated HIC and MIC fields jointly from the unique 27-b portable identity code (PIC), serving as a link end-point identifier. Lastly, we have to note that bit 5 of octet 1 represents the signalling rate (SR) request/response bit, which is used by the calling party to specify the choice of the 66- or 68-b M1 format. Specifically, $SR = 1$ represents the four bit/burst M1 signalling format. The first 6 octets are then protected by the parity check information contained in octets 7 and 8.

29.6 CPP-Initiated Link Setup Procedures

Calls can be initiated at both the CPP and CFP, and the call initation and detection procedures invoked depend on which party initiated the call. Let us first consider calling channel detection at the CFP, which ensues as follows. Under the instruction of the CFP control scheme, the RF synthesizer tunes to a legitimate RF channel and after a certain settling time commences reception. Upon receiving the M3 bursts from the CPP, the automatic gain control (AGC) circuitry adjusts its gain factor, and during the 12-b preamble in the fifth M3 burst, bit synchronization is established. This specific 144-b M3 burst, is transmitted every 14 ms, corresponding to every seventh 144-b burst. Now the CFP is ready to bit-synchronously correlate the received sequences with its locally stored CHMP word in order to identify any CHMP word arriving from the CPP. If no valid CHMP word is detected, the CFP may retune itself to the next legitimate RF channel, etc.

As mentioned, the call identification and link initialization process is shown in the flowchart of Fig. 29.6. If a valid 24-b CHMP word is identified, D-channel frame synchronization can take place using the 16-b SYNCD sequence and the next 8-octet L2 D-channel message delivering the link_request handshake portrayed earlier in Fig. 29.5 and Table 29.2 is decoded by the CFP. The required $16 + 64 = 80$ D bits are accommodated in this scenario by the $4 \cdot 20 = 80$ D bits of the next four 144-b bursts of the M3 structure, where the 20 D bits of the four submultiplex segments are transmitted four times within the same burst before the D message changes. If the decoded LID code of Fig. 29.5 is recognized by the CFP, the link may be reinitialized based on the master's timing information using the M2 burst associated with SYNF and containing the link_grant message addressed to the specific CPP identified by its PID.

Otherwise the CFP returns to its scanning mode and attempts to detect the next CHMP message. The reception of the CFP's 24-b SYNF segment embedded in the M2 message shown previously in Fig. 29.2 allows the CPP to identify the position of the CFP's transmit and receive windows and, hence, the CPP now can respond with another M2 burst within the receive window of the CFP. Following a number of M2 message exchanges, the CFP then sends a L3 message to instruct the CPP to switch to M1 bursts, which marks the commencement of normal voice communications and the end of the link setup session.

29.7 CFP-Initiated Link Setup Procedures

Similar procedures are followed when the CPP is being polled. The CFP transmits the 24-b CHMF words hosted by the 24-b SYN segment of the M2 burst shown in Fig. 29.2 in order to indicate that one or more CPPs are being paged. This process is displayed in the flowchart of Fig. 29.7, as well as in the timing diagram displayed in Fig. 29.8. The M2 D-channel messages convey the identifiers of the polled CPPs.

The CPPs keep scanning all 40 legitimate RF channels in order to pinpoint any 24-b CHMF words. Explicitly, the CPP control scheme notifies the RF synthesizer to retune to the next legitimate RF channel if no CHMF words have been found on the current one. The synthesizer needs a finite time to settle on the new center frequency and then starts receiving again. Observe in Fig. 29.8 that at this stage only the CFP is transmitting the M2 bursts; hence, the uplink-half of the 2-ms TDD frame is unused.

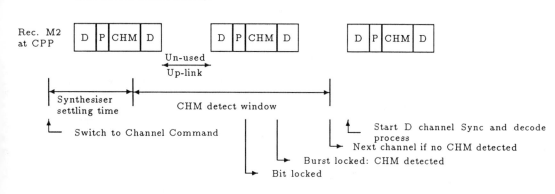

FIGURE 29.8   CT-2 call detection by the CPP.

Since the M2 burst commences with the D-channel bits arriving from the CFP, the CPP receiver's AGC will have to settle during this 16-b interval, which corresponds to about $16 \cdot 1/[72 \text{ kb/s}] \approx 0.22$ ms. Upon the arrival of the 10 alternating one–zero preamble bits, bit synchronization is established. Now the CPP is ready to detect the CHMF word using a simple correlator circuitry, which establishes the appropriate frame synchronization. If, however, no CHMF word is detected within the receive window, the synthesizer will be retuned to the next RF channel, and the same procedure is repeated, until a CHMF word is detected.

When a CHMF word is correctly decoded by the CPP, the CPP is now capable of frame and bit synchronously decoding the D-channel bits. Upon decoding the D-channel message of the M2 burst, the CPP identifier (ID) constituted by the LID and PID segments of Fig. 29.5 is detected and compared to the CPP's own ID in order to decide as to whether the call is for this specific CPP. If so, the CPP ID is reflected back to the CFP along with a SYNP word, which is included in the SYN segment of an uplink M2 burst. This channel scanning and retuning process continues until a legitimate incoming call is detected or the CPP intends to initiate a call.

More precisely, if the specific CPP in question is polled and its own ID is recognized, the CPP sends its poll_response message in three consecutive M2 bursts, since the capacity of a single M2 burst is 32 D bits only, while the handshake messages of Fig. 29.5 and Table 29.2 require 8 octets preceded by a 16-b SYNCD segment. If by this time all paged CPPs have responded, the CFP changes the CHMF word to a SYNF word, in order to prevent activating dormant CPPs who are not being paged. If any of the paged CPPs intends to set up the link, then it will change its poll_response to a L2 link_request message, in response to which the CFP will issue an M2 link_grant message, as seen in Fig. 29.7, and from now on the procedure is identical to that of the CPP-initiated link setup portrayed in Fig. 29.6.

29.8 Handshaking

Having established the link, voice communications is maintained using M1 bursts, and the link quality is monitored by sending handshaking (HS) signalling messages using the D-channel bits. The required frequency of the handshaking messages must be between once every 400 ms and 1000 ms. The CT-2 codewords ID_OK, ID_lost, link_request and link_grant of Table 29.2 all represent valid handshakes. When using M1 bursts, however, the transmission of these 8-octet messages using the 2- or 4-b/2ms D-channel segment must be spread over 16 or 32 M1 bursts, corresponding to 32 or 64 ms.

Let us now focus our attention on the *handshake protocol* shown in Fig. 29.9. Suppose that the CPP's handshake interval of Thtx_p = 0.4 s since the start of the last transmitted handshake has

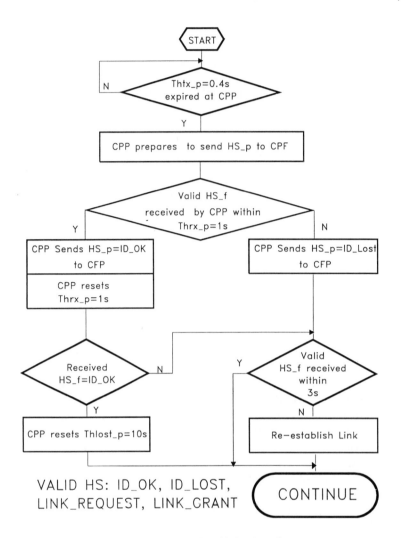

FIGURE 29.9 CT-2 handshake algorithms.

expired, and hence the CPP prepares to send a handshake message HS_p. If the CPP has received a valid HS_f message from the CFP within the last Thrx_p = 1s, the CPP sends an HS_p = ID_OK message to the CFP, otherwise an ID_Lost HS_p. Furthermore, if the valid handshake was HS_f = ID_OK, the CPP will reset its HS_f lost timer Thlost_p to 10 s. The CFP will maintain a 1-s timer referred to as Thrx_f, which is reset to its initial value upon the reception of a valid HS_p from the CPP.

The CFP's actions also follow the structure of Fig. 29.9 upon simply interchanging CPP with CFP and the descriptor _p with _f. If the Thrx_f = 1 s timer expires without the reception of a valid HS_p from the CPP, then the CFP will send its ID_Lost HS_f message to the CPP instead of the ID_OK message and will not reset the Thlost_f = 10 s timer. If, however, the CFP happens to detect a valid HS_p, which can be any of the ID_OK, ID_Lost, link_request and link_grant messages of Table 29.2, arriving from the CPP, the CFP will reset its Thrx_f = 1 s timer and resumes transmitting the ID_OK HS_f message instead of the ID_Lost. Should any of the HS messages go astray for more than 3 s, the CPP or the CFP may try and re-establish the link on the current or another RF channel. Again, although any of the ID_OK, ID_Lost, link_request and link_grant represent valid handshakes, only the reception of the ID_OK HS message is allowed to reset the Thlost = 10 s timer at both the CPP and CFP. If this timer expires, the link will be relinquished and the call dropped.

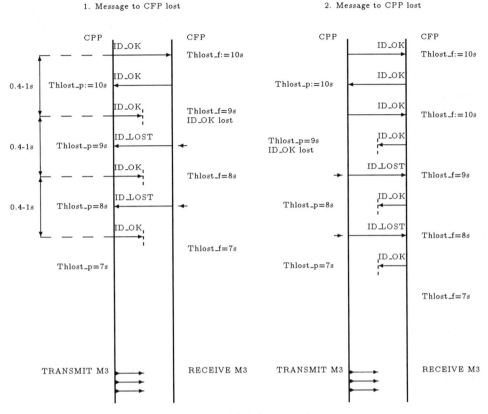

FIGURE 29.10 Handshake loss scenarios.

The handshake mechanism is further augmented by referring to Fig. 29.10, where two different scenarios are examplified, portraying the situation when the HS message sent by the CPP to the CFP is lost or, conversely, that transmitted by the CFP is corrupted.

Considering the first scenario, during error-free communications the CPP sends HS_p = ID_OK, and upon receiving it the CFP resets its Thlost_f timer to 10 s. In due course it sends an HS_f = ID_OK acknowledgement, which also arrives free from errors. The CPP resets the Thlost_f timer to 10 s and, after the elapse of the 0.4–1 s handshake interval, issues an HS_p = ID_OK message, which does not reach the CFP. Hence the Thlost_f timer is now reduced to 9 s and an HS_f = ID_Lost message is sent to the CPP. Upon reception of this, the CPP now cannot reset its Thlost_p timer to 10 s but can respond with an HS_p = ID_OK message, which again goes astray, forcing the CFP to further reduce its Thlost_f timer to 8 s. The CFP issues the valid handshake HS_f = ID_Lost, which arrives at the CPP, where the lack of HS_f = ID_OK reduces Thlost_p to 8 s. Now the corruption of the issued HS_p = ID_OK reduces Thlost_f to 7 s, in which event the link may be reinitialized using the M3 burst. The portrayed second example of Fig. 29.10 can be easily followed in case of the scenario when the HS_f message is corrupted.

29.9 Main Features of the CT-2 System

In our previous discourse we have given an insight in the algorithmic procedures of the CT-2 MPT 1375 recommendation. We have briefly highlighted the four-part structure of the standard dealing with the radio interface, signalling layers 1 and 2, signalling layer 3, and the speech coding issues, respectively. There are forty 100-kHz wide RF channels in the band 864.15–868.15 MHz, and the 72 kb/s bit stream modulates a Gaussian filtered FSK modem. The multiple access technique is

TDD, transmitting 2-ms duration, 144-b M1 bursts during normal voice communications, which deliver the 32-kb/s ADPCM-coded speech signal. During link establishment the M2 and M3 bursts are used, which were also portrayed in this treatise, along with a range of handshaking messages and scenarios.

Defining Terms

AFC: Automatic frequency correction
CAI: Common air interface
CFP: Cordless fixed part
CHM: Channel marker sequence
CHMF: CFP channel marker
CHMP: CPP channel marker
CPP: Cordless portable part
CT: Cordless telephone
DCA: Dynamic channel allocation
DCW: Data code word
DECT: Digital European cordless telecommunications system
FT: Frame format type bit
GFSK: Gaussian frequency shift keying
GP: Guard period
HIC: Handset identification code
HS: Handshaking
ID: Identifier
L2:   Signalling layer 2
L3:   Signalling layer 3
LAN: Local area network
LID: Link identification
LOS: Line of sight
LS: Link status
M1:   Multiplex one burst format
M2:   Multiplex two burst format
M3:   Multiplex three burst format
MIC: Manufacturer identification code
MPT-1375:   British CT2 standard
PCN: Personal communications network
PIC: Portable identification code
PLMR: Public land mobile radio
SNR: Signal-to-noise ratio
SR: Signalling rate bit
SYN: Synchronization sequence
SYNCD: 16-b D-channel frame synchronization word
TDD: Time division duplex multiple access scheme
TP: Telepoint

References

Tuttlebee, W.H.W. ed. 1990. *Cordless Telecommunication in Europe*, Springer–Verlag.
Steele, R. ed. 1992. *Mobile Radio Communications*, Pentech Press, London.
Ochsner, H. 1990. The digital European cordless telecommunications specification DECT. In *Cordless telecommunication in Europe*, ed. W.H.W. Tuttlebee, pp. 273–285. Springer–Verlag.

Asghar, S. 1995. Digital European cordless telephone (DECT), In *The Mobile Communications Handbook,* Chap. 30, CRC Press, Inc. Boca Raton, FL.

Steedman, R.A.J. 1990. The Common Air Interface MPT 1375. In *Cordless Telecommunication in Europe,* ed. W.H.W. Tuttlebee, pp. 261–272, Springer–Verlag.

Gardiner, J.G. 1990. Second generation cordless (CT-2) telephony in the UK: telepoint services and the common air-interface, Elec. & Comm. Eng. J. (April):71–78.

Jabbari, B. 1995. Dynamic channel assignment, In *The Mobile Communications Handbook,* Chap. 21, CRC Press, Inc. Boca Raton, FL.

Hanzo, L. 1995. The Pan-European mobile radio system, In *The Mobile Communications Handbook,* Chap. 25, CRC Press, Inc. Boca Raton, FL.

30

Digital European Cordless Telephone

Saf Asghar
Advanced Micro Devices, Inc.

30.1 Introduction

Cordless technology, in contrast to cellular radio, primarily offers access technology rather than fully specified networks. The digital European cordless telecommunications (DECT) standard, however, offers a proposed network architecture in addition to the air interface physical specification and protocols but without specifying all of the necessary procedures and facilities. During the early 1980s a few proprietary digital cordless standards were designed in Europe purely as coexistence standards. The U.K. government in 1989 issued a few operator licenses to allow public-access cordless known as telepoint. Interoperability was a mandatory requirement leading to a common air interface (CAI) specification to allow roaming between systems. This particular standard (CT2/CAI), has been described in the previous chapter. The European Telecommunications Standards Institute (ETSI) in 1988 took over the responsibility for DECT. After formal approval of the specifications by the ETSI technical assembly in March 1992, DECT became a European telecommunications standard, ETS300-175 in August 1992. DECT has a guaranteed pan-European frequency allocation, supported and enforced by European Commission Directive 91/297. The CT2 specification has been adopted by ETSI alongside DECT as an interim standard I-ETSI 300 131 under review.

0-8493-8573-3/96/$0.00+$.50
© 1996 by CRC Press, Inc.

30.2 Application Areas

Initially, DECT was intended mainly to be a private system, to be connected to a private automatic branch exchange (PABX) to give users mobility, within PABX coverage, or to be used as a single cell at a small company or in a home. As the idea with telepoint was adopted and generalized to public access, DECT became part of the public network. DECT should not be regarded as a replacement of an existing network but as created to interface seamlessly to existing and future fixed networks such as public switched telephone network (PSTN), integrated services digital network (ISDN), global system for mobile communications (GSM), and PABX. Although telepoint is mainly associated with CT2, implying public access, the main drawback in CT2 is the ability to only make a call from a telepoint access point. Recently there have been modifications made to the CT2 specification to provide a structure that enables users to make and receive calls. The DECT standard makes it possible for users to receive and make calls at various places, such as airport/railroad terminals, and shopping malls. Public access extends beyond telepoint to at least two other applications: replacement of the wired local loop, often called cordless local loop (CLL), (Fig. 30.1) and neighborhood access, Fig. 30.2. The CLL is a tool for the operator of the public network. Essentially, the operator will install a multiuser base station in a suitable campus location for access to the public network at a subscriber's telephone hooked up to a unit coupled to a directional antenna. The advantages of CLL are high flexibility, fast installation, and possibly lower investments. CLL does not provide mobility. Neighborhood access is quite different from CLL. Firstly, it offers mobility to the users and, secondly, the antennas are not generally directional, thus requiring higher field strength (higher output power or more densely packed base stations). It is not difficult to visualize that CLL systems could be merged with neighborhood access systems in the context of establishments, such as supermarkets, gas stations, shops etc., where it might be desirable to set up a DECT system for their own use and at the same time also provide access to customers. The DECT standard already includes signalling for authentication, billing, etc. DECT opens possibilities for a new operator structure, with many diversified architectures connected to a global network operator (Fig. 30.3). DECT is designed to have extremely high capacity. A small size is used, which may seem an expensive approach for covering large areas. Repeaters placed at strategic locations overcome this problem.

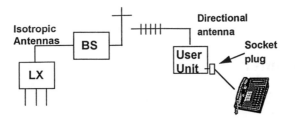

FIGURE 30.1

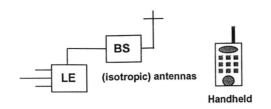

FIGURE 30.2

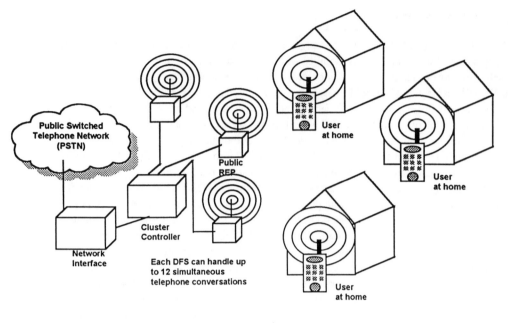

FIGURE 30.3

30.3 DECT/ISDN Interworking

From the outset, a major objective of the DECT specification was to ensure that ISDN services were provided through the DECT network. Within the interworking profile two configurations have been defined: DECT end system and DECT intermediate system. In the end system the ISDN is terminated in the DECT fixed system (DFS). The DFS and the DECT portable system (DPS) may be seen as a ISDN terminal equipment (TE1). The DFS can be connected to an S, S/T, or a P interface. The intermediate system is fully transparent to the ISDN. The S interface is regenerated even in the DPS. Both configurations have the following services specified: 3.1-kHz telephony, i.e, standard telephony; 7-kHz telephony; i.e, high-quality audio; video telephony; group III fax, modems, X.25 over the ISDN; and telematic services, such as group IV fax, telex, and videotex.

30.4 DECT/GSM Interworking

Groupe Speciale Mobile (GSM) is a pan-European standard for digital cellular radio operational throughout the European community. ETSI has the charter to define an interworking profile for GSM and DECT. The profile describes how DECT can be connected to the fixed network of GSM and the necessary air interface functions. The users obviously benefit from the mobility functions of GSM giving DECT a wide area mobility. The operators will gain access to another class of customer. The two systems when linked together will form the bridge between cordless and cellular technologies. Through the generic access profile ETSI will specify a well-defined level of interoperability between DECT and GSM. The voice coding aspect in both of these standards is different; therefore, this subject will be revisited to provide a sensible compromise.

30.5 DECT Data Access

The DECT standard is specified for both voice and data applications. It is not surprising that ETSI confirmed a role for DECT to support cordless local area network (LAN) applications. A new technical committee, ETSI RES10 has been established to specify the high performance European radio LAN similar to IEEE 802.11 standard in the U.S. (Table 30.1).

TABLE 30.1   DECT Characteristics

Parameters	DECT
Operating frequency, MHz	1880–1990 (Europe)
Radio carrier spacing, MHz	1.728
Transmitted data rate, Mb/s	1.152
Channel assignment method	DCA
Speech data rate, kb/s	32
Speech coding technique	ADPCM G.721
Control channels	In-call-embedded (various logical channels C, P, Q, N)
In-call control channel data rate, kb/s	4.8 (plus 1.6 CRC)
Total channel data rate, kb/s	41.6
Duplexing technique	TDD
Multiple access-TDMA	12 TDD timeslots
Carrier usage-FDMA/MC	10 carriers
Bits per TDMA timeslot, b	420 (424 including the 2 field)
Timeslot duration (including guard time), μs	417
TDMA frame period, ms	10
Modulation technique	Gaussian filtered FSK
Modulation index	0.45–0.55
Peak output power, mW	250
Mean output power, mW	10

30.6 How DECT functions

DECT employs frequency division multiple access (FDMA), time division multiple access (TDMA), and time division duplex (TDD) technologies for transmission. Ten carrier frequencies in the 1.88- and 1.90-GHz band are employed in conjunction with 12 time slots per carrier TDMA and 10 carriers per 20 MHz of spectrum FDMA. Transmission is through TDD. Each channel has 24 time slots, 12 for transmission and 12 for receiving. A transmission channel is formed by the combination of a time slot and a frequency. DECT can, therefore, handle a maximum of 12 simultaneous conversations. TDMA allows the same frequency to use different time slots. Transmission takes place for 10 ms, and during the rest of the time the telephone is free to perform other tasks, such as channel selection. By monitoring check bits in the signalling part of each burst, both ends of the link can tell if reception quality is satisfactory. The telephone is constantly searching for a channel for better signal quality, and this channel is accessed in parallel with the original channel to ensure a seamless changeover. Call handover is also seamless, each cell can handle up to 12 calls simultaneously, and users can roam around the infrastructure without the risk of losing a call. Dynamic channel assignment (DCA) allows the telephone and base station to automatically select a channel that will support a new traffic situation, particularly suited to a high-density office environment.

30.7 Architectural Overview

Baseband Architecture

A typical DECT portable or fixed unit consists of two sections: a baseband section and a radio frequency section. The baseband partitioning includes voice coding and protocol handling (Fig. 30.4).

Voice Coding and Telephony Requirements

This section addresses the audio aspects of the DECT specification. The CT2 system as described in the previous chapter requires adaptive differential pulse code modulation (ADPCM) for voice coding. The DECT standard also specifies 32-kb/s ADPCM as a requirement. In a mobile environment it is debatable whether the CCITT G.721 recommendation has to be mandatory. In the handset or the mobile it would be quite acceptable in most cases to implement a compatible or a less

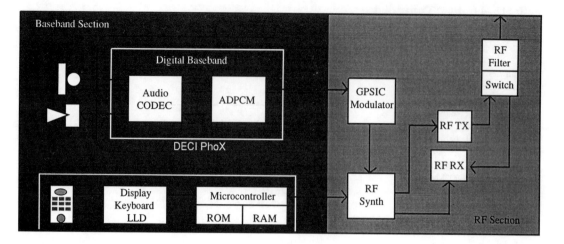

FIGURE 30.4

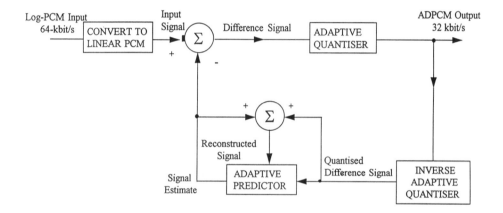

FIGURE 30.5   ADPCM encoder.

complex version of the recommendation. We are dealing with an air interface and communicating with a base station that in the residential situation terminates with the standard POTS line, hence compliance is not an issue. The situation changes in the PBX, however, where the termination is a digital line network. DECT is designed for this case, hence compliancy to the voice coding recommendation becomes important. Adhering to this strategy for the base station and the handset has some marketing advantages.

G.721 32-kb/s ADPCM from its inception was adopted to coexist with G.711 64-kb/s pulse code modulation (PCM) or work in tandem, the primary reason being an increase in channel capacity. For modem type signalling, the algorithm is suboptimal in handling medium-to-high data rates, which is probably one of the reasons why there really has not been a proliferation of this technology in the PSTN infrastructure. The theory of ADPCM transcoding is available in books on speech coding techniques, e.g., O'Shaughnessy, 1987.

The ADPCM transcoder consists of an encoder and a decoder. From Figs. 30.5 and 30.6 it is apparent that the decoder exists in the encoder structure. A benefit derived from this structure allows for efficient implementation of the transcoder.

The encoding process takes a linear speech input signal (the CCITT specification relates to a nonwireless medium such as a POTS infrastructure), and subtracts its estimate derived from earlier input signals to obtain a difference signal. This difference signal is 4-b coded with a 16-level adaptive

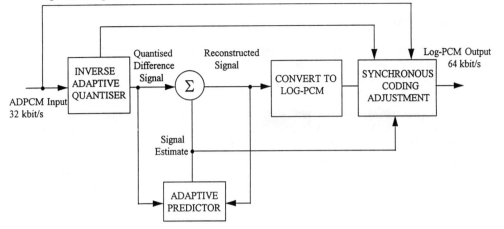

FIGURE 30.6 ADPCM decoder.

quantizer every 125 μs, resulting in a 32-kb/s bit stream. The signal estimate is constructed with the aid of the inverse adaptive quantizer that forms a quantized difference signal that added to the signal estimate is also used to update the adaptive predictor. The adaptive predictor is essentially a second-order recursive filter and a sixth-order nonrecursive filter,

$$S_0(k) = \sum_{i=1}^{2} a_i(k-1)\varepsilon_r(k-i) + \sum_{i=1}^{6} b_i(k-1)d_q(k-i) \qquad (30.1)$$

where coefficients a and b are updated using gradient algorithms.

As suggested, the decoder is really a part of the encoder, that is, the inverse adaptive quantizer reconstructs the quantized difference signal, and the adaptive predictor forms a signal estimate based on the quantized difference signal and earlier samples of the reconstructed signal, which is also the sum of the current estimate and the quantized difference signal as shown in Fig. 30.6. Synchronous coding adjustment tries to correct for errors accumulating in ADPCM from tandem connections of ADPCM transcoders.

ADPCM is basically developed from PCM. It has good speech reproduction quality, comparable to PSTN quality, which therefore led to its adoption in CT2 and DECT.

Telephony Requirements

A general cordless telephone system would include an acoustic interface, i.e., microphone and speaker at the handset coupled to a digitizing compressor/decompressor analog to uniform PCM to ADPCM at 32 kb/s enabling a 2:1 increase in channel capacity as a bonus. This digital stream is processed as described in earlier sections to be transmitted over the air interface to the base station where the reverse happens, resulting in a linear or a digital stream to be transported over the land-based network. The transmission plans for specific systems have been described in detail in Tuttlebee ed., 1995.

An important subject in telephony is the effect of network echoes [Weinstein, 1977]. Short delays are manageable even if an additional delay of, say, less than 15 μs is introduced by a cordless handset. Delays of a larger magnitude, in excess of 250 μs (such as satellite links [Madsen and Fague, 1993]), coupled to cordless systems can cause severe degradation in speech quality and transmission; a small delay introduced by the cordless link in the presence of strong network echoes is undesirable. The DECT standard actually specifies the requirement for network echo control. Additional material can be obtained from the relevant CCITT documents [CCITT, 1984–1985].

DECT Protocol Model

This section provides an overview of the software layer entities and the message interfaces between the software layers for the DECT common interface software package (Fig. 30.7).

The functionality of the DECT protocol is described in the ETSI specifications ETS 300 175-1–ETS 300 175-5.

The DECT protocol model is based on the International Standards Organization (ISO) open systems interconnection (OSI) seven-layer model. The complete DECT air interface corresponds to the first three ISO OSI layers; however, DECT defines four layers of protocol: physical, medium access control, data link, and network (Fig. 30.8).

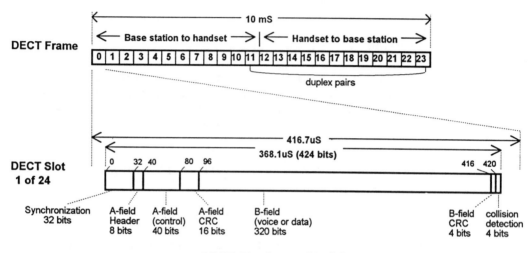

DECT Hardware Model

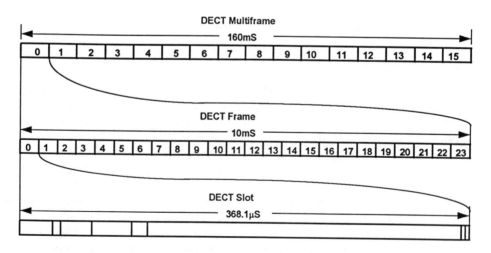

Multiframe Structure

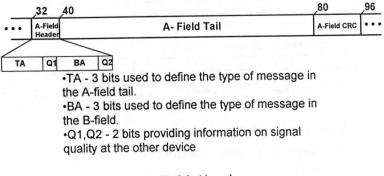

•TA - 3 bits used to define the type of message in the A-field tail.
•BA - 3 bits used to define the type of message in the B-field.
•Q1,Q2 - 2 bits providing information on signal quality at the other device

A-Field Header

- TA=000: Ct packet 0 - Higher layer control information
- TA=001: Ct packet 1 - Higher layer control information
- TA=010: Nt on connectionless bearer - Base station identification
- TA=011: Nt - Identification message for traffic bearers
- TA=100: Qt - Slot, frame, multiframe synchronization information
- TA=101: Escape
- TA=110: Mt - MAC layer control
- TA=111(PP):   Mt for first portable part transmission

    TA=111(RFP): Pt - Paging message

A-Field Tail Formats

DECT Multiframe - RFP to PP

0	1	2	3	4	5	6	7	8	9	10	11	12	13	14	15

160mS

Pt	Mt	Pt	Mt	Pt	Mt	Pt	Mt	Qt	Mt	Pt	Mt	Pt	Mt		Mt
	Ct		Ct		Ct		Ct		Ct		Ct		Ct		Ct
Nt	Nt	Nt	Nt	Nt	Nt	Nt	Nt		Nt	Nt	Nt	Nt	Nt	Nt	Nt

DECT Multiframe - PP to RFP

0	1	2	3	4	5	6	7	8	9	10	11	12	13	14	15

160mS

Mt		Mt		Mt		Mt		Mt		Mt		Mt		Mt	
Ct		Ct		Ct		Ct		Ct		Ct		Ct		Ct	
Nt	Nt	Nt	Nt	Nt	Nt	Nt	Nt	Nt	Nt	Nt	Nt	Nt	Nt	Nt	Nt

T-MUX Algorithm

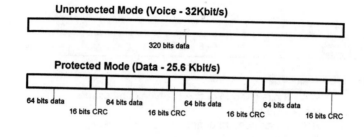

Unprotected Mode (Voice - 32Kbit/s)

320 bits data

Protected Mode (Data - 25.6 Kbit/s)

64 bits data   16 bits CRC   64 bits data   16 bits CRC   64 bits data   16 bits CRC   64 bits data   16 bits CRC

- **Unprotected mode used for voice information**
- **Protected mode used for transferring higher layer control information.**

Protected/Unprotected B-Field

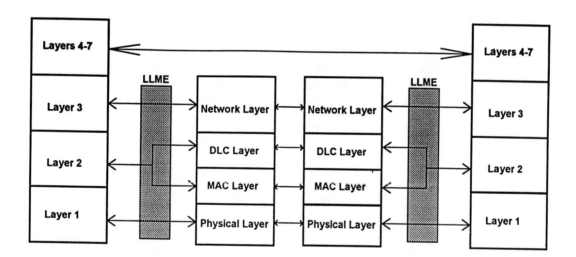

FIGURE 30.7   DECT protocol structure.

The complete DECT air interface corresponds to the first three ISO OSI layers; however, since the OSI Layer does not adequately consider multiple access to one transmission medium, the DECT structure uses four layers for the node-to-node communication: physical, medium access control, data link, and network layers (Fig. 30.9).

These first protocol layers serve to support the creation of a functional data link through the cordless network whereas layers 4, 5, 6, and 7 are concerned with supporting communications between the end users/networks.

Physical Layer

Divides the radio spectrum into physical channels using TDMA operation on ten RF carriers.

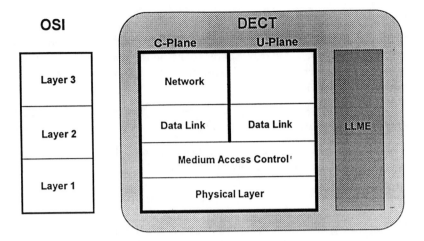

FIGURE 30.8

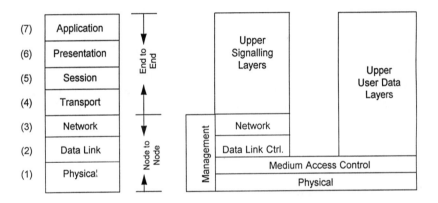
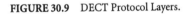

FIGURE 30.9 DECT Protocol Layers.

MAC Layer

The MAC layer selects the physical channels and then establishes or releases connections on those channels. It also multiplexes/demultiplexes control information in slot-sized packets.

These functions provide three services: broadcast service, connection oriented service, and a connectionless service. The broadcast service multiplexes broadcast information into the A-field, and this field appears as part of all active transmissions. In the absence of user traffic, at least one physical channel broadcasts.

Data Link Control (DLC) Layer

The DLC layer provides a reliable data link to the network layer. The DECT DLC layer separates the operation into two planes: the C plane and the U plane. The C plane is common to all applications and provides reliable links for the transmission of internal control signalling and limited user information traffic. It uses LAPC for full error control. The U plane provides a family of alternative services optimized to the specific application. i.e., transparent unprotected service for voice, circuit mode, and packet mode.

Network Layer

This is the main signalling layer of the protocol and is similar to the ISDN layer 3 protocol. It supports the establishment, maintenance, and release of the connections. Some network layer services are: call control, supplementary services, connection oriented message service, Connection less message service, and mobility management.

Lower Layer Management Entity (LLME)

The LLME interfaces with all of the previous layers to provide procedures that concern more than one layer. Most of these procedures have only local significance.

Physical Layer

The physical layer is responsible for the segmentation of the transmission media into physical channels using TDMA on ten carriers between 1880 and 1900 MHz. Each carrier contains a TDMA structure defined as 24 timeslots in a 10-ms frame. These timeslots can be used to transmit data packets.

   Each burst contains a 32-b synchronization field and a data field. The synchronization field is used for clock and packet synchronization. The data field is received by the MAC layer.

MAC Interface: Digital-Service Access Point (D-SAP)

The physical layer communicates to the MAC layer via the D-SAP digital-service access point (D-SAP). The D-SAP is mainly used to exchange D fields. D-field segments may be passed in either direction. The following primitives are exchanged though the D-SAP:

   PL-TX
   PL-RX
   PL-FREQ-ADJ

LLME Interface: PM-SAP

The primitives passed through PM-SAP are mainly used to invoke and control physical layer processes. The following primitives are exchanged through PM-SAP:

   PL-ME-SYNC
   PL-ME-SIG_STR
   PL-ME-TIME_ADJ

Prior to the use of a DECT physical channel, the receiver measures the strength of signals on that physical channel using the PL-ME_STR primitive. Using these signal measurements, the LLME produces two ordered lists: least interfered channels and channels with greatest field strength (PP only).

Physical Layer Procedures

- Addition of the synchronization field and transmission
- Packet reception and removal of synchronization field
- Signal strength measurement
- Synchronization pulse detection
- Timing adjustment
- Frequency adjustment

- Transmission and reception of the Z field
- Sliding collision detection

Medium Access Layer

The medium access layer specifies three groups of MAC services: the broadcast message control service (BMC), the connectionless message control service (CMC), and the multibearer control service (MBC). It also specifies the logical channels that are used by the MAC services and how they are multiplexed and mapped into the service data units that are interchanged with the physical layer.

The MAC layer controls the reception of data for short, half, and full slots, and the transmission of full slots by issuing primitives to the physical layer. Full slots are numbered $K = 0$–23, slot numbers (frame timing) are only defined in a special Q-channel message that is transmitted at a low rate by all FPs.

The MAC layer superimposes a multiframe structure on the TDMA frame. This is a time division multiplex of 16 frames. The multiframe numbering is defined the same way for FPs and PPs. Frame numbers (multiframe timing) are not included in a transmission and must be interpolated from a multiframe marker included in frame 8 in all FP transmissions. The MAC layer software should support MAC messages.

MAC Primitives

The MAC layer interfaces with the physical layer, the data link layer, and the LLME using four types of primitives: request (Req), indicate (Ind), response (Res), and confirm (Cfm). A Cfm primitive only occurs at a confirmation of an action initiated by a Req primitive. A Res primitive can only follow a Ind primitive.

Connection Oriented Primitives

MAC-CON: Connection setup
MAC-MOD: Connection modification
MAC-CO_DTR: CO data transmit ready
MAC-CO_DATA: CO data transfer
MAC-RES_DLC: Restart DLC
MAC-DIS: Connection release
MAC-BW: MAC bandwidth
MAC-ENC_KEY: Load encryption key
MAC-ENC_EKS: Enable/disable encryption

Connectionless and Broadcast Primitives

MAC-PAGE: Paging
MAC-DOWN_CON: Downlink connection
MAC-UP_CON: Uplink connection

Low-Level Management Entity

MAC-ME-CON: Connection setup
MAC-ME-CON_ALL: Connection setup allowed
MAC-ME-REL: Bearer release
MAC-ME-REL_REP: MBC release support
MAC-ME-RFP_PRELOAD: FP information preloading

MAC-ME-PT_PRELOAD: PT information preloading
MAC-ME-INFO: Systems information output
MAC-ME-EXT: Extended system info
MAC-ME-CHANMAP: Channel map
MAC-ME-STATUS: Status report
MAC-ME-ERROR: Error reports

MAC Procedures

Broadcast and Connectionless Procedures

- Downlink broadcast and connectionless procedures
- Uplink Connectionless procedures

Connection/Oriented Service Procedures

- C/O connection setup
- C/O connection modification
- C/O connection release
- C/O bearer setup
- C/O bearer handover
- C/O bearer release
- C/O data transfer
- MOD-2 protected I-channel operation
- Higher layer unprotected information (IN) and MAC error detection services (IP).

LLME Procedures

- Broadcasting
- Extended system information
- PP states and transitions
- Physical channel selection
- In-connection quality control
- Maximum allowed system load at RFPs
- PMID and FMID definitions
- RFP idle receiver scan sequence
- PT fast setup receiver scan sequence

Data Link Layer

As indicated previously, the data link control layer contains two independent planes of protocol: the C plane and the U plane. The C plane is the control plane of the DECT protocol stacks, and the U plane is the user plane of the DECT protocol stacks. This plane contains most of the end-to-end and (external) user information and control.

C Plane

The C plane is common to all applications and provides reliable links for the transmission of internal control signalling and limited user information traffic. It uses LAPC for full error control. The DLC C plane provides two independent services: the data link service (LAPC + Lc) and the broadcast service (Lb). Each of these services is completely independent and is accessed through independent SAPs.

C Plane Data Link Services

These services are provided by two protocol entities called LAPC and Lc. These entities separate the link access protocol functions from the lower link control functions. Each independent data link has an associated instance of these entities. The data link service is accessed via S-SAP.

C Plane Broadcast Service

This service contains only one instance of the Lb lower entity. This entity provides a restricted broadcast service in the down link direction and uses the dedicated MAC broadcast service. The broadcast service is accessed via B-SAP.

U-Plane Services

These services are application dependent. Each U plane is divided into an upper (LUx) entity and a lower (FBx) entity. The upper entity contains all of the procedures, and the lower entity buffers and fragments the U-plane frames from the MAC layer.

The following LUx members are defined by the protocol:

- LU1: Transparent Unprotected service (TRUP)
- U2: Frame relay service (FREL)
- LU3: Frame Switching service (FSW)
- LU4: Forward error correction service (FEC)
- LU5: Basic rate adaption service (BRAT)
- LU6: Secondary rate adaption service (SRAT)
- LU7-LU15: Reserved
- LU16: Escape for nonstandard family (ESC)

Lower Layer Management Entity

The LLME Provides coordination and control for the C-plane and U-plane processes. The LLME controls the routing of the C-plane and U-plane frames from the available MAC connections and controls the opening and closing (handover) of the MAC connections in response to service demands.

Data Link Control Primitives

These primitives describe the DLC interactions with other layers.

Primitives to the Network Layer via S-SAP

DL-ESTABLISH
DL-RELEASE
DL-DATA
DL-UNIT-DATA
DL-SUSPEND
DL-RESUME
DL-ENC_KEY
DL-ENCRYPT

Primitives to the Network Layer via B-SAP

DL-BROADCAST
DL-EXPEDITED

Network Layer

The DECT protocol specifies the C plane of the Network Layer. The C-plane contains all of the internal signalling information.

Entities

The network layer protocols are grouped as:

- Call control (CC) entity
- Supplementary services: call independent supplementary services (CISS) entity
- Connection oriented message service (COMS) entity
- Connectionless message service (CLMS) entity
- Mobility management (MM) entity
- Link control entity (LCE)

Call control (CC). This is the main service instance, and provides a set of procedures to establish, maintain, and release circuit switched services, as well as support for all call related signalling.

Supplementary services: CRSS and CISS. Supplementary services provide additional capabilities to be used with bearer services and teleservices. Two types of supplementary services are defined: call related supplementary services (CRSS) and call independent supplementary services (CISS). CRSS are explicitly associated with a single instance of a CC; CISS may refer to all CC instances.

Connection oriented message service (COMS). COMS offers point-to-point connection oriented packet service. This service only supports packet mode calls and offers a faster and simpler call establishment than the CC entity.

Connectionless message service (CLMS). CLMS offers a connectionless point-to-point or point-to-multipoint service.

Mobility manager (MM). This entity handles functions necessary for the secure provision of DECT services and supports, in particular, incoming calls. The MM procedures are described in seven groups:

- Identity procedures
- Authentication procedure
- Location procedure
- Access rights procedure
- Key allocation procedure
- Parameter retrieval procedure
- Ciphering related procedure

Link control entity (LCE). This is the lowest entity in the network layer. It performs the following tasks:

- Supervises the lower layer link states for every data link endpoint in the C plane
- Downlink routing
- Uplink routing
- Queuing of messages to all C-plane data link endpoints
- Creates and manages the LCD-REQUEST-PAGING messages (B-SAP)
- Queues and submits other messages to B-SAP
- Assigns new data link endpoint identifiers (DLEI)
- Assigns layer 2 instances to existing data link endpoints
- Reports data link failures to all layer 3 instances that are using the link

LLME

All of the network layer entities interface to the LLME which provides coordination of the operation between different network layer entities and also between the network and lower layers. The LLME interfaces with the physical, MAC, DLC, and NWK layers to provide procedures that concern more than one layer. Most of these procedures have only local significance.

LLME Physical Layer: PM-SAP

The primitives passed through PM-SAP are mainly used to invoke and control physical layer processes. The following primitives are exchanged through PM-SAP: PL-ME-SYNC, PL-ME-SIG_STR, and PL-ME-TIME_ADJ. Using these primitives the LLME implements procedures to produce: the list of quietest channels, the list of physical channels with greatest field strength (PP only), and timing information.

Prior to the use of a DECT physical channel, the receiver measures the strength of signals on that physical channel using the PL-ME_STR primitive. Using these signal measurements, the LLME produces two ordered lists: a list of least interfered channels and a list channels with greatest field strength (PP only). Using the PL-ME-SYNC primitives and interworking with higher layer detection of slot numbers, the LLME extracts timing information to establish the slot and frame timing.

MAC Layer

- Creation, maintenance and release of bearers, by activating and deactivating pairs of physical channels.
- Physical channels management, including the choice of free physical channels and the assessment of quality of received signals.

Connection and control.

MAC-ME-CON: Connection setup
MAC-ME-CON_ALL: Connection setup allowed
MAC-ME-REL: Bearer release
MAC-ME-REL_REP: MBC release support

System information and identities.

MAC-ME-RFP_PRELOAD: FP information preloading
MAC-ME-PT_PRELOAD: PT information preloading
MAC-ME-INFO: Systems information output
MAC-ME-EXT: Extended system information
MAC-ME-CHANMAP: Channel map
MAC-ME-STATUS: Status report
MAC-ME-ERROR: Error reports

Procedures.

- Broadcasting
- Extended system information
- PP states and transitions
- Physical channel selection
- In-connection quality control
- Maximum allowed system load at RFPs

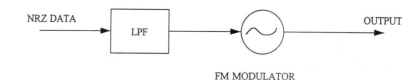

FM MODULATOR

FIGURE 30.10   Premodulation baseband-filtered MSK.

- PMID and FMID definitions
- RFP idle receiver scan sequence
- PT fast setup receiver scan sequence

Data Link Layer (DLC)

DLC covers routing of C-plane and U-plane data to suitable connections and connection management, which includes the establishment and release of connections in response to network layer demands. It also provides coordination and control for the C-plane and U-plane processes. The LLME controls the routing of the C-plane and U-plane frames from the available MAC connections and controls the opening and closing (handover) of the MAC connections in response to service demands.

Network Layer

The network layer provides service negotiation and mapping. All of the network layer entities interface to the LLME, which provides coordination of the operation between different network layer entities and also between the network and lower layers.

Modulation Method

The modulation method for DECT is Gaussian filtered frequency shift keying (GFSK) with a nominal deviation of 288 kHz [Madsen and Fague, 1993]. The BT, i.e., Gaussian filter bandwidth to bit ratio, is 0.5 and the bit rate is 1.152 Mb/s. Specification details can be obtained from the relevant ETSI documents listed in the reference section.

Digital transmission channels in the radio frequency bands, including the DECT systems, present serious problems of spectral congestion and introduce severe adjacent/cochannel interference problems. There were several schemes employed to alleviate these problems: new allocations at high frequencies, use of frequency-reuse techniques, efficient source encoding, and spectrally efficient modulation techniques.

Any communication system is governed by mainly two criteria, transmitted power and channel bandwidth. These two variables have to be exploited in an optimum manner in order to achieve maximum bandwidth efficiency, defined as the ratio of data rate to channel bandwidth (units of bits/Hz/s) [Pasupathy, 1979]. GMSK/GFSK has the properties of constant envelope, relatively narrow bandwidth, and coherent detection capability. Minimum shift keying (MSK) can be generated directly from FM, i.e., the output power spectrum of MSK can be created by using a premodulation low-pass filter. To ensure that the output power spectrum is constant, the low-pass filter should have a narrow bandwidth and sharp cutoff, low overshoot, and the filter output should have a phase shift $\pi/2$, which is useful for coherent detection of MSK; see Fig. 30.10.

Properties of GMSK satisfy all of these characteristics. We replace the low-pass filter with a premodulation Gaussian low-pass filter [Murota and Hirade, 1981]. As shown in Fig. 30.11, it is relatively simple to modulate the frequency of the VCO directly by the baseband Gaussian pulse stream, however, the difficulty lies in keeping the center frequency within the allowable value. This becomes more apparent when analog techniques are employed for generating such signals. A possible solution to this problem in the analog domain would be to use a phase-lock loop (PLL)

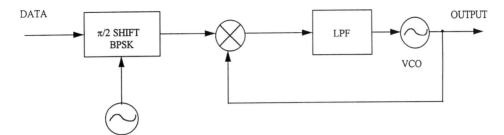

FIGURE 30.11 PLL-type GMSK modulator.

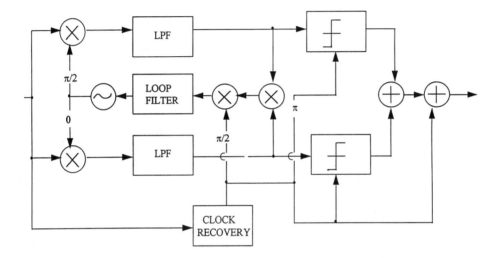

FIGURE 30.12 Costa Loop.

modulator with a precise transfer function. It is desirable these days to employ digital techniques, which are far more robust in meeting the requirements talked about earlier. This would suggest an orthogonal modulator with digital waveform generators [de Jager and Dekker, 1978].

The demodulator structure in a GMSK/GFSK system is centered around orthogonal coherent detection, the main issue being recovery of the reference carrier and timing. A typical method, is described in de Buda, 1972, where the reference carrier is recovered by dividing by four the sum of the two discrete frequencies contained in the frequency doubler output, and the timing is recovered directly from their difference. This method can also be considered to be equivalent to the Costas loop structure as shown in Fig. 30.12.

In the following are some theoretical and experimental representations of the modulation technique just described. Considerable literature is available on the subject of data and modulation schemes and the reader is advised to refer to Pasupathy, 1979, and Murota and Hirade, 1981, for further access to relevant study material.

Radio Frequency Architecture

We have discussed the need for low power consumption and low cost in designing cordless telephones. These days digital transmitter/single conversion receiver techniques are employed to provide highly accurate quadrature modulation formats and quadrature down conversion schemes that allow a great deal of flexibility to the baseband section. Generally, one would have used digital

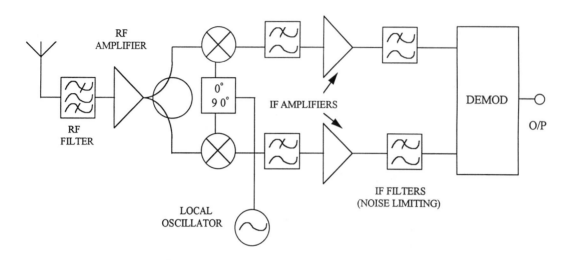

FIGURE 30.13   Direct conversion receiver architecture.

signal processors to perform most of the demodulation functions at the cost of high current consumption. With the advent of application specific signal processing, solutions with these techniques have become more attractive.

From a system perspective, range, multipath, and voice quality influence the design of a DECT phone. A high bit rate coupled with multipath reflections in an indoor environment makes DECT design a challenging task. The delay spread (multipath) can be anywhere in the 100–200 ns range, and a DECT bit time is 880 ns. Therefore, a potential delay spread due to multipath reflections is 1–20% of a bit time. Typically, antenna diversity is used to overcome such effects.

DECT employs a TDMA/TDD method for transmission, which simplifies the complexity of the radio frequency end. The transmitter is on for 380 ms or so. The receiver is also only on for a similar length of time.

A single conversion radio architecture requires fast synthesizer switching speed in order to transmit and receive on as many as 24 timeslots per frame. In this single conversion transmitter structure, the synthesizer has to make a large jump in frequency between transmitting and receiving, typically in the order of 110 MHz. For a DECT transceiver, the PLL synthesizer must have a wide tuning bandwidth at a high-frequency reference in addition to good noise performance and fast switching speed. The prescaler and PLL must consume as low a current as possible to preserve battery life.

In the receive mode the RF signal at the antenna is filtered with a low-loss antenna filter to reduce out-of-band interfering signals. This filter is also used on the transmit side to attenuate harmonics and reduce wideband noise. The signal is further filtered, shaped, and down converted as shown in the Fig. 30.13. The signal path really is no different from most receiver structures. The challenges lie in the implementation, and this area has become quite a competitive segment, especially in the semiconductor world.

The direct conversion receiver usually has an intermediate frequency nominally at zero frequency, hence the term zero IF. The effect of this is to fold the spectrum about zero frequency, which results in the signal occupying only one-half the bandwidth. The zero IF architecture possesses several advantages over the normal superheterodyne approach. First, selectivity requirements for the RF filter are greatly reduced due to the fact that the IF is at zero frequency and the image response is coincident with the wanted signal frequency. Second, the choice of zero frequency means that the

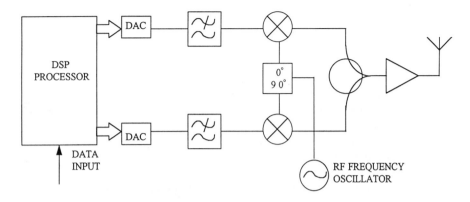

FIGURE 30.14   Transmit section.

bandwidth for the IF paths is only half the wanted signal bandwidth. Third, channel selectivity can be performed simply by a pair of low-bandwidth low-pass filters.

For the twin IF chains of a direct conversion receiver, automatic gain control (AGC) is always required due the fact that each IF channel can vary between zero and the envelope peak at much lower rates than the highest signal bandwidth frequency. An additional requirement in newer systems is received signal strength indication (RSSI) to measure the signal or interference level on any given channel.

Moving on to the transmitter architecture (shown in Fig. 30.14 is a typical I-Q system), it is safe to say that the task of generating an RF signal is much simpler than receiving it. A transmitter consists of three main components: a final frequency generator, a modulator, and the power amplifier. These components can all be combined in common circuits, i.e., frequency synthesizer with inbuilt modulator. The problem of generating a carrier at a high frequency is largely one of frequency control. The main approach for accurately generating an output frequency from a crystal reference today is the PLL, and there is considerable literature available on the subject [Gardner, 1979]. In the modulation stage, depending upon the tightness of the phase accuracy specification of a cordless system, it may be necessary to apply tight control on the modulation index to ensure that the phase path of the signal jumps exactly in 90° increments. Description of an I-Q modulator is also given in a previous chapter.

Defining Terms

AGC: Automatic gain control
ARQ: Automatic repeat request
AWGN: Additive white Gaussian noise
BABT: British approvals board for telecommunications
Base Station: The fixed radio component of a cordless link. This may be single-channel (for domestic) or multi-channel (for Telepoint and business)
BER: Bit error rate (or ratio)
CCITT: Comitè Consultatif International des Tèlègraphes et Tèlèphones, part of the ITU
CEPT: Conference of European Posts and Telecommunications Administrations
CPFSK: Continuous phase frequency shift keying
CPP: Cordless portable part, the cordless telephone handset carried by the user
CRC: Cyclic redundancy check
CT2: Second generation cordless telephone-digital
D Channel: Control and information data channel (16 kb/s in ISDN)

DCT: Digital cordless telephone
DECT: Digital European cordless telecommunications
DLC: Data link control layer, protocol layer in DECT
DSP: Digital signal processing
DTMF: Dual tone multiple frequency (audio tone signalling system)
ETSI: European Telecommunications Standards Institute
FDMA: Frequency division multiple access
FSK: Frequency shift keying
GMSK: Gaussian filtered minimum shift keying
ISDN: Integrated services digital network
ITU: International Telecommunications Union
MPT 1375:   U.K. standard for common air interface (CAI) digital cordless telephones
MSK: Minimum shift keying
PSK: Phase shift keying
RES 3:   Technical subcommittee, radio equipment and systems 3 of ETSI, responsible for the specification of DECT
RSSI: Received signal strength indication
SAW: Surface acoustic wave
TDD: Time division duplex
TDMA: Time division multiple access

References

Cheer, A.P. 1985. Architectures for digitally implemented radios, IEE Colloquium on Digitally Implemented Radios, London.

Comité Consultatif International des Télégraphes et Téléphones. 1984, "32 kbits/sec Adaptive Differential Pulse Code Modulation (ADPCM)," CCITT Red Book, Fascicle III.3, Rec G721.

Comité Consultatif International des Télégraphes et Téléphones. 1984–1985. *General Characteristics of International Telephone Connections and Circuits*, CCITT Red Book, Vol. 3, Facicle III.1, Rec. G101–G181.

de Buda, R. 1972. Coherent demodulation of frequency shifting with low deviation ratio, *IEEE Trans.* COM-20 (June):466–470.

de Jager, F. and Dekker, C.B. 1978. Tamed frequency modulation. A novel method to achieve spectrum ecomony in digital transmission, *IEEE Trans. in Comm.* COM-20 (May):534–542.

Dijkstra, S. and Owen, F. 1994. The case for DECT, *Mobile Comms Int.* :60–65.

European Telecommunications Standards Inst. 1992. RES-3 DECT Ref. Doc. ETS 300 175-1 (Overview). Oct. ETSI Secretariat, Sophia Antipolis Ceder, France.

European Telecommunications Standards Inst. 1992. RES-3 DECT Ref. Doc. ETS 300 175-2 (Physical Layer) Oct. ETSI Secretariat, Sophia Antipolis Ceder, France.

European Telecommunications Standards Inst. 1992. RES-3 DECT Ref. Doc. ETS 300 175-3 (MAC Layer) Oct. ETSI Secretariat, Sophia Antipolis Ceder, France.

European Telecommunications Standards Inst. 1992. RES-3 DECT Ref. Doc. ETS 300 175-4 (Data Link Control Layer) Oct. ETSI Secretariat, Sophia Antipolis Ceder, France.

European Telecommunications Standards Inst. 1992. RES-3 DECT Ref. Doc. ETS 300 175-5 (Network Layer) Oct. ETSI Secretariat, Sophia Antipolis Ceder, France.

European Telecommunications Standards Inst. 1992. RES-3 DECT Ref. Doc. ETS 300 175-6 (Identities and Addressing) Oct. ETSI Secretariat, Sophia Antipolis Ceder, France.

European Telecommunications Standards Inst. 1992. RES-3 DECT Ref. Doc. ETS 300 175-7 (Security Features) Oct. ETSI Secretariat, Sophia Antipolis Ceder, France.

European Telecommunications Standards Inst. 1992. RES-3 DECT Ref. Doc. ETS 300 175-8 (Speech Coding & Transmission) Oct. ETSI Secretariat, Sophia Antipolis Ceder, France.

European Telecommunications Standards Inst. 1992. RES-3 DECT Ref. Doc: ETS 300 175-9 (Public Access Profile) Oct. ETSI Secretariat, Sophia Antipolis Ceder, France.

European Telecommunications Standards Inst. 1992. RES-3 DECT Ref. Doc. ETS 300 176 (Approval Test Spec) Oct. ETSI Secretariat, Sophia Antipolis Ceder, France.

Gardner, F.M. 1979. *Phase Lock Techniques*, Wiley-Interscience, New York.

Madsen, B. and Fague, D. 1993. Radios for the future: designing for DECT. *RF Design*. (April):48–54.

Murota, K. and Hirade, K. 1981,GMSK modulation for digital mobile telephony, *IEEE Trans.*, COM-29(7), 1044–1050.

Olander, P. 1994. DECT a powerful standard for multiple applications, *Mobile Comms Int.*. 14–16.

O'Shaughnessy, D. 1987. *Speech Communication*, Addison-Wesley. Reading, MA.

Pasupathy, S. 1979. Minimum shift keying: spectrally efficient modulation. *IEEE Comm. Soc. Mag.* 17(4):14–22.

Tuttlebee, W.H.W., ed., 1995. *Cordless Telecommunications Worldwide*, Springer–Verlag.

Weinstein, S.B. 1977. Echo cancellation in the telephone network, *IEEE Comm. Soc. Mag.* 15(1):9–15.

31

The RACE Program

Stanley Chia
BT Laboratories

31.1 Introduction

The objective of the European Union's program for research and development in advanced communications technologies in Europe (RACE) is to enable the development of advanced techniques and technologies that will facilitate the introduction of innovative telecommunications services and improve quality and cost effectiveness of traditional ones. They must meet the significant demand for fast data transmission between computers and contend with the rapid growth in visual communications, thus offering the capabilities of telecommunications to a wider community with better adaptation to human communication needs. With the collaborative research activities being prenormative and precompetitive, the aim is to create a series of common architectures, open interfaces, and standardized protocols in order to ensure system interoperability.

The RACE program started in 1987 as a ten-year program to be carried out in two phases, RACE I and RACE II, preceded by a definition phase. The main goal of the program is to contribute to the introduction of **integrated broadband communications** (IBC) into community-wide services, taking into account the evolving integrated service digital network and national introduction strategies. During the progress of the program, it becomes obvious that telecommunications and advanced information services will play a key role in world socio-economic development as we approach the 21st century. This offers tremendous market opportunities for new services and infrastructure implementation, and, thus, accelerates the maturity of the integrated broadband communication concept and technologies. It is also noted that an underlying objective of RACE is to "promote the competitiveness of the Community's telecommunications industry, operators and service providers in order to make available to final users the services which will sustain the competitiveness of the European economy and contribute to maintaining and creating employment in Europe." The program has, indeed, created large-scale awareness of the market opportunities within the industry that will accompany the implementation of the next generation of telecommunications services and systems in Europe. In addition, it is contributing to the creation of awareness among users of the advantage that application of advanced communications will deliver.

Up to now mobile services are divided essentially into four groups, cellular, paging, cordless, and private mobile radio, each providing services using a number of different networks and incompatible systems. The standardization of systems, interfaces to users and service providers, and the decreasing cost of terminals and networks have created economies of scale and high-usage levels. This rapid market penetration, however, will lead to capacity shortage by the turn of the century. A third generation mobile system is envisaged by the European Union to accommodate the requirements of higher capacity, service integration, and universal personal telecommunications. This third generation mobile system will provide seamless connections between mobile and fixed network environments. A prerequisite for this is a flexible air interface that can adapt to meet varying service and operational requirements as may be found within domestic and business customer premises networks and public networks. In the mid-1980s work began internationally, in the International Telecommunication Union-Radio (ITU-R, previously CCIR) Interim Working Party 8/13 (now Task Group 8/1), on systems that would supersede existing cellular and cordless systems. In Europe, the work of the RACE mobile definition phase project and then the full mobile project in the RACE I program laid conceptual foundations for a **universal mobile telecommunications system** (UMTS). Subsequently, this evolved into a major project line in the RACE II program and has since begun to be standardized within the European Telecommunication Standard Institute (ETSI).

31.2 RACE I Mobile Project

The RACE I program is designed to evaluate the options and clear the path for standardization. Specifically, the RACE I mobile project has contributed to the development of third generation mobile telecommunication systems, conceived as the universal mobile telecommunication system and mobile broadband system (MBS), intended to realize true personal mobile radio communications, at the least, from anywhere within Europe, and allow people to communicate freely with each other in domestic or business environments, urban or rural settings, and fixed locations or roving vehicles; see Fig 31.1. The project has paved the way for the creation of the ETSI subtechnical committee SMG-5 charged to provide standardization for the UMTS. The project has also made significant

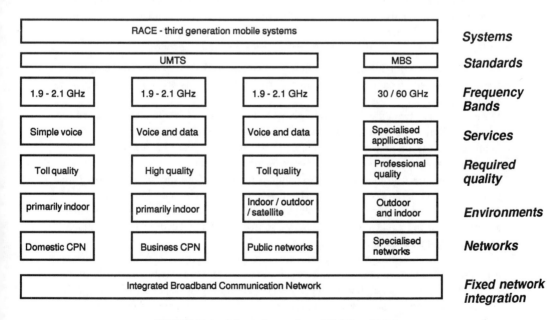

FIGURE 31.1   Schematic overview of UMTS and MBS.

contributions towards the allocation of the radio spectrum requirements for future public land mobile telecommunications systems. It was on the basis of the RACE work that the **WARC'92** decided to allocate 230 MHz in the 2-GHz band that will replace the current pan-European Global System for Mobile Communication (GSM) system by the early part of the 21st century. In addition, the RACE I project also provided the technologies and service foundations for mobile broadband systems operating in the vicinity of 60 GHz.

The UMTS is designed to be a universal multiservice, multifunction digital system, evolving from currently operational or planned second generation systems. The provision of user data rate of up to 2 Mb/s will be achieved through the UMTS in the 2-GHz frequency band and will allow seamless interface between the mobile and the fixed network. This will also allow integrated service digital network (ISDN) services in the fixed network to be extended into the mobile domain. For user data rates in excess of 2 Mb/s, the use of the 60-GHz band will be required to provide sufficient bandwidth and to share the broadband transport infrastructure and functionality directly. In brief, the RACE I mobile project has made significant contributions to the creation of a system description document for the third generation mobile system, the definition of a flexible air interface, and the key functions of the fixed network, supporting the allocation of suitable frequencies and establishing common functional specifications for the UMTS as the basis for the contribution to standards. Indeed, the project has provided a firm foundation for the second phase of the project: RACE II.

31.3 RACE II Mobile Project Line

The RACE II program is structured into project lines containing groups of related projects in specific subject areas. For mobile applications, the mobile and personal communications project line is directly related to the implementation of third generation wireless communications. The European Commission recognized that "recent market developments and initiatives taking place in Japan and North America, indicate that in order to meet the challenges of globalization and international competition, very significant efforts should be devoted in particular to the development of mobile/personal communications networks, systems, products and services, as a means of capturing a meaningful percentage of world market." This provided a strong impetus to the European Union to direct the development not only of a Pan-European market but also of the world markets. The primary objective of the project line, particularly for UMTS, is the contribution to standardization activities. This emphasis is justified by the fact that mobile standards are becoming increasingly complex and can only be achieved efficiently through close collaboration of operators and manufacturers. Activities in support of standards development are as follows:

1. The study of network architecture, signalling protocol, network management, mobility management, radio and fixed network resource management, security aspects, and billing and accounting.
2. The evaluation of two different multiple access techniques for the radio interface based on novel spread spectrum techniques and the enhancement of the relatively mature time division multiple access technology through computer simulation and real-time testbeds with a set of common assumptions and performance assessment scenarios.
3. The study of the integration of UMTS satellite components into the fixed infrastructure, including radio interface compatibility as well as signalling and interworking issues.
4. The research and development of technologies, models, and tools in areas including adaptive antenna design and applications, modulation, source and channel coding, terminal design, and radio-wave propagation characteristics.
5. The study of the marketing aspects, service creation, quality of service, service requirements, evolution and implementation strategies, operational and functional requirements in anticipation of the increasing number and types of services currently envisaged for UMTS, and the reduction of transmission rates due to more efficient source coding and data compression techniques.

The time frame for the UMTS activities is designed to be compatible with the expected availability of a standard around 1998, which will allow the commercial exploitation to take place by the turn of the century. By contrast, the standardization of mobile broadband systems is likely to take place over a longer period of time and the current activities are focused on the development of the enabling technologies and the evaluation of the economic feasibility of mobile broadband services.

The mobile and personal communication project line is divided into a number of projects in order to evaluate the market and technical requirements, as well as the physical implementation aspects, of the UMTS. Specifically fixed network issues are studied within the MONET project, whereas multiple access techniques are addressed by the CODIT and ATDMA projects. The latter two projects investigate the relative merits of spread spectrum multiple access and time division multiple access (TDMA), respectively. The issue of integrating satellite mobile components of the UMTS into the terrestrial network is addressed by SIANT. This project will interact with MONET on the fixed network aspect and CODIT and ATDMA on air interface definition issues. The MAVT project is aimed at developing a mobile audio-visual terminal demonstrator to support bandwidth efficient video services over mobile radio channels. It evaluates different source and channel coding techniques for low-bit-rate image and speech transmission for future mobile terminals. The MBS project, which studies the requirement of supporting high-bit-rate services using millimeter-wave technology, will have strong interaction with the MONET project on network architecture and evolution, as well as CODIT and ATDMA projects on air interface and cellular coverage definition. The project TSUNAMI, which researches into smart antenna technologies for universal advanced mobile infrastructure, is considering the benefits that adaptive antenna technology can bring to both air interface techniques, as well as developing appropriate component technologies. Finally, the GIRAFE project is expected to have close interaction with MAVT, CODIT, ATDMA, and SIANT on terminal technology issues. In the following the key project areas are described in more details.

Code Division Multiple Access

The overall objective of the CODIT project is to explore the potential of code division multiple access for a future high-capacity UMTS. A system concept based on CDMA is established, together with advanced radio frequency technologies and cellular architecture and techniques. A system demonstrator comprising test mobile stations, radio base stations, and a radio network controller is to be designed and built in order to validate the system concept through laboratory bench test, field trials, and computer simulation.

The CODIT system is proposed to be able to handle a large number of users in pico-, micro-, and macro-cells with simple spectrum management, which would accommodate multiple operators as well as private networks. Operating environments will be for both indoor and outdoor with a high grade of service. Low-power terminals will be supported for high-quality voice services. It will also be able to support variable data rates for advanced data service and will be compatible with existing data networks.

It is noted that UMTS will need to accommodate different bit rates with instantaneous data rates as high as 2 Mb/s. Following the evaluation of different spread spectrum multiple access schemes, a direct-sequence CDMA (DS-CDMA) scheme with three different chip rates is adopted. The CODIT system will, therefore, be able to support three different bandwidths of approximately 1, 5, and 20 MHz. For each bandwidth, there is an associated chip rate. The radio access scheme, which is based on direct sequence code division multiple access, allows flexibility when choosing the appropriate coding of information at any given time instant. During the course of a frame, which has a duration of 10 ms, the coding is not changed. The selected parameters, however, can change from frame to frame, which is utilized by the speech codec for adaptation. The speech codec, which is based on a codebook excited linear prediction algorithm, can adaptively vary its rate from 16 kb/s down to 4 kb/s. In order to minimize the requirements of synchronization and avoid code planning, user information is spread with pseudonoise (PN) sequences. Only one long PN sequence is adopted, and each link is assigned a unique phase shift. In addition, the system is unsynchronized in the sense that mobile stations connected to a base station are not synchronized, and base stations are not synchronized among themselves. An exception is during soft handover when synchronous reception at the mobile station is required to combine two signals from two different base stations arriving within the time window of the Rake receiver of a mobile station. With measurements from the mobile station, the time of transmission from the base stations is adjusted on a call-by-call basis in order to achieve synchronous reception at the former.

A computer simulation of the system in accordance with the proposed air interface has been set up and is used to verify the specification and produce radio interface performance. Simulation results have shown that system capacity is dependent on the environment in which the system operates and is typically lower in picocellular and microcellular than in macrocellular environments. It is, therefore, important that capacity is to be boosted in microcell and picocell environments in order to ensure that DS-CDMA is to be an all round solution for UMTS service provision. This can be achieved by the use of antenna diversity, macrodiversity techniques, and the use of special spreading codes. Furthermore, the positioning of base stations for a DS-CDMA system could affect capacity. The reduction in cellular system capacity from hexagonally distributed base stations to randomly distributed base stations is about 15%. Macrodiversity, however, can be used to improve both uplink and downlink capacity, improve signal quality, and enable a make-before-break soft handover routine at the expense of more modems in the base station and more infrastructure.

Advanced Time Division Multiple Access (ATDMA)

The main objective of the ATDMA project is to develop and quantify the potential of a time division multiple access (TDMA) system for UMTS. Given that the maximum capabilities of TDMA are far from being attained by present systems (e.g., GSM and DECT), the target of the project is to develop a set of techniques that can improve the overall system capacity by an order of magnitude, while at the same time improving the quality of service, grade of service, and range of services. The goals of the ATDMA project are to compare the performance of an improved TDMA system with other radio access techniques and to evaluate the system architecture developed and implemented in the testbed. The real-time testbed is intended to provide a signal processing platform for the implementation of transport techniques and a protocol test platform for the control techniques, where the latter is aimed at assessing the effectiveness of the link adaptation concept.

The project is structured into four technical areas: dealing with techniques studies, channel characterization, testbed implementation, and system performance evaluation. System evaluation will be performed using both real-time testbed and computer simulations. The real-time system demonstrator comprising two mobile stations, two radio base stations, a network emulator, and a radio channel simulator will be implemented to evaluate the performance of the radio frequency subsystem, transport and low-level control subsystem. This testbed will support a range of service demonstrations, including advanced video and voice coding technologies. The services will operate under different environments taking into account the needs of multiple operators and low-cost terminals. To achieve this, the radio control issues, such as channel allocation strategies, resource reservation techniques, signalling channel assignments, and response time of resource management mechanisms, are to be optimized in accordance with allowable interference and noise levels, transmission bandwidth, modulation schemes, forward error correction codes, robustness of the source, and channel coding, etc.

The impact of duplexing on the division of the UMTS frequency band and the system complexity and flexibility were examined resulting in the conclusion that both time and frequency division duplex must be employed. The transport interface will support two carrier symbol rates of 450 kb/s for macrocells and 1.8 Mb/s for both microcells and picocells. In addition, it will support the following: a common modulation scheme for all environments, a single demodulator-channel estimator, a prefilter equalizer, flexible burst and frame structures, a speech codec with gross rate of 13 kb/s, including error protection and net speech rate 6.4 and 9.6 kb/s, and adaptive error protection schemes.

The key to the capacity gain is by the dynamic utilization of radio resource so as to optimize the system performance from a global perspective. This will be realised via the combined use of packet access and a generic air interface that supports three cell types, picocell, microcell, and macrocell, and incorporates static and dynamic adaptation strategies. Specifically, static adaptation will enable system parameter changes to occur on a medium-to-long term basis. This form of adaptation will be used at call setup and during interoperator or intercell type handover to set the air interface's initial conditions. Dynamic adaptation will be performed during each call in order to cope with the time-varying radio-wave propagation characteristics of an environment and the traffic loading on the terminal and the base station. From the transport and control plane point of view, two adaptation mechanisms can be differentiated. The transport plane consists of transport techniques (including modulation, equalization, burst and frame structure, etc.) and error protection schemes (including automatic repeat request, forward error correction, etc.) which could be modified dynamically. Evidently, the forward error correction code rate, slot allocation, and interleaving depth could be mutually interchanged to minimize the required average resource allocation while still maintaining a rated quality of service and the need to minimize the overall spectrum demand. The dynamic adaptation can also be realised in the control plane. This includes the use of adaptive power control, handover algorithm optimization, packet transmission with bandwidth allocation on demand for mixed voice and data services, and link adaptation, which dynamically adapts the parameters of the air interface during a call in order to cope with varying propagation, interference, quality of service, and traffic loading conditions. Thus, if the rated quality of service objectives can be relaxed during overloading conditions, then this mechanism can offer a form of soft capacity.

Mobile Network

The main aim of the MONET project is to develop network standards for UMTS that will integrate the infrastructure for mobile and fixed communications, as well as offer the same range of services as provided by the fixed networks. The underlying challenge is to define a fixed infrastructure capable of supporting a huge volume of mobile connected traffic over a wide geographical area. To achieve this, it is essential that new concepts for handover, call handling, location management, security, telecommunication management, and database and base station interconnection are developed in order to minimize the signalling load due to the mobility of the user. Furthermore, the UMTS network architecture has to be aligned with the integrated broadband communication network so that intelligent network (IN) and universal personal telecommunications features can be exploited.

The design of a network to support a distributed system like UMTS is a complex undertaking. The design methodology is based on the ITU-T (previously CCITT) three-stage methodology. As the starting pointing for the construction of the functional architecture and network architecture, functional models of the most prominent mobility procedures (e.g., location management, handover, and call handling) are analyzed. A functional architecture will be defined covering the domestic, business, and public environments. The mapping of functional entities to the network elements may be different for different types of environment. A broad range of possibilities for environments exists ranging from a single location area to more advanced ones comprising several location areas and paging areas. Other possible arrangements may depend on the size of cells and the expected traffic volumes. A network architecture will be defined for the domestic, business and public environments. The network architecture identifies the physical entities of the system and the physical interfaces between them; only interfaces that are relevant for standardization processes are identified. The functional model will be used for the functional descriptions of the mobility procedures. These detailed descriptions will allow the precise definition of the requirements on the functional entities of the model in terms of capabilities.

The UMTS protocol framework defines the overall structure of the application layer for UMTS signalling protocols. It contains a list of all application layer signalling protocols to be used and the application entities used to host the protocols. The design of UMTS application layer signalling protocols follows a number of new paths. Two new features are protocol reuse and protocol flexibility. Current signalling protocols have been designed for specific systems with little or no consideration given to the reuse of the resulting protocols in other systems. In addition, current signalling protocols are very tightly coupled to a single physical network topology. In UMTS, however, potentially very different topologies could be used in different environments. Thus, protocols must not be optimized for a single topology nor for a single allocation of functions. To facilitate the flexible reuse of protocols between different systems, a clear separation between the system functionality and the supporting protocol must be made. Thus, a protocol should only reflect the communication aspects of the functionality it supports. As a consequence, system specification should only indicate how relevant protocols are to be used whereas the protocol specification itself is provided separately.

A major goal with respect to the implementation of UMTS is to allow its integration into networks for fixed communication. It is because UMTS and B-ISDN are likely to mature in the same time frame that integration of UMTS into B-ISDN is logical. The B-ISDN is expected to be present in major parts of the world during the introduction of UMTS to the market. This integration scenario, thus, underlines the idea of UMTS to be a world standard: UMTS will be the wireless access for the B-ISDN. For techno-economical reasons, it is also considered important to allow integration of UMTS into networks for fixed communication and to reuse existing and forthcoming infrastructure as much as possible. It also enables UMTS to progress rapidly such that it can exploit the advanced signalling possibilities and services of the B-ISDN. In addition the B-ISDN protocols and functions will be used where possible complete with UMTS specific protocols and functions. Regarding the available bandwidth, B-ISDN will not impose limitations on the service provisioning by UMTS. As far as the integration process is concerned, promising developments in the intelligent network arena could facilitate a convenient route. As a word of caution, although there is a strong emphasis on integration aspects in the UMTS requirements, it still might require a fall-back option to implement UMTS as a stand-alone system (especially in areas where no B-ISDN is available). This scenario requires the UMTS to be able to operate optionally with no constraints from existing or forthcoming systems. The result is that both the radio access part and the fixed part of the network are UMTS specific, together with new UMTS specific network components, such as base stations, switches, and control equipment.

The intelligent network concept plays an important role in existing and future telecommunication networks, including mobile systems, in Europe. Originally, the IN was conceived as a means for service providers to provide rapid creation and introduction of new services upon existing fixed networks. For each mobility procedure, a part of the functions can be implemented in IN whereas the remaining part is implemented in the backbone network. In ETSI IN capability set one, service control and data functions are centralized. The need for future development of IN towards wider service coverage with interconnection of different IN networks has already been recognized and this requires decentralized service control. The additional signalling efficiency requirements of handover, and the structure of distributed databases associated with UMTS also reinforced the need for distribution of service control and data functions. Up to now IN has only addressed services that are applied to calls that are in the process of setting up, in-progress, or clearing down. Hence there is great emphasis on the basic call state model in the switch. Mobility procedures, however, like location updating can occur while there is no active call. Developments are necessary in order to manage these noncall related functions. In the current IN, only one service at a time can be activated from the basic call. In a mobile system, however, several services and mobility procedures can be active in parallel. For example, a location updating or a handover might occur during invocation of another IN service. To support this, a more advanced mechanism for interaction between basic call control and service control is required.

Security is an indispensable feature of any wireless communication system. The development of a security architecture for UMTS has taken into account two main factors: The first is that security requirements, which have an equivalent in existing systems, can be fulfilled in a novel and more efficient way due to the advances in the field of data and network security and the supporting technology. It is now possible to use a wider range of security mechanisms in order to realize the basic security services and the associated key management. The second factor is that UMTS offers new features, which distinguish it from existing systems. These new features also lead to new security requirements and demand new security solutions. It is generally assumed that a UMTS user is represented by a smart card in security procedures. This smart card plays a role in most of the user related features and is called a subscription identification device. These user related features are 1) multiuser terminals, 2) direct support to UPT, 3) new forms of payment, 4) user anonymity, 5) incontestable charging, 6) detection of modification, 7) authentication of a service provider by the user, and 8) authentication of a network operator by the user. Network related features are 1) authentication during handover, 2) incontestable charging between network operator and service provider, 3) restricted availability of authentication information to visited network operator, 4) security in the fixed network, and 5) mobile customer premises network (CPN).

Subscriber accounting can be divided into three subprocesses: usage metering, charging, and billing. Metering of the use of public UMTS resources is done in the originating and terminating public UMTS. The charges for a UMTS call are based on an origin–destination relationship, not on the actual route used. All public UMTS users associated with a subscriber have a home operator. Since the originating network knows the terminating network, the charges for the first part of the connection can be deduced from a so-called mobile originating usage record. As the called party might be roaming, however, the charges for the last part of the connection are to be deduced from a so-called mobile terminating usage record. Furthermore, as part of the charging process is responsible for collecting usage records, all usage records of a particular call that are registered by visited public UMTS operators must eventually be transferred to the home UMTS operator of the involved users. This is done only in the case of billing the subscriber. A UMTS subscriber shall be billed by his home operator only. It is preferred that all UMTS services used are specified in one bill. Call charges are billed either to the home operator of a subscriber or directly to the user.

Interoperator accounting again is proposed to follow the three stages of usage metering, charging, and billing. Unlike subscriber accounting, usage metering information is to be based on the actual route of a call. Interoperator accounting must be verifiable. This implies that when traffic over a given connection or link between two operators is to be charged, usage metering needs to be done on both outgoing and incoming traffic. Since part of the charging process is responsible for collecting usage records, all usage records that are registered by visited UMTS operators must eventually be sent to the home UMTS operators or to a clearinghouse. The billing process compiles the usage records and clears with other operators. This can be done directly among operators or through a clearinghouse.

Mobile Audio-Visual Terminal

The main objective of the MAVT project is to develop video and audio coding schemes for the transmission of multimedia services in a mobile environment taking into account user and service requirements, network and channel characteristics, as well as terminal architecture. A demonstrator will be implemented with low-bit-rate video and audio coding algorithms in accordance with the ISO MPEG4 coding standard. The demonstrator will be interfaced with the CODIT and ATDMA testbed for performance evaluation.

Following the development of the $p \times 8$ kb/s source coding scheme and the low-bit-rate audio algorithm, the possibility of transmission of video data via mobile radio channels has been confirmed. New algorithms based on current standards [ITU-T Rec H.61—Codec for audio visual services at $n \times 384$ kbits, the International Standards Organization, Moving Picture Experts Group (MPEG),

International Standards Organization, Joint Photographic Experts Group (JPEG)] are developed by optimizing existing low-bit-rate coding schemes in terms of video and audio coding delay. Rate compatible punctured convolutional codes, together with schemes that enable flexible exchange between source and channel data rate, as well as combined source/channel coding methods are evaluated.

Mobile Broadband System

The MBS project addresses the system concepts, techniques, and technologies required for the realization of a mobile broadband system. It also identifies the potential market and economical issues relating to the widespread introduction of the corresponding systems and services. The project also aims at demonstrating the industrial capability to produce the subsystems required by future high-data rate (154 Mb/s) mobile communication systems in a cost-effective manner. In addition, the system aspects, radio access schemes, network management issues, integration with Integrated Broadband Communications Networks (IBCN), broadband wireless local area networks (LANs) and multimedia applications are to be studied. It is clear that MBS will be a mobile extension to the B-ISDN and that it will encompass services to be provided by other systems such as HIPERLAN (High performence radio local area network). Even though MBS will be able to support low-bit-rate services, it will not be a replacement of UMTS but an enhancement, providing its applications mainly to the professional users.

From the study, it was identified that forward mobile controlled handover is the most appropriate scheme to support continuous mobility for mobile broadband services. Studies on antenna configuration suitable for elongated street cells have indicated the use of a single radiator solution based on a dielectric antenna structure as the best solution. The use of adaptive beam forming antennas in order to improve spectrum efficiency and transmission quality is potentially feasible.

Smart Antenna Technology for Advanced Mobile Infrastructure

The TSUNAMI project addresses the development of adaptive antenna component technologies at frequencies and bandwidths appropriate for UMTS. Components of interest include array antennas, radio frequency, intermediate frequency and digital beam forming networks, digital signal processor, and efficient adaptive array control algorithms. The improvement of mobile system performance through the increase in antenna directivity will also be addressed.

Adaptive antenna component specifications and prototypes are generated using a top down approach. The primary focus will be the use of adaptive antennas at the base stations. This will consist of a balanced combination of requirement definition, traffic analysis, propagation analysis, and measurement and performance evaluation using field trials and computer simulations. Radio planning and traffic engineering models will be developed based on the coverage characteristics of adaptive base station antennas.

Integration of Satellite in Future Mobile Network

The SIANT project is designed for the evaluation and identification of the requirements for the integration of satellites into the UMTS. The project is charged to optimize the whole mobile network taking into account the terrestrial backbone network defined by the MONET project. The project is further responsible for providing a set of recommendations on service aspects, and radio aspects, network aspects, and security aspects for the satellite integration of the UMTS.

The project sets out to define mobile services to be offered by considering terrestrial UMTS and MBS services. The result is a set of operational and functional requirements satisfying user needs, as well as network interoperability constraints. The definition of the air interface will take into account the features derived either from ATDMA or CODIT projects. Key issues to be addressed are the requirements and limitations for the integration of satellite systems, optimization of radio resource management, orbit selection and satellite architecture, technology aspects for the terminals, and the air interface definition.

Radio Front End

The GIRAFE project is to investigate the application of microelectronics integration and packaging techniques to the radio front end for mobile and wireless telephones of different standards at 1.5–2.3 GHz by creating a library of basic radio frequency building blocks, including transmit mixer, receive mixer, phase shifter, voltage control oscillator, and fully integrated phase lock loop in a high-frequency silicon bipolar process. It will also investigate and develop novel techniques that will have a major impact on low-cost, high-volume packaging and external passive inductive components for radio frequency integrated front end applications.

Key issues to be addressed are 1) multisystem requirement specification, 2) low-voltage, low-power radio frequency designs, 3) low-cost external planar passive inductive components for matching and filtering, and 4) low-cost, high-volume radio frequency packaging techniques. The goals are to compile a set of common functional specifications for several mobile and wireless standards and to develop a library of low-voltage, low-power, highly integrable basic radio frequency building blocks. In addition it is also expected to develop external radio frequency planar inductive passive components and low-cost, high-volume radio frequency packaging techniques.

31.4 Beyond the RACE Program

It is recognized by the European Union that there is a continuing need for communication research and development in Europe. Future work will build on the achievements of the RACE program and contribute further to the success of European activities in the area. Based on the view of the RACE management committee and other expressed opinions of leading experts in the field, a new program is considered by the European Union to follow on from RACE, taking into account the changed situation in view of the contribution from RACE. This new program is entitled research and development in advanced communications technologies and services (ACTS) [Dasilva and Fernandes, 1995]. This will benefit the well-established practice of collaboration on a European level. An important distinction between ACTS and RACE is that the former focuses on operational trials and the rationale for performing all of the research and development, whereas the latter focuses on technology and service development. The ACTS program is budgeted for resource of in excess of 100 million ECU, thereby giving a new impetus to the further development of mobile and personal communication systems.

Defining Terms

Universal mobile telecommunications system: The European version of the third generation mobile system that is being standardized by the European Telecommunications Standard Institution.

WARC'92: The World Administration Radio Congress, which identified the radio spectrum for future public land mobile telecommunications systems internationally in 1992.

Integrated broadband communications: A global concept that covers all kinds of communications and technical and operational means to offer services. Its related infrastructure and services will be offered by network operators and service providers using mobile, terrestrial, broadcast, and satellite transmissions, with a range of equipment from different manufacturers adapted for business and civilian application.

References

Commission of the European Communities. 1994. Research and technology development in advanced communications technologies in Europe-RACE 1994. Directorate General XIII. Annual Rept. Feb. Brussels, Belgium.

Commission of the European Communities. 1994. Research and technology development in advanced communications technologies in Europe-RACE 1994. Sec. 2.1, p. 3. The RACE Programme-Objectives.

Commission of the European Communities. 1994. Research and technology development in advanced communications in Europe-RACE 1994. Sec. 6.3.2.1, p. 43. "Project Line 3-Mobile and Personal Communications Overview of Project Line, Introduction."

Commission of the European Communities. 1994. Research and technology development in advanced communications technologies in Europe-RACE 1994. Sec.2.1, p. 3. The RACE Programme–Objectives.

Commission of the European Communities. 1994. *Proceeding of RACE Mobile Telecommunications Workshop*, Amsterdam. Directorate General XIII. Brussels. Vol. 1 and 2, May.

Commission of the European Communities. 1993. *Proceeding of RACE Mobile Telecommunications Workshop*, Metz, France. Directorate General XIII. Brussels. June.

Chia, S. 1992. The universal mobile telecommunication system. *IEEE Comm. Mag.* Dec., 30(2):54–62.

Chia, S.T.S. and Grillo, D. 1992. UMTS—mobile communications beyond the year 2000: requirements, architecture and system options. *IEE Electronics and Comm. Eng. J.* Oct., 4(5):331–340.

Hsing, T.R., Chen, C.T., and Bellisio, J.A. 1995. Video communications and services in the copper loop. *IEEE Com. Mag.* Jan., 31(1):62–68.

Dasilva, J. and Fernandes, B. E. 1995. The European research programme for advanced mobile systems. *IEEE Personal Comm. Mag.* Feb., 2(1):14–19.

1993. Special issue on the European path toward UMTS, *IEEE Personal Comm. Mag.* Feb., 2(1).

Further Information

The *IEE Electronics Communication Engineering Journal* frequently publishes review papers submitted by the RACE projects. These papers are normally identified by RACE section in each issue. Another active forum for detailed information of the RACE projects is the Subtechnical Committee, Special Mobile Group 5 of the European Telecommunications Standards Institute (ETSI), which is charged to study universal mobile telecommunications system matters. This body meets quarterly with a significant amount of contributions submitted by RACE participants as temporary documents to support standardization. These temporary documents can be obtained by applying directly to the ETSI. Finally, the European Union Directorate General XIII RACE Office in Brussels publishes annual reports and workshop digest on all RACE projects.

32

Half-Rate Standards

Wai-Yip Chan
Illinois Institute of Technology

Ira Gerson
*Motorola Corporate Systems
Research Laboratories*

Toshio Miki
*NTT Mobile Communication
Network, Inc.*

32.1 Introduction

A half-rate speech coding standard specifies a procedure for digital transmission of speech signals in a digital cellular radio system. The speech processing functions that are specified by a half-rate standard are depicted in Fig. 32.1. An input speech signal is processed by a *speech encoder* to generate a digital representation at a *net bit rate* of R_s bits per second. The encoded bit stream representing the input speech signal is processed by a *channel encoder* to generate another bit stream at a *gross bit rate* of R_c bits per second, where $R_c > R_s$. The channel encoded bit stream is organized into data frames, and each frame is transmitted as payload data by a radio-link access controller and modulator. The net bit rate R_s counts the number of bits used to describe the speech signal, and the difference between the gross and net bit rates ($R_c - R_s$) counts the number of error protection bits needed by the *channel decoder* to correct and detect transmission errors. The output of the channel decoder is given to the *speech decoder* to generate a *quantized* version of the speech encoder's input signal. In current digital cellular radio systems that use time-division multiple access (TDMA), a voice connection is allocated a fixed transmission rate (i.e., R_c is a constant). The operations performed by the speech and channel encoders and decoders and their input and output data formats are governed by the half-rate standards.

Globally, three major TDMA cellular radio systems have been developed and deployed. The initial digital speech services offered by these cellular systems were governed by *full-rate standards*. Because of the rapid growth in demand for cellular services, the available transmission capacity in some areas is frequently saturated, eroding customer satisfaction. By providing essentially the same voice quality but at half the gross bit rates of the full-rate standards, half-rate standards can readily double the number of callers that can be serviced by the cellular systems. The gross bit rates of the full-rate and half-rate standards for the European Groupe Speciale Mobile (GSM), Japanese Personal Digital Cellular[1] (PDC), and North American cellular (IS-54) systems are listed

[1] Personal Digital Cellular was formerly Japanese Digital Cellular (JDC)

FIGURE 32.1   Digital speech transmission for digital cellular radio.   Boxes with solid outlines represent processing modules that are specified by the half-rate standards.

TABLE 32.1   Gross Bit Rates Used for Digital Speech Transmission in Three TDMA Cellular Radio Systems

Standard Organization and Digital Cellular System	Gross Bit Rate, b/s	
	Full Rate	Half Rate
European Telecommunications Standards Institute (ETSI), GSM	22,800	11,400
Research & Development Center for Radio Systems (RCR), PDC	11,200	5,600
Telecommunication Industries Association (TIA), IS-54	13,000	6,500

in Table 32.1. The three systems were developed and deployed under different time tables. Their disparate full- and half-bit rates partly reflect this difference. At the time of writing (January, 1995), the European and the Japanese systems have each selected an algorithm for their respective half-rate **codec**. Standardization of the North American half-rate codec has not reached a conclusion as none of the candidate algorithms has fully satisfied the standard's requirements. Thus, we focus here on the Japanese and European half-rate standards and will only touch upon the requirements of the North American standard.

32.2 Speech Coding for Cellular Mobile Radio Communications

Unlike the relatively benign transmission media commonly used in the public-switched telephone network (PSTN) for analog and digital transmission of speech signals, mobile radio channels are impaired by various forms of fading and interference effects. Whereas proper engineering of the radio link elements (modulation, power control, diversity, equalization, frequency allocation, etc.) ameliorates fading effects, burst and isolated bit errors still occur frequently. The net effect is such that speech communication may be required to be operational even for bit-error rates greater than 1%. In order to furnish reliable voice communication, typically half of the transmitted payload bits are devoted to error correction and detection.

It is common for low-bit-rate speech codecs to process samples of the input speech signal one frame at a time, e.g., 160 samples processed once every 20 ms. Thus, a certain amount of time is required to gather a block of speech samples, encode them, perform channel encoding, transport the encoded data over the radio channel, and perform channel decoding and speech synthesis. These processing steps of the speech codec add to the overall end-to-end transmission delay. Long transmission delay hampers conversational interaction. Moreover, if the cellular system is inter-connected with the PSTN and a four-wire to two-wire (analog) circuit conversion is performed in the network, feedbacks called *echoes* may be generated across the conversion circuit. The echoes can be heard by the originating talker as a delayed and distorted version of his/her speech and can be quite annoying. The annoyance level increases with the transmission delay and may necessitate (at additional costs) the deployment of **echo cancellers**.

A consequence of user mobility is that the level and other characteristics of the acoustic back-ground noise can be highly variable. Though acoustic noise can be minimized through suitable acoustic transduction design and the use of adaptive filtering/cancellation techniques [Ohya, Suda, and Miki, 1994; Suda, Ikeda, and Ikedo, 1994; Gibson, Koo, and Gray, 1991], the speech encoding algorithm still needs to be robust against background noise of various levels and kinds (e.g., babble, music, noise bursts, and colored noise).

Processing complexity directly impacts the viability of achieving a circuit realization that is compact and has low-power consumption, two key enabling factors of equipment portability for the end user. Factors that tend to result in low complexity are fixed-point instead of floating-point computation, lack of complicated arithmetic operations (division, square roots, transcendental functions), regular algorithm structure, small data memory, and small program memory. Since, in general, better speech quality can be achieved with increasing speech and channel coding delay and complexity, the digital cellular mobile-radio environment imposes conflicting and challenging requirements on the speech codec.

32.3 Codec Selection and Performance Requirements

The half-rate speech coding standards are drawn up through competitive testing and selection. From a set of candidate codec algorithms submitted by contending organizations, the one algorithm that meets basic selection criteria and offers the best performance is selected to form the standard. The codec performance measures and codec testing and selection procedures are set out in a test plan under the auspices of the organization (Table 32.1) responsible for the standardization process [see, e.g., Telecommunication Industries Association, 1993]. Major codec characteristics evaluated are speech quality, delay, and complexity. The full-rate codec is also evaluated as a *reference codec*, and its evaluation scores form part of the selection criteria for the codec candidates.

The speech quality of each candidate codec is evaluated through listening tests. To conduct the tests, each candidate codec is required to process speech signals and/or encoded bit streams that have been preprocessed to simulate a range of operating conditions: variations in speaker voice and level, acoustic background noise type and level, channel error rate, and stages of **tandem coding**. During the tests, subjects listen to processed speech signals and judge their quality levels or annoyance levels on a five-point opinion scale. The opinion scores collected from the tests are suitably averaged over all trials and subjects for each test condition [see Jayant and Noll, 1984, for mean opinion score (MOS) and degradation mean opinion score]. The categorical opinion scales of the subjects are also calibrated using *modulated noise reference units* (*MNRUs*) [Dimolitsas, Corcoran, and Baraniecki, 1994]. Modulated noise better resembles the distortions created by speech codecs than noise that is uncorrelated with the speech signal. Modulated noise is generated by multiplying the speech signal with a noise signal. The power level of the resultant product signal is scaled to a desired level and then added to the uncoded (clean) speech signal. The ratio between the power level of the speech signal and that of the modulated noise is expressed in decibels and given the notation *dBQ*. Under each test condition, subjects are presented with speech signals processed by the codecs as well as speech signals corrupted by modulated noise. Through presenting a range of modulated-noise levels, the subjects' opinions are calibrated on the dBQ scale. Thereafter, the mean opinion scores obtained for the codecs can also be expressed on that scale.

For each codec candidate, a profile of scores is compiled, consisting of speech quality scores, delay measurements, and complexity estimates. Each candidate's score profile is compared with that of the reference codec, ensuring that basic requirements are satisfied [see, e.g., Masui and Oguchi, 1993]. An overall figure of merit for each candidate is also computed from the profile. The candidates, if any, that meet the basic requirements then compete on the basis of maximizing the figure of merit.

Basic performance requirements for each of the three half-rate standards are summarized in Table 32.2. In terms of speech quality, the GSM and PDC half-rate codecs are permitted to under-perform their respective full-rate codecs by no more than 1 dBQ averaging over all test conditions and no more than 3 dBQ within each test condition. More stringently, the North American half-rate codec is required to furnish a speech-quality profile that is statistically equivalent to that of the North American full-rate codec as determined by a specific statistical procedure for multiple comparisons [TIA, 1993]. Since various requirements on the half-rate standards are set relative to their full-rate counterparts, an indication of the *relative* speech quality between the three half-rate standards can be deduced from the test results of De Martino [1993] comparing the three full-rate

TABLE 32.2   Basic Performance Requirements for the Three Half-Rate Standards

Digital Cellular Systems	Basic performance requirements		
	Min. Speech Quality, dBQ Rel. to Full Rate	Max. Delay, ms	Max. Complexity Rel. to Full Rate
Japanese (PDC)	−1 average, −3 maximum	94.8	3×
European (GSM)	−1 average, −3 maximum	90	4×
North American (IS-54)	Statistically equivalent	100	4×

codecs. The maximum delays in Table 32.2 apply to the total of the delays through the speech and channel encoders and decoders (Fig. 32.1). Codec complexity is computed using a formula that counts the computational operations and memory usage of the codec algorithm. The complexity of the half-rate codecs is limited to 3 or 4 times that of their full-rate counterparts.

32.4 Speech Coding Techniques in the Half-Rate Standards

Existing half-rate and full-rate standard coders can be characterized as *linear-prediction based analysis-by-synthesis* (LPAS) speech coders [Gersho, 1994]. LPAS coding entails using a time-varying all-pole filter in the decoder to synthesize the quantized speech signal. A short segment of the signal is synthesized by driving the filter with an *excitation* signal that is either *quasiperiodic* (for *voiced* speech) or *random* (for *unvoiced* speech). In either case, the excitation signal has a *spectral envelope* that is relatively flat. The synthesis filter serves to shape the spectrum of the excitation input so that the spectral envelope of the synthesized output resembles the filter's magnitude frequency response. The magnitude response often has prominent peaks; they render the *formants* that give a speech signal its phonetic character. The synthesis filter has to be adapted to the current frame of input speech signal. This is accomplished with the encoder performing a linear prediction (LP) analysis of the frame: the inverse of the all-pole synthesis filter is applied as an LP *error filter* to the frame, and the values of the filter parameters are computed to minimize the energy of the filter's output error signal. The resultant filter parameters are quantized and conveyed to the decoder for it to update the synthesis filter.

Having executed an LP analysis and quantized the synthesis filter parameters, the LPAS encoder performs analysis-by-synthesis (ABS) on the input signal to find a suitable excitation signal. An ABS encoder maintains a *copy* of the decoder. The encoder examines the possible outputs that can be produced by the decoder copy in order to determine how best to instruct (using transmitted information) the actual decoder so that it would output (synthesize) a good approximation of the input speech signal. The decoder copy tracks the state of the actual decoder, since the latter evolves (under ideal channel conditions) according to information received from the encoder. The details of the ABS procedure vary with the particular excitation model employed in a specific coding scheme. One of the earliest seminal LPAS schemes is *code excited linear prediction* (*CELP*) [Gersho, 1994]. In CELP, the excitation signal is obtained from a **codebook** of *code vectors*, each of which is a candidate for the excitation signal. The encoder searches the codebook to find the one code vector that would result in a best match between the resultant synthesis output signal and the encoder's input speech signal. The matching is considered best when the energy of the difference between the two signals being matched is minimized. A *perceptual weighting filter* is usually applied to the difference signal (prior to energy integration) to make the minimization more relevant to human perception of speech fidelity. Regions in the frequency spectrum where human listeners are more sensitive to distortions are given relatively stronger weighting by the filter and vice versa. For instance, the concentration of spectral energy around the formant frequencies gives rise to stronger *masking* of coder noise (i.e. rendering the noise less audible) and, therefore, weaker weighting can be applied to the formant frequency regions. For masking to be effective, the weighting filter has to be adapted to the time-varying speech spectrum. Adaptation is achieved usually by basing the weighting filter parameters on the synthesis filter parameters.

The CELP framework has evolved to form the basis of a great variety of speech coding algorithms, including all existing full-and half-rate standard algorithms for digital cellular systems. We outline next the basic CELP encoder-processing steps, in a form suited to our subsequent detailed descriptions of the PDC and GSM half-rate coders. These steps have accounted for various computational efficiency considerations and may, therefore, deviate from a conceptual functional description of the encoder constituents.

1. LP analysis on the current frame of input speech to determine the coefficients of the all-pole synthesis filter;
2. quantization of the LP filter parameters;
3. determination of the open-loop **pitch period** or lag;
4. adapting the perceptual weighting filter to the current LP information (and also pitch information when appropriate) and applying the adapted filter to the input speech signal;
5. formation of a filter cascade (which we shall refer to as *perceptually weighted synthesis filter*) consisting of the LP synthesis filter, as specified by the quantized parameters in step 2, followed by the perceptual weighting filter;
6. subtraction of the *zero-input response* of the perceptually weighted synthesis filter (the filter's decaying response due to past input) from the perceptually weighted input speech signal obtained in step 4;
7. an *adaptive codebook* is searched to find the most suitable periodic excitation, i.e., when the perceptually weighted synthesis filter is driven by the best code vector from the adaptive code-book, the output of the filter cascade should best match the difference signal obtained in step 6;
8. one or more nonadaptive excitation codebooks are searched to find the most suitable random excitation vectors that, when added to the best periodic excitation as determined in step 7 and with the resultant sum signal driving the filter cascade, would result in an output signal best matching the difference signal obtained in step 6.

Steps 1–6 are executed once per frame. Steps 7 and 8 are executed once for each of the *subframes* that together constitute a frame. Step 7 may be skipped depending on the pitch information from step 3, or if step 7 were always executed, a *nonperiodic excitation* decision would be one of the possible outcomes of the search process in step 7. Integral to steps 7 and 8 is the determination of gain (scaling) parameters for the excitation vectors. For each frame of input speech, the filter and excitation and gain parameters determined as outlined are conveyed as encoded bits to the speech decoder.

In a properly designed system, the data conveyed by the channel decoder to the speech decoder should be free of errors most of the time, and the speech signal synthesized by the speech decoder would be identical to that as determined in the speech encoder's ABS operation. It is common to enhance the quality of the synthesized speech by using an adaptive *postfilter* to attenuate coder noise in the perceptually sensitive regions of the spectrum. The postfilter of the decoder and the perceptual weighting filter of the encoder may seem to be functionally identical. The weighting filter, however, influences the selection of the best excitation among available choices, whereas the postfilter actually shapes the spectrum of the synthesized signal. Since postfiltering introduces its own distortion, its advantage may be diminished if tandem coding occurs along the end-to-end communication path. Nevertheless, proper design can ensure that the net effect of postfiltering is a reduction in the amount of audible codec noise [Chen and Gersho, 1995]. Excepting postfiltering, all other speech synthesis operations of an LPAS decoder are (effectively) duplicated in the encoder (though the converse is not true). Using this fact, we shall illustrate each coder in the sequel by exhibiting only a block diagram of its encoder or decoder but not both.

32.5 Channel Coding Techniques in the Half-Rate Standards

Crucial to the maintainence of quality speech communication is the ability to transport coded speech data across the radio channel with minimal errors. Low-bit-rate LPAS coders are particularly sensitive to channel errors; errors in the bits representing the LP parameters in one frame, for instance, could result in the synthesis of nonsensical sounds for longer than a frame duration. The error rate of a digital cellular radio channel with no channel coding can be catastropically high for LPAS coders. The amount of tolerable transmission delay is limited by the requirement of interactive communication and, consequently, *forward error control* is used to remedy transmission errors. "Forward" means that channel errors are remedied in the receiver, with no additional information from the transmitter and, hence, no additional transmission delay. To enable the channel decoder to correct channel errors, the channel encoder conveys more bits than the amount generated by the speech encoder. The additional bits are for error *protection*, as errors may or may not occur in any particular transmission epoch. The ratio of the number of encoder input (information) bits to the number of encoder output (code) bits is called the (channel) *coding rate*. This is a number no more than one and generally decreases as the error protection power increases. Though a lower channel coding rate gives more error protection, fewer bits will be available for speech coding. When the channel is in good condition and, hence, less error protection is needed, the received speech quality could be better if bits devoted to channel coding were used for speech coding. On the other hand, if a high channel coding rate were used, there would be uncorrected errors under poor channel conditions and speech quality would suffer. Thus, when nonadaptive forward error protection is used over channels with nonstationary statistics, there is an inevitable tradeoff between quality degradation due to uncorrected errors and that due to expending bits on error protection (instead of on speech encoding).

Both the GSM and PDC half-rate coders use *convolutional coding* [Proakis, 1995] for error correction. Convolutional codes are sliding or sequential codes. The encoder of a rate $m/n, m < n$ convolutional code can be realized using m shift registers. For every m information bits input to the encoder (one bit to each of the m shift registers), n code bits are output to the channel. Each code bit is computed as a modulo-2 sum of a subset of the bits in the shift registers. Error protection overhead can be reduced by exploiting the unequal sensitivity of speech quality to errors in different positions of the encoded bit stream. A family of *rate-compatible punctured convolutional codes* (RCPCCs) [Hagenauer, 1988] is a collection of related convolutional codes; all of the codes in the collection except the one with the lowest rate are derived by *puncturing* (dropping) code bits from the convolutional code with the lowest rate. With an RCPCC, the channel coding rate can be varied on the fly (i.e., variable-rate coding) while a sequence of information bits is being encoded through the shift registers, thereby imparting on different segments in the sequence different degrees of error protection.

For decoding a convolutional coded bit stream, the *Viterbi algorithm* [Proakis, 1995] is a computationally efficient procedure. Given the output of the demodulator, the algorithm determines the most likely sequence of data bits sent by the channel encoder. To fully utilize the error correction power of the convolutional code, the amplitude of the demodulated *channel symbol* can be quantized to more bits than the minimum number required, i.e., for subsequent *soft decision decoding*. The minimum number of bits is given by the number of channel-coded bits mapped by the modulator onto each channel symbol; decoding based on the minimum-rate bit stream is called *hard decision* decoding. Although soft decoding gives better error protection, decoding complexity is also increased.

Whereas convolutional codes are most effective against randomly scattered bit errors, errors on cellular radio channels often occur in bursts of bits. These bursts can be broken up if the bits put into the channel are rearranged after demodulation. Thus, in *block interleaving*, encoded bits are read into a matrix by row and then read out of the matrix by column (or vice versa) and then passed on to the modulator; the reverse operation is performed by a *deinterleaver* following demodulation. Interleaving increases the transmission delay to the extent that enough bits need to be collected in order to fill up the matrix.

Owing to the severe nature of the cellular radio channel and limited available transmission capacity, uncorrected errors often remain in the decoded data. A common countermeasure is to append an error detection code to the speech data stream prior to channel coding. When residual channel errors are detected, the speech decoder can take various remedial measures to minimize the

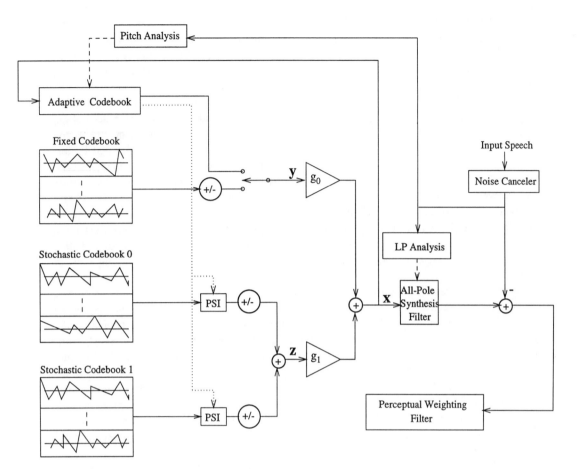

FIGURE 32.2 Basic structure of the PSI-CELP encoder.

negative impact on speech quality. Common measures are repetition of speech parameters from the most recent good frames and gradual muting of the possibly corrupted synthesized speech.

The PDC and GSM half-rate standard algorithms together embody some of the latest advances in speech coding techniques, including: *multimodal coding* where the coder configuration and bit allocation change with the type of speech input; *vector quantization* (*VQ*) [Gersho and Gray, 1991] of the LP filter parameters; higher precision and improved coding efficiency for pitch-periodic excitation; and postfiltering with improved tandeming performance. We next explore the more distinctive features of the PDC and GSM speech coders.

32.6 The Japanese Half-Rate Standard

An algorithm was selected for the Japanese half-rate standard in April 1993, following the evaluation of 12 submissions in a first round, and four final candidates in a second round [Masui and Oguchi, 1993]. The selected algorithm, called pitch synchronous innovation CELP[2] (PSI-CELP), met all of the basic selection criteria and scored the highest among all candidates evaluated. A block diagram of the PSI-CELP encoder is shown in Fig. 32.2, and bit allocations are summarized in Table 32.3. The complexity of the coder is estimated to be approximately 2.4 times that of the PDC full-rate

---

[2]There were two candidate algorithms named PSI-CELP in the PDC half-rate competition. The algorithm described here was contributed by NTT Mobile Communications Network, Inc. (NTT DoCoMo).

TABLE 32.3 Bit Allocations for the PSI-CELP Half-Rate PDC Speech Coder

Parameter	Bits	Error Protected Bits
LP synthesis filter	31	15
Frame energy	7	7
Periodic excitation	8×4	8×4
Stochastic excitation	10×4	0
Gain	7×4	3×4
Total	138	66

coder. The frame size of the coder is 40 ms, and its subframe size is 10 ms. These sizes are longer than those used in most existing CELP-type standard coders. However, LP analysis is performed twice per frame in the PSI-CELP coder.

A distinctive feature of the PSI-CELP coder is the use of an adaptive noise canceller [Ohya, Suda, and Miki, 1994; Suda, Ikeda, and Ikedo, 1994] to suppress noise in the input signal prior to coding. The input signal is classified into various modes, depending on the presence or absence of background noise and speech and their relative power levels. The current active mode determines whether *Kalman filtering* [Gibson, Koo, and Gray, 1991] is applied to the input signal and whether the parameters of the Kalman filter are adapted. Kalman filtering is applied when a significant amount of background noise is present or when both background noise and speech are strongly present. The filter parameters are adapted to the statistics of the speech and noise signals in accordance with whether they are both present or only noise is present.

The LP filter parameters in the PSI-CELP coder are encoded using VQ. A tenth-order LP analysis is performed every 20 ms. The resultant filter parameters are converted to 10 *line spectral frequencies* (LSFs).[3] The LSF parameters have a naturally increasing order, and together are treated as the ordered components of a vector. Since the speech spectral envelope tends to evolve slowly with time, there is intervector dependency between adjacent LSF vectors that can be exploited. Thus, the two LSF vectors for each 40-ms frame are paired together and jointly encoded. Each LSF vector in the pair is split into three subvectors. The pair of subvectors that cover the same vector component indexes are combined into one composite vector and vector quantized. Altogether, 31 b are used to encode a pair of LSF vectors. This three-way *split VQ*[4] scheme embodies a compromise between the prohibitively high complexity of using a large vector dimension and the performance gain from exploiting intra- and intervector dependency.

The PSI-CELP encoder uses a perceptual weighting filter consisting of a cascade of two filter sections. The sections exploit the pitch-harmonic structure and the LP spectral-envelope structure of the speech signal, respectively. The pitch-harmonic section has four parameters, a pitch lag and three coefficients, whose values are determined from an analysis of the periodic structure of the input speech signal. Pitch-harmonic weighting reduces the amount of noise in between the pitch harmonics by aggregating coder noise to be closer to the harmonic frequencies of the speech signal. In high-pitched voice, the harmonics are spaced relatively farther apart, and pitch-harmonic weighting becomes correspondingly more important.

The excitation vector x (Fig. 32.2) is updated once every subframe interval (10 ms) and is constructed as a *linear combination* of two vectors

$$x = g_0 y + g_1 z \qquad (32.1)$$

where g_0 and g_1 are scalar gains, y is labeled as the *periodic* component of the excitation and z as the *stochastic* or *random* component. When the input speech is voiced, the ABS operation would find a value for y from the *adaptive codebook* (Fig. 32.2). The codebook is constructed out of past samples of the excitation signal x; hence, there is a feedback path into the adaptive codebook in Fig. 32.2. Each code vector in the adaptive codebook corresponds to one of the 192 possible pitch lag L values available for encoding; the code vector is populated with samples of x beginning with the Lth sample backward in time. L is not restricted to be an integer, i.e., *fractional pitch period* is permitted.

---

[3] Also known as line spectrum pairs (LSPs).

[4] Matrix quantization is another possible description.

Successive values of L are more closely spaced for smaller values of L; short, medium, and long lags are quantized to one-quarter, one-half, and one sampling-period resolution, respectively. As a result, the *relative* quantization error in the encoded pitch frequency (which is the reciprocal of the encoded pitch lag) remains roughly constant with increasing pitch frequency. When the input speech is unvoiced, y would be obtained from the fixed codebook (Fig. 32.2). To find the best value for y, the encoder searches through the aggregate of 256 code vectors from both the adaptive and fixed codebooks. The code vector that results in a synthesis output most resembling the input speech is selected. The best code vector thus chosen also implicitly determines the voicing condition (voiced/unvoiced) and the pitch lag value L^* most appropriate to the current subframe of input speech. These parameters are said to be determined in a *closed-loop* search.

The stochastic excitation z is formed as a sum of two code vectors, each selected from a *conjugate codebook* (Fig. 32.2) [Ohya, Suda, and Miki, 1994]. Using a pair of conjugate codebooks each of size 16 code vectors (4 b) has been found to improve robustness against channel errors, in comparison with using one single codebook of size 256 code vectors (8 b). The synthesis output due to z can be decomposed into a sum of two orthorgonal components, one of which points in the same direction as the synthesis output due to the periodic excitation y and the other component points in a direction orthogonal to the synthesis output due to y. The latter synthesis output component of z is kept, whereas the former component is discarded. Such decomposition enables the two gain factors g_0 and g_1 to be separately quantized. For voiced speech, the conjugate code vectors are preprocessed to produce a set of *pitch synchronous innovation* (PSI) vectors. The first L^* samples of each code vector are treated as a fundamental period of samples. The fundamental period is replicated until there are enough samples to populate a subframe. If L^* is not an integer, interpolated samples of the code vectors are used (upsampled versions of the code vectors can be precomputed). PSI has been found to reinforce the periodicity and substantially improve the quality of synthesized voiced speech.

The postfilter in the PSI-CELP decoder has three sections, for enhancing the formants, the pitch harmonics, and the high frequencies of the synthesized speech, respectively. Pitch-harmonic enhancement is applied only when the adaptive codebook has been used. Formant enhancement makes use of the decoded LP synthesis filter parameters, whereas a refined pitch analysis is performed on the synthesized speech to obtain the values for the parameters of the pitch-harmonic section of the postfilter. A first-order high-pass filter section compensates for the low-pass spectral tilt [Chen and Gersho, 1995] of the formant enhancement section.

Of the 138 speech data bits generated by the speech encoder every 40-ms frame, 66 b (Table 32.3) receive error protection and the remaining 72 speech data bits of the frame are not error protected. An error detection code of 9 *cyclic redundancy check* (CRC) bits is appended to the 66 b and then submitted to a rate 1/2, punctured convolutional encoder to generate a sequence of 152 channel coded bits. Of the unprotected 72 b, the 40 b that index the excitation codebooks (Table 32.3) are remapped or *pseudo-Gray coded* [Zeger and Gersho, 1990] so as to equalize their channel error sensitivity. As a result, a bit error occuring in an index word is likely to cause about the same amount of degradation regardless of the bit error position in the index word. For each speech frame, the channel encoder emits 224 b of payload data. The payload data from two adjacent frames are interleaved before transmission over the radio link.

Uncorrected errors in the most critical 66 b are detected with high probability as a CRC error. A finite state machine keeps track of the recent history of CRC errors. When a sequence of CRC errors is encountered, the power level of the synthesized speech is progressively suppressed, so that muting is reached after four consecutive CRC errors. Conversely, following the cessation of a sequence of CRC errors, the power level of the synthesized speech is ramped up gradually.

32.7 The European GSM Half-Rate Standard

A *vector sum excited linear prediction* (VSELP) coder, contributed by Motorola, Inc., was selected in January 1994 by the main GSM technical committee as a basis for the GSM half-rate standard. The

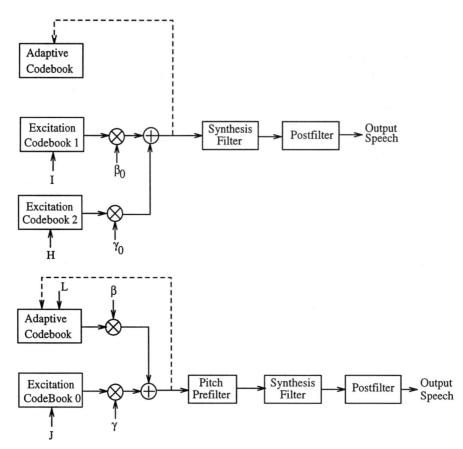

FIGURE 32.3 Basic structure of the GSM VSELP decoder. Top is for mode 0 and bottom is for modes 1, 2, and 3.

standard was finally approved in January 1995. VSELP is a generic name for a family of algorithms from Motorola; the North American full-rate and the Japanese full-rate standards are also based on VSELP. All VSELP coders make use of the basic idea of representing the excitation signal by a linear combination of *basis vectors* [Gerson and Jasiuk, 1990]. This representation renders the excitation codebook search procedure very computationally efficient. A block diagram of the GSM half-rate decoder is depicted in Fig. 32.3 and bit allocations are tabulated in Table 32.4. The coder's frame size is 20 ms, and each frame comprises four subframes of 5 ms each. The coder has been optimized for execution on a processor with 16-b word length and 32-b accumulator. The GSM standard is a *bit exact* specification: in addition to specifying the codec's processing steps, the numerical formats and precisions of the codec's variables are also specified.

The synthesis filter coefficients in GSM VSELP are encoded using the *fixed point lattice technique* (FLAT) [Gerson et al., 1994] and vector quantization. FLAT is based on the *lattice filter* representation of the linear prediction error filter. The tenth-order lattice filter has 10 stages, with the ith stage, $i \in \{1, \ldots, 10\}$, containing a a *reflection coefficient* parameter r_i. The lattice filter has an *order-recursion* property such that the best prediction error filters of all orders less than ten are all embedded in the best tenth-order lattice filter. This means that once the values of the lower order reflection coefficients have been optimized, they do not have to be reoptimized when a higher order predictor is desired; in other words, the coefficients can be optimized sequentially from low to high orders. On the other hand, if the lower order coefficients were suboptimal (as in the case when the coefficients

TABLE 32.4 Bit Allocations for the VSELP Half-Rate GSM Coder

Parameter	Bits/subframe	Bits/frame
LP synthesis filter		28
Soft interpolation		1
Frame energy		5
Mode selection		2
Mode 0		
Excitation code I	7	28
Excitation code H	7	28
Gain code G_s, P_0	5	20
Mode 1, 2, and 3		
Pitch lag L (first subframe)		8
Difference lag (subframes 2, 3, 4)	4	12
Excitation code J	9	36
Gain code G_s, P_0	5	20
Total		112

are quantized), the higher order coefficients could still be selected to minimize the prediction *residual* (or error) energy at the output of the higher order stages; in effect, the higher order stages can compensate for the suboptimality of lower order stages.

In the GSM VSELP coder, the ten reflection coefficients $\{r_1, \ldots, r_{10}\}$ that have to be encoded for each frame are grouped into three coefficient vectors $v_1 = [r_1 r_2 r_3]$, $v_2 = [r_4 r_5 r_6]$, $v_3 = [r_7 r_8 r_9 r_{10}]$. The vectors are quantized sequentially, from v_1 to v_3, using a b_i-bit VQ codebook C_i for v_i, where b_i, $i = 1, 2, 3$ are 11, 9, and 8 b, respectively. The vector v_i is quantized to minimize the prediction error at the energy output of the jth stage of the lattice filter where r_j is the highest order coefficient in the vector v_i. The computational complexity associated with quantizing v_i is reduced by searching only a small subset of the code vectors in C_i. The subset is determined by first searching a *prequantizer* codebook of size c_i bits, where c_i, $i = 1, \ldots, 3$ are 6, 5, and 4 b, respectively. Each code vector in the prequantizer codebook is associated with $2^{b_i - c_i}$ code vectors in the target codebook. The subset is obtained by pooling together all of the code vectors in C_i that are associated with the top few best matching prequantizer code vectors. In this way, a factor of reduction in computational complexity of nearly $2^{b_i - c_i}$ is obtained for the quantization of v_i.

The half-rate GSM coder changes its configuration of excitation generation (Fig. 32.3) in accordance with a *voicing mode* [Gerson and Jasiuk, 1992]. For each frame, the coder selects one of four possible voicing modes depending on the values of the *open-loop* pitch-prediction gains computed for the frame and its four subframes. Open loop refers to determining the pitch lag and the pitch-predictor coefficient(s) via a direct analysis of the input speech signal or, in the case of the half-rate GSM coder, the perceptually weighted (LP-weighting only) input signal. Open-loop analysis can be regarded as the opposite of closed-loop analysis, which in our context is synonymous with ABS. When the pitch-prediction gain for the frame is weak, the input speech signal is deemed to be unvoiced and mode 0 is used. In this mode, two 7-b *trained* codebooks (excitation codebooks 1 and 2 in Fig. 32.3) are used, and the excitation signal for each subframe is formed as a linear combination of two code vectors, one from each of the codebooks. A trained codebook is one designed by applying the coder to a representative set of speech signals while optimizing the codebook to suit the set. Mode 1, 2, or 3 is chosen depending on the strength of the pitch-prediction gains for the frame and its subframes. In these modes, the excitation signal is formed as a linear combination of a code vector from an 8-b adaptive codebook and a code vector from a 9-b trained codebook (Fig. 32.3). The code vectors that are summed together to form the excitation signal for a subframe are each scaled by a gain factor (β and γ in Fig. 32.3). Each mode uses a gain VQ codebook specific to that mode.

As depicted in Fig. 32.3, the decoder contains an adaptive pitch prefilter for the voiced modes and an adaptive postfilter for all modes. The filters enhance the perceptual quality of the decoded speech and are not present in the encoder. It is more conventional to locate the pitch prefilter as a section of the postfilter; the distinctive placement of the pitch prefilter in VSELP was chosen to reduce artifacts caused by the time-varying nature of the filter. In mode 0, the encoder uses an LP spectral weighting filter in its ABS search of the two excitation codebooks. In the other modes, the encoder uses a pitch-harmonic weighting filter in cascade with an LP spectral weighting filter for searching excitation codebook 0, whereas only LP spectral weighting is used for searching the adaptive codebook. The pitch-harmonic weighting filter has two parameters, a pitch lag and a coefficient, whose values are determined in the aforementioned open-loop pitch analysis.

A code vector in the 8-b adaptive codebook has a dimension of 40 (the duration of a subframe) and is populated with past samples of the excitation signal beginning with the Lth sample back from the present time. L can take on one of 256 different integer and fractional values. The best adaptive code vector for each subframe can be selected via a complete ABS; the required exhaustive search of the adaptive codebook is, however, computationally expensive. To reduce computation, the GSM VSELP coder makes use of the aforementioned open-loop pitch analysis to produce a list of *candidate lag values.* The open-loop pitch-prediction gains are ranked in decreasing order, and only the lags corresponding to top-ranked gains are kept as candidates. The final decisions for the four L values of the four subframes in a frame are made jointly. By assuming that the four L values can not vary over the entire range of all possible 256 values in the short duration of a frame, the L of the first subframe is coded using 8 b, and the L of each of the other three subframes is coded *differentially* using 4 b. The 4 b represent 16 possible values of deviation relative to the lag of the previous subframe. The four lags in a frame trace out a *trajectory* where the change from one time point to the next is restricted; consequently, only 20 b are needed instead of 32 b for encoding the four lags. Candidate trajectories are constructed by linking top ranked lags that are commensurate with differential encoding. The best trajectory among the candidates is then selected via ABS.

The trained excitation codebooks of VSELP have a special vector sum structure that facilitates fast searching [Gerson and Jasiuk, 1990]. Each of the 2^b code vectors in a b-bit trained codebook is formed as a linear combination of b *basis vectors.* Each of the b scalar weights in the linear combination is restricted to have a binary value of either 1 or -1. The 2^b code vectors in the codebook are obtained by taking all 2^b possible combinations of values of the weights. A substantial storage saving is incurred by storing only b basis vectors instead of 2^b code vectors. Computational saving is another advantage of the vector-sum structure. Since filtering is a linear operation, the synthesis output due to each code vector is a linear combination of the synthesis outputs due to the individual basis vectors, where the same weight values are used in the output linear combination as in forming the code vector. A vector sum codebook can be searched by first performing synthesis filtering on its b basis vectors. If, for the present subframe, another trained codebook (mode 0) or an adaptive codebook (mode 1, 2, 3) had been searched, the filtered basis vectors are further orthogonalized with respect to the signal synthesized from that codebook, i.e., each filtered basis vector is replaced by its own component that is orthogonal to the synthesized signal. Further complexity reduction is obtained by examining the code vectors in a sequence such that two successive code vectors differ in only one of the b scalar weight values; that is, the entire set of 2^b code vectors is searched in a *Gray coded* sequence. With successive code vectors differing in only one term in the linear combination, it is only necessary in the codebook search computation to progressively track the difference [Gerson and Jasiuk, 1990].

The total energy of a speech frame is encoded with 5 b (Table 32.4). The two gain factors (β and γ in Fig. 32.3) for each subframe are computed after the excitation codebooks have been searched and are then transformed to parameters G_s and P_0 to be vector quantized. Each mode has its own 5-b gain VQ codebook. G_s represents the energy of the subframe relative to the total frame energy, and P_0 represents the fraction of the subframe energy due to the first excitation source (excitation codebook 1 in mode 0, or the adaptive codebook in the other modes).

An *interpolation bit* (Table 32.4) transmitted for each frame specifies to the decoder whether the LP synthesis filter parameters for each subframe should be obtained from interpolating between the decoded filter parameters for the current and the previous frames. The encoder determines the value of this bit according to whether interpolation or no interpolation results in a lower prediction residual energy for the frame. The postfilter in the decoder operates in concordance with the actual LP parameters used for synthesis.

The speech encoder generates 112 b of encoded data (Table 32.4) for every 20-ms frame of the speech signal. These bits are processed by the channel encoder to improve, after channel decoding at the receiver, the uncoded bit-error rate and the detectability of uncorrected errors. Error detection coding in the form of 3 CRC bits is applied to the most critical 22 data bits. The combined 25 b plus an additional 73 speech data bits and 6 *tail bits* are input to an RCPCC encoder (the tail bits serve to bring the channel encoder and decoder to a fixed terminal state at the end of the payload data stream). The 3 CRC bits are encoded at rate 1/3 and the other 101 b are encoded at rate 1/2, generating a total of 211 channel coded bits. These are finally combined with the remaining 17 (uncoded) speech data bits to form a total of 228 b for the payload data of a speech frame. The payload data from two speech frames are interleaved for transmission over four timeslots of the GSM TDMA channel.

With the Viterbi algorithm, the channel decoder performs soft decision decoding on the demodulated and deinterleaved channel data. Uncorrected channel errors may still be present in the decoded speech data after Viterbi decoding. Thus, the channel decoder classifies each frame into three integrity categories: bad, unreliable, and reliable, in order to assist the speech decoder in undertaking error concealment measures. A frame is considered bad if the CRC check fails or if the received channel data is close to more than one candidate sequence. The latter evaluation is based on applying an adaptive threshold to the metric values produced by the Viterbi algorithm over the course of decoding the most critical 22 speech data bits and their 3 CRC bits. Frames that are not bad may be classified as unreliable, depending on the metric values produced by the Viterbi algorithm and on channel reliability information supplied by the demodulator.

Depending on the recent history of decoded data integrity, the speech decoder can take various error concealment measures. The onset of bad frames is concealed by repetition of parameters from previous reliable frames, whereas the persistence of bad frames results in power attenuation and ultimately muting of the synthesized speech. Unreliable frames are decoded with normality constraints applied to the energy of the synthesized speech.

32.8 Conclusions

The half-rate standards employ some of the latest techniques in speech and channel coding to meet the challenges posed by the severe transmission environment of digital cellular radio systems. By halving the bit rate, the voice transmission capacity of existing full-rate digital cellular systems can be doubled. Although advances are still being made that can address the needs of quarter-rate speech transmission, much effort is currently devoted to enhancing the speech quality and robustness of full-rate (GSM and IS-54) systems, aiming to be closer to *toll quality*. On the other hand, the imminent introduction of competing wireless systems that use different modulation schemes [e.g., coded division multiple access (CDMA)] and/or different radio frequencies [e.g., personal communications systems (PCS)] is poised to alleviate congestion in high-user-density areas.

Defining Terms

Codebook: An ordered collection of all possible values that can be assigned to a scalar or vector variable. Each unique scalar or vector value in a codebook is called a *codeword*, or *code vector* where appropriate.

Codec: A contraction of *(en)coder–decoder*, used synonymously with the word *coder*. The encoder and decoder are often designed and deployed as a pair. A half-rate standard codec by definition performs speech as well as channel coding.

Echo canceller: A signal processing device that given the source signal causing the echo signal, generates an estimate of the echo signal and subtracts the estimate from the signal being interfered with by the echo signal. The device is usually based on a discrete-time adaptive filter.

Pitch period: The period of a voiced speech waveform that can be regarded as periodic over a short-time interval (quasiperiodic). The reciprocal of pitch period is *pitch frequency* or simply, *pitch.*

Tandem coding: More than one encoder–decoder pair in an end-to-end transmission path. In cellular radio communications, having a radio link at each end of the communication path could subject the speech signal to two passes of speech encoding–decoding. In general, repeated encoding and decoding increases the distortion.

Acknowledgment

The authors would like to thank Erdal Paksoy and Mark A. Jasiuk for their valuable comments.

References

Chen, J.-H. and Gersho, A. 1995. Adaptive postfiltering for quality enhancement of coded speech. *IEEE Trans. Speech & Audio Proc.* 3(1) 59–71.

De Martino, E. 1993. Speech quality evaluation of the European, North-American and Japanese speech codec standards for digital cellular systems. In *Speech and Audio Coding for Wireless and Network Applications,* ed. B.S. Atal, V. Cuperman, and A. Gersho, pp. 55–58, Kluwer Academic Publishers, Norwell, MA.

Dimolitsas, S., Corcoran, F.L., and Baraniecki, M.R. 1994. Transmission quality of North American cellular, personal communications, and public switched telephone networks. *IEEE Trans. Veh. Tech.* 43(2):245–251.

Gersho, A. 1994. Advances in speech and audio compression. *Proc. IEEE.* 82(6) 900–918.

Gersho, A. and Gray, R.M. 1991. *Vector Quantization and Signal Compression,* Kluwer Academic Publishers, Norwell, MA.

Gerson, I.A. and Jasiuk, M.A. 1990. Vector sum excited linear prediction (VSELP) speech coding at 8 kbps. In *Proceedings, IEEE Intl. Conf. Acoustics, Speech, & Sig. Proc.* pp. 461–464, April.

Gerson, I.A. and Jasiuk, M.A. 1992. Techniques for improving the performance of CELP—type speech coders. *IEEE J. Sel. Areas Comm.* 10(5):858–865.

Gerson, I.A., Jasiuk, M.A., Nowack, J.M., Winter, E.H., and Müller, J.-M. 1994. Speech and channel coding for the half-rate GSM channel. In *Proceedings, ITG-Report 130 on Source and Channel Coding,* pp. 225–232. Munich, Germany, Oct.

Gibson, J.D., Koo, B., and Gray, S.D. 1991. Filtering of colored noise for speech enhancement and coding. *IEEE Trans. Sig. Proc.* 39(8):1732–1742.

Hagenauer, J. 1988. Rate-compatible punctured convolutional codes (RCPC codes) and their applications. *IEEE Trans. Comm.* 36(4):389–400.

Jayant, N.S. and Noll, P. 1984. *Digital Coding of Waveforms,* Prentice-Hall, Englewood Cliffs, NJ.

Masui, F. and Oguchi, M. 1993. Activity of the half rate speech codec algorithm selection for the personal digital cellular system. *Tech. Rept. of IEICE.* RCS93-77(11):55–62 (in Japanese).

Ohya, T., Suda, H., and Miki, T. 1994. 5.6 kbits/s PSI-CELP of the half-rate PDC speech coding standard. In *Proceedings, IEEE Veh. Tech. Conf.* pp. 1680–1684, June.

Proakis, J.G. 1995. *Digital Communications,* 3rd ed. McGraw-Hill, New York.

Suda, H., Ikeda, K., and Ikedo, J. 1994. Error protection and speech enhancement schemes of PSI-CELP, *NTT R&D.* (Special issue on PSI-CELP speech coding system for mobile communications), 43(4):373–380 (in Japanese).

Telecommunication Industries Association (TIA). 1993. Half-rate speech codec test plan V6.0. TR45.3.5/93.05.19.01.

Zeger, K. and Gersho, A. 1990. Pseudo-Gray coding. *IEEE Trans. Comm.* 38(12):2147–2158.

Further Information

Additional technical information on speech coding can be found in the books, periodicals, and conference proceedings that appear in the list of references. Other relevant publications not represented in the list are *Speech Communication*, Elsevier Science Publishers; *Advances in Speech Coding*, B. S. Atal, V. Cuperman, and A, Gersho, eds., Kluwer Academic Publishers; and *Proceedings of the IEEE Workshop on Speech Coding*.

33

Modulation Methods

Gordon L. Stüber
Georgia Institute of Technology

33.1 Introduction

Modulation is the process where the message information is added to the radio carrier. Most first generation cellular systems such as the advanced mobile telephone system (AMPS) use analog **frequency modulation** (**FM**), because analog technology was very mature when these systems were first introduced. Digital modulation schemes, however, are the obvious choice for future wireless systems, especially if data services such as wireless multimedia are to be supported. Digital modulation can also improve spectral efficiency, because digital signals are more robust against channel impairments. Spectral efficiency is a key attribute of wireless systems that must operate in a crowded radio frequency spectrum.

To achieve high spectral efficiency, modulation schemes must be selected that have a high **bandwidth efficiency** as measured in units of bits per second per Hertz of bandwidth. Many wireless communication systems, such as cellular telephones, operate on the principle of frequency reuse, where the carrier frequencies are reused at geographically separated locations. The link quality in these systems is limited by cochannel interference. Hence, modulation schemes must be identified that are both bandwidth efficient and capable of tolerating high levels of cochannel interference. More specifically, digital modulation techniques are chosen for wireless systems that satisfy the following properties.

Compact Power Density Spectrum: To minimize the effect of adjacent channel interference, it is desirable that the power radiated into the adjacent channel be 60–80 dB below that in the desired channel. Hence, modulation techniques with a narrow main lobe and fast rolloff of sidelobes are desirable.

Good Bit-Error-Rate Performance: A low-bit-error probability should be achieved in the presence of cochannel interference, adjacent channel interference, thermal noise, and other channel impairments, such as fading and intersymbol interference.

0-8493-8573-3/96/$0.00+$.50
© 1996 by CRC Press, Inc.

Envelope Properties: Portable and mobile applications typically employ nonlinear (class C) power amplifiers to minimize battery drain. Nonlinear amplification may degrade the bit-error-rate performance of modulation schemes that transmit information in the amplitude of the carrier. Also, spectral shaping is usually performed prior to up-conversion and nonlinear amplification. To prevent the regrowth of spectral sidelobes during nonlinear amplification, the input signal must have a relatively constant envelope.

A variety of digital modulation techniques are currently being used in wireless communication systems. Two of the more widely used digital modulation techniques for cellular mobile radio are $\pi/4$ phase-shifted quadrature **phase shift keying** ($\pi/4$-QPSK) and **Gaussian minimum shift keying** (**GMSK**). The former is used in the North American IS-54 digital cellular system and Japanese Personal Digital Cellular (PDC), whereas the latter is used in the global system for mobile communications (GSM system). This chapter provides a discussion of these and other modulation techniques that are employed in wireless communication systems.

33.2 Basic Description of Modulated Signals

With any modulation technique, the bandpass signal can be expressed in the form

$$s(t) = \text{Re}\{v(t)e^{j2\pi f_c t}\} \qquad (33.1)$$

where $v(t)$ is the complex envelope, f_c is the carrier frequency, and $\text{Re}\{z\}$ denotes the real part of z. For digital modulation schemes $v(t)$ has the general form

$$v(t) = A \sum_k b(t - kT, x_k) \qquad (33.2)$$

where A is the amplitude of the carrier $x_k = (x_k, x_{k-1}, \ldots, x_{k-K})$ is the data sequence, T is the symbol or baud duration, and $b(t, x_i)$ is an equivalent shaping function usually of duration T. The precise form of $b(t, x_i)$ and the memory length K depends on the type of modulation that is employed. Several examples are provided in this chapter where information is transmitted in the amplitude, phase, or frequency of the bandpass signal.

The **power spectral density** of the bandpass signal $S_{ss}(f)$ is related to the power spectral density of the complex envelope $S_{vv}(f)$ by

$$S_{ss}(f) = \frac{1}{2}[S_{vv}(f - f_c) + S_{vv}(f + f_c)] \qquad (33.3)$$

The power density spectrum of the complex envelope for a digital modulation scheme has the general form

$$S_{vv}(f) = \frac{A^2}{T} \sum_m S_{b,m}(f)e^{-j2\pi f mT} \qquad (33.4)$$

where

$$S_{b,m}(f) = \frac{1}{2}E[B(f, x_m)B^*(f, x_0)] \qquad (33.5)$$

$B(f, x_m)$ is the Fourier transform of $b(t, x_m)$, and $E[\cdot]$ denotes the expectation operator. Usually symmetric signal sets are chosen so that the complex envelope has zero mean, i.e., $E[b(t, x_0)] = 0$. This implies that the power density spectrum has no discrete components. If, in addition, x_m and

x_0 are independent for $|m| > K$, then

$$S_{vv}(f) = \frac{A^2}{T} \sum_{|m|<K} S_{b,m}(f) e^{-j2\pi f mT} \qquad (33.6)$$

33.3 Analog Frequency Modulation

With analog frequency modulation the complex envelope is

$$v(t) = A \exp\left[j2\pi k_f \int_0^t m(\tau)\, d\tau \right] \qquad (33.7)$$

where $m(t)$ is the modulating waveform and k_f in Hz/v is the frequency sensitivity of the FM modulator. The bandpass signal is

$$s(t) = A \cos\left[2\pi f_c t + 2\pi k_f \int_0^t m(t)\, dt \right]. \qquad (33.8)$$

The instantaneous frequency of the carrier $f_i(t) = f_c + k_f m(t)$ varies linearly with the waveform $m(t)$, hence the name frequency modulation. Notice that FM has a constant envelope making it suitable for nonlinear amplification. However, the complex envelope is a nonlinear function of the modulating waveform $m(t)$ and, therefore, the spectral characteristics of $v(t)$ cannot be obtained directly from the spectral characteristics of $m(t)$.

With the sinusoidal modulating waveform $m(t) = A_m \cos(2\pi f_m t)$ the instantaneous carrier frequency is

$$f_i(t) = f_c + \Delta_f \cos(2\pi f_m t) \qquad (33.9)$$

where $\Delta_f = k_f A_m$ is the peak frequency deviation. The complex envelope becomes

$$v(t) = \exp\left[2\pi \int_0^t f_i(t)\, dt \right]$$
$$= \exp[2\pi f_c t + \beta \sin(2\pi f_m t)] \qquad (33.10)$$

where $\beta = \Delta_f / f_m$ is called the modulation index. The bandwidth of $v(t)$ depends on the value of β. If $\beta < 1$, then narrowband FM is generated, where the spectral widths of $v(t)$ and $m(t)$ are about the same, i.e., $2f_m$. If $\beta \gg 1$, then wideband FM is generated, where the spectral occupancy of $v(t)$ is slightly greater than $2\Delta_f$. In general, the approximate bandwidth of an FM signal is

$$W \approx 2\Delta_f + 2f_m = 2\Delta_f\left(1 + \frac{1}{\beta}\right) \qquad (33.11)$$

which is a relation known as Carson's rule. Unfortunately, typical analog cellular radio systems use a modulation index in the range $1 \lesssim \beta \lesssim 3$ where Carson's rule is not accurate. Furthermore, the message waveform $m(t)$ is not a pure sinusoid so that Carson's rule does not directly apply.

In analog cellular systems the waveform $m(t)$ is obtained by first companding the speech waveform and then hard limiting the resulting signal. The purpose of the limiter is to control the peak frequency deviation Δ_f. The limiter introduces high-frequency components that must be removed with a low-pass filter prior to modulation. To estimate the bandwidth occupancy, we first determine the ratio of the frequency deviation Δ_f corresponding to the maximum amplitude of $m(t)$, and the highest frequency component B that is present in $m(t)$. These two conditions are the most extreme cases, and the resulting ratio, $D = \Delta_f / B$, is called the *deviation ratio*. Then replace β by D and f_m by B in Carson's rule, giving

$$W \approx 2\Delta_f + 2B = 2\Delta_f \left(1 + \frac{1}{D} \right) \qquad (33.12)$$

This approximation will overestimate the bandwidth requirements. A more accurate estimate of the bandwidth requirements must be obtained from simulation or measurements.

33.4 Phase Shift Keying (PSK) and $\pi/4$-QPSK

With **phase shift keying** (PSK), the equivalent shaping function in Eq. (33.2) has the form

$$b(t, \mathbf{x}_k) = \psi_T(t) \exp \left[j \frac{\pi}{M} x_k h_s(t) \right], \qquad \mathbf{x}_k = x_k \qquad (33.13)$$

where $h_s(t)$ is a phase shaping pulse, $\psi_T(t)$ an amplitude shaping pulse, and M the size of the modulation alphabet. Notice that the phase varies linearly with the symbol sequence $\{x_k\}$, hence the name phase shift keying. For a modulation alphabet size of M, $x_k \in \{\pm 1, \pm 3, \ldots, \pm(M-1)\}$. Each symbol x_k is mapped onto $\log_2 M$ source bits. A QPSK signal is obtained by using $M = 4$, resulting in a transmission rate of 2 b/symbol.

Usually, the phase shaping pulse is chosen to be the rectangular pulse $h_s(t) = u_T(t) \overset{\Delta}{=} u(t) - u(t - T)$, where $u(t)$ is the unit step function. The amplitude shaping pulse is very often chosen to be a square root raised cosine pulse, where the Fourier transform of $\psi_T(t)$ is

$$\Psi_T(f) = \begin{cases} \sqrt{T} & 0 \le |f| \le (1-\beta)/2T \\ \sqrt{\dfrac{T}{2} \left[1 - \sin \dfrac{\pi T}{\beta} \left(f - \dfrac{1}{2T} \right) \right]} & (1-\beta)/2T \le |f| \le (1+\beta)/2T \end{cases} \qquad (33.14)$$

The receiver implements the same filter $\Psi_R(f) = \Psi_T(f)$ so that the overall pulse has the raised cosine spectrum $\Psi(f) = \Psi_R(f)\Psi_T(f) = |\Psi_T(f)|^2$. If the channel is affected by flat fading and additive white Gaussian noise, then this partitioning of the filtering operations between the transmitter and receiver will optimize the signal to noise ratio at the output of the receiver filter at the sampling instants. The rolloff factor β usually lies between 0 and 1 and defines the **excess bandwidth** $100\beta\%$. Using a smaller β results in a more compact power density spectrum, but the link performance becomes more sensitive to errors in the symbol timing. The IS-54 system uses $\beta = 0.35$, while PDC uses $\beta = 0.5$.

The time domain pulse corresponding to Eq. (33.14) can be obtained by taking the inverse Fourier transform, resulting in

$$\psi_T(t) = 4\beta \frac{\cos[(1+\beta)\pi t/T] + \sin[(1-\beta)\pi t/T](4\beta t/T)^{-1}}{\pi\sqrt{T}[1 - 16\beta^2 t^2/T^2]} \qquad (33.15)$$

A typical square root raised cosine pulse with a rolloff factor of $\beta = 0.5$ is shown in Fig. 33.1. Strictly speaking the pulse $\psi_T(t)$ is noncausal, but in practice a truncated time domain pulse is used. For example, in Fig. 33.1 the pulse is truncated to $6T$ and time shifted by $3T$ to yield a causal pulse.

Unlike conventional QPSK that has four possible transmitted phases, $\pi/4$-QPSK has eight possible transmitted phases. Let $\theta(n)$ be the transmitted carrier phase for the nth epoch, and let

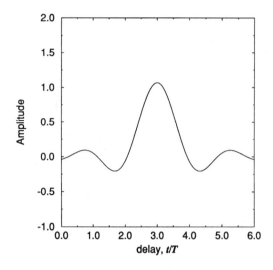

FIGURE 33.1 Square root raised cosine pulse with rolloff factor $\beta = 0.5$.

$\Delta\theta(n) = \theta(n) - \theta(n-1)$ be the differential carrier phase between epochs n and $n-1$. With $\pi/4$-QPSK, the transmission rate is 2 b/symbol and the differential phase is related to the symbol sequence $\{x_n\}$ through the mapping

$$\Delta\theta(n) = \begin{cases} -3\pi/4, & x_n = -3 \\ -\pi/4, & x_n = -1 \\ \pi/4, & x_n = +1 \\ 3\pi/4, & x_n = +3 \end{cases} \tag{33.16}$$

Since the symbol sequence $\{x_n\}$ is random, the mapping in Eq. (33.16) is arbitrary, except that the phase differences must be $\pm\pi/4$ and $\pm3\pi/4$. The phase difference with the given mapping can be written in the convenient algebraic form

$$\Delta\theta(n) = x_n\frac{\pi}{4} \tag{33.17}$$

which allows us to write the equivalent shaping function of the $\pi/4$-QPSK signal as

$$b(t, \underline{x}_k) = \psi(t)\exp\left\{j\left[\theta(k-1) + x_k\frac{\pi}{4}\right]\right\}$$
$$= \psi_T(t)\exp\left[j\frac{\pi}{4}\left(\sum_{n=-\infty}^{k-1} x_n + x_k\right)\right] \tag{33.18}$$

The summation in the exponent represents the accumulated carrier phase, whereas the last term is the phase change due to the kth symbol. Observe that the phase shaping function is the rectangular pulse $u_T(t)$. The amplitude shaping function $\psi_T(t)$ is usually the square root raised cosine pulse in Eq. (33.15).

The phase states of QPSK and $\pi/4$-QPSK signals can be summarized by the signal space diagram in Fig. 33.2 that shows the phase states and allowable transitions between the phase states. However, it does not describe the actual phase trajectories. A typical diagram showing phase trajectories with

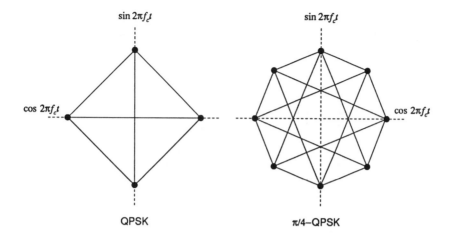

FIGURE 33.2   Signal-space constellations for QPSK and $\pi/4$-DQPSK.

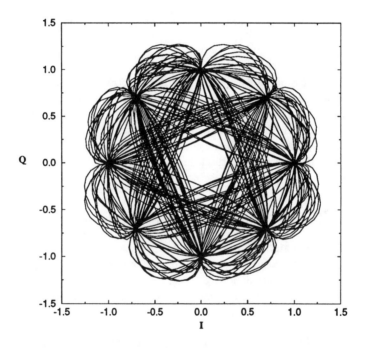

FIGURE 33.3   Phase diagram of $\pi/4$-QPSK with square root raised cosine pulse; $\beta = 0.5$.

square root raised cosine pulse shaping is shown in Fig. 33.3. Note that the phase trajectories do not pass through the origin. This reduces the envelope fluctuations of the signal making it less susceptible to amplifier nonlinearities and reduces the dynamic range required of the power amplifier.

The power density spectrum of QPSK and $\pi/4$-QPSK depends on both the amplitude and phase shaping pulses. For the rectangular phase shaping pulse $h_s(t) = u_T(t)$, the power density spectrum of the complex envelope is

$$S_{vv}(f) = \frac{A^2}{T}|\Psi_T(f)|^2 \tag{33.19}$$

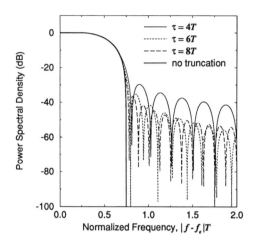

FIGURE 33.4 Power density spectrum of truncated square root raised cosine pulse with various truncation lengths; $\beta = 0.5$.

With square root raised cosine pulse shaping, $\Psi_T(f)$ has the form defined in Eq. (33.14). The power density spectrum of a pulse $\tilde{\psi}_T(t)$ that is obtained by truncating $\psi_T(t)$ to length τ can be obtained by writing $\tilde{\psi}_T(t) = \psi_T(t)\text{rect}(t/\tau)$. Then $\tilde{\Psi}_T(f) = \Psi_T(f) * \tau\text{sinc}(f\tau)$, where $*$ denotes the operation of convolution, and the power density spectrum is again obtained by applying Eq. (33.19). Truncation of the pulse will regenerate some side lobes, thus causing adjacent channel interference. Figure 33.4 illustrates the power density spectrum of a truncated square root raised cosine pulse for various truncation lengths τ.

33.5 Continuous Phase Modulation (CPM) and MSK

Continuous phase modulation (CPM) refers to a broad class of frequency modulation techniques where the carrier phase varies in a continuous manner. A comprehensive treatment of CPM is provided in [Anderson, Aulin, and Sundberg, 1986]. CPM schemes are attractive because they have constant envelope and excellent spectral characteristics. The complex envelope of any CPM signal is

$$v(t) = A \exp\left[j2\pi k_f \int_{-\infty}^{t} \sum_n x_n h_s(\tau - nT)\, d\tau \right] \qquad (33.20)$$

The instantaneous frequency deviation from the carrier is

$$f_{\text{dev}}(t) = k_f \sum_n x_n h_s(t - nT) \qquad (33.21)$$

where k_f is the peak frequency deviation. If the frequency shaping pulse $h_s(t)$ has duration T, then the equivalent shaping function in Eq. (33.2) has the form

$$b(t, \mathbf{x}_k) = \exp\left\{ j\left[\beta(T) \sum_{n=-\infty}^{k-1} x_n + x_k \beta(t) \right] \right\} u_T(t) \qquad (33.22)$$

where

$$\beta(t) = \begin{cases} 0, & t < 0 \\ \dfrac{\pi h}{\int_0^T h_s(\tau)\,d\tau} \displaystyle\int_0^t h_s(\tau)\,d\tau, & 0 \le t \le T \\ \pi h, & t \ge T \end{cases} \qquad (33.23)$$

is the phase shaping pulse, and $h = \beta(T)/\pi$ is called the modulation index.

Minimum shift keying (MSK) is a special form of binary CPM ($x_k \in \{-1, +1\}$) that is defined by a rectangular frequency shaping pulse $h_s(t) = u_T(t)$, and a modulation index $h = 1/2$ so that

$$\beta(t) = \begin{cases} 0, & t < 0 \\ \pi t/2T, & 0 \le t \le T \\ \pi/2, & t \ge T \end{cases} \qquad (33.24)$$

Therefore, the complex envelope is

$$v(t) = A \exp\left(j\frac{\pi}{2} \sum_{n=-\infty}^{k-1} x_n + \frac{\pi}{2} x_k \frac{t - kT}{T} \right) \qquad (33.25)$$

A MSK signal can be described by the phase trellis diagram shown in Fig. 33.5 which plots the time behavior of the phase

$$\theta(t) = \frac{\pi}{2} \sum_{n=-\infty}^{k-1} x_n + \frac{\pi}{2} x_k \frac{t - kT}{T} \qquad (33.26)$$

The MSK bandpass signal is

$$\begin{aligned} s(t) &= A \cos\left(2\pi f_c t + \frac{\pi}{2} \sum_{n=-\infty}^{k-1} x_n + \frac{\pi}{2} x_k \frac{t - kT}{T} \right) \\ &= A \cos\left[2\pi\left(f_c + \frac{x_k}{4T} \right)t - \frac{k\pi}{2} x_k + \frac{\pi}{2} \sum_{n=-\infty}^{k-1} x_n \right] \qquad kT \le t \le (k+1)T \end{aligned}$$

$$(33.27)$$

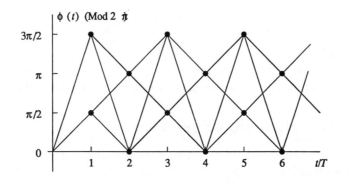

FIGURE 33.5   Phase-trellis diagram for MSK.

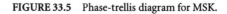

From Eq. (33.27) we observe that the MSK signal has one of two possible frequencies $f_L = f_c - 1/4T$ or $f_U = f_c + 1/4T$ during each symbol interval. The difference between these frequencies is $f_U - f_L = 1/2T$. This is the minimum frequency difference between two sinusoids of duration T that will ensure orthogonality with coherent demodulation [Proakis, 1995], hence the name minimum shift keying. By applying various trigonometric identities to Eq. (33.27) we can write

$$s(t) = A\big[x_k^I \psi(t - k2T)\cos(2\pi f_c t) - x_k^Q \psi(t - k2T - T)\sin(2\pi f_c t)\big],$$

$$kT \le t \le (k+1)T \quad (33.28)$$

where

$$x_k^I = -x_{k-1}^Q x_{2k-1}$$

$$x_k^Q = x_k^I x_{2k}$$

$$\psi(t) = \cos\left(\frac{\pi t}{2T}\right), \qquad -T \le t \le T$$

Note that the x_k^I and x_k^Q are independent binary symbols that take on elements from the set $\{-1, +1\}$, and the half-sinusoid amplitude shaping pulse $\psi(t)$ has duration $2T$ and $\psi(t - T) = \sin(\pi t/2T), 0 \le t \le 2T$. Therefore, MSK is equivalent to offset quadrature amplitude shift keying (OQASK) with a half-sinusoid amplitude shaping pulse.

To obtain the power density spectrum of MSK, we observe from Eq. (33.28) that the equivalent shaping function of MSK has the form

$$b(t, \mathbf{x}_k) = x_k^I \psi(t) + j x_k^Q \psi(t - T) \quad (33.29)$$

The Fourier transform of Eq. (33.29) is

$$B(f, \mathbf{x}_k) = \big(x_k^I + j x_k^Q e^{-j2\pi fT}\big)\Psi(f) \quad (33.30)$$

Since the symbols x_k^I and x_k^Q are independent and zero mean, it follows from Eqs. (33.5) and (33.6) that

$$S_{vv}(f) = \frac{A^2 |\Psi(f)|^2}{2T} \quad (33.31)$$

Therefore, the power density spectrum of MSK is determined solely by the Fourier transform of the half-sinusoid amplitude shaping pulse $\psi(t)$, resulting in

$$S_{vv}(f) = \frac{16A^2 T}{\pi^2}\left[\frac{\cos 2\pi fT}{1 - 16f^2 T^2}\right]^2 \quad (33.32)$$

The power spectral density of MSK is plotted in Fig. 33.8. Observe that an MSK signal has fairly large sidelobes compared to $\pi/4$-QPSK with a truncated square root raised cosine pulse (c.f., Fig. 33.4).

33.6 Gaussian Minimum Shift Keying

MSK signals have all of the desirable attributes for mobile radio, except for a compact power density spectrum. This can be alleviated by filtering the modulating signal $x(t) = \sum_n x_n u_T(t - nT)$ with a low-pass filter prior to frequency modulation, as shown in Fig. 33.6. Such filtering removes the higher frequency components in $x(t)$ and, therefore, yields a more compact spectrum. The low-pass

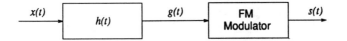

FIGURE 33.6 Premodulation filtered MSK.

filter is chosen to have 1) narrow bandwidth and a sharp transition band, 2) low-overshoot impulse response, and 3) preservation of the output pulse area to ensure a phase shift of $\pi/2$.

GMSK uses a low-pass filter with the following transfer function:

$$H(f) = A \exp\left\{-\left(\frac{f}{B}\right)^2 \frac{\ln 2}{2}\right\} \tag{33.33}$$

where B is the 3-dB bandwidth of the filter and A a constant. It is apparent that $H(f)$ is bell shaped about $f = 0$, hence the name Gaussian MSK. A rectangular pulse $\text{rect}(t/T) = u_T(t + T/2)$ transmitted through this filter yields the frequency shaping pulse

$$h_s(t) = A\sqrt{\frac{2\pi}{\ln 2}}(BT) \int_{t/T-1/2}^{t/T+1/2} \exp\left\{-\frac{2\pi^2(BT)^2 x^2}{\ln 2}\right\} dx \tag{33.34}$$

The phase change over the time interval from $-T/2 \le t \le T/2$ is

$$\theta\left(\frac{T}{2}\right) - \theta\left(\frac{-T}{2}\right) = x_0\beta_0(T) + \sum_{\substack{n=-\infty \\ n\neq 0}}^{\infty} x_n\beta_n(T) \tag{33.35}$$

where

$$\beta_n(T) = \frac{\pi h}{\int_{-\infty}^{\infty} h_s(v)\, dv} \int_{-T/2-nT}^{T/2-nT} h_s(v)\, dv \tag{33.36}$$

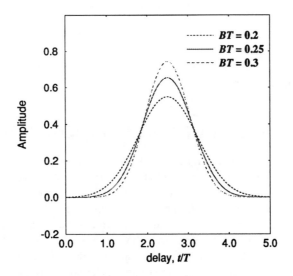

FIGURE 33.7 GMSK frequency shaping pulse for various normalized filter bandwidths BT.

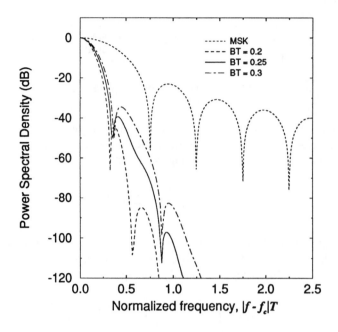

FIGURE 33.8   Power density spectrum of MSK and GMSK.

The first term in Eq. (33.35) is the desired term, and the second term is the intersymbol interference (ISI) introduced by the premodulation filter. Once again, with GMSK $h = 1/2$ so that a total phase shift of $\pi/2$ is maintained.

Notice that the pulse $h_s(t)$ is noncausal so that a truncated pulse must be used in practice. Figure 33.7 plots a GMSK frequency shaping pulse that is truncated to $\tau = 5T$ and time shifted by $2.5T$, for various normalized filter bandwidths BT. Notice that the frequency shaping pulse has a duration greater than T so that ISI is introduced. As BT decreases, the induced ISI is increased. Thus, whereas a smaller value of BT results in a more compact power density spectrum, the induced ISI will degrade the bit-error-rate performance. Hence, there is a tradeoff in the choice of BT. Some studies have indicated that $BT = 0.25$ is a good choice for cellular radio systems [Murota, Kinoshita, and Hirade, 1981].

The power density spectrum of GMSK is quite difficult to obtain, but can be computed by using published methods [Garrison, 1975]. Fig. 33.8 plots the power density spectrum for $BT = 0.2$, 0.25, and 0.3, obtained from Wesolowski, 1994. Observe that the spectral sidelobes are greatly reduced by the Gaussian low-pass filter.

33.7 Orthogonal Frequency Division Multiplexing (OFDM)

Orthogonal frequency division multiplexing (OFDM) is a modulation technique that has been recently suggested for use in cellular radio [Birchler and Jasper, 1992], digital audio broadcasting [Le Floch, Halbert-Lassalle, and Castelain, 1989], and digital video broadcasting. The basic idea of OFDM is to transmit blocks of symbols in parallel by employing a (large) number of orthogonal subcarriers. With block transmission, N serial source symbols each with period T_s are converted into a block of N parallel modulated symbols each with period $T = NT_s$. The block length N is chosen so that $NT_s \gg \sigma_\tau$, where σ_τ is the rms delay spread of the channel. Since the symbol rate on each subcarrier is much less than the serial source rate, the effects of delay spread are greatly reduced. This has practical advantages because it may reduce or even eliminate the need for equalization. Although the block length N is chosen so that $NT_s \gg \sigma_\tau$, the channel dispersion will still cause consecutive blocks to overlap. This results in some residual ISI that will degrade the performance. This residual ISI can be eliminated at the expense of channel capacity by using guard intervals between the blocks that are at least as long as the effective channel impulse response.

The complex envelope of an OFDM signal is described by

$$v(t) = A \sum_{k} \sum_{n=0}^{N-1} x_{k,n} \phi_n(t - kT) \tag{33.37}$$

where

$$\phi_n(t) = \exp\left\{ j \frac{2\pi \left(n - \frac{N-1}{2} \right) t}{T} \right\} U_T(t), \qquad n = 0, 1, \ldots, N - 1 \tag{33.38}$$

are orthogonal waveforms and $U_T(t)$ is a rectangular shaping function. The frequency separation of the subcarriers, $1/T$, ensures that the subcarriers are orthogonal and phase continuity is maintained from one symbol to the next, but is twice the minimum required for orthogonality with coherent detection. At epoch k, N-data symbols are transmitted by using the N distinct pulses. The data symbols $x_{k,n}$ are often chosen from an M-ary **quadrature amplitude modulation** (M-QAM) constellation, where $x_{k,n} = x_{k,n}^I + j x_{k,n}^Q$ with $x_{k,n}^I, x_{k,n}^Q \in \{\pm 1, \pm 3, \ldots, \pm(N-1)\}$ and $N = \sqrt{M}$.

A key advantage of using OFDM is that the modulation can be achieved in the discrete domain by using either an inverse discrete Fourier transform (IDFT) or the more computationally efficient inverse fast Fourier transform (IFFT). Considering the data block at epoch $k = 0$ and ignoring the frequency offset $\exp\{-j[2\pi(N-1)t/2T]\}$, the complex low-pass OFDM signal has the form

$$v(t) = \sum_{n=0}^{N-1} x_{0,n} \exp\left\{ \frac{j2\pi nt}{NT_s} \right\}, \qquad 0 \le t \le T \tag{33.39}$$

If this signal is sampled at epochs $t = kT_s$, then

$$v^k = v(kT_s) = \sum_{n=0}^{N-1} x_{0,n} \exp\left\{ \frac{j2\pi nk}{N} \right\}, \qquad k = 0, 1, \ldots, N - 1 \tag{33.40}$$

Observe that the sampled OFDM signal has duration N and the samples $v^0, v^1, \ldots, v^{N-1}$ are just the IDFT of the data block $x_{0,0}, x_{0,1}, \ldots, x_{0,N-1}$. A block diagram of an OFDM transmitter is shown in Fig. 33.9.

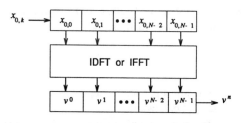

FIGURE 33.9 Block diagram of OFDM transmitter using IDFT or IFFT.

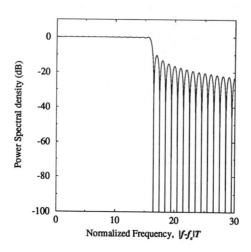

FIGURE 33.10 Power density spectrum of OFDM with $N = 32$.

The power spectral density of an OFDM signal can be obtained by treating OFDM as independent modulation on subcarriers that are separated in frequency by $1/T$. Because the subcarriers are only separated by $1/T$, significant spectral overlap results. Because the subcarriers are orthogonal, however, the overlap improves the spectral efficiency of the scheme. For a signal constellation with zero mean and the waveforms in Eq. (33.38), the power density spectrum of the complex envelope is

$$S_{vv}(f) = \frac{A^2}{T}\sigma_x^2 \sum_{n=0}^{N-1} \left| \text{sinc}\left[fT - \left(n - \frac{N-1}{2} \right) \right] \right|^2 \qquad (33.41)$$

where $\sigma_x^2 = \frac{1}{2}E[|x_{k,n}|^2]$ is the variance of the signal constellation. For example, the complex envelope power spectrum of OFDM with $N = 32$ subcarriers is shown in Fig. 33.10.

33.8 Conclusions

A variety of modulation schemes are employed in wireless communication systems. Wireless modulation schemes must have a compact power density spectrum, while at the same time providing a good bit-error-rate performance in the presence of channel impairments such as cochannel interference and fading. The most popular digital modulation techniques employed in wireless systems are GMSK in the European GSM system, $\pi/4$-QPSK in the North American IS-54 and Japanese PDC systems, and OFDM in digital audio broadcasting systems.

Defining Terms

Bandwidth efficiency: Transmission efficiency of a digital modulation scheme measured in units of bits per second per Hertz of bandwidth.

Power spectral density: Relative power in a modulated signal as a function of frequency.

Frequency modulation: Modulation where the instantaneous frequency of the carrier varies linearly with the data signal.

Phase shift keying: Modulation where the instantaneous phase of the carrier varies linearly with the data signal.

Excess bandwidth: Percentage of bandwidth that is in excess of the minimum of $1/2T$ (T is the baud or symbol duration) required for data communication.

Continuous phase modulation: Frequency modulation where the phase varies in a continuous manner.

Minimum shift keying: A special form of continuous phase modulation having linear phase trajectories and a modulation index of $1/2$.

Gaussian minimum shift keying: MSK where the data signal is prefiltered with a Gaussian filter prior to frequency modulation.

Orthogonal frequency division multiplexing: Modulation by using a collection of low-bit-rate orthogonal subcarriers.

Quadrature amplitude modulation: Modulation where information is transmitted in the amplitude of the cosine and sine components of the carrier.

References

Anderson, J.B., Aulin, T., and Sundberg, C.-E. 1986. *Digital Phase Modulation*, Plenum Press, New York.

Birchler, M.A., and Jasper, S.C, 1992. A 64 kbps digital land mobile radio system employing M-16QAM. *Proc. 5th Nordic Sem. Dig. Mobile Radio Commun.* Dec.:237–241.

Le Floch, B., Halbert-Lassalle, R., and Castelain, D. 1989. Digital sound broadcasting to mobile receivers, *IEEE Trans. Consum. Elec.* 35:(Aug.).

Garrison, G.J. 1975. A power spectral density analysis for digital FM, *IEEE Trans. Commun.*, COM-23(Nov.):1228–1243.

Wesolowski, K. 1994. Private Communication.

Murota, K., Kinoshita, K., and Hirade, K. 1981. Spectral efficiency of GMSK land mobile radio. *Proc. ICC'81.* (June):23.8.1.

Murota, K. and Hirade, K. 1981. GMSK modulation for digital mobile radio telephony, *IEEE Trans. Commun.*, COM-29(July):1044–1050.

Proakis, J.G. 1989. *Digital Communications*, 2nd ed. McGraw-Hill, New York.

Further Information

A good discussion of digital modem techniques is presented in *Advanced Digital Communications*, edited by K. Feher, Prentice-Hall, 1987.

Proceedings of various IEEE conferences such as the Vehicular Technology Conference, International Conference on Communications, and Global Telecommunications Conference, document the lastest development in the field of wireless communications each year.

Journals such as the *IEEE Transactions on Communications* and *IEEE Transactions on Vehicular Technology* report advances in wireless modulation.

34

Wireless LANs

Suresh Singh
University of South Carolina

34.1 Introduction

A proliferation of high-performance portable computers combined with end-user need for communication is fueling a dramatic growth in wireless **local area network** (LAN) technology. Users expect to have the ability to operate their portable computer globally while remaining connected to communications networks and service providers. Wireless LANs and cellular networks, connected to high-speed networks, are being developed to provide this functionality.

Before delving deeper into issues relating to the design of wireless LANs, it is instructive to consider some scenarios of user mobility.

1. A simple model of user mobility is one where a computer is physically moved while retaining network connectivity at either end. For example, a move from one room to another as in a hospital where the computer is a hand-held device displaying patient charts and the nurse using the computer moves between wards or floors while accessing patient information.

2. Another model situation is where a group of people (at a conference, for instance) set up an ad-hoc LAN to share information as in Fig. 34.1.

3. A more complex model is one where several computers in constant communication are in motion and continue to be networked. For example, consider the problem of having robots in space collaborating to retrieve a satellite.

A great deal of research has focused on dealing with physical layer and **medium access control** (MAC) layer protocols. In this chapter we first summarize standardization efforts in these areas. The remainder of the chapter is then devoted to a discussion of networking issues involved in wireless LAN design. Some of the issues discussed include routing in wireless LANs (i.e., how does data find its destination when the destination is mobile?) and the problem of providing service guarantees

0-8493-8573-3/96/$0.00+$.50

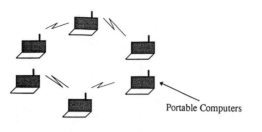

FIGURE 34.1   Ad-hoc wireless LAN.

to end users (e.g., error-free data transmission or bounded delay and bounded bandwidth service, etc.).

34.2 Physical Layer Design

Two media are used for transmission over wireless LANs, infrared and radio frequency. RF LANs are typically implemented in the industrial, scientific, and medical (ISM) frequency bands 902–928 MHz, 2400–2483.5 MHz and 5725–5850 MHz. These frequencies do not require a license allowing the LAN product to be portable, i.e., a LAN can be moved without having to worry about licensing.

IR and RF technologies have different design constraints. IR receiver design is simple (and thus inexpensive) in comparison to RF receiver design because IR receivers only detect the amplitude of the signal not the frequency or phase. Thus, a minimal of filtering is required to reject interference. Unfortunately, however, IR shares the electromagnetic spectrum with the sun and incandescent or fluorescent light. These sources of modulated infrared energy reduce the signal-to-noise ratio of IR signals and, if present in extreme intensity, can make the IR LANs inoperable. There are two approaches to building IR LANs.

1. The transmitted signal can be focused and aimed. In this case the IR system can be used outdoors and has an area of coverage of a few kilometers.
2. The transmitted signal can be bounced off the ceiling or radiated omnidirectionally. In either case, the range of the IR source is 10–20 m (i.e., the size of one medium-sized room).

RF systems face harsher design constraints in comparison to IR systems for several reasons. The increased demand for RF products has resulted in tight regulatory constraints on the allocation and use of allocated bands. In the U.S., for example, it is necessary to implement spectrum spreading for operation in the ISM bands. Another design constraint is the requirement to confine the emitted spectrum to a band, necessitating amplification at higher carrier frequencies, frequency conversion using precision local oscillators, and selective components. RF systems must also cope with environmental noise that is either naturally occurring, for example, atmospheric noise or man made, for example, microwave ovens, copiers, laser printers, or other heavy electrical machinery. RF LANs operating in the ISM frequency ranges also suffer interference from amateur radio operators.

Operating LANs indoors introduces additional problems caused by multipath propagation, Rayleigh fading, and absorption. Many materials used in building construction are opaque to IR radiation resulting in incomplete coverage within rooms (the coverage depends on obstacles within the room that block IR) and almost no coverage outside closed rooms. Some materials, such as white plasterboard, can also cause reflection of IR signals. RF is relatively immune to absorption and reflection problems. Multipath propagation affects both IR and RF signals. The technique to alleviate the effects of multipath propagation in both types of systems is the same use of aimed (directional) systems for transmission enabling the receiver to reject signals based on their angle of incidence. Another technique that may be used in RF systems is to use multiple antennas. The phase difference between different paths can be used to discriminate between them.

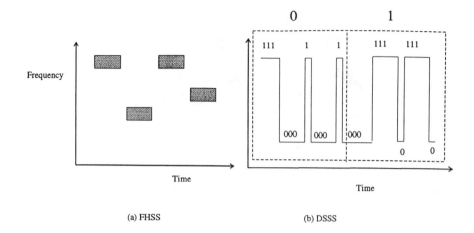

Time

(a) FHSS   (b) DSSS

FIGURE 34.2   Spread spectrum.

Rayleigh fading is a problem in RF systems. Recall that Rayleigh fading occurs when the difference in path length of the same signal arriving along different paths is a multiple of half a wavelength. This causes the signal to be almost completely canceled out at the receiver. Because the wavelengths used in IR are so small, the effect of Rayleigh fading is not noticeable in those systems. RF systems, on the other hand, use wavelengths of the order of the dimension of a laptop. Thus, moving the computer a small distance could increase/decrease the fade significantly.

Spread spectrum transmission technology is used for RF-based LANs and it comes in two varieties: direct-sequence spread spectrum (DSSS) and frequency-hopping spread spectrum (FHSS). In a FHSS system, the available band is split into several channels. The transmitter transmits on one channel for a fixed time and then hops to another channel. The receiver is synchronized with the transmitter and hops in the same sequence; see Fig. 34.2(a). In DSSS systems, a random binary string is used to modulate the transmitted signal. The relative rate between this sequence and user data is typically between 10 and 100; see Fig. 34.2(b).

The key requirements of any transmission technology is its robustness to noise. In this respect DSSS and FHSS show some differences. There are two possible sources of interference for wireless LANs: the presence of other wireless LANs in the same geographical area (i.e., in the same building, etc.) and interference due to other users of the ISM frequencies. In the latter case, FHSS systems have a greater ability to avoid interference because the hopping sequence could be designed to prevent potential interference. DSSS systems, on the other hand, do exhibit an ability to recover from interference because of the use of the spreading factor [Fig. 34.2(b)].

It is likely that in many situations several wireless LANs may be collocated. Since all wireless LANs use the same ISM frequencies, there is a potential for a great deal of interference. To avoid interference in FHSS systems, it is necessary to ensure that the hopping sequences are orthogonal. To avoid interference in DSSS systems, on the other hand, it is necessary to allocate different channels to each wireless LAN. The ability to avoid interference in DSSS systems is, thus, more limited in comparison to FHSS systems because FHSS systems use very narrow subchannels (1 MHz) in comparison to DSSS systems that use wider subchannels (for example, 25 MHz), thus, limiting the number of wireless LANs that can be collocated. A summary of design issues can be found in Bantz and Bauchot, 1994.

34.3   MAC Layer Protocols

MAC protocol design for wireless LANs poses new challenges because of the in-building operating environment for these systems. Unlike wired LANs (such as the ethernet or token ring), wireless LANs operate in strong multipath fading channels where channel characteristics can change in very

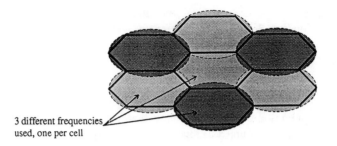

3 different frequencies
used, one per cell

FIGURE 34.3 Cellular structure for wireless LANs (note frequency reuse).

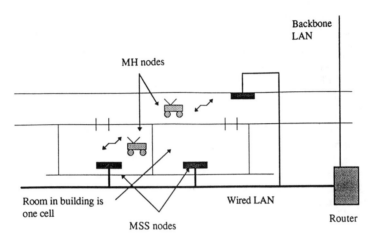

FIGURE 34.4 In-building LAN (made up of several wireless LANs).

short distances resulting in unreliable communication and unfair channel access due to capture. Another feature of the wireless LAN environment is that carrier sensing takes a long time in comparison to wired LANs; it typically takes between 30 and 50 μs (see Chen and Lee, 1994), which is a significant portion of the packet transmission time. This results in inefficiencies if the CSMA family of protocols is used without any modifications.

Other differences arise because of the mobility of users in wireless LAN environments. To provide a building (or any other region) with wireless LAN coverage, the region to be covered is divided into cells as shown in Fig. 34.3. Each cell is one wireless LAN, and adjacent cells use different frequencies to minimize interference. Within each cell there is an access point called a **mobile support station** (MSS) or base station that is connected to some wired network. The mobile users are called **mobile hosts (MH)**. The MSS performs the functions of channel allocation and providing connectivity to existing wired networks; see Fig. 34.4. Two problems arise in this type of an architecture that are not present in wired LANs.

1. The number of nodes within a cell changes dynamically as users move between cells. How can the channel access protocol dynamically adapt to such changes efficiently?
2. When a user moves between cells, the user has to make its presence known to the other nodes in the cell. How can this be done without using up too much bandwidth? The protocol used to solve this problem is called a handoff protocol and works along the following lines: A switching station (or the MSS nodes working together, in concert) collects signal strength information for each mobile host within each cell. Note that if a mobile host is near a

cell boundary, the MSS node in its current cell as well as in the neighboring cell can hear its transmissions and determine signal strengths. If the mobile host is currently under the coverage of MSS M1 but its signal strength at MSS M2 becomes larger, the switching station initiates a handoff whereby the MH is considered as part of M2's cell (or network).

The mode of communication in wireless LANs can be broken in two: communication from the mobile to the MSS (called *uplink* communication) and communication in the reverse direction (called *downlink* communication). It is estimated that downlink communication accounts for about 70–80% of the total consumed bandwidth. This is easy to see because most of the time users request files or data in other forms (image data, etc.) that consume much more transmission bandwidth than the requests themselves. In order to make efficient use of bandwidth (and, in addition, guarantee service requirements for real-time data), most researchers have proposed that the downlink channel be controlled entirely by the MSS nodes. These nodes allocate the channel to different mobile users based on their current requirements using a protocol such as **time division multiple access** (TDMA). What about uplink traffic? This is a more complicated problem because the set of users within a cell is dynamic, thus making it infeasible to have a static channel allocation for the uplink. This problem is the main focus of MAC protocol design.

What are some of the design requirements of an appropriate MAC protocol? The IEEE 802.11 recommended standard for wireless LANs has identified almost 20 such requirements, some of which are discussed here (the reader is referred to Chen, 1994, for further details). Clearly any protocol must maximize throughput while minimizing delays and providing fair access to all users. In addition to these requirements, however, mobility introduces several new requirements.

1. The MAC protocol must be independent of the underlying physical layer transmission technology adopted (be it DSSS, FHSS or IR).
2. The maximum number of users can be as high as a few hundred in a wireless LAN. The MAC protocol must be able to handle many users without exhibiting catastrophic degradation of service.
3. The MAC protocols must provide secure transmissions because the wireless medium is easy to tap.
4. The MAC protocol needs to work correctly in the presence of collocated networks.
5. It must have the ability to support ad-hoc networking (as in Fig. 34.1).
6. Other requirements include the need to support priority traffic, preservation of packet order, and an ability to support multicast.

Several contention-based protocols currently exist that could be adapted for use in wireless LANs. The protocols currently being looked by IEEE 802.11 include protocols based on **carrier sense multiple access** (CSMA), polling, and TDMA. Protocols based on **code division multiple access** (CDMA) and **frequency division multiple access** (FDMA) are not considered because the processing gains obtained using these protocols are minimal while, simultaneously, resulting in a loss of flexibility for wireless LANs.

It is important to highlight an important difference between networking requirements of ad-hoc networks (as in Fig. 34.1) and networks based on cellular structure. In cellular networks, all communication occurs between the mobile hosts and the MSS (or base station) within that cell. Thus, the MSS can allocate channel bandwidth according to requirements of different nodes, i.e., we can use centralized channel scheduling for efficient use of bandwidth. In ad-hoc networks there is no such central scheduler available. Thus, any multiaccess protocol will be contention based with little explicit scheduling. In the remainder of this section we focus on protocols for cell-based wireless LANs only.

All multiaccess protocols for cell-based wireless LANs have a similar structure; see Chen, 1994.

1. The MSS announces (explicitly or implicitly) that nodes with data to send may contend for the channel.

2. Nodes interested in sending data contend for the channel using protocols such as CSMA.
3. The MSS allocates the channel to successful nodes.
4. Nodes transmit packets (contention-free transmission).
5. MSS sends an explicit acknowledgment (ACK) for packets received.

Based on this model we present three MAC protocols.

Reservation-TDMA (R-TDMA)

This approach is a combination of TDMA and some contention protocol (see PRMA in Goodman, 1990). The MSS divides the channel into slots (as in TDMA), which are grouped into frames. When a node wants to transmit it needs to reserve a slot that it can use in every consecutive frame as long as it has data to transmit. When it has completed transmission, other nodes with data to transmit may contend for that free slot. There are four steps to the functioning of this protocol.

a. At the end of each frame the MSS transmits a feedback packet that informs nodes of the current reservation of slots (and also which slots are free). This corresponds to steps 1 and 3 from the preceding list.
b. During a frame, all nodes wishing to acquire a slot transmit with a probability ρ during a free slot. If a node is successful it is so informed by the next feedback packet. If more than one node transmits during a free slot, there is a collision and the nodes try again during the next frame. This corresponds to step 2.
c. A node with a reserved slot transmits data during its slot. This is the contention-free transmission (step 4).
d. The MSS sends ACKs for all data packets received correctly. This is step 5.

The R-TDMA protocol exhibits several nice properties. First and foremost, it makes very efficient use of the bandwidth, and average latency is half the frame size. Another big benefit is the ability to implement power conserving measures in the portable computer. Since each node knows when to transmit (nodes transmit during their reserved slot only) it can move into a power-saving mode for a fixed amount of time, thus increasing battery life. This feature is generally not available in CSMA-based protocols. Furthermore, it is easy to implement priorities because of the centralized control of scheduling. One significant drawback of this protocol is that it is expensive to implement (see Barke and Badrinath, 1994).

Distributed Foundation Wireless MAC (DFWMAC)

The CSMA/CD protocol has been used with great success in the ethernet. Unfortunately, the same protocol is not very efficient in a wireless domain because of the problems associated with cell interference (i.e., interference from neighboring cells), the relatively large amount of time taken to sense the channel (see Glisic, 1991) and the hidden terminal problem (see Tobagi and Kleinrock, 1975a and 1975b). The current proposal is based on a CSMA/collision avoidance (CA) protocol with a four-way handshake; see Fig. 34.5.

The basic operation of the protocol is simple. All MH nodes that have packets to transmit compete for the channel by sending ready to transmit (RTS) messages using nonpersistent CSMA. After a station succeeds in transmitting a RTS, the MSS sends a clear to transmit (CTS) to the MH. The MH transmits its data and then receives an ACK. The only possibility of collision that exists is in the RTS phase of the protocol and inefficiencies occur in the protocol, because of the RTS and CTS stages. Note that unlike R-TDMA it is harder to implement power saving functions. Furthermore, latency is dependent on system load making it harder to implement real-time guarantees. Priorities are also not implemented. On the positive side, the hardware for this protocol is very inexpensive.

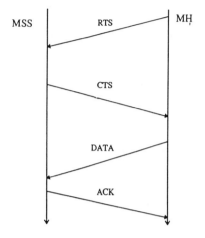

FIGURE 34.5 CSMA/CA and four-way handshaking protocol.

Randomly Addressed Polling (RAP)

In this scheme, when a MSS is ready to collect uplink packets it transmits a READY message. At this point all nodes with packets to send attempt to grab the channel as follows.

a. Each MH with a packet to transmit generates a random number between 0 and P.

b. All active MH nodes simultaneously and orthogonally transmit their random numbers (using CDMA or FDMA). We assume that all of these numbers are received correctly by the MSS. Remember that more than one MH node may have selected the same random number.

c. Steps a and b are repeated L times.

d. At the end of L stages, the MSS determines a stage (say, k) where the total number of distinct random numbers was the largest. The MSS polls each distinct each random number in this stage in increasing order. All nodes that had generated the polled random number transmit packets to the MSS.

e. Since more than one node may have generated the same random number, collisions are possible. The MSS sends a ACK or NACK after each such transmission. Unsuccessful nodes try again during the next iteration of the protocol.

The protocol is discussed in detail in Chen and Lee, 1993 and a modified protocol called GRAP (for group RAP) is discussed in Chen, 1994. The authors propose that GRAP can also be used in the contention stage (step 2) for TDMA- and CSMA-based protocols.

34.4 Network Layer Issues

An important goal of wireless LANs is to allow users to move about freely while still maintaining all of their connections (network resources permitting). This means that the network must route all packets destined for the mobile user to the MSS of its current cell in a transparent manner. Two issues need to be addressed in this context.

- How can users be addressed?
- How can active connections for these mobile users be maintained?

Ioanidis, Duchamp, and Maguire, 1991 propose a solution called the IPIP (IP-within-IP) protocol. Here each MH has a unique **internet protocol** (IP) address called its home address. To deliver a packet to a remote MH, the source MSS first broadcasts an address resolution protocol (ARP) request to all other MSS nodes to locate the MH. Eventually some MSS responds. The source MSS then encapsulates each packet from the source MH within another packet containing the IP address of the MSS in whose cell the MH is located. The destination MSS extracts the packet and delivers it to the MH. If the MH has moved away in the interim, the new MSS locates the new location of the MH and performs the same operation. This approach suffers from several problems as discussed in Teraoka and Tokoro, 1993. Specifically, the method is not scaleable to a network spanning areas larger than a campus for the following reasons.

1. IP addresses have a prefix identifying the campus subnetwork where the node lives; when the MH moves out of the campus, its IP address no longer represents this information.
2. The MSS nodes serve the function of routers in the mobile network and, therefore, have the responsibility of tracking all of the MH nodes globally causing a lot of overhead in terms of message passing and packet forwarding; see Ghai and Singh, 1994.

Teraoka and Tokoro, 1993, have proposed a much more flexible solution to the problem called virtual IP (VIP). Here every mobile host has a virtual IP address that is unchanging regardless of the location of the MH. In addition, hosts have physical network addresses (traditional IP addresses) that may change as the host moves about. At the transport layer, the target node is always specified by its VIP address only. The address resolution from the VIP address to the current IP address takes place either at the network layer of the same machine or at a gateway. Both the host machines and the gateways maintain a cache of VIP to IP mappings with associated timestamps. This information is in the form of a table called *address mapping table* (AMT). Every MH has an associated *home gateway*. When a MH moves into a new subnetwork, it is assigned a new IP address. It sends this new IP address and its VIP address to its home gateway via a *VipConn* control message. All intermediate gateways that relay this message update their AMT tables as well. During this process of updating the AMT tables, all packets destined to the MH continue to be sent to the old location. These packets are returned to the sender, who then sends them to the home gateway of the MH. It is easy to see that this approach is easily scaleable to large networks, unlike the IPIP approach.

Alternative View of Mobile Networks

The approaches just described are based on the belief that mobile networks are merely an extension of wired networks. Other authors [Singh, 1995] disagree with this assumption because there are fundamental differences between the mobile domain and the fixed wired network domain. Two examples follow.

1. The available bandwidth at the wireless link is small; thus, end-to-end packet retransmission for transmission control protocol (TCP)-like protocols (implemented over datagram networks) is a bad idea. This leads to the conclusion that transmission within the mobile network must be connection oriented. Such a solution, using virtual circuits (VC), is proposed in Ghai and Singh, 1994.
2. The bandwidth available for a MH with open connections changes dynamically since the number of other users present in each cell varies randomly. This is a feature not present in fixed high-speed networks where, once a connection is set up, its bandwidth does not vary much. Since bandwidth changes are an artifact of mobility and are dynamic, it is necessary to deal with the consequences (e.g., buffer overflow, large delays, etc.) locally to both, i.e., shield fixed network hosts from the idiosyncrasies of mobility as well as to respond to changing bandwidth quickly (without having to rely on end-to-end control). Some other differences are discussed in Singh, 1995.

A Proposed Architecture

Keeping these issues in mind, a more appropriate architecture has been proposed in Ghai and Singh, 1994, and Singh, 1995. Mobile networks are considered to be different and separate from wired networks. Within a mobile network is a three-layer hierarchy; see Fig. 34.6. At the bottom layer are the MHs. At the next level are the MSS nodes (one per cell). Finally, several MSS nodes are controlled by a **supervisor host (SH)** node (there may be one SH node per small building). The SH nodes are responsible for flow control for all MH connections within their domain; they are also responsible for tracking MH nodes and forwarding packets as MH nodes roam. In addition, the SH nodes serve as a *gateway* to the wired networks. Thus, any connection setup from a MH to a fixed host is broken in two, one from the MH to the SH and another from the SH to the fixed host. The MSS nodes in this design are simply connection endpoints for MH nodes. Thus, they are simple devices that implement the MAC protocols and little else. Some of the benefits of this design are as follows.

1. Because of the large coverage of the SH (i.e., a SH controls many cells) the MH remains in the domain of one SH much longer. This makes it easy to handle the consequences of dynamic bandwidth changes locally. For instance, when a MH moves into a crowded cell, the bandwidth available to it is reduced. If it had an open ftp connection, the SH simply buffers undelivered packets until they can be delivered. There is no need to inform the other endpoint of this connection of the reduced bandwidth.

2. When a MH node sets up a connection with a service provider in the fixed network, it negotiates some quality of service (QOS) parameters such as bandwidth, delay bounds, etc. When the MH roams into a crowded cell, these QOS parameters can no longer be met because the available bandwidth is smaller. If the traditional view is adopted (i.e., the mobile networks are extensions of fixed networks) then these QOS parameters will have to be renegotiated each time the bandwidth changes (due to roaming). This is a very expensive proposition because of the large number of control messages that will have to be exchanged. In the approach of Singh, 1995, the service provider will never know about the bandwidth changes since it deals only with the SH that is accessed via the wired network. The SH bears the responsibility of handling bandwidth changes by either buffering packets until the bandwidth available to the MH increases (as in the case of the ftp example) or it could discard a fraction of real-time packets (e.g., a voice connection) to ensure delivery of most of the packets within their deadlines. The SH could also instruct the MSS to allocate a larger amount of bandwidth to the MH when the number of buffered packets becomes large. Thus, the service provider in the fixed network is shielded from the mobility of the user.

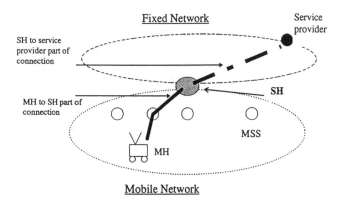

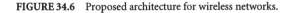

FIGURE 34.6   Proposed architecture for wireless networks.

Networking Issues

It is important for the network to provide connection-oriented service in the mobile environment (as opposed to connectionless service as in the internet) because bandwidth is at a premium in wireless networks, and it is, therefore, inadvisable to have end-to-end retransmission of packets (as in TCP). The proposed architecture is well suited to providing connection-oriented service by using VCs.

In the remainder of this section we look at how virtual circuits are used within the mobile network and how routing is performed for connections to mobile hosts. Every connection set up with one or more MH nodes as a connection endpoint is routed through the SH nodes and each connection is given a unique VC number. The SH node keeps track of all MH nodes that lie within its domain. When a packet needs to be delivered to a MH node, the SH first buffers the packet and then sends it to the MSS at the current location of the MH or to the predicted location if the MH is currently between cells. The MSS buffers all of these packets for the MH and transmits them to the MH if it is in its cell. The MSS discards packets after transmission or if the SH asks it to discard the packets. Packets are delivered in the correct order to the MH (without duplicates) by having the MH transmit the expected sequence number (for each VC) during the initial handshake (i.e., when the MH first enters the cell). The MH sends ACKs to the SH for packets received. The SH discards all packets that have been acknowledged. When a MH moves from the domain of SH1 into the domain of SH2 while having open connections, SH1 continues to forward packets to SH2 until either the connections are closed or until SH2 sets up its own connections with the other endpoints for each of MH's open connections (it also gives new identifiers to all these open connections). The detailed protocol is presented in Ghai and Singh, 1994.

The SH nodes are all connected over the fixed (wired) network. Therefore, it is necessary to route packets between SH nodes using the protocol provided over the fixed networks. The VIP protocol appears to be best suited to this purpose. Let us assume that every MH has a globally unique VIP address. The SHs have both a VIP as well as a fixed IP address. When a MH moves into the domain of a SH, the IP address affixed to this MH is the IP address of the SH. This ensures that all packets sent to the MH are routed through the correct SH node. The SH keeps a list of all VIP addresses of MH nodes within its domain and a list of open VCs for each MH. It uses this information to route the arriving packets along the appropriate VC to the MH.

34.5 Transport Layer Design

The transport layer provides services to higher layers (including the application layer), which include connectionless services like UDP or connection-oriented services like TCP. A wide variety of new services will be made available in the high-speed networks, such as continuous media service for real-time data applications such as voice and video. These services will provide bounds on delay and loss while guaranteeing some minimum bandwidth.

Recently variations of the TCP protocol have been proposed that work well in the wireless domain. These proposals are based on the traditional view that wireless networks are merely extensions of fixed networks. One such proposal is called I-TCP [Barke and Badrinath, 1994] for indirect TCP. The motivation behind this work stems from the following observation. In TCP the sender times out and begins retransmission after a timeout period of several hundred milliseconds. If the other endpoint of the connection is a mobile host, it is possible that the MH is disconnected for a period of several seconds (while it moves between cells and performs the initial greeting). This results in the TCP sender timing out and transmitting the same data several times over, causing the effective throughput of the connection to degrade rapidly. To alleviate this problem, the implementation of I-TCP separates a TCP connection into two pieces—one from the fixed host to another fixed host that is near the MH and another from this host to the MH (note the similarity of this approach with the approach in Fig. 34.6). The host closer to the MH is aware of mobility and has a larger timeout period. It serves as a type of gateway for the TCP connection because it sends ACKs back to the sender before receiving ACKs from the MH. The performance of I-TCP is far superior to traditional TCP for the mobile networks studied.

In the architecture proposed in Fig. 34.6, a TCP connection from a fixed host to a mobile host would terminate at the SH. The SH would set up another connection to the MH and would have the responsibility of transmitting all packets correctly. In a sense this is a similar idea to I-TCP except that in the wireless network VCs are used rather than datagrams. Therefore, the implementation of TCP service is made much easier.

A problem that is unique to the mobile domain occurs because of the unpredictable movement of MH nodes (i.e., a MH may roam between cells resulting in a large variation of available bandwidth in each cell). Consider the following example. Say nine MH nodes have opened 11-kb/s connections in a cell where the available bandwidth is 100 kb/s. Let us say that a tenth mobile host M10, also with an open 11-kb/s connection, wanders in. The total requested bandwidth is now 110 kb/s while the available bandwidth is only 100 kb/s. What is to be done? One approach would be to deny service to M10. However, this seems an unfair policy. A different approach is to penalize all connections equally so that each connection has 10-kb/s bandwidth allocated.

To reduce the bandwidth for each connection from 11 kb/s to 10 kb/s, two approaches may be adopted:

1. Throttle back the sender for each connection by sending control messages.
2. Discard 1-kb/s data for each connection at the SH. This approach is only feasible for applications that are tolerant of data loss (e.g., real-time video or audio).

The first approach encounters a high overhead in terms of control messages and requires the sender to be capable of changing the data rate dynamically. This may not always be possible; for instance, consider a teleconference consisting of several participants where each mobile participant is subject to dynamically changing bandwidth. In order to implement this approach, the data (video or audio or both) will have to be compressed at different ratios for each participant, and this compression ratio may have to be changed dynamically as each participant roams. This is clearly an unreasonable solution to the problem. The second approach requires the SH to discard 1-kb/s of data for each connection. The question is, how should this data be discarded? That is, should the 1 kb of discarded data be consecutive (or clustered) or uniformly spread out over the data stream every 1 s ? The way in which the data is discarded has an effect on the final perception of the service by the mobile user. If the service is audio, for example, a random uniform loss is preferred to a clustered loss (where several consecutive words are lost). If the data is compressed video, the problem is even more serious because most random losses will cause the encoded stream to become unreadable resulting in almost a 100% loss of video at the user.

A solution to this problem is proposed in Seal and Singh, 1995, where a new sublayer is added to the transport layer called the *Loss profile transport sublayer (LPTSL)*. This layer determines how data is to be discarded based on special transport layer markers put by application calls at the sender and based on negotiated loss functions that are part of the QOS negotiations between the SH and service provider. Figure 34.7 illustrates the functioning of this layer at the service provider, the SH, and the MH. The original data stream is broken into *logical segments* that are separated by markers (or flags). When this stream arrives at the SH, the SH discards entire logical segments (in the case of compressed video, one logical segment may represent one frame) depending on the bandwidth available to the MH. The purpose of discarding entire logical segments is that discarding a part of such a segment of data makes the rest of the data within that segment useless—so we might as well discard the entire segment. Observe also that the flags (to identify logical segments) are inserted by the LPTSL via calls made by the application layer. Thus, the transport layer or the LPTSL does not need to know encoding details of the data stream. This scheme is currently being implemented at the University of South Carolina by the author and his research group.

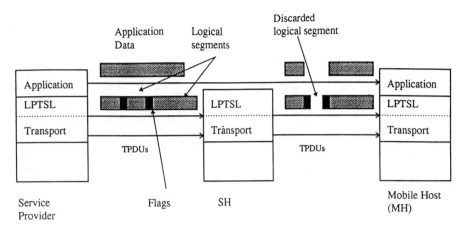

FIGURE 34.7 LPTSL, an approach to handle dynamic bandwidth variations.

34.6 Conclusions

The need for wireless LANs is driving rapid development in this area. The IEEE has proposed standards (802.11) for the physical layer and MAC layer protocols. A great deal of work, however, remains to be done at the network and transport layers. There does not appear to be a consensus regarding subnet design for wireless LANs. Our work has indicated a need for treating wireless LAN subnetworks as being fundamentally different from fixed networks, thus resulting in a different subnetwork and transport layer designs. Current efforts are underway to validate these claims.

Defining Terms

Carrier-sense multiple access (CSMA): Protocols such as those used over the ethernet.
Medium access control (MAC): Protocols arbitrate channel access between all nodes on a wireless LAN.
Mobile host (MH) nodes: The nodes of wireless LAN.
Supervisor host (SH): The node that takes care of flow-control and other protocol processing for all connections.

References

Barke, A. and Badrinath, B.R. 1994. I-TCP: indirect TCP for mobile hosts. Tech. Rept. DCS-TR-314, Dept. Computer Science, Rutgers Univ. Piscataway, NJ.

Bantz, D.F. and Bauchot, F.J. 1994. Wireless LAN design alternatives. *IEEE Network.* 8(2):43–53.

Chen, K.-C. and Lee, C.H. 1993. RAP: a novel medium access control protocol for wireless data networks. *Proc. IEEE GLOBECOM'93*, IEEE Press, Piscataway, NJ 08854. 1713–1717.

Chen, K.-C. 1994. Medium access control of wireless LANs for mobile computing. *IEEE Network*, 8(5):50–63.

Glisic, S.G. 1991. 1-Persistent carrier sense multiple access in radio channel with imperfect carrier sensing. *IEEE Trans. on Comm.* 39(3):458–464.

Goodman, D.J. 1990. Cellular packet communications. *IEEE Trans. on Comm.* 38(8):1272–1280.

Ioanidis, J., Duchamp, D., and Maguire, G.Q. 1991. IP-based protocols for mobile internetworking. *Proc. of ACM SIGCOMM'91*, ACM Press, New York, NY 10036 (Sept.):235–245.

Ghai, R. and Singh, S. 1994. An architecture and communication protocol for picocellular networks. *IEEE Personal Comm. Mag.* 1(3):36–46.

Singh, S. 1995. Quality of service guarantees in mobile computing. *J. of Computer Comm.*, to appear.

Seal, K. and Singh, S. 1995. Loss profiles: a quality of service measure in mobile computing. *J. Wireless Networks*, submitted for publication.

Teraoka, F. and Tokoro, M. 1993. Host migration transparency in IP networks: the VIP approach. *Proc. of ACM SIGCOMM*, ACM Press, New York, NY 10036 (Jan.):45–65.

Tobagi, F. and Kleinrock, L. 1975a. Packet switching in radio channels: Part I carrier sense multiple access models and their throughput delay characteristic. *IEEE Trans. on Comm.*, 23(12):1400–1416.

Tobagi, F. and Kleinrock, L. 1975b. Packet switching in radio channels: Part II the hidden terminal problem in CSMA and busy-one solution. *IEEE Trans. on Comm.*, 23(12):1417–1433.

Further Information

A good introduction to physical layer issues is presented in Bantz, 1994 and MAC layer issues are discussed in Chen, 1994. For a discussion of network and transport layer issues, see Singh, 1995 and Ghai and Singh, 1994.

35

Wireless Data

Allen H. Levesque
GTE Laboratories, Inc.

Kaveh Pahlavan
Worcester Polytechnic Institute

35.1 Introduction

Wireless data services and systems represent a rapidly growing and increasingly important segment of the communications industry. Whereas the wireless data industry is becoming increasingly diverse, one can identify two mainstreams that relate directly to users' requirement for data services. On one hand, there are requirements for relatively low-speed data services provided to mobile users over wide geographical areas, as provided by private mobile data networks and by data services carried on common-carrier cellular telephone networks. On the other hand, there are requirements for high-speed data services in local areas, as provided by cordless private branch exchange (PBX) systems and wireless local area networks (LANs), as well as by the emerging personal communications services (PCS). Personal communications services are treated in Chapter 18 and wireless LANs are treated in Chapter 34. In this chapter we mainly address wide-area wireless data systems, commonly called *mobile data systems*, and briefly touch upon data services to be incorporated into the emerging digital cellular systems.

Mobile data systems provide a wide variety of services for both business users and public safety organizations. Basic services supporting most businesses include electronic mail, enhanced paging, modem and facsimile transmission, remote access to host computers and office LANs, and information broadcast services. Public safety organizations, particularly law-enforcement agencies, are making increasing use of wireless data communications over traditional VHF and UHF radio dispatch networks and over public cellular telephone networks. In addition, there are wireless services supporting vertical applications that are more or less tailored to the needs of specific companies or

0-8493-8573-3/96/$0.00+$.50

industries, such as transaction processing, computer-aided delivery dispatch, customer service, fleet management, and emergency medical services. Work currently in progress to develop the national Intelligent vehicle highway system (IVHS) includes the definition of a wide array of new traveler services, many of which will be supported by standardized mobile data networks.

Much of the growth in use of wireless data services has been spurred by the rapid growth of the paging service industry and increasing customer demand for more advanced paging services, as well as the desire to increase work productivity by extending to the mobile environment the suite of digital communications services readily available in the office environment. There is also a desire to make more cost-efficient use of the mobile radio and cellular networks already in common use for mobile voice communications by incorporating efficient data transmission services into these networks. The services and networks that have evolved to date represent a variety of specialized solutions and, in general, they are not interoperable with each other. As the wireless data industry expands, there is an increasing demand for an array of attractively priced standardized services and equipment accessible to mobile users over wide geographic areas. Thus, we see the growth of nationwide privately operated service networks as well as new data services built upon the first and second generation cellular telephone networks. The establishment of new PCS systems in the 2-GHz bands will further extend this evolution.

In this chapter we describe the principal existing and evolving wireless data networks and the related standards activities now in progress. We begin with a discussion of the technical characteristics of wireless data networks.

35.2 Characteristics of Wireless Data Networks

From the perspective of the data user, the basic requirement for wireless data service is convenient, reliable, low-speed access to data services over a geographical area appropriate to the user's pattern of daily business operation. By low speed we mean data rates comparable to those provided by standard data modems operating over the public switched telephone network (PSTN). This form of service will support a wide variety of short-message applications, such as notice of electronic mail or voice mail, as well as short file transfers or even facsimile transmissions that are not overly lengthy. The user's requirements and expectations for these types of services are different in several ways from the requirements placed on voice communication over wireless networks. In a wireless voice service, the user usually understands the general characteristics and limitations of radio transmission and is tolerant of occasional *signal fades* and brief dropouts. An overall level of acceptable voice quality is what the user expects. In a data service, the user is instead concerned with the accuracy of delivered messages and data, the time-delay characteristics of the service network, the ability to maintain service while traveling about, and, of course, the cost of the service. All of these factors are dependent on the technical characteristics of wireless data networks, which we discuss next.

Radio Propagation Characteristics

The chief factor affecting the design and performance of wireless data networks is the nature of radio propagation over wide geographic areas. The most important mobile data systems operate in various land–mobile radio bands from roughly 100 to 200 MHz, the specialized mobile radio (SMR) band around 900 MHz, and the cellular telephone bands at 824–894 MHz. In these frequency bands, radio transmission is characterized by distance-dependent field strength, as well as the well-known effects of *multipath fading*, signal shadowing, and signal blockage. The signal coverage provided by a radio transmitter, which in turn determines the area over which a mobile data receiving terminal can receive a usable signal, is governed primarily by the *power–distance relationship*, which gives signal power as a function of distance between transmitter and receiver. For the ideal case of single-path transmission in free space, the relationship between transmitted power P_t and received power P_r is given by

$$P_r/P_t = G_tG_r(\lambda/4\pi d)^2 \qquad (35.1)$$

where G_t and G_r are the transmitter and receiver antenna gains, respectively, d is the distance between the transmitter and the receiver, and λ is the wavelength of the transmitted signal. In the mobile radio environment, the power-distance relationship is in general different from the free-space case just given. For propagation over an Earth plane at distances much greater than either the signal wavelength or the antenna heights, the relationship between P_t and P_r is given by

$$P_r/P_t = G_tG_r(h_1^2h_2^2/d^4) \qquad (35.2)$$

where h_1 and h_2 are the transmitting and receiving antenna heights. Note here that the received power decreases as the fourth power of the distance rather than the square of distance seen in the ideal free-space case. This relationship comes from a propagation model in which there is a single signal reflection with phase reversal at the Earth's surface, and the resulting received signal is the vector sum of the direct line-of-sight signal and the reflected signal. When user terminals are used in mobile situations, the received signal is generally characterized by rapid fading of the signal strength, caused by the vector summation of reflected signal components, the vector summation changing constantly as the mobile terminal moves from one place to another in the service area. Measurements made by many researchers show that when the fast fading is averaged out, the signal strength is described by a Rayleigh distribution having a log-normal mean. In general, the power-distance relationship for mobile radio systems is a more complicated relationship that depends on the nature of the terrain between transmitter and receiver.

Various propagation models are used in the mobile radio industry for network planning purposes, and a number of these models are described in Bodson, McClure, and McConoughey eds., 1984. Propagation models for mobile communications networks must take account of the terrain irregularities existing over the intended service area. Most of the models used in the industry have been developed from measurement data collected over various geographic areas. A very popular model is the *Longley–Rice model* [Rice et al., 1967; Longley and Rice, 1968]. Many wireless networks are concentrated in urban areas. A widely used model for propagation prediction in urban areas is one usually referred to as the *Okumura–Hata model* [Okumura et al., 1968; Hata, 1980]. The Longley–Rice and Okumura–Hata propagation models are discussed in Chapter 22 of this handbook.

By using appropriate propagation prediction models, one can determine the range of signal coverage for a base station of given transmitted power. In a wireless data system, if one knows the level of received signal needed for satisfactory performance, the area of acceptable performance can, in turn, be determined. Cellular telephone networks utilize base stations that are typically spaced 1–5 mi apart, though in some mid-town areas, spacings of 1/2 mi or less are now being used. In packet-switched data networks, higher power transmitters are used, spaced about 5–15 mi apart.

An important additional factor that must be considered in planning a wireless data system is the in-building penetration of signals. Many applications for wireless data services involve the use of mobile data terminals inside buildings, for example, for trouble-shooting and servicing computers on customers' premises. Another example is wireless communications into hospital buildings in support of emergency medical services. It is usually estimated that in-building signal penetration losses will be in the range of 15–30 dB. Clearly, received signal strengths can be satisfactory in the outside areas around a building but totally unusable inside the building. This becomes an important issue when a service provider intends to support customers using mobile terminals inside buildings.

One important consequence of the rapid fading experienced on mobile channels is that errors tend to occur in bursts, causing the transmission to be very unreliable for short intervals of time. Another problem is signal dropouts that occur, for example, when a data call is handed over from one base station to another, or when the mobile user moves into a location that severely blocks the signal. Because of this, mobile data systems employ various error-correction and error-recovery techniques to insure accurate and reliable delivery of data messages.

35.3 Market Issues

Although the market for personal computers (PCs) is not growing as it has in past years, the market for portable computers such as laptops, pen-pads, and notebook computers is growing rapidly. Of greater importance to the wireless data communication industry, the market for networked portables is growing much faster than the market for portable computing. Wireless is the communication method of choice for portable terminals. Mobile data communication services discussed here provide a low-speed solution for wide area coverage. For high-speed and local communications, a portable terminal with wireless access can bring the processing and database capabilities of a large computer directly to specific locations for short periods of time, thus opening a horizon for new applications. For example, one can take portable terminals into classrooms for instructional purposes, or to hospital beds or accident sites for medical diagnosis.

35.4 Modem Services Over Cellular Networks

A simple form of wireless data communication now in common use is data transmission using modems or facsimile terminals over analog cellular telephone links. In this form of communication, the mobile user simply accesses a cellular channel just as he would in making a standard voice call over the cellular network. The user then operates the modem or facsimile terminal just as would be done from office to office over the PSTN. A typical connection is shown in Fig. 35.1, where the mobile user has a lap-top computer and portable modem in the vehicle, communicating with another modem and computer in the office. Typical users of this mode of communication include service technicians, real estate agents, and traveling sales people. In this form of communication, the network is not actually providing a data service but simply a voice link over which the data modem or fax terminal can interoperate with a corresponding data modem or fax terminal in the office or service center. The connection from the mobile telephone switching office (MTSO) is a standard landline connection, exactly the same as is provided for an ordinary cellular telephone call. Many portable modems and fax devices are now available in the market and are sold as elements of the so-called mobile office for the traveling business person. Law enforcement personnel are also making increasing use of data communication over cellular telephone and dispatch radio networks to gain rapid access to databases for verification of automobile registrations and drivers' licenses. Portable devices are currently available that operate at transmission rates up to 9.6 or 14.4 kb/s. Error-correction modem protocols such as MNP-10, V.34, and V.42 are used to provide reliable delivery of data in the error-prone wireless transmission environment.

In another form of mobile data service, the mobile subscriber uses a portable modem or fax terminal as already described but now accesses a modem provided by the cellular service operator

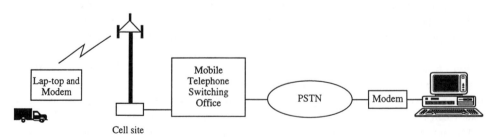

FIGURE 35.1   Modem operation over an analog cellular voice connection.

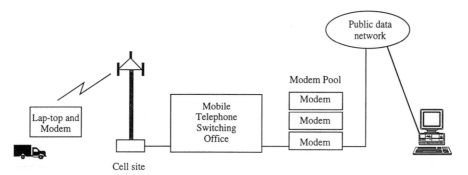

FIGURE 35.2 Cellular data service supported by modem pools in the network.

as part of a *modem pool*, which is connected to the MTSO. This form of service is shown in Fig. 35.2. The modem pool might provide the user with a choice of several standard modem types. The call connection from the modem pool to the office is a digital data connection, which might be supported by any of a number of public packet data networks, such as those providing X.25 service. Here, the cellular operator is providing a special service in the form of modem pool access, and this service in general carries a higher tariff than does standard cellular telephone service, due to the operator's added investment in the modem pools. In this form of service, however, the user in the office or service center does not require a modem but instead has a direct digital data connection to the desk-top or host computer.

Each of the types of wireless data transmission just described is in effect an appliqué onto an underlying cellular telephone service and, therefore, has limitations imposed by the characteristics of the underlying voice connection. That is, the cellular segment of the call connection is a circuit-mode service, which might be cost effective if the user needs to send long file transfers or fax transmissions but might be relatively costly if only short messages are to be transmitted and received. This is because the subscriber is being charged for a circuit-mode connection, which stays in place throughout the duration of the communication session, even if only intermittent short messages exchanges are needed. The need for systems capable of providing cost-effective communication of relatively short message exchanges led to the development of wireless packet data networks, which we describe next.

35.5 Private Data Networks

Here we describe two packet data networks that provide mobile data services to users in major metropolitan areas throughout the United States.

ARDIS

ARDIS is a two-way radio service developed as a joint venture between IBM and Motorola and first implemented in 1983. In mid-1994, IBM sold its interest in ARDIS to Motorola. The ARDIS network consists of four network control centers with 32 network controllers distributed through 1250 base station in 400 cities in the U.S. The service is suitable for two-way transfers of data files of size less than 10 kilobytes, and much of its use is in support of computer-aided dispatching, such as is used by field service personnel, often while they are on customers' premises. Remote users access the system from laptop radio terminals, which communicate with the base stations. Each of the ARDIS base stations is tied to one of the 32 radio network controllers, as shown in Fig. 35.3. The backbone of the network is implemented with leased telephone lines. The four ARDIS hosts, located in Chicago, New York, Los Angeles, and Lexington, KY, serve as access points

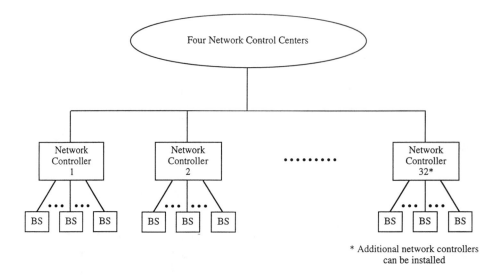

FIGURE 35.3   ARDIS network architecture.

for a customer's mainframe computer, which can be linked to an ARDIS host using async, bisync, SNA, or X.25 dedicated circuits.

The operating frequency band is 800 MHz, and the RF links use separate transmit and receive frequencies spaced by 45 MHz. The system was initially implemented with an RF channel data rate 4800 b/s per 25-kHz channel, using the MDC-4800 protocol. This has been upgraded to 19.2 kb/s, using the RD-LAP protocol, which provides a user data rate of about 8000 b/s. The system architecture is cellular, with cells overlapped to increase the probability that the signal transmission from a portable transmitter will reach at least one base station. The base station power is 40 W, which provides line-of-sight coverage up to a radius of 10–15 miles. The portable units operate with 4 W of radiated power. The overlapping coverage, combined with designed power levels, and error-correction coding in the transmission format, insures that the ARDIS can support portable communications from inside buildings, as well as on the street. This capability for in-building coverage is an important characteristic of the ARDIS service. The modulation technique is frequency-shift keying (FSK), the access method is frequency division multiple access (FDMA), and the transmission packet length is 256 bytes.

Although the use of overlapping coverage, almost always on the same frequency, provides reliable radio connectivity, it poses the problem of interference when signals are transmitted simultaneously from two adjacent base stations. The ARDIS network deals with this by turning off neighboring transmitters, for 0.5–1 s, when an outbound transmission occurs. This scheme has the effect of constraining overall network capacity.

The laptop portable terminals access the network using a random access method called data sense multiple access (DSMA) [Pahlavan and Levesque, 1995]. A remote terminal listens to the base station transmitter to determine if a busy bit is on or off. When the busy bit is off, the remote terminal is allowed to transmit. If two remote terminals begin to transmit at the same time, however, the signal packets may collide, and retransmission will be attempted, as in other contention-based multiple access protocols. The busy bit lets a remote user know when other terminals are transmitting and, thus, reduces the probability of packet collision.

MOBITEX

The MOBITEX system is a nationwide, interconnected trunked radio network developed by Ericsson and Swedish Telecom. The first MOBITEX network went into operation in Sweden in 1986, and networks have either been implemented or are being deployed in 13 countries. A MOBITEX operations association oversees the open technical specifications and coordinates software and hardware developments [Khan and Kilpatrick, 1995]. In the U.S., MOBITEX service was introduced by RAM Mobile Data in 1991 and now covers 7500 cities and towns, with automatic roaming across all service areas. By locating its base stations close to major business centers, the RAM Mobile system provides a degree of in-building signal coverage. Although the MOBITEX system was designed to carry both voice and data service, the U.S. and Canadian networks are used to provide data service only. MOBITEX is an intelligent network with an open architecture that allows establishing virtual networks. This feature facilitates the mobility and expandability of the network [Kilpatrick, 1992; Parsa, 1992].

The MOBITEX network architecture is hierarchical, as shown in Fig. 35.4. At the top of the hierarchy is the network control center (NCC), from which the entire network is managed. The top level of switching is a national switch (MHX1) that routes traffic between service regions. The next level comprises regional switches (MHX2s), and below that are local switches (MOXs), each of which handles traffic within a given service area. At the lowest level in the network, multichannel trunked-radio base stations communicate with the mobile and portable data sets. MOBITEX uses packet-switching techniques, as does ARDIS, to allow multiple users to access the same channel at the same time. Message packets are switched at the lowest possible network level. If two mobile users in the same service area need to communicate with each other, their messages are relayed through the local base station, and only billing information is sent up to the network control center.

The base stations are laid out in a grid pattern using the same frequency reuse rules as are used for cellular telephone networks. In fact, the MOBITEX system operates in much the same way as a cellular telephone system, except that handoffs are not managed by the network. That is, when a radio connection is to be changed from one base station to another, the decision is made by the mobile terminal, not by a network computer as in cellular telephone systems.

To access the network, a mobile terminal finds the base station with the strongest signal and then registers with that base station. When the mobile terminal enters an adjacent service area, it automatically re-registers with a new base station, and the user's whereabouts are relayed to the

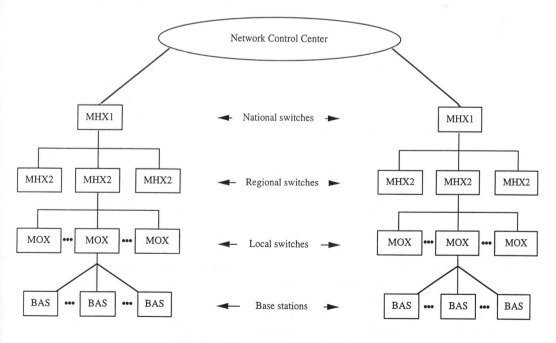

FIGURE 35.4   MOBITEX network architecture.

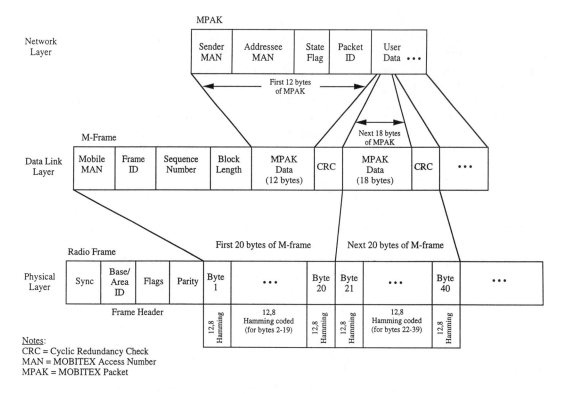

FIGURE 35.5　MOBITEX packet and frame structure at three layers of the protocol stack.

higher level network nodes. This provides automatic routing of messages bound for the mobile user, a capability known as *roaming*. The MOBITEX network also has a store-and-forward capability.

The mobile units transmit at 896–901 MHz and the base stations at 935 to 940 MHz. The system uses dynamic power setting, in the range of 100 mW–10 W for mobile units and 100 mW–4 W for portable units. The Gaussian minimum shift keying (GMSK) modulation technique is used, with $BT = 0.3$ and noncoherent demodulation. The transmission rate is 8000 b/s half-duplex in 12.5-kHz channels, and the service is suitable for file transfers up to 20 kilobytes. The MOBITEX system uses a proprietary network-layer protocol called MPAK, which provides a maximum packet size of 512 bytes and a 24-b address field. Forward-error correction, as well as retransmissions, are used to ensure the bit-error-rate quality of delivered data packets. Fig. 35.5 shows the packet structure at various layers of the MOBITEX protocol stack. The system uses the reservation-slotted ALOHA (R-S-ALOHA) random access method.

35.6 Cellular Data Networks and Services

Cellular Digital Packet Data (CDPD)

The cellular digital packet data (CDPD) system was designed to provide packet data services as an overlay onto the existing analog cellular telephone network, which is called advanced mobile phone service (AMPS). CDPD was developed by IBM in collaboration with the major cellular carriers. These companies will cover 95% of the U.S., including all major urban areas. Any cellular carrier owning a license for AMPS service is free to offer its customers CDPD service without any need for further licensing. A basic goal of the CDPD system is to provide data services on a noninterfering basis with the existing analog cellular telephone services using the same 30-kHz channels. This is accomplished in either of two ways. First, one or a few AMPS channels in each cell site can be devoted to CDPD service. Second, CDPD is designed to make use of a cellular channel that is temporarily not being used for voice traffic and to move to another channel when the current channel is allocated to voice service. The compatibility of CDPD with the existing cellular telephone system allows it to be installed in any AMPS cellular system in North America, providing data services that are not dependent on support of a digital cellular standard in the service area. The participating companies issued release 1.0 of the CDPD specification in July 1993, and release 1.1 was issued in late 1994 [CDPD, 1994]. At this writing (mid-1995), CDPD service is implemented in more than a dozen major market areas, and deployment in an additional 55 markets is planned by the end of 1995. Intended applications for CDPD service include: electronic mail, field support servicing, package delivery tracking, inventory control, credit card verification, security reporting, vehicle theft recovery, traffic and weather advisory services, and a potentially wide range of information retrieval services.

Although CDPD cannot increase the number of channels usable in a cell, it can provide an overall increase in user capacity if data users use CDPD instead of voice channels. This capacity increase would result from the inherently greater efficiency of a connectionless packet data service relative to a connection-oriented service, given bursty data traffic. That is, a packet data service does not require the overhead associated with setup of a voice traffic channel in order to send one or a few data packets. In the following paragraphs we briefly describe the CDPD network architecture and the principles of operation of the system. Our discussion follows Quick and Balachandran, 1993, closely.

The basic structure of a CDPD network (Fig. 35.6) is similar to that of the cellular network with which it shares transmission channels. Each mobile end system (M-ES) communicates with a mobile data base station (MDBS) using the protocols defined by the air-interface specification,

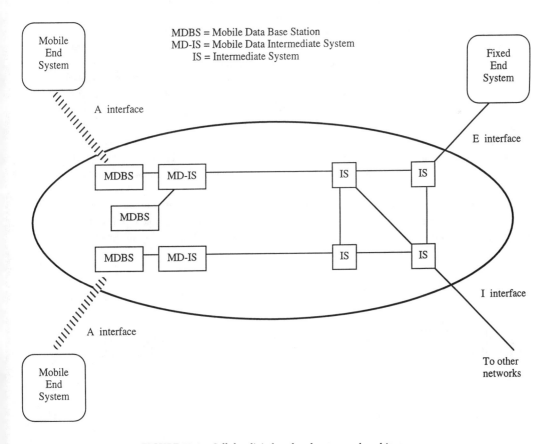

FIGURE 35.6 Cellular digital packet data network architecture.

to be described subsequently. The MDBSs are expected to be collocated with the cell equipment providing cellular telephone service to facilitate the channel-sharing procedures. All of the MDBSs in a service area will be linked to a mobile data intermediate system (MD-IS) by microwave or wireline links. The MD-IS provides a function analogous to that of the mobile switching center (MSC) in a cellular telephone system. The MD-IS may be linked to other MD-ISs and to various services provided by end systems outside the CDPD network. The MD-IS also provides a connection to a network management system and supports protocols for network management access to the MDBSs and M-ESs in the network.

Service endpoints can be local to the MD-IS or remote, connected through external networks. A MD-IS can be connected to any external network supporting standard routing and data exchange protocols. A MD-IS can also provide connections to standard modems in the PSTN by way of appropriate modem interworking functions (modem emulators). Connections between MD-ISs allow routing of data to and from M-ESs that are roaming, that is, operating in areas outside their home service areas. These connections also allow MD-ISs to exchange information required for mobile terminal authentication, service authorization, and billing.

CDPD employs the same 30-kHz channelization as used in existing AMPS cellular systems throughout North America. Each 30-kHz CDPD channel will support channel transmission rates up to 19.2 kb/s. Degraded radio channel conditions, however, will limit the actual information payload throughput rate to lower levels, typically 5–10 kb/s, and will introduce additional time delay due to the error-detection and retransmission protocols.

The CDPD radio link physical layer uses GMSK modulation at the standard cellular carrier frequencies, on both forward and reverse links. The Gaussian pulse shaping filter is specified to have bandwidth-time product $B_bT = 0.5$. The specified B_bT product assures a transmitted waveform with bandwidth narrow enough to meet adjacent-channel interference requirements, while keeping the intersymbol interference small enough to allow simple demodulation techniques. The choice of 19.2 kb/s as the channel bit rate yields an average power spectrum that satisfies the emission requirements for analog cellular systems and for dual-mode digital cellular systems.

The forward channel carries data packets transmitted by the MDBS, whereas the reverse channel carries packets transmitted by the M-ESs. In the forward channel, the MDBS forms data frames by adding standard high level data link control (HDLC) terminating flags and inserted zero bits, and then segments each frame into blocks of 274 b. These 274 b, together with an 8-b *color code* for MDBS and MD-IS identification, are encoded into a 378-b coded block using a (63, 47) Reed–Solomon code over a 64-ary alphabet. A 6-b synchronization and flag word is inserted after every 9 code symbols. The flag words are used for reverse link access control. The forward link block structure is shown in Fig. 35.7.

In the reverse channel, when an M-ES has data frames to send, it formats the data with flags and inserted zeros in the same manner as in the forward link. That is, the reverse link frames are segmented and encoded into 378-b blocks using the same Reed–Solomon code as in the forward

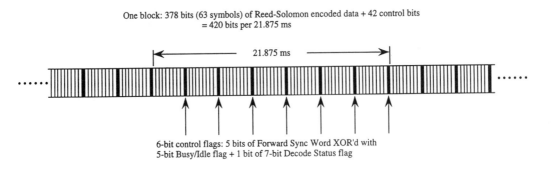

One block: 378 bits (63 symbols) of Reed-Solomon encoded data + 42 control bits
= 420 bits per 21.875 ms

21.875 ms

6-bit control flags: 5 bits of Forward Sync Word XOR'd with
5-bit Busy/Idle flag + 1 bit of 7-bit Decode Status flag

FIGURE 35.7   Cellular digital packet data forward link block structure.

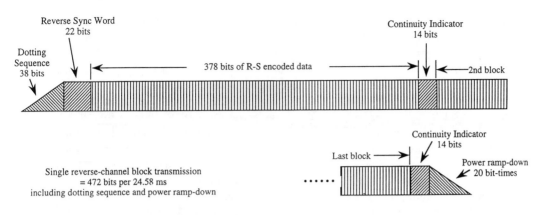

FIGURE 35.8   Cellular digital packet data reverse link block structure.

channel. The M-ES may form up to 64 encoded blocks for transmission in a single reverse channel transmission burst. During the transmission, a 7-b transmit continuity indicator is interleaved into each coded block and is set to all ones to indicate that more blocks follow, or all zeros to indicate that this is the last block of the burst. The reverse channel block structure is shown in Fig. 35.8.

The media access control (MAC) layer in the forward channel is relatively simple. The receiving M-ES removes the inserted zeros and HDLC flags and reassembles data frames that were segmented into multiple blocks. Frames are discarded if any of their constituent blocks are received with uncorrectable errors.

On the reverse channel (M-ES to MDBS), access control is more complex, since multiple M-ESs must share the channel. CDPD uses a multiple access technique called digital sense multiple access (DSMA), which is closely related to the carrier sense multiple access/collision detection (CSMA/CD) access technique.

The network layer and higher layers of the CDPD protocol stack are based on standard ISO and internet protocols. It is expected that the earliest CDPD products will use the internet protocols.

The selection of a channel for CDPD service is accomplished by the radio resource management entity in the MDBS. Through the network management system, the MDBS is informed of the channels in its cell or sector that are available either as dedicated data channels or as potential CDPD channels when they are not being used for analog cellular service, depending on which channel allocation method is implemented. For the implementation in which CDPD service is to use channels of opportunity, there are two ways in which the MDBS can determine whether the channels are in use. If a communication link is provided between the analog system and the CDPD system, the analog system can inform the CDPD system directly about channel usage. If such a link is not available, the CDPD system can use a forward power monitor (sniffer antenna) to detect channel usage on the analog system. Circuitry to implement this function can be built into the cell sector interface.

Digital Cellular Data Services

In response to the rapid growth in demand for cellular telephone service throughout the U.S and Canada, the Cellular Telecommunications Industry Association (CTIA) and the Telecommunications Industry Association (TIA) have been developing standards for new digital cellular systems to replace the existing AMPS cellular system. Two air-interface standards have now been published. The IS-54 standard specifies a three-slot TDMA system, and the I-95 standard specifies a CDMA spread spectrum system. In both systems, a variety of data services are being planned.

The general approach taken in the definition of IS-95 data services has been to base the services on standard data protocols, to the greatest extent possible [Tiedemann, 1993]. The previously-specified physical layer of the IS-95 protocol stack was adopted for the physical layer of the data services, with an appropriate radio link protocol (RLP) overlaid. The current standardization effort is directed to defining three primary services: 1) asynchronous data, 2) group-3 facsimile, and 3) packet data service carried over a circuit-mode connection. Later, attention will be given to other services, including synchronous data and contention-based packet data service.

IS-95 asynchronous data will be structured as a circuit-switched service. For circuit-switched connections, a dedicated path is established between the data devices for the duration of the call. It is used for connectivity through the PSTN requiring point-to-point communications to the common PC or fax user. There are several applications that fall into this category. For file transfer involving PC-to-PC communications the asynchronous data service is the desired cellular service mode. The service will employ an RLP to protect data from transmission errors caused by radio channel degradations at the air interface. The RLP employs automatic repeat request (ARQ), forward error correction (FEC), and flow control. Flow control and retransmission of data blocks with errors are used to provide an improved error performance in the mobile segment of the data connection at the expense of variations in throughput and delay. Typical raw channel rates for digital cellular transmission are measured at approximately a 10^{-2} bit-error rate. Acceptable data transmission, however, usually requires a bit-error rate of approximately 10^{-6} and achieving this requires the design of efficient ARQ and error-correction codes to deal with error characteristics in the mobile environment.

At this writing, the TIA CDMA data services task group has developed a service description for asynchronous data service standards. Work on defining a synchronous data service is underway, whereas work on contention-based packet data service standards has not yet begun. In parallel with the CDMA data services effort, anther TIA task group, TR45.3.2.5, has been defining standards for digital data services for the TDMA digital cellular standard IS-54 [Sacuta, 1992; Weissman, Levesque, and Dean, 1993]. As with the IS-95 data services effort, initial priority is being given to standardizing circuit-mode asynchronous data and group-3 facsimile services. As of this writing, a standard for asynchronous data and group 3 facsimile service has been completed [TIA, 1994].

35.7 Other Planned Systems

Trans-European Trunked Radio (TETRA)

As is the case in the U.S. and Canada, there is interest in Europe in establishing fixed wide-area standards for mobile data communications. Whereas the Pan-European standard for digital cellular, termed Global Systems for Mobile communications (GSM), will provide an array of data services, data will be handled as a circuit-switched service, consistent with the primary purpose of GSM as a voice service system. Therefore, the European Telecommunications Standards Institute (ETSI) has begun developing a public standard for trunked radio and mobile data systems. The standards, which are known generically as Trans-European Trunked Radio (TETRA), are the responsibility of the ETSI RES 6 subtechnical committee [Haine, Martin, and Goodings, 1992].

TETRA is being developed as a family of standards. One branch of the family is a set of radio and network interface standards for trunked voice (and data) services. The other branch is an air-interface standard optimized for wide-area packet data services for both fixed and mobile subscribers and supporting standard network access protocols. Both versions of the standard will use a common physical layer, based on $\pi/4$ dfferential quadrature phase shift keying ($\pi/4$-DQPSK) modulation operating at a channel rate of 36 kb/s in each 25-kHz channel.

Figure 35.9 is a simplified model of the TETRA network, showing the three interfaces at which services will be defined. The U_m interface is the radio link between the base station (BS) and the mobile station (MS). At the other side of the network is the fixed network access point (FNAP) through which mobile users gain access to fixed users. Fixed host computers and fixed data networks typically use standardized interfaces and protocols, and it is intended that the mobile segments of the TETRA network will utilize these same standards. Finally, there is the interface to the mobile data

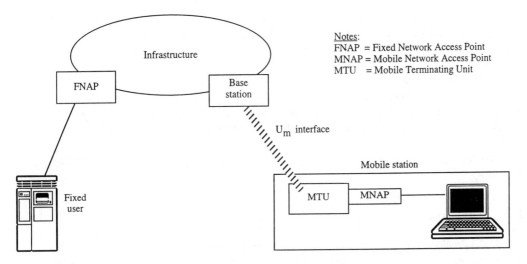

Notes:
FNAP = Fixed Network Access Point
MNAP = Mobile Network Access Point
MTU   = Mobile Terminating Unit

FIGURE 35.9   TETRA network interfaces.

user, the mobile network access point (MNAP). This is shown in the figure as a physical interface between a mobile terminating unit (MTU) and a data terminal. The MNAP, however, may be only a logical interface and not a physical interface, if the MS is an integrated device with no external data port. It is envisioned that all three interfaces, FNAP, MNAP, and the air interface U_m, will support data services with packet-mode protocols.

It is planned that TETRA will provide both connection-oriented and connectionless data services. Although work is still in progress on defining protocols for the three interfaces in the TETRA network model, some broad decisions have been made. It has been decided that the physical data port at the mobile station will be a true X.25 interface, which means that any attached data terminal

TABLE 35.1   Characteristics and Parameters of Five Mobile Data Services

System:	ARDIS	MOBITEX	CDPD	IS-95[b]	TETRA[b]
Frequency band					
Base to mobile, (MHz).	(800 band,	935–940[a]	869–894	869–894	(400 and
Mobile to base, (MHz).	45-kHz sep.)	896–901	824–849	824–849	900 Bands)
RF channel spacing	25 kHz (U.S.)	12.5 kHz	30 kHz	1.25 MHz	25 kHz
Channel access/ multiuser access	FDMA/ DSMA	FDMA/ dynamic- R-S-ALOHA	FDMA/ DSMA	FDMA/ CDMA-SS	FDMA/ DSMA & SAPR[c]
Modulation method	FSK, 4-FSK	GMSK	GMSK	4-PSK/DSSS	$\pi/4$-QDPSK
Channel bit rate, kb/s	19.2	8.0	19.2	9.6	36
Packet length	Up to 256 bytes (HDLC)	Up to 512 bytes	24–928 b	(Packet service-TBD)	192 b (short) 384 b (long)
Open architecture	No	Yes	Yes	Yes	Yes
Private or Public Carrier	Private	Private	Public	Public	Public
Service Coverage	Major metro. areas in U.S.	Major metro. areas in U.S.	All AMPS areas	All CDMA cellular areas	European trunked radio
Type of coverage	In-building & mobile	In-building & mobile	Mobile	Mobile	Mobile

[a] Frequency allocation in the U.S. in the U.K., 380–450 MHz band is used.
[b] IS-95 and TETRA data services standardization in progress.
[c] Slotted-ALOHA packet reservation

must implement the X.25 protocol. This will provide true peer-to-peer communication between the mobile and fixed ends of the call connection. For connectionless data services, the protocol definition is still incomplete, but it is envisioned that the protocols will be based on ISO standards for connectionless-mode network service [ISO, 1987].

The protocol for the radio link has not yet been defined, but it will certainly employ a combination of forward-error correction (FEC) coding, CRC error-detection coding, and an ARQ scheme. It has been reported that the TETRA standard is being designed to accommodate two popular forms of multiuser access, slotted Aloha (with and without packet reservation), and data-sense multiple access.

Table 35.1 compares the chief characteristics and parameters of the five wireless data services described.

35.8 Conclusions

Mobile data radio systems have grown out of the success of the paging-service industry and the increasing customer demand for more advanced services. Today 100,000 customers are using mobile data services and the industry expects 13 million users by the year 2000. This could be equivalent to 10–30% of the revenue of the cellular radio industry. Today, mobile data services provide length-limited wireless connections with in-building penetration to portable users in metropolitan areas. The future direction is toward wider coverage, higher data rates, and capability for transmitting longer data files.

References

Bodson, D., McClure, G.F., and McConoughey, S.R. eds. 1984. *Land-Mobile Communications Engineering*, Selected Reprint Ser., IEEE Press, New York.

CDPD Industry Coordinator. 1994. Cellular Digital Packet Data Specification, Release 1.1, November 1994, Pub. CDPD, Kirkland, WA.

Haine, J.L., Martin, P.M., and Goodings, R.L.A. 1992. A European standard for packet-mode mobile data, *Proceedings of Personal, Indoor, and Mobile Radio Conference (PIMRC'92)*, Boston, MA. Pub. IEEE, New York.

Hata, M. 1980. Empirical formula for propagation loss in land-mobile radio services. *IEEE Trans. on Vehicular Tech.* 29(3):317–325.

International Standards Organization (ISO), 1987. Protocol for providing the connectionless-mode network service. Pub. ISO 8473.

Khan, M. and Kilpatrick, J. 1995. MOBITEX and mobile data standards. *IEEE Comm. Maga.*, 33(3):96–101.

Kilpatrick, J.A., 1992. Update of RAM Mobile Data's packet data radio service. *Proceedings of the 42nd IEEE Vehicular Technology Conference (VTC'92)*, Denver, CO: 898–901, Pub. IEEE, New York.

Longley, A.G. and Rice, P.L. 1968. Prediction of tropospheric radio transmission over irregular terrain. A. computer method—1968, Environmental Sciences and Services Administration Tech. Rep. ERL 79-ITS 67, U.S. Government Printing Office, Washington, DC.

Okumura, Y., Ohmori, E., Kawano, T., and Fukuda, K., 1968. Field strength and its variability in VHF and UHF land-mobile service. *Review of the Electronic Communication Laboratory*, 16:825–873.

Pahlavan, K. and Levesque, A.H. 1994. Wireless data communications. *Proceedings of the IEEE*, 82(9):1398–1430.

Pahlavan, K. and Levesque, A.H. 1995. *Wireless Information Networks*, J. Wiley & Sons, New York.

Parsa, K. 1992. The MOBITEX packet-switched radio data system. *Proceedings of the Personal, Indoor and Mobile Radio Conference (PIMRC'92)*, Boston, MA: 534–538. Pub. IEEE, New York.

Quick, R.R., Jr. and Balachandran, K. 1993. Overview of the cellular packet data (CDPD) system. *Proceedings of the Personal, Indoor and Mobile Radio Conference (PIMRC'93)*, Yokoham, Japan: 338–343. Pub. IEEE, New York.

Rice, P.L., Longley, A.G., Norton, K.A., and Barsis, A.P. 1967. Transmission loss predictions for tropospheric communication circuits. National Bureau of Standards, Tech. Note 101, Boulder, CO.

Sacuta, A. 1992. Data standards for cellular telecommunications—a service-based approach. *Proceedings of the 42nd IEEE Vehicular Technology Conference*, Denver CO: 263–266. Pub. IEEE, New York.

Telecommunications Industry Association. 1994. Async data and fax. Project No. PN-3123, and Radio link protocol 1. Project No. PN-3306, Nov. 14. Issued by TIA, Washington, DC.

Tiedemann, E. 1993. Data services for the IS-95 CDMA standard. presented at Personal, Indoor and Mobile Radio Conf. PIMRC'93. Yokohama, Japan.

Weissman, D., Levesque, A.H., and Dean, R.A. 1993. Interoperable wireless data. *IEEE Comm. Mag.* 31(2):68–77.

Further Information

Pahlavan and Levesque, 1994 provides a comprehensive survey of the wireless data field as of mid-1994. The monthly journals *IEEE Communications Magazine* and *IEEE Personal Communications Magazine*, and the bimonthly journal *IEEE Transactions on Vehicular Technology* report advances in many areas of mobile communications, including wireless data. For subscription information contact: IEEE Service Center, 445 Hoes Lane, P. O. Box 1331, Piscataway, NJ, 08855-1131. Phone (800)678-IEEE.

Index

Page on which term is defined is indicated in bold.

Page on which term is defined is indicated in bold.

Page on which term is defined is indicated in bold.

Page on which term is defined is indicated in bold.

Page on which term is defined is indicated in bold.